Interdisciplinary Applied Mathematics

Interdisciplinary Applied Mathematics

Volume 14

Problems in engineering, computational science, and the physical and biological sciences are using increasingly sophisticated mathematical techniques. Thus, the bridge between the mathematical sciences and other disciplines is heavily traveled. The correspondingly increased dialog between the disciplines has led to the establishment of the series: *Interdisciplinary Applied Mathematics*.

The purpose of this series is to meet the current and future needs for the interaction between various science and technology areas on the one hand and mathematics on the other. This is done, firstly, by encouraging the ways that mathematics may be applied in traditional areas, as well as point towards new and innovative areas of applications; and, secondly, by encouraging other scientific disciplines to engage in a dialog with mathematicians outlining their problems to both access new methods and suggest innovative developments within mathematics itself.

The series will consist of monographs and high-level texts from researchers working on the interplay between mathematics and other fields of science and technology.

Akira Okubo, with Simon A. Levin

Diffusion and Ecological Problems: Modern Perspectives

Second Edition

With 114 Illustrations

 Springer

Simon A. Levin
Department of Ecology and
  Evolutionary Biology
Princeton University
Princeton, NJ 08544
USA
slevin@eno.princeton.edu

Editors

S.S. Antman
Department of Mathematics
and
Institute for Physical Science and Technology
University of Maryland
College Park, MD 20742-4015
USA

J.E. Marsden
Control and Dynamical Systems
  Applied Mathematics
California Institute of Technology
Pasadena, CA 91125
USA

L. Sirovich
Division of Mail Code 107-81
Brown University
Providence, RI 02912
USA

S. Wiggins
Control and Dynamical Systems
Mail Code 107-81
California Institute of Technology
Pasadena, CA 91125
USA

Mathematics Subject Classification (2000): 92A17, 35K55

Library of Congress Cataloging-in-Publication Data
Diffusion and ecological problems : modern perspectives / [edited by] Akira Okubo,
Simon A. Levin.—2nd ed.
    p. cm. — (Interdisciplinary applied mathematics ; 14)
    Rev. ed. of: Diffusion and ecological problems / Akira Okubo. 1980.
    Includes bibliographical references.

    1. Ecology—Mathematical models. 2. Diffusion—Mathematical models.
3. Biogeography—Mathematical models. I. Okubo, Akira. II. Levin, Simon A.
III. Okubo, Akira. Diffusion and ecological problems. IV. Interdisciplinary applied
mathematics ; v. 14.
QH541.15.M3 O38   2001
577'.01'5118—dc21   00-052258

Printed on acid-free paper.

Printed in the United States of America.

9 8 7 6 5 4 3 2 1

ISBN 978-1-4419-3151-1

Springer-Verlag New York Berlin Heidelberg
A member of BertelsmannSpringer Science+Business Media GmbH

Dedicated to J.G. Skellam

Preface to the Second Edition

The story of this edition is a testament to an almost legendary figure in theoretical ecology and to the influence his work and charisma has had on the field. It is also a story that can only be told by a trip back in time, to the genesis of the First Edition and before.

Akira Okubo and I were students together, but never knew it at the time. He was a graduate student at The Johns Hopkins University, where I was an undergraduate in mathematics. We both studied "modern physics," taught by Dino Franco Rasetti, and we decided years later that we must have been in the same class. Akira was then a chemical oceanographer, but ship time and his stomach did not agree. So he turned to theory, and the rest is history. His impact has been phenomenal, and the First Edition of this book was his most influential work. Building on his famous work with dye-diffusion experiments, he turned his attention to organisms and created a unique melding of ideas from physics and biology.

In the early 1970s, Lee Segel and I began to work on problems of planktonic patchiness, following some pathways that were simultaneously being explored by Akira (on diffusive instabilities). This brought Akira and me together, and he presented me with a copy of his 1975 book, *Ecology and Diffusion*, published in Japanese by Tsukji Shokan, Tokyo. I could understand all of the Greek in the book, but none of the Japanese. Still, I recognized enough that was familiar to know that this was an important book, but one whose influence was likely to be a bit limited if it remained only in Japanese. I was at the time Editor of the Biomathematics Series for Springer-Verlag, and I encouraged Akira to produce an updated version of his work for us, in large part so that I could read it. He readily accepted.

The next several years produced unexpected benefits for me. As Akira developed his chapters and sent them to me for comments and editorial suggestions, he began frequent pilgrimages from the Marine Sciences Research Center at Stony Brook to Cornell, where I was a faculty member. This led to fruitful collaborations between us, while he became a de facto member of our family, and an additional adviser for many of my students.

As I read each chapter, I was amazed by the novel insights, the originality, and the new vistas opened for me.

The success of *Diffusion and Ecological Problems* stimulated much research on diffusion and other spatial models in ecology over the past two decades, and it became clear than an updated version, greatly expanded, was needed. Akira set to work on this as his last project but never finished. When he passed away in 1996, he left a mass of notes, mostly cryptic ones to himself, about changes he planned to make. His close friend, the late Keiko Parker, assembled these from among the papers he left, and sent them to me with the simple statement that Akira asked her to do so and that I would know what to do with them. Considering the state of the notes, this was a daunting challenge, but one that I could not duck. Fortunately, Akira had many friends. My job was reduced primarily to one of editor.

I approached Achi Dosanjh, Mathematics Editor of Springer, who was enthusiastic about the idea of a Second Edition of this popular book. I laid out plans for the project to friends and colleagues of Akira and found more who were willing to pitch in than I could use. The idea was that each chapter would be adopted by someone expert in the area, who would amend what Akira had written 20 years ago, guided by the notes he left, and then add new material where appropriate to trace the influence of the original book in later work. Each chapter posed unique challenges, and the results I feel sure would have gratified Akira.

It will be clear in the reading that some branches of the subject have developed much more than others; this of course is to be expected. The early chapters provided the classical background for the subject and have changed only in small doses. Some of the later chapters summarize extremely active areas of research and have expanded substantially. Furthermore, the coverage is somewhat idiosyncratic, as Akira would have wanted. For example, no effort has been made to provide extensive coverage of the great advances in metapopulation theory and interacting particle systems, though these are introduced. Diffusion remains the integrative theme.

One piece remains unchanged—Akira's original preface. As Editor, I resisted suggestions that it was too personal and not sufficiently scientific. A reading of it will make clear that it is a gem, capturing a view of Akira's personality and humanity that could emerge in no other way.

This book is a gift from Akira's friends and colleagues to the memory of our Sensei. Akira, we hope you like it. It was a work of love.

Princeton, New Jersey   *Simon A. Levin*
October 2001

Preface to the First Edition

This book is an extended version of the Japanese edition, *Ecology and Diffusion*, published in 1975 by Tsukiji Shokan, Tokyo. All the chapters are revised and up-to-date; the last chapter is completely rewritten. These changes reflect an increased awareness of the importance of spatial processes such as diffusion in population ecology. The number of references cited in this book is almost double that of the original Japanese text.

This book surveys a wide variety of mathematical models of diffusion in the ecological context. The introductory chapter provides a brief history of diffusion problems in ecology together with a discussion on the use of mathematical models. The material presented throughout this book is almost entirely concerned with deterministic differential equation models. Chapter 2 presents the basics of diffusion ranging from a simple random walk model to environmental turbulent diffusion. Chapters 3 and 4 deal with the application of physical diffusion models to the dispersal of passive abiotic and biotic properties in ecosystems. A detailed discussion of mathematical modeling for animal dispersal is presented in Chapter 5; in Chapter 6 some examples of modeling animal diffusion with or without biological processes are given.

The tone of the book changes slightly in Chapter 7, which deals with the dynamics of animal grouping such as insect swarms and fish schools. Here two counteracting processes, advection and diffusion, are recognized to be important in modeling.

Chapter 8 is concerned with animal movements in home ranges, and Chapter 9 examines the role of advection and diffusion in the patchy distribution of organisms. The book ends with Chapter 10, which presents an extensive discussion of mathematical models of dispersing populations with intra- and inter-species interactions; some topics of interest are traveling waves of dispersing populations, models for density-dependent dispersal, and diffusion-induced instability in interacting populations.

The book is written with the primary intent of providing scientists, particularly physicists but also biologists, with some background of the mathematics and physics of diffusion and how they can be applied to ecological

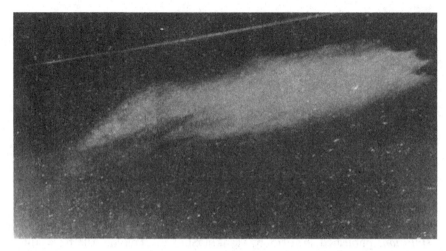

A swarm of desert locusts, from the air. Swarm covering 1 km^2 with locusts flying mainly below 20 m. Kenya, 13 January 1953. Photography by H. J. Sayer (from Rainey, 1958).

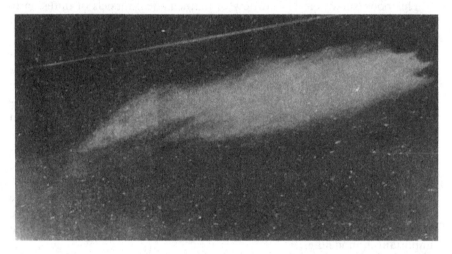

A patch of tracer dye in the sea, from the air. Patch covering 1 km^2 with most dye staying upper 5 m of water. Sea off Argentine coast, 27 July 1964. Photography by B. Katz.

problems. The secondary intent is to provide a specialized textbook for graduate students, who are interested in mathematical ecology. The reader is assumed to have a basic knowledge of probability and differential equations. Although this book is a volume in the *Biomathematics Series*, the use of mathematics is limited to the extent that mathematical models contribute to the interpretation of actual data or to gaining insight into the role of diffusion in model ecosystems.

A great deal of good luck was necessary for the completion of this book; in fact, seven "good lucks" combined. For the first part of the good luck, Dr. J. G. Skellam, "father of ecological diffusion," to whom this book is dedicated, encouraged a physical oceanographer who was greatly inspired by the pioneer work of Skellam and who was madly determined to study rodent dispersal on land. Without Dr. Skellam's interest in my approach, this book would probably not have been written.

For the second part of the good luck, Dr. H. C. Chiang, eminent entomologist, has been so generous as to allow me to work with him on midge swarming. I consider myself very lucky to have an association with Dr. Chiang. The tiny midges have educated me in the danger of building mathematical models for biological problems on sand. Without Dr. Chiang's interdisciplinary mind, certain pertinent chapters would have been missing in this book.

For the third part of good luck, I work at the Marine Sciences Research Center under the direction of Dr. J. R. Schubel, who is a most broadminded leader. Obviously, it is sometimes very difficult to justify an oceanographer who works with midges in a corn field of Minnesota. Fortunately enough, Dr. Schubel has a sense of scientific humor and a perception of how knowledge in one field can add to that in another. "Who knows? Many millions of years from now, those darned insects might invade deep into the last remaining sanctuary on the earth (i.e., the ocean)!" Without Dr. Schubel's understanding, I would have had to resign from MSRC to complete this book. (What would happen if I ask him to go to the Serengeti National Park to study wildebeest swarming, I now wonder.)

For the fourth part of good luck, I am indebted to Dr. S. A. Levin and Mr. K. Yano. Mr. Yano was the science editor of Tsukiji Shokan who kindly encouraged an amateur "ecologist" to write a book on ecological modeling. The result was my Japanese text in 1975. Unfortunately, the Shokan went bankrupt immediately afterward; I felt bad about it, wondering whether there was a causative relationship. Meanwhile, Dr. Levin, a distinguished mathematical ecologist, became aware of the book and made a generous offer to publish an English translation of the book as a Lecture Note in *Biomathematics* for which he serves as the managing editor. As days passed, the idea of translation evolved into a new book. Dr. Levin kindly reviewed all my drafts and gave constructive comments and criticism. Nonetheless, I may not have taken into account all the comments that he provided. Any mistakes are all my own. Without Dr. Levin and Mr. Yano, this book would probably not have been born.

The fifth part of good luck is attributed to Dr. G. N. Parker, who helped me in translating the original Japanese text and later in improving my English text. He is one of the rarest American scientists; one who can speak, read, and write Japanese as fluently as the natives. This gift is due partly to his intellectual Japanese wife and partly to his own talent. Throughout the preparation of this book, Dr. and Mrs. Parker encouraged me very much; often they whipped the lazy horse to keep running. Without their push this book would have been delayed at least five years.

The sixth part of good luck is very much biological; animals must eat. In the isolated village of Stony Brook, a determined nondriver was suffering from malnutrition. Then a miracle happened. After a random walk, or rather "drunkard walk," I discovered a cozy Japanese restaurant in Port Jefferson, least expected on this "remote" island. Thanks to the kindness and warm hospitality (e.g., free warm *sake*) of the Hashimotos, the hungry mouse narrowly escaped starvation, and the fire of life kept burning. Without the Hashimotos this book would have been written in a hospital bed.

Finally, the seventh part of good luck is due to an enormous amount of help and encouragement received from many other people, including friends and colleagues.

To all of these people I am very grateful. Also, I should like to express my thanks to Eileen Quinn for her skillful and patient typing of the final manuscript, to Joanne D'Amico and Deborah Ulrich for their efficient typing of the drafts, to Deborah Bray for indexing, and to Dean Loose, Marie Eisel, and Carol Cassidy for the illustrations; last of all, to Chiyo for her companionship in the long, long agonizing period of writing.

Stony Brook, New York
November 1978

Akira Okubo

Contents

Contributors

Josef David Ackerman
Environmental Studies Programme
University of Northern British
  Columbia
3333 University Way
Prince George, BC V2N 4Z9
Canada

Robert A. Armstrong
Marine Sciences Research Center
State University of New York
Stony Brook, NY 11794
USA

Leah Edelstein-Keshet
Department of Mathematics
University of British Columbia
1984 Mathematics Rd. #121
Vancouver, BC V6T 1Z2
Canada

Louis Gross
The Institute for Environmental
  Modeling
University of Tennessee
Knoxville, TN 37996
USA

Daniel Grünbaum
School of Oceanography
University of Washington
Seattle, WA 98195
USA

Alan Hastings
Department of Environmental
  Science and Policy
University of California, Davis
Davis, CA 95616
USA

Peter Kareiva
National Marine Fisheries Service
Route: F/NWC
Seattle, WA 98112
USA

Simon A. Levin
Department of Ecology and
  Evolutionary Biology
Princeton University
Princeton, NJ 08544
USA

James G. Mitchell
Department of Biology
Flinders University
Adelaide, SA 5001
Australia

Thomas Powell
Department of Integrative Biology
University of California, Berkeley
Berkeley, CA 95616
USA

Dennis P. Swaney
Boyce Thompson Institute
Cornell University
Ithaca, NY 14583
USA

Jeannette Yen
Marine Sciences Research Center
State University of New York
Stony Brook, NY 11794
USA

ERRATA

Diffusion and Ecological Problems: Modern Perspectives

Second Edition

Akira Okubo
Simon A. Levin

Page x: The top photograph, depicting a swarm of desert locusts, should be as follows:

Pages xiii and xix: Josef Ackerman's middle name is Daniel, not David.

1
Introduction: The Mathematics of Ecological Diffusion

Akira Okubo

1.1   A History of Research on Diffusion in Ecology

The mathematical treatment of the essential problem of ecology, the investigation of the interrelations between the distribution and quantity of organisms and their environment, has come into existence only recently. The remarks of Pielou (1969) in her book, *An Introduction to Mathematical Ecology*, merit consideration:

> The fact that ecology is essentially a mathematical subject is becoming ever more widely accepted.—One should not, however, draw the conclusion that mathematical ecology, as a subject, is a unified whole; far from it. Anyone who is dismayed by the seemingly fragmentary nature of the work described in this book must console himself with the thought that the subject is still in its early stages and that the welding together of its disconnected parts is a challenging job yet to be done.

Mathematical ecology has its roots in population ecology, which treats the increase and fluctuation of populations. It was along these lines that Lotka (1924) and Volterra (1926) established their original works on the expression of predator–prey and competing species relations in terms of simultaneous nonlinear differential equations, making the first breakthrough in modern mathematical ecology. One of the earlier treatises of this initial era was the Russian geophysicist Kostitzin's (1937, 1939) monumental work, *Mathematical Biology*. The model ecosystem of Lotka–Volterra has since been extended and developed, and the original deterministic model has been complemented by a probabilistic model; this discipline is still one of the mainstreams of mathematical ecology (Goel et al., 1971; May, 1973, 1976; Maynard Smith, 1974).

The goal of the Lotka–Volterra-type model is, briefly, the quantification of the interaction within and between species and their inorganic environment, and the investigation of the temporal variation of groups of individuals of various species. However, spatial variation is not considered. It is nevertheless true that time and space are inseparable "sister coordinates," and only when populations of organisms are considered in both time and space can the ecological situation be understood.

The fact that the majority of models in mathematical ecology treat only temporal variation may be due to some degree of partiality but is rather more likely caused by the difficulty of mathematical expression. The time rate of change of the number of individuals N in a population may be expressed as the derivative with respect to t, dN/dt. The equations of the ecosystem are then established by equating this derivative to another relation expressing the effect of species interaction on population. Such a simple analysis is not possible when spatial variations are considered. What is directly related to species interaction is the net *population flux* through an arbitrary infinitesimal piece of space rather than the spatial rate of change of the population itself, and thus a proper expression is unattainable without knowledge of the mechanism of movement of the organisms.

Consider the inorganic world; where a spatial inhomogeneity occurs in temperature, heat flows from regions of high temperature to regions of low temperature (conduction), and where an inhomogeneity occurs in the concentration of matter, matter flows from high to low concentration (diffusion). Also, when a space-occupying medium (e.g., air) itself flows, heat and matter are transported accordingly (advection or convection). The spatial gradients of the fluxes of heat or matter caused by spatial inhomogeneity specify the local rate of change of that heat or matter. While recognizing that the flow of populations of organisms represents a process vastly different from the flow of heat or mass, it is natural to utilize the thoroughly investigated techniques of the inorganic world as a point of departure for the mathematics of spatial variation in populations.

Such thinking had already begun to develop in the random walk theories of organisms due to Pearson and Blakeman (1906) and Brownlee (1911). It developed further with the models of the spread of genes and epidemics of Fisher (1937), Kolmogorov et al. (1937), Kendall (1948), and others. About that time Terada (1933) had indicated that to a certain extent even humans could be viewed as a group of inorganic particles and treated with the methods of statistical physics.

With the work of Skellam (1951), the basics of the theory of random dispersion of biological populations took shape. Skellam's method involved applying the analytical expression for molecular diffusion directly to ecological problems, relating it to the interaction among and between species. His treatise constitutes one of the classic works on biological dispersion.

However, as Skellam (1951, 1972, 1973) has suggested, the process of biological diffusion cannot be said to be purely random; rather, as can be seen in the activities of most animals, a special portion of space is preferred for use, and there is an element of choice in location. The motion of animal populations does not simply consist of spreading. Animals often concentrate together to form "groups": insect swarms, fish schools, bird flocks, buffalo herds, etc. In such cases, an effect that opposes diffusion occurs due to behavioral patterns and interaction between individuals. One of the important features that distinguishes the movements of animals from the random motion of inorganic material is this delicate balance between "spreading" and "concentrating." Animal dispersion cannot be analyzed without an understanding of these processes. Although the random walk forms our point of departure, we cannot simply rest peacefully with it. In addition, species propagation and intra- and interspecies relations work in the direction of the formation of spatially concentrated patterns of organisms. When, on the other hand, too much crowding occurs, population pressure causes the commencement of organism dispersion. Here again the endless interplay between order and disorder occurs.

Airborne bacteria, fungi, pollen, etc., have been of continuing interest to botanists; wind transport and atmospheric diffusion have been considered as processes that contribute to the determination of spore diffusion. The work of researchers in meteorology and fluid mechanics concerning atmospheric turbulent diffusion has made possible the quantification of spore diffusion, embodied in the investigations of such researchers as Rombakis (1947), Schrödter (1960), and Gregory (1961). Recently a new field, "aerobiology" (Inoue, 1974), involving the study of all objects suspended in the atmospheric that play a role in biological processes, has been founded. This field amalgamates the knowledge of physics, biology, chemistry, and mathematics.

The movement of many organisms, e.g., plankton, that are suspended in water is controlled by the motion of the water of the surrounding environment, and thus lake or ocean dispersion cannot be ignored in the consideration of organism dispersion. Even when the inorganic environment of the ocean appears uniform, horizontal density variations of plankton have been observed to exist, and such plankton "patchiness" has been of interest to biological oceanographers (Cassie, 1963; Steele, 1976a). While a tendency for the formation of patches exists, the effect of oceanic diffusion is to supply a mechanism opposing patch formation so that a balance can be maintained. However, when diffusion is combined with intra- and interspecies relations, instability of the ecological system may arise (Segel and Jackson, 1972; Levin, 1974). This can be realized from an application of the Turing (1952) effect, familiar in the dynamic theory of morphogenesis.

Experimental investigation of the phenomenon of animal dispersion developed first from insects, easily obtained in large numbers. Since the famous experiments of Dobzhansky and Wright (1943, 1947) on the release of Drosophila flies, a variety of excellent research has been conducted. Included are the works of such Japanese entomologists as Watanabe et al. (1952), Kono (1952), Ito (1952), and Morisita (1952, 1954, 1971), who contributed to the recognition of the concept of biological diffusion by including interactive forces between dispersing individuals in a quantification of "density-dependent dispersion." They emphasized that a model of dispersion must consider the forces operating between population individuals, and it cannot be limited to the simple random walk. These forces at times cause popula-

tions to "flow" in an orderly fashion, and one method of accounting for
them is their inclusion as advection in the diffusion equation. Some effort
has recently been made to construct advection–diffusion models that take
into account these behavioral aspects of animals (see, e.g., Shigesada and
Teramoto, 1978).

There are two viewpoints from which to investigate the motion of populations; the Lagrangian viewpoint involves identifying (marking) each individual and following the subsequent motion; in the Eulerian viewpoint the
flow of population individuals past a fixed point is observed. These two
viewpoints represent concepts originating in fluid mechanics, but observation
of the Lagrangian flow of fluid particles is decidedly more difficult than that
of Eulerian flow and can, in fact, be impossible in some cases. The basic
nature of Lagrangian observation is, however, much more suited to the
study of biological organisms. This fact allows for a glimmer of optimism to
fall on the otherwise nearly intractable study of animal motion; the development of telemetry techniques together with the computer has opened a
new horizon of research (Siniff and Jessen, 1969). In the future it can be expected that a quantitative knowledge of animal diffusion that can actually be
applied to the study of ecology will be based on such experimental research.

1.2   The Value of Mathematical Models

A mathematical treatment is indispensable if the dynamics of ecosystems are
to be analyzed and predicted quantitatively. The method is essentially the
same as that used in such fields as classical and quantum mechanics, molecular biology, and biophysics. In a quantification of "macrobiology," however, one has no access to such established principles as Newton's equations
of motion or the Schrödinger wave equation. If only diffusion is considered,
it may be noted that although molecular diffusion obeys Fick's law, turbulent diffusion in that part of the ecosystem known as the natural environment has found no firm basis comparable to Fick's law (Corrsin, 1974).

It must be realized that quantum mechanics, as well established as it is
now, was as vague as mist in its early stages. In that mist, Planck's quantum
hypothesis and Rutherford–Nagaoka's atomic model based on a solar-
system analogy were born. Bohr formulated the model and was able to explain the spectral lines of hydrogen by introducing the quantum hypothesis.
Thus, the method of (i) constructing a model based on analogy, (ii) formulating it, and (iii) solving it by introducing a hypothesis or hypotheses
that characterize the subject under consideration provides a very effective
approach toward systemization in fields where Baedekers do not exist. Of
course, one must not forget that analogies are no better than analogies,
models nothing more than models, and hypotheses simply hypotheses. The
mere fact that a mathematical model agrees well with a small amount of

data does not suffice, as the agreement could be coincidental. Moreover, models should not be confused with fundamental equations or laws. Only those hypotheses that have withstood large amounts of critical scrutiny can be elevated to the status reserved for laws.

In this regard it is natural that there should be ecologists who hardly recognize the value of mathematical models, and such is a proper attitude toward this field. I would like to assign, on the other hand, at least some value to mathematical models. It may be optimistic, but I feel that through trial and error the use of mathematical models in the field of ecological diffusion will eventually lead to the establishment of laws and basic equations. Even in physics, one half-century elapsed between the works of Galileo and Newton, and more than ten years were needed from Bohr's beginning to the solidification of quantum mechanics; so in the extremely complex field of ecology, the goal certainly cannot be reached in a ten-year time span. Certainly, however, if nothing is attempted, no results will be obtained at all.

Even while recognizing the value of mathematical models, an overemphasis is dangerous. Mathematical models appear to have a particular faddish appeal, especially to some young researchers, but science is never developed from fads alone. One must not become enamored of mathematical models; there is no mystique associated with them. Especially in ecology, physics and mathematics must be considered as tools rather than sources of knowledge, tools that are effective but nonetheless dangerous if misused.

Mathematical models can be rather broadly categorized into two types. One may be referred to as "educational," and the other as "practical" in nature. Educational models are based on a small number of simple assumptions and are analytically tractable. An example is the Lotka–Volterra predator–prey model. The method employed by such models involves the investigation of one or two essential processes that have been isolated from the complexities of the totality of ecological relationships. For this reason there exists a danger that the analysis may become divorced from reality. The real virtue of these models lies rather in the fact that they provide a process for gaining insight, expressing ideas, and eventually extending to more complex models. In this book, emphasis has been placed on educational models. This is necessary as an appetizer preceding more practical models, for indigestion is likely to result if one begins to gobble the latter before having savored the former well.

Practical models are based on realistic assumptions and thus involve the parameterization of interrelationships of large numbers of variables, and often the formulation of numerous (e.g., 20) equations containing numerous (e.g., 172) parameters is required (Kowal, 1971). In such cases analytical treatment becomes impossible, and one must rely on computer calculations. Here again the computer serves merely as a tool. Also, the deduction of basic and general rules from these complex models is more difficult than one might suppose. As Otake (1970) has observed,

The more complex a model becomes, the more parameters are involved. As models become more and more complex, it becomes progressively more difficult to demonstrate statistically that they do not in fact describe reality. Every time a new parameter is introduced in the model, a degree of freedom is lost, and our error terms in statistical tests lose strength. (Watt, 1968)

Finally, the reader may find papers by Steele (1976b) and Conway (1977) on mathematical modeling in ecology both "educational" und useful.

1.3   Deterministic Versus Stochastic Methods

Mathematical models can also be delineated into those of either a deterministic or stochastic nature. Roughly speaking, educational models often employ deterministic methods, while practical models tend to be stochastic in approach. Strictly speaking, nearly all biological processes are stochastic. However, a greater portion of the individuals involved in a process may be said to follow a single deterministic path *on the average*.

As an example, consider a population subject to exponential growth. According to deterministic theory, the rate of population increase, dN/dt, is proportional to the number of individuals at that time, $N(t)$. If λ denotes the coefficient of growth (pure birthrate), the following then holds:

$$\frac{dN}{dt} = \lambda N. \tag{1.1}$$

Solving (1.1) under the initial condition, $N = N_0$ at $t = 0$, we obtain

$$N = N_0 e^{\lambda t}. \tag{1.2}$$

The change in population is uniquely given by (1.2), and if λ is held constant for each run we will get the same result *deterministically*.

However, the process of population growth is usually stochastic, and at a given instant the growth does not necessarily take place at a rate λN, as shown in experiments with yeast. Thus, one must consider the probability $p(N,t)$ that at a given time the population will be N. According to Pielou (1969),

$$p(N,t) = (N-1)!/(N_0-1)!(N-N_0)!e^{-\lambda N_0 t}(1-e^{-\lambda t})^{N-N_0} \tag{1.3}$$

where $\lambda \Delta t$ represents the probability that in a small time interval Δt one individual will give birth to another individual. In this sense, λ should be considered as an average rate of growth. According to (1.3), the population at a given time varies from experiment to experiment even though N_0 and λ remain constant. Figure 1.1 illustrates the difference in the two methods (Skellam, 1955).

How can the difference between the two methods be resolved? Since the

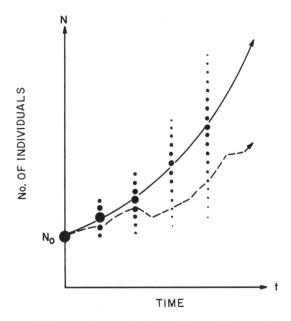

FIGURE 1.1. Time variations of population undergoing an exponential growth. Deterministic model (———→) vs. stochastic model (– – –→) (from Skellam, 1955).

probability density function is given by (1.3), we can calculate the expected value (probability average of population), $\bar{N}$,

$$\bar{N} = \sum_{N=0}^{\infty} Np(N, t) = N_0 e^{\lambda t},$$

and this coincides with the deterministic value. In other words, the deterministic method expresses the average state of the actual stochastic process; this statement is true in the case of exponential growth but is not necessarily true in general.

One then may inquire as to what extent this average state represents reality, i.e., actual experimental results. To answer this question, we calculate the standard deviation (s.d.) of N about its average value:

$$\text{s.d.} = (e^{\lambda t} - 1)^{1/2} \bar{N}^{1/2} = N_0^{1/2} e^{\lambda t} (1 - e^{-\lambda t})^{1/2}.$$

Taking the ratio of the s.d. with the average value,

$$\text{s.d.}/\bar{N} = (1 - e^{-\lambda t})^{1/2}/N_0^{1/2} \rightarrow 1/N_0^{1/2} (t \rightarrow \infty). \tag{1.4}$$

That is, the relative error of the deterministic method varies inversely with $N_0^{1/2}$ after a sufficient amount of time has passed. The larger the (initial) population, the better the agreement is between experimental values (which are not necessarily equal to the average value) and values obtained from the deterministic method. This result applies not only to simple exponential growth but also to the general case.

However, cases exist where a critical difference arises between the two methods. Consider simple birth and death processes occurring simultaneously, with the birthrate λ exceeding the death rate μ. According to the deterministic theory, $N = N_0 e^{(\lambda-\mu)t}$, so that the population must always increase. According to the stochastic theory, on the other hand, the possibility that the population becomes extinct also exists. The probability of extinction after a sufficient time has elapsed is given by $(\mu/\lambda)^{N_0}$ (Pielou, 1969). Since $\lambda > \mu$, when N_0 is sufficiently large, this probability is extremely small, but it is not zero. The existence of the possibility of extinction can in no way be obtained from the deterministic theory. Here the basic difference between the two methods appears, and in certain cases the deterministic method must be discarded. Clearly, the stochastic method is considerably less tractable.

As illustrated in Chap. 2, diffusion is also a stochastic process, and thus the above discussion must be generalized to consider the probability density function $p(S, \mathbf{x}, t)$ of finding the population density S at time t and at place $\mathbf{x}$. However, an evaluation of this function for diffusion in ecosystems is exceedingly difficult at present. Perhaps one must be content with considering only the change in time and space of the average population density, $\bar{S}$. In effect this implies a deterministic viewpoint. Thus, intra- and interspecific interactions should also be expressed in deterministic terms. Unfortunately, we encounter many cases in which the population density of the organisms concerned is not sufficiently large to validate the use of a deterministic method. This point has not yet been analyzed in detail for the case of ecological diffusion.

Note. In the present consideration of the deterministic population model, we tacitly assume that population growth is a continuous process and generations overlap. For many species such as certain insects, population growth takes place at discrete intervals of time, and generations are completely nonoverlapping; the appropriate mathematical description for this growth process is in terms of nonlinear *difference* equations rather than *differential* equations. Such nonlinear difference equations, even if simple and *deterministic* with respect to their characteristic parameters, can exhibit a remarkable spectrum of dynamic behavior including *apparent random* (chaotic) fluctuations. Thus, arbitrarily close initial population sizes can lead to entirely different patterns of population growth as time progresses. In fact, the dynamical fluctuations of the system are in many respects indistinguishable from the sample realizations of a random process. This rich dynamical structure of nonlinear difference equations has recently become a subject of considerable mathematical and ecological interest (Lorenz, 1964; May, 1975, 1976; May and Oster, 1976; Pielou, 1977; Li and Yorke, 1975; Yamaguti, 1978).

Editor's Note (Simon A. Levin)

This chapter is such a beautiful introduction to Akira Okubo's philosophy of science, and philosophy of modeling, that changing it in any way could only diminish it. Of course, Akira's observations about "recent" attention to nonlinear difference equations could be expanded, since the subject indeed has received great attention in the years since the First Edition was published. This chapter, however, is not the appropriate place for such discussions, which do receive attention later in the book.

Akira's notes to himself in guiding revision included, among a few other references, the work of Denis Mollison (1984) on model simplification and Stephen Wolfram on endogenous randomness (Wolfram, 1985). Mollison (1977a, b) also wrote seminal papers on spatial stochastic processes, discussed in the First Edition of this book, and Wolfram (1986) pioneered the use of cellular automata as models of self-organization. The exploration of interacting particle models—that is, stochastic cellular automata—as models for spatial phenomena has seen extensive development in the last decade and provides an important complement to diffusion models (Durrett and Levin, 1994a, b). So too have patch and metapopulation models (Levin and Paine, 1974; Gyllenberg et al., 1997; Hanski, 1991; Chesson, 1986), representing a tremendous growth industry especially in conservation biology, as well as point process models that describe the dynamics of forests and other communities of sessile organisms (e.g., Pacala et al., 1993). These topics are touched upon in this book, but a fuller survey of them can be found in Tilman and Kareiva (1997).

2
The Basics of Diffusion

Akira Okubo and Simon A. Levin

The concept of diffusion may be viewed naively as the tendency for a group of particles initially concentrated near a point in space to spread out in time, gradually occupying an ever larger area around the initial point. Herein the term "particles" refers not only to physical particles, but to biological population individuals or to any other identifiable units as well. Furthermore, the term "space" does not refer only to ordinary Euclidean space but can also be an abstract space (such as ecological niche space).

However, such a definition of diffusion is liable to invite confusion. For example, consider a number of particles released simultaneously from a point on a plane, each heading straight with its own speed and direction. Clearly, the particles will spread and occupy ever-increasing areas. However, such a process is not called diffusion.

Diffusion is a phenomenon by which the particle group as a whole spreads according to the irregular motion of each particle. Rephasing, when the microscopic irregular motion of each particle gives rise to a regularity of motion of the total particle group (macroscopic regularity), the phenomenon of diffusion arises. (A consideration of the long-term statistical trend of the irregular motion of a single particle also leads to the concept of diffusion.) In this fashion, one is naturally led to the concept of randomness.

2.1 Random Variables

When the individual values that a variable takes are specified only in terms of probabilities, the variable is termed a *random*, or *stochastic*, variable. For example, the result of casting a die is a random variable; it is not possible to predict which face will appear with each cast, but if the die is unbiased, the chance that each face will appear is 1/6. The term "randomness" is invariably defined in terms of probability. In such a fashion, randomness is seen to be neither haphazard nor unruly; rather it possesses a statistical regularity of its own.

In terms of randomness, diffusion can then be defined to be *a basically irreversible phenomenon by which matter, particle groups, population, etc., spread out within a given space according to individual random motion.* The term "irreversible" must be interpreted in a probabilistic sense. Its use does not imply that once a particle group has spread, it cannot return to its original concentrated configuration, again by a solely random process. Rather, as the number of diffusing particles increases, the probability of returning to the original state decreases and in the limit diffusion becomes strictly irreversible. Furthermore, even if the number of particles is small, a statistical average illustrates that diffusion proceeds. For example, even though a single particle may instantaneously return to its origin at one time, its average distance, i.e., the root mean square deviation, from the origin, increases with time.

However, it goes without saying that in certain instances nonrandom processes may occur simultaneously in such a way that the effect of diffusion is negated, and as a result antidiffusion (concentration) is observed.

2.2   The Random Walk and Diffusion

The point of departure for the theory of diffusion is the random walk, or random flight, model. Let us consider the one-dimensional space illustrated in Fig. 2.1. Each individual moves a short distance λ to the right or left in a short time τ. This motion is taken to be completely random (isotropic), and the probability of moving to either the right or left is $1/2$. After time τ, a given population spreads, one-half moving a distance λ to the right of the origin and one-half moving the same distance to the left. We do not know which individual will belong to which half. Continuing, in the next time interval τ, each individual moves a distance λ to the right or left, each alternative with a probability of $1/2$, independently of the previous motion. Thus, $1/4$ of the population finds itself a distance 2λ to the right of the origin, $1/4$ is the same distance to the left, and $1/2$ returns to the origin. The population as a whole exhibits spreading. We should like to know the form that the spatial distribution of population takes after time $n\tau$ has passed.

Clearly, after time $n\tau$ has passed, an individual occupies one of the points $-n, -n+2, \ldots, n-2, n$. However, the spatial population distribution is not uniform.

For example, an individual occupying the point farthest to the right, n, has arrived there by repeating a motion to the right n times successively; the

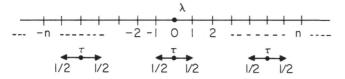

FIGURE 2.1. Random walk model (one-dimensional case).

probability of this occurrence becomes exceedingly small as n becomes large. We shall consider the probability of an individual arriving at a point m steps to the right of the origin after time $n\tau$ has passed. (It is assumed that $n \pm m$ is even.) If, to reach that point, an individual has gone a steps to the right and b steps to the left, then $a - b = m$ and $a + b = n$. Hence, $a = (n+m)/2$ and $b = (n-m)/2$. The number of possible paths by which the individual can arrive at m is known from combinatorial theory to be

$$\frac{n!}{a!b!} = \frac{n!}{\{(n+m)/2\}!\{(n-m)/2\}!}.$$

Now the total number of possible paths of n steps is 2^n, and thus the probability $p(m,n)$ that a particle will arrive at point m after n steps is

$$p(m,n) = \left(\frac{1}{2}\right)^n \frac{n!}{\{(n+m)/2\}!\{(n-m)/2\}!}. \tag{2.1}$$

Equation (2.1) is Bernoulli's distribution (also called the binomial distribution).

When n becomes extremely large, this distribution converges to the Gaussian distribution (also called the normal distribution). That is,

$$\lim_{n\to\infty} p(m,n) = (2/\pi n)^{1/2} \exp(-m^2/2n). \tag{2.2}$$

In the above, if $m\lambda = x$, $n\tau = t$, and (x,t) are considered to be continuous variables, the probability density function of the continuous random walk is

$$p(x,t) = \frac{1}{2(\pi Dt)^{1/2}} \exp(-x^2/4Dt). \tag{2.3}$$

Here D is defined to be the limit of $\lambda^2/2\tau$ as λ and τ become small and is called the coefficient of diffusion, or simply diffusivity. As discussed in detail in Chap. 5, D is obtained as λ and τ approach zero in such a way that the ratio of λ^2 and τ becomes constant in the limit. Herein if λ/τ is formally set to c (speed of walk), we can write $D = 1/2c\lambda = 1/2c^2\tau$, but in the limit c approaches infinity.

Distributions (2.2) and (2.3) apply only under the conditions $n \gg 1$, $m \gg 1$, or in other words, when the time of observation t is much greater than the duration time τ of each random step, and when the scale of observation x is much greater than the length λ of each random step. The results of a comparison of Eqs. (2.1) and (2.2) are seen in Table 2.1, from which it is clear that the Bernoulli distributions can be approximated by the Gaussian distributions for $n > 6$.

The above approach is essentially a Lagrangian one, in which individual particles are followed. An alternative, more general approach is to derive a recursion equation for $p(x,t)$, the probability that an individual is at point x at time t, or equivalently, the fraction of a population of individuals that are at x at time t. Because of the independence of individual motions in this

TABLE 2.1. Comparison between random walk model [Eq. (2.1)] and diffusion model [Eq. (2.2)]

p		0	±2	±4	±6	±8	±10	±12	±14
					m				
$n = 2$	Eq. (2.1)	0.50	0.25	0					
	Eq. (2.2)	0.562	0.21	0.01	0				
$n = 4$	Eq. (2.1)	0.375	0.25	0.063	0				
	Eq. (2.2)	0.400	0.24	0.054	0.004	0			
$n = 6$	Eq. (2.1)	0.3125	0.234	0.094	0.016	0			
	Eq. (2.2)	0.326	0.234	0.086	0.016	0.0016	0		
$n = 8$	Eq. (2.1)	0.273	0.219	0.109	0.03	0.004	0		
	Eq. (2.2)	0.282	0.220	0.104	0.03	0.005	0.0005	0	
$n = 10$	Eq. (2.1)	0.246	0.205	0.117	0.044	0.0097	0.001	0	
	Eq. (2.2)	0.252	0.206	0.114	0.042	0.010	0.002	0.0002	0

simple model, the population description is identical to the probability distribution that would apply to repeated independent realizations of a random process. The approach, however, can be extended to more general situations, as we will see in Chapters 5 and 7.

Given spatial step size λ and temporal step size τ, the recursion equation on p takes the form

$$p(x, t + \tau) = (p(x + \lambda) + p(x - \lambda))/2 \qquad (2.3.1)$$

because of the assumption that all individuals move every time step. To arrive at a continuous approximation to this formulation, one must allow λ and τ to shrink to 0 in a manner such that the limit

$$D = \lim_{(\lambda, \tau \to 0)} \lambda^2/2\tau \qquad (2.3.2)$$

exists (see, for example, Murray, 1989), which leads to the limiting diffusion equation

$$\partial p/\partial t = D\partial^2 p/\partial x^2 \qquad (2.3.3)$$

It is easily confirmed that (2.3) solves this equation, corresponding to the particular solution with all of the mass initially concentrated at the origin (δ-function). More generally, however, (2.3.3) describes the time evolution of any initial population distribution.

The situation just described is the simplest possible: Every individual moves one spatial step in each time step, and with probability 1/2 to the left and probability 1/2 to the right. This can be easily generalized to allow individuals not to move at all (the essential effect is simply to change the time scale) or to move to the left and right with different probabilities (introducing an advection term in the limiting equation), or it can be generalized to higher dimensions. We will encounter all of these modifications in later chapters from other perspectives. For a fuller treatment of the recursion approach to these situations, the reader is referred to the excellent book of Murray (1989). Density-dependent random walks have been considered in more detail by Aronson (1985), Flierl et al. (1999), Grünbaum (1992), Turchin (1989), and Yamazaki and Okubo (1995), among others. See also Chapters 5 and 7.

The constant-step random walk settles down to its limiting behavior relatively quickly compared, for example, to walks with exponentially distributed step sizes. The slower approach for the latter is a consequence of the need to smooth out the large fluctuations in step size inherent in the exponential distribution. See Rapaport (1985).

Random walks in two dimensions are recurrent; this is not true in higher dimensions (Feller, 1968). Put another way, this means that two particles undergoing random walks in a plane become arbitrarily close eventually with probability 1, but this does not hold for particles in three dimensions. Thus, in the acroporiid and some faviid coral species, gametes are discharged as buoyant egg–sperm bundles that rise to the surface and then break apart (Harrison et al., 1984; Cox and Sethian, 1985), in order to translate the three-dimensional problem into a two-dimensional one.

2.3   Fick's Equation of Diffusion

The classic theory of diffusion was founded more than one hundred years ago by the physiologist A. Fick. According to Fick's law, the amount of transport of matter in the x direction across a unit normal area in a unit time, i.e., the flux J_x, is proportional to the gradient of the concentration of matter. Thus, $J_x = -D\partial C/\partial x$, where C is the concentration of matter and D is the diffusivity. The minus sign indicates that diffusion occurs from high concentration to low concentration. Using this law, we can obtain Fick's equation of diffusion:

$$\frac{\partial C}{\partial t} = -\frac{\partial J_x}{\partial x} = \frac{\partial}{\partial x}\left(D\frac{\partial C}{\partial x}\right). \tag{2.4}$$

As a special case, when D is constant, Eq. (2.4) can be solved subject to the conditions that initially M particles in a unit area are concentrated at $x = 0$, resulting in

$$C(x,t) = \frac{M}{2(\pi Dt)^{1/2}} \exp(-x^2/4Dt). \tag{2.5}$$

If D is defined appropriately, Eq. (2.5) provides basically the same result as that of the random walk, Eq. (2.3). Thus, Fickian diffusion is applicable only to a diffusion phenomenon that corresponds to the random walk when λ and τ are small compared with, respectively, x and t.

Problems involving diffusion can be handled from both the probabilistic considerations of Sect. 2.2 and the use of the diffusion equation of this section. The former can be applied generally to include cases less restricted than those where the Fickian diffusion equations apply; however, the difficulty of mathematical formulations tends to increase as more complicated problems are considered. On the other hand, the equation of diffusion belongs to the family of parabolic linear partial differential equations, and its solutions for a broad range of problems have been investigated (e.g., Carslaw and Jaeger, 1959). This allows for extensions to more complex cases. However, the direct importation of a diffusion equation that applies strictly to such microscopic phenomena as the diffusion of molecular solutes to problems involving macroscopic diffusions, especially animal diffusion, is dangerous. With due caution regarding this, the central purpose of the present volume is to study ecological diffusion according to a diffusion equation appropriate for the problem.

However successful diffusion models have been in the physical sciences, it must be recognized that there are situations where the naive approach will not work well. Mollison (1977a, b) has emphasized this for situations in which long-distance moves are possible: A modified formulation is needed especially where it is important to recognize that individuals are packaged in discrete units, not as infinitesimals. Durrett and Levin (1994a) show how diffusion approximations can sometimes be obtained for such situations and how they differ from the naive formulation.

2.4   Turbulence and Turbulent Diffusion

When a fluid flows in an orderly fashion, the flow is called *laminar*; when the fluid flows in an irregular fashion with mixing, the flow is called *turbulent* (Fig. 2.2). Flows in the atmosphere and oceans are generally in the turbulent

a)
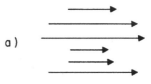

b)

FIGURE 2.2. Environmental fluid flows. (a) laminar flow, (b) turbulent flow.

state. Turbulence must be viewed as a type of random motion since it consists of many superposed whirls ("eddies") moving in a fashion that is spatially and temporally extremely complicated.

The motion of a small particle placed in an environmental fluid is influenced by the turbulence and accordingly is random. When the motion possesses randomness, it is accompanied by diffusion. We call the diffusion due to the turbulence of environmental fluids *turbulent diffusion* in order to distinguish it from the molecular diffusion associated with the random molecular motion that constitutes heat.

Turbulent diffusion is macroscopic in nature. When the diffusion of ecological systems is considered, another type of diffusion, due to the random motion of the organisms themselves, appears in addition to turbulent diffusion. This is referred to herein as *biodiffusion*.

Typical mean time scales of molecular motion are 10^{-10} sec (air) and 10^{-12} sec (water), and typical mean length scales are 10^{-5} cm (air) and 10^{-7} cm (water). Since these values are considerably smaller than our typical time and length scales of observation, molecular diffusion obeys Fick's law. However, the typical time and length scales of turbulence in such environmental flows as occur in the atmosphere, in lakes, or in the ocean are of the same order of magnitude as typical time and length scales of observation (Batchelor, 1950; Csanady, 1973).

For this reason we have no guarantee that a simple analogy to Fick's law is applicable to turbulent diffusion. Similarly, attempts to apply the random walk model to turbulent diffusion tread on thin ice. In the first place, the turbulent motion of particles is continuous, and motions separated by small time intervals are correlated with each other (Taylor, 1921); thus, this class of motions belongs to the category of correlated random walks.

Some characteristic features of turbulent diffusion are the following:

i. Since in the turbulent case elements of flow that are of much larger scale than the distances of molecular motion participate in mixing, turbulent diffusion is much more effective than molecular diffusion, at least with regard to macroscopic dispersion.

ii. Turbulence is composed of a wide range of eddy sizes, from small to large. Thus, the scale of the eddies participating in diffusion varies with the scale of the phenomenon, ℓ. As ℓ becomes larger, more and larger eddies participate in diffusion, and the effective diffusivity, A, increases. While it is difficult to provide a rigorous definition of the effective diffusivity, a definition, useful for factual diffusion experiments using tracers, can be given in terms of the observed time rate of change of the variance, σ^2, of the concentration distribution. That is, $A \equiv (1/2)\, d\sigma^2/dt$. If the diffusion obeys Fick's law and if the diffusivity D is constant, then A is equal to D regardless of the scale of the phenomenon. Both theoretically and experimentally, we have found that A varies with $\ell^{4/3}$ in the atmosphere, the

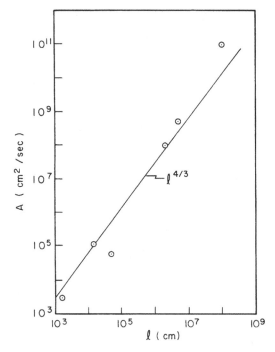

FIGURE 2.3. Relationship between horizontal diffusivity (A) and the scale of phenomenon (ℓ) in the atmosphere (from Richardson, 1926).

ocean, and lakes (Figs. 2.3 and 2.4). To be precise, the 4/3 power does not necessarily always apply but is quite sufficient as a general approximation.

iii. With the exception of very small-scale eddies (typically less than 10 cm), the turbulence associated with natural environments is nonisotropic, with horizontal scales that greatly exceed vertical scales. Accordingly, the horizontal diffusivity is several orders of magnitude larger than the vertical diffusivity. In fact, the 4/3 power law applies only to the horizontal diffusivity; in many cases the influence of the fluid density distribution is more important in determining the vertical diffusivity than the scale of the phenomenon.

iv. In a fluid near a solid surface such as in the atmosphere near ground level, the turbulence structure is affected by the boundary. The eddies decrease in size as ground level is approached. Accordingly, the vertical diffusivity also decreases (Fig. 2.5). Furthermore, topographical eddies can be created by the effect of ground objects like trees and buildings (Slade, 1968), and the turbulence structure differs from that over a flat surface (Fig. 2.6).

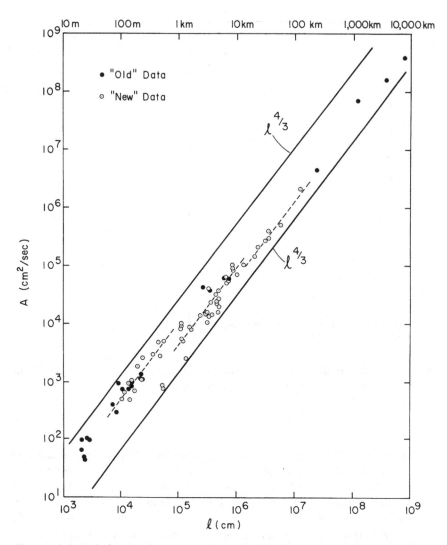

FIGURE 2.4. Relationship between horizontal diffusivity (A) and the scale of phenomenon (ℓ) in the sea (from Okubo, 1971a).

2.5   Diffusion in a Force Field

Thus far we have discussed the process of diffusion alone. However, in reality other processes also come into play, and it becomes necessary to consider the total resulting effect. For example, in the case of the vertical distribution of pollen in the atmosphere, the effect of particles settling under their own weight, in addition to atmospheric turbulent diffusion, must be taken into

MEAN WIND

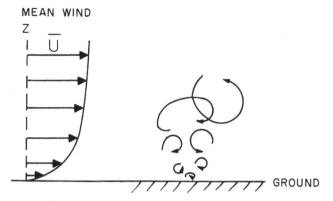

FIGURE 2.5. Wind profile and turbulent eddies near ground level.

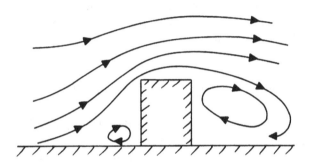

FIGURE 2.6. Topographical eddies due to a ground object.

account. Thus, we deal with diffusion under the influence of the earth's gravitation. The existence of forces acting on individual organisms is an important aspect of biodiffusion. Such forces are of general interpretation and may include the effect of environment on individuals, the attraction or repulsion between individuals, population pressure, and force fields that specify individual behavior.

In general, such a force can be decomposed into random and nonrandom components. Random forces are related to the random motion of the individuals themselves and can be included in the diffusion process. In many cases they are parameterized by the diffusivity.

The force remaining after the exclusion of its random part is not directly related to diffusion, but it brings about a mean (regular) motion of the individuals and acts to change the spatial distribution of the population. In some cases it counteracts diffusion and causes a concentration of population in a specific place (see Sect. 9.3), and in other cases it augments diffusion (see shear diffusion: Sect. 2.6.2). In most cases dispersion of biotic populations cannot be handled solely by a consideration of simple diffusion.

2.6   The Theory of Diffusion in Natural Environments: Physical Diffusion

Research on the subject of physical diffusion in natural environments began with atmospheric studies, starting from the pioneering work of Taylor (1921), Schmidt (1925), and Richardson (1926), and progressing through the efforts of such investigators as Sutton (1953), Frenkiel (1953), Ogura (1952), Inoue (1950, 1951), Gifford (1959), and Pasquill (1962, 1976) to the point where it can be utilized today in problems of practical significance such as air pollution (Seinfeld, 1986). Also, for oceanic and limnological diffusion a wealth of research has been produced since Joseph and Sendner (1958), Ozmidov (1958), Bowles et al. (1958), etc., developed methods unique for oceanography, to be distinguished from the mere imitation of the techniques of atmospheric diffusion. Development of the theoretical model for oceanic diffusion has closely paralleled the development of a field technique using fluorescent dyes (Pritchard and Carpenter, 1960; Carter and Okubo, 1978). Treatments of physical diffusion in a natural environment fall into three main categories.

(i)   The Theory of the Variation of Diffusion Width of a Group of Particles

This treatment takes the standard deviation of the particle distribution as the "diffusion width"; in the case of diffusion from an instantaneous source, the time variation of the spread (diffusion width) is considered, and in the case of release from a continuous source, the variation of spread as a function of distance from the source is considered (Figs. 2.7 and 2.8). From a theoretical standpoint, we deal with the square of the standard deviation, i.e., the (statistical) variance σ^2, rather than the standard deviation itself.

The variance of particle position can be expressed in terms of the correla-

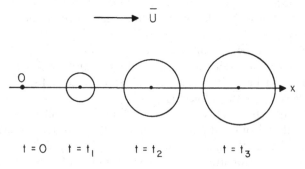

FIGURE 2.7. Time variation of the diffusion width of a patch release from an instantaneous source; $t = 0 < t_1 < t_2 < t_3$.

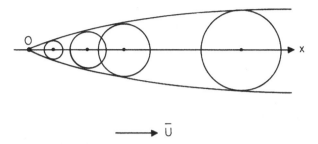

FIGURE 2.8. Variation of the diffusion width of a plume as a function of distance from a continuous source.

tion function or correlation coefficient of particle velocity at two different times (Taylor, 1921). The structure of this function is determined by the structure of the atmospheric or oceanic turbulence. Thus, the diffusion width also depends on the structure of the turbulence; in the initial stages of diffusion, each of the particles may be regarded as dispersing outward with its own initial speed, and the variance increases linearly with the second power of time or distance. In the later stages of diffusion, when the width of a group of particles becomes larger than the largest effective eddies composing the turbulence, the subsequent diffusion resembles that of molecular diffusion. The variance then increases proportionally to time or distance; the coefficient of proportionality is the (constant) diffusivity, but it is many magnitudes of order larger than the molecular diffusivity. (For molecular diffusion, $\sigma^2 = 2Dt$. The value of the molecular diffusivity D is 0.2 cm^2/sec for air and of the order of 10^{-5} cm^2/sec for water.) In actual cases of diffusion, it is common to observe that the final stage has not in fact been reached; thus, the variation of the diffusion width in the intermediate stages becomes the object of our concern, and Fickian diffusion, whereby diffusivity is independent of scale, loses its applicability. This point can also be seen clearly in Figs. 2.3 and 2.4.

(ii)   Particle Distribution Relations Obtained from Diffusion Width and
       Similarity Hypotheses

Let us consider a group of particles instantaneously released from a single point. As time proceeds, the patch spreads. We may postulate that the shape of the particle distribution relative to the center of gravity is preserved. This is known as the *similarity hypothesis of diffusion* (Fig. 2.9). In other words, we assume that, while the concentration at the center of the patch decreases in time, the concentration distribution relative to the center can maintain the same shape when reduced with a spatial scale that varies appropriately with time. It is natural to choose the diffusion width as this spatial measure. As a result, the laws of similarity allow us to express the three-dimensional con-

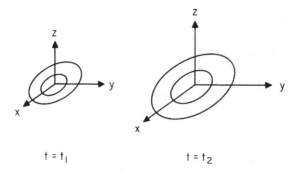

$$t = t_1 \qquad\qquad\qquad t = t_2$$

FIGURE 2.9. Similarity hypothesis of diffusion from an instantaneous source; $t = t_1 < t_2$.

centration distribution as (Okubo, 1970)

$$\bar{S}(x, y, z, t) = \frac{M}{a\sigma_x\sigma_y\sigma_z} F\{x/\sigma_x, y/\sigma_y, z/\sigma_z\}. \qquad (2.6)$$

Here M is the total number of particles, and σ_x, σ_y, and σ_z are, respectively, the standard deviations in the x, y, and z directions and are in general functions of time t. F is a functional form that determines the shape of the distribution, and a is a constant depending on the nature of F. If a Gaussian distribution is assumed for F, we obtain from (2.6),

$$\bar{S}(x, y, z, t) = \frac{M}{(2\pi)^{1/2}\sigma_x\sigma_y\sigma_z} \exp\left\{-\frac{1}{2}(x^2/\sigma_x^2 + y^2/\sigma_y^2 + z^2/\sigma_z^2)\right\}, \quad (2.7)$$

where we can use the diffusion widths discussed in (i) for σ_x, σ_y, and σ_z.

Likewise, the similarity hypothesis applied to continuously released particle plumes allows for delineation of the concentration distribution (Fig. 2.10). Let us assume that there is a mean flow (wind, ocean current) with velocity U and that the plume has reached steady state. The concentration along the central axis, x, decreases with distance. If we assume that the concentration distribution in a plane normal to the x-axis maintains a similar shape, we obtain, analogous to the previous result,

$$\bar{S}(x, y, z) = \frac{Q}{\beta U\sigma_y\sigma_z} G\{y/\sigma_y, z/\sigma_z\} \qquad (2.8)$$

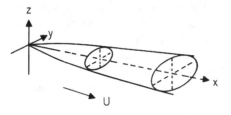

FIGURE 2.10. Similarity hypothesis of diffusion from a continuous source.

(Okubo, 1970). Q is the rate of release of particles, and σ_y and σ_z are, respectively, the standard deviations in the y and z directions and are in general functions of the distance x from the point of release. (The values $\sigma_y(t)$ and $\sigma_z(t)$ of the patch and $\sigma_y(x)$ and $\sigma_z(x)$ of the plume are identical if $t = x/U$.) G is the functional form that determines the cross-sectional shape of the distribution, and β is a constant depending on the nature of G. If a Gaussian distribution is assumed for G, we obtain

$$\bar{S}(x, y, z) = \frac{Q}{2\pi U \sigma_y \sigma_z} \exp\left\{ -\frac{1}{2}(y^2/\sigma_y^2 + z^2/\sigma_z^2) \right\}. \tag{2.9}$$

The above considerations apply to diffusion in a boundary-free environment; with appropriate correction they can be utilized for cases where influence is felt from ground or water surfaces. Furthermore, they can be applied to the case where the source is not a point but is rather of finite extent. Environmental conditions are implicit in the values of σ_x, σ_y, and σ_z, and thus, they must be expressed in terms of parameters that express the nature of the environment. Also, in a rigorous treatment the shapes of F and G must be determined from the structure of the environmental turbulence.

Thus far, we have discussed statistical characteristics of diffusion. These theories are useful for predicting the overall characteristics of diffusing particles. We may not use them, however, to estimate instantaneous properties, say of particle concentrations in a particular spatial region at a particular time (Squires and Eaton, 1991). In ecological problems, we often need to know the instantaneous properties of particle concentrations. To this end, Squires and Eaton simulated the movement of particles in a turbulent flow field. Thus, they found that particle inertia causes aggregation of particles in regions of low vorticity and high strain rate, in agreement with Maxey (1987). Conventional wisdom that the particles are uniformly distributed by turbulence can be grossly in error (see also Sect. 4.1.4).

(iii)   The Equations of Turbulent Diffusion

Let us consider an infinitesimal volume around some point $\mathbf{x}$ in a turbulent environment and express the instantaneous particle concentration within it as $S(\mathbf{x}, t)$. The turbulence causes the value of S to change randomly; thus, we consider an average value $\bar{S}$ obtained by taking the statistical average of the values resulting from an infinite number of experiments in which the macroscopic conditions creating the turbulence do not vary. We then query as to what diffusion equation $\bar{S}$ obeys.

Due to the turbulence, the instantaneous velocity $u(\mathbf{x}, t)$ at a given place also varies randomly. The instantaneous flux of particles in the x direction generated by this velocity is uS. For an infinitesimal volume, we can assume that the diffusion of the instantaneous concentration obeys Fick's law. Therefore, the instantaneous flux in the x direction due to molecular diffusion is given by $-D\partial S/\partial x$. The total instantaneous flux in the x direction is given by the sum of these two,

$$J_x = uS - D\partial S/\partial x.$$

(In Sect. 2.3, flux due to a velocity field was not considered.) The one-dimensional equation of diffusion for instantaneous concentration in an infinitesimal volume can thus be expressed as

$$\frac{\partial S}{\partial t} = -\frac{\partial J_x}{\partial x} = -\frac{\partial}{\partial x}(uS) + \frac{\partial}{\partial x}\left(D\frac{\partial S}{\partial x}\right). \tag{2.10}$$

Now let us decompose the instantaneous quantities u and S into their statistical averages, $\bar{u}$ and $\bar{S}$, and the deviations (due to turbulence) from the averages, u' and S':

$$u = \bar{u} + u';$$
$$S = \bar{S} + S'. \tag{2.11}$$

Substituting (2.11) into (2.10) and averaging (from the definition of statistical averages, $\bar{S}' = 0$, $\bar{\bar{S}} = \bar{S}$), we obtain

$$\frac{\partial \bar{S}}{\partial t} = -\frac{\partial}{\partial x}(\bar{u}\bar{S}) + \frac{\partial}{\partial x}(-\overline{u'S'}) + \frac{\partial}{\partial x}\left(D\frac{\partial \bar{S}}{\partial x}\right). \tag{2.12}$$

In other words, the time rate of change of the average concentration is determined by the sum of the net transport due to the mean velocity (advection), the net transport due to the turbulence (turbulent diffusion), and molecular diffusion. Thus, a direct application of Fickian diffusion to $\bar{S}$ would not allow for the turbulent diffusion term. The turbulent transport term is, moreover, generally very much larger than the molecular diffusion term; however, being composed of the average of the product of the deviations of the velocity and concentration, it is an unknown quantity.

The theory of turbulent transport allows us to make an analogy to the kinetic theory of gases and relate $\overline{u'S'}$ to the gradient of the average concentration $\partial \bar{S}/\partial x$ (for details, see Corrsin, 1974; Ogura, 1955). In other words,

$$\overline{u'S'} = -K\partial \bar{S}/\partial x \tag{2.13}$$

where K is the diffusivity due to turbulence in the x direction and is variously referred to as the *turbulent diffusivity* or the *eddy diffusivity*. Using (2.13) to rewrite (2.12), we obtain

$$\frac{\partial \bar{S}}{\partial t} = -\frac{\partial}{\partial x}(\bar{u}\bar{S}) + \frac{\partial}{\partial x}\left\{(D+K)\frac{\partial \bar{S}}{\partial x}\right\}. \tag{2.14}$$

According to transport theory, K can be expressed as the product of a mixing length ℓ^* and a measure of the intensity of the turbulent velocity fluctuation u^*. For a one-dimensional model,

$$K = \frac{1}{2}\ell^* u^*. \qquad (2.15)$$

Comparing (2.15) to the random walk expression for D, we see that if $\ell^* \leftrightarrow \lambda$ and $u^* \leftrightarrow c$, then $K \leftrightarrow D$. The turbulent diffusivity formally appears to be an extension of the concept of molecular diffusivity; however, since the scale of turbulence ℓ^* and the scale of diffusion ℓ (e.g., the diffusion width) are of the same order of magnitude, there is a fundamental contradiction involved in applying the random walk analogy to turbulent diffusion. Accordingly, the notion that diffusion can be determined in terms of an eddy diffusivity can be regarded as a necessary evil (for details, see Corrsin, 1974).

It is a simple matter to generalize (2.14) to the three-dimensional case. We shall neglect molecular diffusivity due to its small size compared to the turbulent diffusivity and, including the possible effect of internal changes of concentration (due to, for example, chemical decomposition of matter) within an infinitesimal volume, generalize to

$$\frac{\partial \bar{S}}{\partial t} = -\frac{\partial}{\partial x}(\bar{u}\bar{S}) - \frac{\partial}{\partial y}(\bar{v}\bar{S}) - \frac{\partial}{\partial z}(\bar{w}\bar{S}) + \frac{\partial}{\partial x}\left(K_x \frac{\partial \bar{S}}{\partial x}\right)$$

$$+ \frac{\partial}{\partial y}\left(K_y \frac{\partial \bar{S}}{\partial y}\right) + \frac{\partial}{\partial z}\left(K_z \frac{\partial \bar{S}}{\partial z}\right) + \bar{R}. \qquad (2.16)$$

Here $\bar{u}, \bar{v}$, and $\bar{w}$ are, respectively, the average velocities in the x, y, and z direction; K_x, K_y, and K_z are the eddy diffusivities in the x, y, and z directions; and $\bar{R}$ is a nonconservative term expressing internal changes in concentration.

Environmental fluid flow may be regarded as incompressible (Imai, 1970; Batchelor, 1967; Phillips, 1966), and thus, the average velocities obey the condition

$$\frac{\partial \bar{u}}{\partial x} + \frac{\partial \bar{v}}{\partial y} + \frac{\partial \bar{w}}{\partial z} = 0. \qquad (2.17)$$

To solve problems of environmental turbulent diffusion it is necessary to specify $\bar{u}, \bar{v}, \bar{w}$. K_x, K_y, K_z, and $\bar{R}$ appropriately as functions of space and time.

Although diffusion in the natural world takes place usually in two- or three-dimensional space, the discussion throughout this book will mostly be dependent on one-dimensional diffusion models, which are applicable to a linear habitat such as a shoreline or riverbank. The use of the one-dimensional model is due to mathematical simplicity, and most of the results from one-dimensional space should be extendable to higher dimensions. However, care must be exercised due to the fact that in a few cases an essential difference exists between diffusion in one-dimensional space and that of higher dimensions (Widder, 1975).

The biological phenomenon of chemotactic collapse may depend on the dimensionality of space (Childress and Percus, 1981). In particular, for the special model that Childress and Percus investigated, collapse cannot occur in a one-dimensional space, may or may not in two dimensions, and must for three or more dimensions, under a perturbation of sufficiently high symmetry. We thus have another instance of the critical role of dimensionality in determining the qualitative character of nonlinearly driven systems.

2.6.1   Diffusion in Atmospheric Boundary Layers

The region extending from ground level to $\sim$10–100 meters above the earth's surface is known as the atmospheric surface boundary layer, where the average wind velocity increases regularly with height. Many terrestrial organisms inhabit portions of this boundary layer.

When the atmosphere is in a state of neutral stability (Haltiner and Martin, 1957; Shono, 1958), the shear stress (τ_0) is assumed to be constant in the vertical direction, and theoretically the average wind velocity is given by the logarithmic law

$$\bar{u} = u_*/k \ln(z/z_0). \tag{2.18}$$

In the above, $u_* = (\tau_0/\rho)^{1/2}$ is the friction velocity, ρ is the density of air, z_0 is the roughness length, and k is the Von Karman constant, taking the value of 0.4.

The vertical eddy diffusivity is a function of height and is given by

$$K_z = ku_*z. \tag{2.19}$$

Thus, the steady-state diffusion from a point source placed in the surface boundary layer is formulated as

$$\bar{u}(z)\frac{\partial \bar{S}}{\partial x} = \frac{\partial}{\partial y}\left(K_y \frac{\partial \bar{S}}{\partial y}\right) + \frac{\partial}{\partial z}\left(K_z \frac{\partial \bar{S}}{\partial z}\right), \tag{2.20}$$

where the diffusion in the wind direction x has been neglected compared to the advection; this hypothesis is widely used in connection with diffusion from a continuous source. It is a good approximation where there is a prevailing mean flow in the environmental fluid (Frenkiel, 1953).

Equation (2.20) can be solved with the aid of (2.18) and (2.19) (Chatwin, 1968). From a practical point of view, more convenient forms are given by

$$\bar{u} = u_* q(z/z_0)^\alpha,$$

$$K_z = ku_*z_0(z/z_0)^\beta \qquad (\alpha, \beta, q: \text{constant}), \tag{2.21}$$

which can be inserted into (2.20) for a solution (Calder, 1949; Deacon, 1949; Sutton, 1943). The constants α, β, and q take various values depending on atmospheric conditions. The interested reader may consult Sutton's book (1953).

Sutton (1953) further used Eq. (2.9) in a treatment of diffusion from a continuous source near ground level, obtaining semiempirical power-function expressions for σ_y and σ_z in terms of x:

$$\sigma_y^2 = \frac{1}{2}c_y^2 x^{2-n}$$

$$\sigma_z^2 = \frac{1}{2}c_z^2 x^{2-n}.$$

(2.22)

Here n, c_y, and c_z are contants depending on such factors as atmospheric stability and height of the source above ground. If Eqs. (2.22) are inserted into (2.9) and a reflective boundary condition is applied at the ground (see Sect. 4.1), the concentration distribution for a source located at the ground is found to be

$$\bar{S}(x, y, z) = \frac{2Q}{\pi U c_y c_z x^{2-n}} \exp\{-x^{n-2}(y^2/c_y^2 + z^2/c_z^2)\}.$$

(2.23)

According to Sutton (1953), when the atmosphere is neutrally stable, $n = 0.25$, and when the ground is flat grassland, $c_y^2 = 0.4$ cm$^{1/8}$ and $c_z^2 = 0.2$ cm$^{1/8}$. These values provide good agreement with experiments. However, one must remember that they are purely empirical.

Yokoyama (1960) evaluated n, c_y, and c_z from experimental data incorporating a variety of meteorological conditions. He found that the Sutton equations were applicable to moderate wind speeds (1–5 m/sec) and intermediate scales (a few kms to a few tens of kms).

While Sutton's theory of atmospheric diffusion is still widely used in practical problems of prediction for dispersion, its general applicability is very limited. For more advanced diffusion theories, consult Slade (1968), Pasquill (1962, 1976), Frenkiel and Munn (1974), and Csanady (1973).

2.6.2 Oceanographic and Limnological Diffusion

Diffusion in the surface layers of oceans and lakes should essentially be approached by the methods developed for atmospheric boundary layers; however, because such factors as the velocity profile and the variation of the vertical diffusivity are not well known, one must rely on simpler models. We shall first discuss diffusion from an instantaneous source.

The horizontal turbulence is of considerably greater scale than that of the vertical direction; thus, horizontal diffusion becomes the main concern. We assume that the diffusing substance is distributed uniformly within a vertical layer of constant thickness h. We furthermore treat the horizontal diffusion as isotropic, allowing us to postulate the concentration distribution to be a function of distance from the patch center r, and t alone. Thus, the diffusion equation of the patch is found to be

$$\frac{\partial \bar{S}}{\partial t} = \frac{1}{r}\frac{\partial}{\partial r}\left(rK_h\frac{\partial \bar{S}}{\partial r}\right), \tag{2.24}$$

where K_h is the horizontal diffusivity and is in general a function of r and t.

Various solutions result according to the form assumed for K_h (Okubo, 1962). Three representative ones are

$$\bar{S} = \frac{M}{4\pi hKt}\exp\left(-\frac{r^2}{4Kt}\right), \tag{2.25}$$

$$\bar{S} = \frac{M}{2\pi hP^2t^2}\exp\left(-\frac{r}{Pt}\right), \tag{2.26}$$

$$\bar{S} = \frac{M}{6\pi h\gamma^3 t^3}\exp\left(-\frac{r^{2/3}}{\gamma t}\right). \tag{2.27}$$

Equation (2.25) represents the Fickian solution, where $K_h = K$ (constant). In the case of turbulent diffusion, K_h is not constant, but for practical purposes the solution can perhaps be used locally with an appropriate choice of K (see Sect. 4.2). Equation (2.26) is the Joseph and Sendner (1958) solution, in which P is called the "diffusion velocity." It takes a value of 1 ± 0.5 cm/sec for oceanic surface layers and large lakes and reduces to about 0.1 cm/sec for thermoclines or deep waters. Equation (2.27) is the Ozmidov (1958) solution, in which γ is called the "energy dissipation parameter," typically taking values of the order of 10^{-2} cm$^{2/3}$/sec in the oceanic surface layer (Okubo, 1962).

The horizontal variances $\sigma_r^2 \,(= 2\sigma_x\sigma_y)$ corresponding to Eq. (2.25), (2.26), or (2.27) are, respectively,

$$\sigma_r^2 = 4Kt, \tag{2.28}$$

$$\sigma_r^2 = 6P^2t^2, \tag{2.29}$$

$$\sigma_r^2 = 60\gamma^3 t^3, \tag{2.30}$$

from which it is seen that the rate of increase of the variance is slowest for the Fickian model. Data for oceanic and limnological diffusion tend to support either (2.29) or (2.30) but do not agree with (2.28) (Murthy, 1976; Okubo, 1971a). We can calculate an apparent horizontal diffusivity $A = \sigma_r^2/4t$ from the temporal variation of the horizontal variance. A delineation of the relation between this and the diffusion scale $\ell = 3\sigma_r$ for (2.28), (2.29), and (2.30) provides, respectively,

$$A = K \quad \text{(independent of } \ell), \tag{2.31}$$

$$A = 0.204P\ell, \tag{2.32}$$

$$A = 0.226\gamma\ell^{4/3}. \tag{2.33}$$

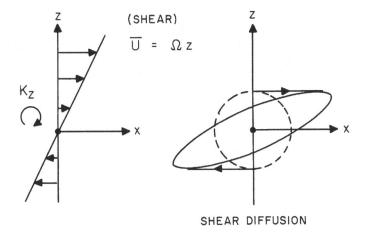

FIGURE 2.11. Shear-diffusion model.

It is thus seen that except for the case of Fickian diffusion, A increases with the scale of diffusion (see Sect. 2.4). In general, horizontal and vertical diffusivities in a lake are smaller than in the sea (Abbott et al., 1984). However, large lakes such as the Great Lakes exhibit diffusivities of the same order as the sea (Murthy and Dunbar, 1981).

The above are models of isotropic horizontal diffusion. However, in reality the shear of the mean flow interacts with diffusion in the direction of the shear to produce a pronounced "shear diffusion" in the direction of flow (Taylor, 1953, 1954; Bowles et al., 1958; Saffman, 1962; Bowden, 1975; Csanady, 1973). This is thought to be the dominant mechanism of diffusion in environmental flows, an idea supported by considerable experimental evidence (Fischer, 1973; Okubo, 1971b).

Shear diffusion can be explained by the use of a simple model (Fig. 2.11). A patch released in a layer where uniform shear given by the velocity profile $\bar{u} = \Omega z$ exists will vertically diffuse a distance of the order of $(K_z t)^{1/2}$, up and down in time t. Accordingly, the vertical velocity difference (effective shear) acting on the patch amounts to $\Omega (K_z t)^{1/2}$, and the patch will disperse horizontally, in time t, a distance $\Omega (K_z t)^{1/2} t$. The horizontal variance is roughly the square of this, or $\Omega^2 K_z t^3$ (according to an exact analysis, $2/3\Omega^2 K_z t^3$). This horizontal variance is the result of interaction between shear and vertical diffusion; the apparent horizontal diffusivity is increased to the order of $\Omega^2 K_z t^2$, and the patch tends to elongate in the direction of flow. The model discussed in regard to diffusion in the atmospheric boundary layer (2.20) is, in fact, an example of shear diffusion.

Shear diffusion also occurs in shallow regions such as estuaries and coastal waters, but since a vertical boundary exists in these cases, the results are somewhat different. The apparent horizontal diffusivity becomes proportional to the product of depth and vertically averaged velocity. Equivalently, it may be expressed as proportional to the product of the frictional velocity and depth. Bowden (1965) has obtained expressions for the horizontal diffusivity due to various velocity distributions.

In the case of rivers, the apparent downstream horizontal diffusivity is generally thought to be determined by shear in the lateral direction (Fischer, 1968, 1973).

3
Passive Diffusion in Ecosystems

Akira Okubo, Josef Daniel Ackerman, and Dennis P. Swaney

Preface: *In writing this revision, we endeavored to include information and directions contained in Akira's notes. He suggested most of the new sections, and we included other new ones based on our understanding of his interests and scientific concerns. We dedicate this effort to a teacher, advisor, mentor, and friend.*

JDA would like to acknowledge the assistance of Julia Bolton for her efforts in many aspects of this work, Trent Hoover for rendering crude sketches into figures, Nicole Bock of the Interlibrary loan office for ordering the mountains of articles, and NSERC for research support. DPS would like to acknowledge the assistance of Karin Limburg, who made helpful comments on sections of the manuscript, and Christoph Humborg, Sven Blomqvist, and Christopher Post, who suggested several references and helped with the literature search.

In this chapter we consider several aspects of passive diffusion in the environment. In this book, *passive diffusion* is defined to be the diffusion of chiefly abiotic objects that are not capable of performing random motion without the help of environmental turbulence. Likewise, *active diffusion* is defined to be the diffusion of objects, chiefly animals, that perform motion by themselves.

Wolfenbarger (1975) introduced the following definition: When transportation is supplied by energy from within the organism, the dispersion is termed "active", and when it is supplied by energy from outside the body, the dispersion is called "passive".

The smaller the organism, the more it is subject to the effect of environmental turbulence. Thus, the diffusion of small animals should be considered as partly passive and partly active. For instance, bacteria and pollen in the air and phytoplankton in the water diffuse almost passively, while many insects in flight undergo varying proportions of passive and active diffusion, according to the degree of movement of the environmental fluid.

Theories of turbulent diffusion are expected to hold approximately for organisms that diffuse in an essentially passive manner. In this case, however, attention must be paid to the fact that, depending on the organism's size, certain components (i.e., small-scale motions of the environmental turbulence) will not be effective agents of dispersion. The effect of falling or rising must be considered when the density of the organism differs significantly from that of the surrounding medium. Moreover, given sufficient time, behavioral responses of small organisms, including phototaxis and geotaxis, may lead to patterns of dispersion that differ from that of the fluid. Remember that dispersion in turbulent fluids is both directional and random: (1) directional in the direction of advection of the fluid; and (2) random or chaotic-like within the eddies associated with the turbulence.

3.1   Diffusion Within and Above Plant Canopies

Plant canopies are of particular importance in the ecology of many organisms, in that they provide structure, habitat, shelter, and resources (Russell et al., 1989). For example, one need only consider that most of the principal agricultural crops are species that form canopies (Heiser, 1990) to understand the implications to human societies. Plant canopies modify local environments and thereby provide the constituent elements for the formation of terrestrial (e.g., grasslands, forests) and aquatic ecosystems (e.g., marshes, weed beds, kelps). While many of the fluid dynamic factors related to diffusion within and above plant canopies (Raupach and Thom, 1981; Raupach et al., 1991; Finnigan, 2000) are similar regardless of the environment, there are sufficient differences such that both terrestrial and aquatic ecosystems are treated separately here.

3.1.1   Terrestrial Plant Canopies

The wind distribution above plant canopies, when the atmosphere is neutrally stable, is represented by a modified logarithmic law (i.e., the law of the wall; refer to Sect. 2.6.1),

$$u(z) = \frac{u_*}{k} \ln\left\{\frac{z-d}{z_0}\right\}, \qquad z > H > d + z_0. \qquad (3.1)$$

This form accounts for the roughness of the surface and displacement of flow due to the presence of plants (Inoue, 1963; Thom, 1975; Grace, 1977); H is the mean plant height, d is called the datum-level displacement, or zero plane displacement, and the other quantities were already defined in Sect. 2.6.1. The wind expressed in (3.1) is a mean quantity; thus, it should have been written $\bar{u}$. The overbar is omitted for simplicity, a convention that is used frequently throughout this book.

Expression (3.1) provides a model for the boundary layer above the plant canopy. The canopy boundary layer can be subdivided into (1) a *roughness sublayer* extending 1 to $2H$ from the top of the canopy, (2) an *inertial* or *logarithmic sublayer* above the smaller *roughness sublayer*, and (3) an *outer sublayer* that extends to the region where u approaches the free-stream velocity within that region of the planetary boundary layer (i.e., that region within ≈ 1 km of the earth's surface). Typically, velocity measurements made within the inertial sublayer can used to estimate u_*, d, and z_0 (e.g., Monteith and Unsworth, 1990).

As a rule of thumb for a typical agricultural crop, the roughness parameter, z_0, is 10% of the length of the surface protuberances, and the zero plane displacement, d, is 60% to 70% of the height of the plant (Grace, 1977). Takeda (1965, 1966) derived a theoretical relation between d and z_0. Later, Maki (1969, 1976) extended the concept to obtain a new relationship, which better fits the data taken from various canopies with different leaf-area indexes. Numerical investigations confirmed these results and included the effects of the density and distribution of vegetation within the canopy on d and z_0 (Shaw and Pereira, 1982). Both d and z_0 show some dependence on wind speed since the canopy is distorted and smoothed with increasing wind speed (Finnigan and Mulhearn, 1978). Some empirically determined values for z_0 range from 0.1 to 9 cm for mowed and 50-cm-long grass canopies, to greater than 120 cm for coniferous forests (Campbell and Norman, 1998).

Within the plant canopy, on the other hand, we must make allowance for the resistance of plant leaves to the wind. By analogy to the law of resistance of an object placed in a field of turbulence (Tani, 1951; Imai, 1970; Prandtl and Tietjens, 1957), the plant resistance per unit volume, F, may be given by

$$F = \frac{1}{2} C_d B \rho u^2, \tag{3.2}$$

where C_d is a resistance coefficient, B is the leaf-area density (i.e., leaf area per unit volume), and ρ is the density of air.

This resistance force balances the vertical variation of the tangential shearing stress due to the wind, τ. We thus have

$$\frac{d\tau}{dz} = \frac{1}{2} C_d B \rho u^2. \tag{3.3}$$

According to turbulent transport theory (Hinze, 1959; Ogura, 1955), the stress can be expressed as

$$\tau = \rho A_v \frac{du}{dz} \tag{3.4}$$

with the use of the eddy viscosity A_v. The mixing-length concept gives

$$A_v = \ell_z^2 \left| \frac{du}{dz} \right|, \tag{3.5}$$

where ℓ_z is the mixing length (Hinze, 1959).

Substituting (3.4) and (3.5) into (3.3), we obtain

$$\frac{d}{dz} \left\{ \ell_z^2 \left(\frac{du}{dz} \right)^2 \right\} = \frac{1}{2} C_d B u^2. \tag{3.6}$$

While a degree of uncertainty remains about whether or not the concept behind (3.3)–(3.5) is valid within terrestrial plant canopies (Saito et al., 1970), some success has been achieved in submerged macrophyte canopies (Ackerman and Okubo, 1993; see below).

Inoue (1963) solved (3.6) assuming that, for plant communities where the foliage distribution is reasonably uniform, ℓ_z is constant except in the immediate vicinity of the ground. The result is

$$u = u_H \exp\left\{-\alpha\left(1 - \frac{z}{H}\right)\right\}, \qquad (3.7)$$

where u_H is the wind velocity at the "height" (top) of the plant canopy $z = H$, and α is an attenuation coefficient defined by $\alpha = H(C_d B/2\ell_z^2)^{1/2}$. Wind profiles within a corn canopy observed by Shaw et al. (1974) agree well with (3.7), with an α-value of about 2.2. The wind velocity within the canopy thus decreases exponentially toward the ground. A logarithmic profile will develop immediately above the ground with new d- and z_0-values characteristic of the soil.

Cionco (1965) computed a mixing-length solution, which showed that the mixing length ℓ_z was nearly constant throughout most of the canopy's vertical extent; this lends support to Inoue's hypothesis. Cionco developed a refined model (Cionco, 1965), which takes into account the variation of ℓ_z in the vicinity of the ground, and found that the simulated canopy wind profiles agree quite well with the observed canopy wind data from a cornfield. Other wind profiles, which are more or less similar to each other, have been proposed. Thus, Landsberg and James (1971) semiempirically derived an analytical form for wind profile which is the same as Thom's (1971). Equation (3.6) can also be applied to the top portion (e.g., within 30%–40%) of canopies with nonuniformly distributed vegetation (e.g., trees; Campbell and Norman, 1998).

Kondo and Akashi (1976) developed a model for two-dimensional horizontal flow in canopy layers which includes the pressure gradient and Coriolis forces.

The structure of *turbulence* in plant canopies has been the object of much study. Some typical velocity profiles obtained from studies within wind tunnels, crops, and forests are presented in Fig. 3.1. The figure shows a point of inflection in the normalized velocity at the top of the canopy ($z/H = 1$). The (1) inflection point, (2) negative skewness in the vertical direction and positive skewness in the horizontal velocity components, and (3) integral length scales on the order of H appear to be consistent features of the turbulence with plant canopies (Finnigan and Brunet, 1995). Historically, it was believed that the high level of turbulence in canopies was the result of the eddy shed downstream of the vegetation. Recent advances in sensors, fluid dynamic modeling including Lagrangian models and higher-order closures of wind field, have led to the conclusion that large-scale, intermittent turbulent eddies are responsible for the transfer of momentum in canopies (Finnigan and

FIGURE 3.1. Velocity profiles through model, cereal, and forest canopies (after Finnigan and Brunet, 1995).

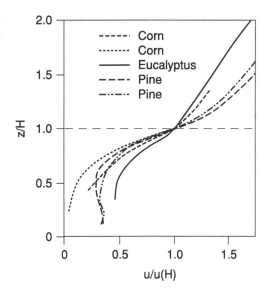

Brunet, 1995). It would appear that turbulence within and just above the canopy is characterized as a region with a broad and continuous eddy spectrum.

Three ranges of eddy vertical length scales (L_w) are of importance: (1) eddies with $L_w \gg H$, the canopy height; (2) eddies with $L_w \sim H$; and (3) eddies with $L_w \ll H$. The largest eddies provide little to the vertical mixing in the canopy, in contrast to eddies of $L_w \sim H$, which contribute most of the momentum transfer. This can best be seen in Fig. 3.2, which depicts the energy spectrum within the canopy as a function of κ, the wavenumber. The principal peak in energy is due to shear production at κ on the order of the reciprocal of the Eulerian length scale (i.e., $1/H$), and these are followed by a cascade, i.e., $\kappa^{-5/3}$, due to viscous dissipation beyond the Kolmogorov microscale ($\lambda = $ mm). The canopy-scale eddies that lead to shear production also cause the *honami* (*ho* = cereal, *nami* = wave; Inoue, 1955) or cowlick patterns observed in cereal fields on windy days. Honami are generated by downward gusts (eddies), whose energy is transferred to the plants through drag, which result in the transfer of momentum (i.e., shear production). After the gust passes, the plants rebound from their deflected position due to the mechanical properties of their tissues and oscillate leading to the waving phenomena (Finnigan, 1979a, b). Honami may contribute to a more rapid dissipation of energy in the canopy (i.e., spectral shortcut in Fig. 3.2) through the splitting of eddies in the wakes of plant materials (i.e., stem wake turbulence) and the oscillatory motion; however, this contributes relatively little to the overall momentum transfer in the canopy (Raupach et al., 1996).

The various, and sometimes deleterious, effects of winds and turbulence

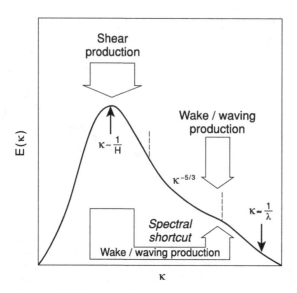

FIGURE 3.2. Schematic diagram of the energy spectrum E through plant canopies as a function of κ, the wavenumber. H: canopy height, λ: Kolmogorov microscale (after Finnigan and Brunet, 1995).

on the mechanics, physiology, ecology, and management of trees have been addressed in a recent multidisciplinary volume (Coutts and Grace, 1995).

Of late, Finnigan and Brunet (1995) suggested that the fluid dynamic interactions—specifically, the active turbulence and coherent motions—near the top of canopies are characteristic of a *plane mixing layer*. The *plane mixing layer* model appears to explain many features associated with canopy turbulence better than boundary layer models (Raupach et al., 1996). Turbulent flow below the top portion of the canopy is less well defined and is a function of the vegetation density profile (i.e., leaf-area density), especially in canopies that are nonuniform in this regard. For more details on the development of ideas related to the turbulence in plant canopies, readers may consult Cionco (1972), Kawatani and Meroney (1970), Baines (1972), Saito et al. (1970), Isobe (1972), Arkin and Perrier (1974), Shaw et al. (1974), Grace (1977), Seginer et al. (1976), Raupach and Thom (1981), Raupach et al. (1991, 1996), Finnigan and Brunet (1995), and Finnigan (2000), among others.

A discussion of the diffusion of CO_2 within canopies may commence with (2.16). For the interior of the plant canopy, away from its periphery, we may ignore the horizontal advection and diffusion of CO_2 and the vertical advective flow. Only the vertical diffusion and nonconservative processes become relevant to the problem. We thus have

$$\frac{\partial S}{\partial t} = \frac{\partial}{\partial z}\left(K_z \frac{\partial S}{\partial z}\right) - (\varepsilon I - r)B, \tag{3.8}$$

where S, ε, I, and r denote, respectively, the CO_2 concentration, photo-synthetic efficiency, incident light intensity, and respiration rate (Inoue, 1965). The last term of (3.8) provides an appropriate expression of $\bar{R}$ in (2.16) for the problem of concern.

For steady-state conditions, (3.8) yields

$$\frac{\partial}{\partial z}\left(K_z \frac{\partial S}{\partial z}\right) = (\varepsilon I - r)B. \tag{3.9}$$

Inoue (1965), assuming that K_z is equal to A_v and that ℓ_z is constant, and using the wind profile (3.7) obtained from (3.5)

$$K_z = K_H \exp\left\{-\alpha\left(1 - \frac{z}{H}\right)\right\} \tag{3.10}$$

where K_H is the value of K_z at the height of the plant canopy, i.e., at $z = H$. Furthermore, if the leaf-area density is invariant with height, Inoue (1965) gives the incident light intensity as

$$I = I_H \exp\left\{-\beta\left(1 - \frac{z}{H}\right)\right\}, \tag{3.11}$$

where I_H is the incident light intensity at height H, and β is a constant of light attenuation that depends on the leaf-area density.

In the upper part of plant canopies, the photosynthetic rate is assumed to overcome the rate of respiration, i.e., $\varepsilon I \gg r$. Neglecting r compared with εI in (3.9)–(3.11), we integrate (3.9) to obtain

$$S = S_H - \frac{H^2 \varepsilon B I_H}{\beta(\beta - \alpha)K_H}\left[1 - \exp\left\{-(\beta - \alpha)\left(1 - \frac{z}{H}\right)\right\}\right],$$

where S_H is the CO_2 concentration at height H.

In the lower part of plant canopies, on the other hand, the respiration overcomes the photosynthesis, i.e., $r \gg \varepsilon I$, and we may neglect the term εI. We also neglect the height variation of K_z, i.e., $K_z = $ constant. The solution of (3.9) is then given by

$$S = S_0 - \frac{J_0}{K_z}z - \frac{rB}{2K_z}z^2,$$

where S_0 is the concentration of CO_2 at the ground, $z = 0$, and J_0 is the flux of CO_2 through the ground due to soil respiration: $J_0 = -(K_z \, \partial S/\partial z)_{z=0}$. (Note that (3.9) in combination with (3.10) and (3.11) can be integrated in general without Inoue's assumption.)

Figure 3.3 shows examples of vertical profiles of CO_2 within crop canopies. The concentration of CO_2 increases upward in the upper part of the canopy, and it also increases downward in the lower part of the canopy. The increase toward the ground is due both to plant respiration and to the supply

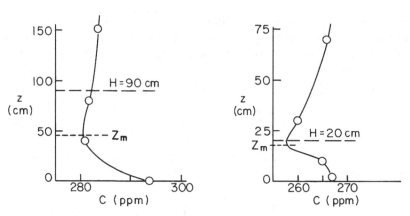

FIGURE 3.3. Vertical profiles of carbon dioxide within and above crop canopies. C: CO_2 concentration, z: height, H: plant height, z_m: height of minimum CO_2 (Inoue, 1965).

of CO_2 from soil respiration. The minimum concentration of CO_2 is found to be within the canopy at $z = z_m$. The compensation height, $z = z_c$, where the photosynthetic assimilation balances the rate of respiration, is seen to be below the height of minimum concentration (Noble, 1983).

There are one direct and two indirect methods for determining CO_2 fluxes (or other scalars) above and within the top portions of uniform canopies (Monteith and Unsworth, 1990). The direct method known as *eddy correlation* involves considerable sophistication in the simultaneous measurement of velocity and CO_2 fluctuations (i.e., turbulence) at small spatial and temporal scales (e.g., Katul et al., 1998). The indirect methods include (1) the *aerodynamic method* proposed by Inoue et al. (1958), which uses measurements of CO_2 concentrations at two heights and an integral exchange coefficient that depends on the wind profile and the stability of a column of air to determine the flux, and (2) the *Bowen ratio method*, which uses an energy balance model (e.g., heat balance) to estimate the flux. The aerodynamic method is advantageous in estimating the CO_2 flux, and accordingly the photosynthetic fixation of CO_2 by plants, without artificial disturbances to the plant and environment. Since the introduction of the aerodynamic method, Japanese scientists at the National Institute of Agricultural Sciences in Tokyo have made much progress in developing more advanced methods (Inoue et al., 1968; Uchijima, 1970; Uchijima et al., 1970; Uchijima and Inoue, 1970). Barring limitations in the measurement of some types of scalars (Pearcy et al., 1989), recent advances in electronics, remote sensing, and statistical measurements of turbulence may lead to dominance of the direct method (i.e., eddy correlation; Finnigan and Brunet, 1995; Raupach et al., 1996).

3.1.2   Aquatic Plant Canopies

A number of different experimental and theoretical approaches have been applied to the study of water flow above and within aquatic plant or macro-phyte canopies. This is due, in part, to the taxonomic (e.g., algae, pterido-phytes, angiosperms) and morphological diversity of macrophytes, which include (1) emergent canopies where the top of the canopy is exposed to the atmosphere (e.g., rushes and marsh grasses), (2) submerged canopies of marine and freshwater plants (e.g., pondweeds and seagrasses), and (3) submerged kelp forests. It is intriguing to postulate analogies between pondweeds/seagrasses and cereals, and between kelps and trees, but the fluid media pose important constraints (Denny, 1993) related directly to the attenuation and, therefore, the acquisition of light (Niklas, 1997). By neces-sity, terrestrial plants are composed of mechanically reinforced structures that act against gravity, whereas aquatic plants are flexible organisms that use specialized gas-filled tissues (e.g., lacuna, aerenchyma, pneumatocysts) as a means of buoyancy. Notwithstanding these differences in growth form, in general, the study of water flow and turbulence above and within macro-phyte canopies is less developed than what was observed above for terrestrial plant canopies.

We will begin our discussion of flow within and above macrophyte can-opies by examining the kelps, which can form forests many tens of meters tall (up to 47.5 m!). Kelps create important ecosystems in coastal temperate waters. One of the largest kelp forests (*Macrocystic pyrifera*) is about 7 km long and 1 to 1.5 km wide adjacent to Point Loma (San Diego), California (Fig. 3.4). Ocean currents introduce nutrient, plankton, larvae, and other waterborne materials into the kelp ecosystem. Jackson and Winant (1983) demonstrated that the currents measured in the kelp forest were approxi-

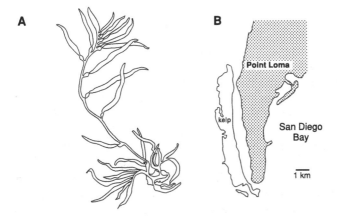

FIGURE 3.4. Kelp forest ecosystems in southern California. A—sketch of the giant kelp, *Macrocystis*. B—Map of the Point Loma forest (after Jackson, 1998).

mately one-third slower than those measured at similar locations outside the forest. The distance over which the longshore current was found to penetrate the kelp canopy (i.e., assuming flow within the kelp) was determined by balancing the advective momentum of the flow with the drag of the plants using

$$u\frac{\partial u}{\partial x} = -(C_d Dq)u^2,$$

where C_d is the drag coefficient (0.5), D is the diameter of the kelp stipe (0.2 m), and q is the plant density (0.1 kelp/m^2) (Jackson and Winant, 1983). The solution subject to $u = u_0$, the free-stream velocity, at $x = 0$, the leading edge is

$$u = u_0 \exp(-(C_d Dq)x),$$

with a scale length of $\approx (C_d Dq)^{-1}$ or approximately 100 m. In other words, the longshore currents are expected to penetrate on the order of 100 m into the kelp forest, which is too little to be of importance for fluid exchange given the 7-km length of the forest.

Jackson (1998) determined the relative importance of the cross-shore currents from the variance in the velocities measured in the cross-shore direction and the first empirical orthogonal function (EOF; a form of principal component analysis) of velocity ($e_{u,1}$). Since the cross-shore velocity was linearly related to the cross-shore distance ($y = -1000$ m), i.e., $u_1 = a(t)y$, it was possible to express u_1 as a function of $e_{u,1}$:

$$u_1(t) = \frac{\sigma}{L} e_{u,1}(t) y,$$

where σ is the standard deviation of velocity and $y = L$. An integration of $u = dy/dt$ leads to the following result:

$$y = y_0 e^{\sigma/L} \int_0^t e_{u,1}(t)\, dt,$$

which, when solved using field data, indicates that currents frequently penetrate more than 400 m into the Point Loma kelp forest (Fig. 3.4) on a daily basis (Jackson, 1998). In other words, the cross-shore exchange of water is substantial given the width of the forest (1 km) and therefore is of greater importance than the longshore exchange.

In addition to the reduction in flow into the kelp forest due to drag, which is about 10 times higher in the kelp forest compared to outside it, Jackson (1984, 1988) demonstrated that the size of the forest is comparable to the wavelengths of suprainertial-frequency Kelvin waves. This leads to a reduction and dampening of coastal-trapped waves in the region. Jackson (1998) also noted that the kelp forests dampen high-frequency waves and surface gravity waves and slow low-frequency waves. The flexibility of the kelp and their response to the currents are likely responsible for this reduction in turbulence (see below). Koehl (1986) recognized that kelps have an advantage of being both flexible and long relative to other subtidal and intertidal organisms in energetic environments where fluid dynamic forces can be damaging or fatal (see Denny, 1988). Long and flexible organisms may be able to reduce the drag they experience by moving with the fluid (i.e., to "go with the flow") and thereby decreasing the water velocity relative to their tissues. While this concept has merit, it does not account for the dynamic loading (i.e., dynamic drag, acceleration reaction: virtual buoyancy and mass) due to the acceleration and deceleration of kelp over the period of a wave. Denny et al. (1998) recently examined the effects of dynamic loading of kelps and other organisms in energetic tidal environments through numerical and empirical studies. They introduced the *Jerk number* (J), a dimensionless number, which is the ratio of the maximum inertial force acting on a moving system to the maximum hydrodynamic force that a stationary object encounters:

$$J = \frac{\sqrt{km}}{K_D u_{x,m}},$$

where k is stiffness of the material ($k \approx 3EI/L^3$, where EI is the flexural stiffness and L is the length), m is the mass, K_D is the drag divided by the square of the velocity ($K_D = 1/2C_d\rho A$, where A is the area), and $u_{x,m}$ is the maximum horizontal velocity of the fluid relative to the bottom. It is possible to estimate the overall forces experienced by an organism when J is combined with the dimensionless frequency (f), which is the ratio of the frequency of oscillation of the wave force (ω) to the natural frequency of the organism's movements,

$$f = \frac{\omega}{\sqrt{km}}.$$

Going with the flow reduces the overall force when J and f are small, whereas under large J and f, flexibility is predicted to increase the overall force experienced (Denny et al., 1998). For example, when this type of approach was applied to the understory kelps *Eisenia* and *Pterygophora*, going with the flow was predicted to be advantageous for plants greater than 1 m tall, at depths deeper than 10 m and under waves less than 2 m in height (Gaylord and Denny, 1997).

There are a number of other coastal plants of smaller scale, which are cosmopolitan in the range of coastal habitats in which they are found (den Hartog, 1970; Dawes, 1998). The seagrasses and their freshwater relatives are the aquatic analogue of terrestrial cereals. These aquatic plants create coastal and inland ecosystems of ecological and economic importance, where water velocities and turbulence are reduced, and where sedimentation is enhanced (see Ackerman and Okubo, 1993). In general, there has been greater emphasis on flow in seagrass canopies; they will be presented here,

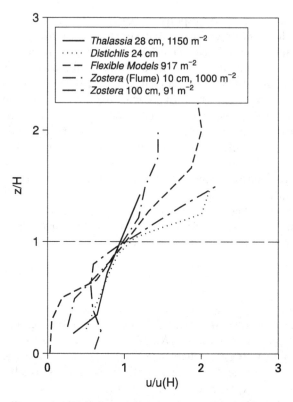

FIGURE 3.5. Velocity profiles through model, marsh, and seagrass canopies. H: canopy height, z: height, u: velocity.

but readers may also consult Dawson and Robinson (1984), Marshall and Westlake (1990), Sand-Jensen and Mebus (1996), Lopez and Garcia (1997), and Sand-Jensen and Pedersen (1999).

Flow is reduced at the top of seagrass canopies (Fig. 3.5), but friction velocities (u_*) may be ten times that of nonvegetative areas (Gambi et al., 1990). Estimated friction factors for the top of the canopy were found to decrease with increasing canopy Reynolds numbers (Re) for seagrasses with ribbon-shaped leaves (Fonseca and Fisher, 1986), and the sedimentation of clays and organic materials was also found to decrease under higher Re (Fonseca et al., 1983). Strong unidirectional flows may lead to a reduction in the canopy friction due to the horizontal deflection of leaves (Fonseca et al., 1982; Ackerman, 1986). Flow within a 1-m-tall eelgrass (*Zostera marina*) canopy at 5.5-m depth closely matched the vegetative profile of the plants (Fig. 3.5) (Ackerman and Okubo, 1993). Ackerman and Okubo (1993) estimated the eddy viscosity (K) and mixing lengths within the canopy by combining (1) the canopy drag $[F(z)]$ as a function of resistance exerted by the leaves on the fluid [cf. (3.2)],

$$F(z) = \frac{1}{2} C(z) \rho u^2 L(z),$$

where C is the drag coefficient and L is the leaf-area index, with (2) the shear stress (τ) from turbulent transport theory [cf. (3.3)],

$$\tau = \rho K \left(\frac{du}{dz} \right).$$

In this case,

$$K(z) = 1 \left/ \left(\frac{du}{dz} \right) \frac{1}{2} \int_0^z C(z) L(z) u(z)^2 \, dz, \right.$$

provided vertical eddy viscosities on the order of 10^{-5} to 10^{-4} m^2/s, which are less than one-quarter the magnitude predicted from models that do not account for vegetation (i.e., $K = \kappa u_* z$, where κ is the von Karman constant). These results were consistent with particle diffusivities measured in the same canopy (Ackerman, 1989). Worcester (1995) estimated horizontal K on the order of 10^{-3} to 10^{-2} m^2/s in shallow seagrass canopies, where the canopy height (H) was similar to the water depth (Z), under slow flow conditions ($u < 5$ cm/s) using a dye-tracking technique. The lack of difference in K estimated in the canopy and in nonvegetative sites draws Worcester's (1995) results into question because they were obtained so close to the free surface, where thermal convective and wind-driven movements would enhance mixing.

The effect of external turbulence on canopy flow was also found to be important by Ackerman and Okubo (1993), who referred to the wavelike oscillations of an eelgrass canopy as *monami* (*mo* = aquatic plant, *nami* = wave). Monami are mechanically different from the honami observed in cereals. In the former case, water deflects the flexible plants horizontally, and the buoyancy created by gas-filled lacunae leads to the return to the vertical (i.e., a hydroelastic response). In the latter case, winds deflect the stiff stems horizontally by loading the panicles, and the stiff elastic tissues in the stem return the plants to the vertical (i.e., a mechanical response). While monami effect the physiology of aquatic plants, Grizzle et al. (1996) also demonstrated that monami affect the settlement of bivalve larvae.

As in the case of terrestrial canopies, shear production is of importance to the overall canopy turbulence in model canopies when flow is unconfined (e.g., $Z/H > 1$) (Nepf and Vivoni, 1999). When the flow is confined by the free surface (i.e., $Z/H \leq 1$), pressure-gradient flow and wake turbulence are of greater importance to the overall turbulence. Energy spectra within seagrass canopies have been found to be consistent with the plant movements (i.e., monami) in that fundamental frequencies were distinct from external conditions (Ackerman and Okubo, 1993). Koch (1996) and Koch and Gust (1999) found that seagrass canopies attenuate wave energy, but

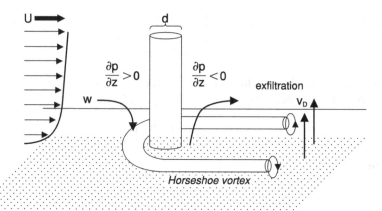

FIGURE 3.6. Secondary flows around aquatic plants. U: horizontal velocity, w: vertical velocity, V_D: exfiltration velocity, p: pressure, z: height, d: diameter.

they did not report distinct plant responses in the velocity spectra. However, careful scrutiny of their results reveals distinct frequencies in the canopy (i.e., fluctuations of 2.5 and 5.5 s) that are consistent with monami (i.e., too long for wake turbulence). Given these results, it is evident that seagrasses reduce and modify the flow within their canopies. Care must be taken, however, in the comparison of results from different studies given differences in vegetative profiles and canopy-to-water depth ratios (i.e., Z/H).

Vertical secondary flows also occur in macrophyte canopies (Nepf and Koch, 1999). These flows are due to the pressure difference upstream and downstream of cylindrical obstructions (diameter d), which lead to downward flows on the upstream face and upward flows on the downstream face of the obstruction (Fig. 3.6). Given hydrostatic flows in the boundary layer, the secondary flow in the vertical $[w(z)]$ is given by

$$w(z) = u_* d \frac{\ln(z/z_0)}{8\kappa^3 z}.$$

The upward-oriented flows downstream of the plant models are thought to be of importance for the exfiltration of nutrient and other materials from the sediments. Nepf and Koch (1999) estimated the exfiltration velocity (V_D) from Darcy's equation

$$V_D = K_{\mathrm{hyd}} \frac{\partial p / \partial s}{\rho g} = \frac{K_{\mathrm{hyd}} u_{z \sim 0}^2}{g d},$$

where K_{hyd} is the hydraulic conductivity, $\partial p / \partial s$ is the pressure gradient between the extremes of pressure along the obstruction, and $u_{z \sim 0}$ is the near-bottom velocity. Results of theoretical and empirical models were consistent using these relationships.

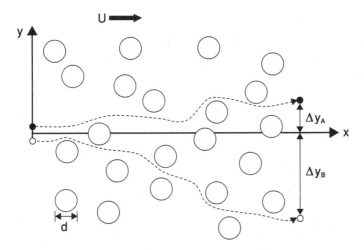

FIGURE 3.7. Mechanical turbulence in emergent plant canopies. Dashed lines represent different fluid paths. (After Nepf, 1999).

As described above, confined flow in macrophyte canopies is characterized by stem wake turbulence rather than shear production. This also applies to the more mechanically rigid emergent plants found in marshes and salt marshes (e.g., sedges, rushes, and grasses). Flow conditions in emergent plant canopies include the reduction of velocity, which matches the vegetative density profile (see Fig. 3.5), and the production of small-scale turbulence by the vortices shed in the downstream wakes (Leonard and Luther, 1995) when stem Re > 200 (Nepf, 1999). This leads to a nonlinear response in the turbulence as velocity is increased. In addition, there is significant anisotropy in the turbulence, with horizontal turbulence accounting for four times that of the vertical component in model canopies (Nepf et al., 1997).

Diffusion in emergent plant canopies was found to be a function of the approaching flow (U) and the plant density (q). Nepf (1999) recognized that *mechanical diffusion* (D_{mech}), which causes the fluid to move laterally due to the physical obstruction of the flow by stems (Fig. 3.7), is also present. Mechanical diffusion increases the total diffusion (D_{total}) over that of molecular diffusion (D_{mol}) when the stem Re < 200, and while much smaller in scale than the turbulent diffusion (D_{turb}) in the stem wakes, D_{mech} does contribute to the D_{total} when stem Re > 200. In both of these situations, an increase in the population density (= fractional volume of the flow occupied by plants; ad, where a is the stem density) leads to increased D_{mech} and D_{turb}. The D_{total} scales through the addition of D_{turb} and D_{mech} according to

$$D_{total} \approx E_k^{1/2} l + [ad] U d,$$

where E_k is the turbulent kinetic energy, and l is the mixing length. In turn, E_k scales with the addition of the turbulence due to shear production and stem wake production according to

$$E_k \sim (1 - ad)C_B U^2 + [\bar{C}_D \, ad]^{2/3} U^2,$$

where C_B is the bed drag coefficient, and $\bar{C}_D$ is the bulk drag coefficient (Nepf, 1999). These scaling relationships provide an opportunity to assess the effect of plant spacing and density on the diffusion within emergent plant canopies, which should be of considerable importance given the ecological relevance of emergent macrophyte canopies (i.e., marshes and wetlands) and the increasing emphasis of wetlands in environmental engineering.

The nature of the motion of fluids immediately around the leaves and fronds of macrophytes has been of interest because it controls mass transfer in aquatic ecosystems. Anderson and Charters (1982) examined the flow through the bushy intertidal algae *Gelidium*, which was found to reduce the velocity and suppress the turbulence of the approaching flow. Associated with these changes was the generation of stem wake turbulence generated at velocities ranging from 6 to 12 cm/s, depending on the diameter and spatial density of the branches.

As mentioned above, the transition in flow induced by the branches of a marine plant is probably a phenomenon of considerable adaptive significance because the turbulence generated by the plant itself, or by neighboring plants, may be the only turbulence that is of the right scale to enhance nutrient uptake and affect the exchange of dissolved gases and solutes. Wheeler (1980) noted an analogous phenomenon in the giant kelp (*Macrocystis pyrifera*), where turbulent boundary layers were observed on fronds at low flows (e.g., 1 cm/s) as a result of the rugosities and spines on the surface of fronds. Hurd and Stevens (1997) confirmed this finding in their examination of the flow around ten types of marine algae. The transition to turbulence occurred at relatively low flows (<3 cm/s), although the transition was much higher (e.g., 12–14 cm/s) on bushy algae like *Gelidium* as a consequence of the stem wake turbulence.

The transport of inorganic carbon within the viscous sublayer (δ_V) of the boundary layer directly next to macrophytes can be modeled using Fick's law (see Sect. 2.4),

$$J = -D\frac{\partial C}{\partial z} \quad \text{or} \quad J = \frac{D}{\delta}(C_\infty - C_0),$$

where J is the flux, D is the molecular diffusivity, C is the concentration (in the free stream C_∞ and at the substratum C_0), z is distance, and δ is the boundary layer thickness (e.g., Wheeler, 1980). The mass transport of inorganic carbon may be limited by δ if the uptake by the plant exceeds the delivery (Wheeler, 1988; Neushul et al., 1992; Falkowski and Raven, 1997). Such *diffusional* or *mass transport stress* occurs at velocities <10 cm/s (Wheeler, 1988; Hurd et al., 1996) equivalent to leaf and frond Re ≪ 10^4 (Ackerman, 1998a), which may also lead to ^{13}C isotope enrichment in their tissues (France and Holmquist, 1997; Keough et al., 1998). This carbon isotope enrichment has been directly related to boundary layer resistance (Smith and Walker, 1980) although the hypothesis has not been examined explicitly. Higher velocities lead to enhanced productivity due to increased nutrient flux (Wheeler, 1988; Ackerman and Okubo, 1993; Hurd et al., 1996; Stevens and Hurd, 1997). At extremely high velocities, nutrient uptake may be limited by a threshold dictated by enzymatic function (Wheeler, 1980; Koch, 1994) and/or drag-induced damage to tissues (Koehl, 1986). Ackerman (1998a) reviewed the effect of velocity on photosynthesis and growth of macrophytes. He found that the *velocity effect* also occurred for other eco-physiological processes in a diverse range of aquatic plants and animals.

Stevens and Hurd (1997) modeled the boundary layers around macrophytes under steady and oscillatory flows. They questioned the use of the flat plate approximation (e.g., Wheeler, 1980) since it ignores the complex morphologies of macrophytes and leads to an underestimation of the diffusional boundary layer thickness (δ_D), which lies within the viscous sublayer of the boundary layer. When realistic measures of u_*/U obtained from direct measurements on macrophytes (e.g., Koehl and Alberte, 1988) are used, advection may balance diffusion, i.e., $\delta_D \to \delta_V$. Moreover, under oscillatory flow, periodic removal of the nutrient gradient (ΔC) by external shear at frequencies of $1/t'$, where t' is the time scale for replenishment, can increase the time-averaged flux ($\bar{J}$)

$$\bar{J} = D\frac{\Delta C}{\delta_D} + \frac{2\Delta C\delta_V}{t'}\sum_{n=1}^{\infty}\left(\frac{1 - e^{-Dn^2\pi^2 t'/\delta_v^2}}{n^2\pi^2}\right)$$

tenfold over the expectation from the equilibrium flux

$$J_v = D\frac{\Delta C}{\delta_v}.$$

The strong influence of macrophyte morphology on u_*/U in these models indicates the need for further laboratory and field study.

3.2   Diffusion of Nutrients in the Sea

Nutrients are chemical elements within marine ecosystems that are directly responsible for the activity of primary producers. At the largest scales, nutrient transport is driven by the advective "oceanic conveyor belt" that constitutes the general circulation (Gordon, 1986; Broecker, 1991). In the open ocean, mesoscale eddies (≈ 100 km) are also significant, episodic drivers of nutrient transport (Falkowski et al., 1991; McGillicuddy et al., 1998; Oschlies and Garcon, 1998). Vertical advection is generally impeded by a strong density gradient (pycnocline) that exists between a well-mixed surface layer and the deep ocean (except in the upwelling areas of coastal regions and at the equator). Turbulent diffusion transports nutrients (e.g., NO_3) across the pycnocline to phytoplankton in the surface layer. In addition to the physical processes of advection and diffusion, the distribution of nutrients is controlled by the nonconservative terms in the balance equation. These include the uptake of nutrients by phytoplankton in the upper mixed layer, much of which lies in the euphotic zone, and the remineralization of organic matter (the remains of planktonic organisms) in the deeper layers of water. This process of nutrient sequestration into organic "bundles" which then sink into deeper layers where they are remineralized is sometimes referred to as a "biological pump" of carbon and other nutrients (cf. Long-hurst, 1991). Thus, the concentration of nutrients such as NO_3, PO_4, and SiO_3 is small in the surface layer, increases with depth, reaches a maximum at a deep layer, and decreases toward the bottom of the ocean.

If the near-surface nutrient profile is described by a simple advection–diffusion model for the vertical distribution of nutrients, assuming steady state and constant values of vertical diffusivity, K_z and vertical advection, w, we find from (2.16) that

$$K_z \frac{d^2 S}{dz^2} - w \frac{dS}{dz} + R = 0, \tag{3.12}$$

where S is the concentration of nutrients and the z-axis is taken downward from the sea surface. Equation (3.12) deals only with the gross features of nutrient distribution, disregarding, for instance, the seasonal variation of nutrients in the upper layer.

Munk (1966), Wyrtki (1962), and Tsunogai (1972a) found (3.12) to be applicable to distributions in deep layers, at depths of perhaps 1 to 5 km, of the interior ocean such as the Pacific, where the estimated values $w = -1.4 \times 10^{-5}$ cm/s and $K_z = 1.3$ cm^2/s are considered reasonable. In other words, the deep water of the interior oceans, while being mixed and diffuse, gradually rises at a speed of 4.4 m per year. The upward flux of mass in the Pacific then amounts to 6×10^{32} g per year. This flux is supplied by the Antarctic Bottom Water, which flows through the South Pacific, thus completing the general circulation.

In deep water the regeneration of nutrients (R) from the remains of organisms (remineralization) causes R to be positive. Thus, Riley (1951), Wyrtki (1962), Tsunogai (1972a), Grill (1970), and Shaffer (1996) all assumed

$$R = R_0 e^{-\alpha z} \tag{3.13}$$

as an analytical expression for the nonconservative term. [Okubo (1954, 1956) obtained a purely theoretical profile for R, the form of which is close to an exponential function. See also Lerman and Lal (1977) and Pond et al., (1971).] Power-law relationships derived from sediment trap data have also been used (Martin et al., 1987). The downward decrease of R is likely due to the facts that the remineralization process starts at an upper layer and that the organic remains sink as they decompose. For the special case of $\alpha = 0$, the regeneration rate, $R = R_0$ (= constant).

FIGURE 3.8. Vertical distribution of dissolved oxygen (O_2), and inorganic phosphorus (P) in the Coral Sea. Observed (———) and theoretical (- - - -) distributions (from Wyrtki, 1962).

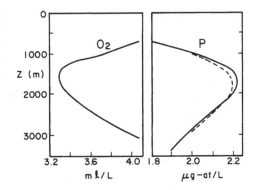

Substituting (3.13) into (3.12) and solving the resulting equation, we obtain

$$S = A + Be^{wz/K_z} + \frac{R_0 e^{-\alpha z}}{K_z \alpha^2 (1 + w/K_z \alpha)}, \qquad (3.14)$$

with constants A and B to be determined from boundary conditions. Thus, if we know the boundary values $S = S_1$ at $z = z_1$ (upper layer) and $S = S_2$ at $z = z_2$ (lower layer), we can evaluate A and B.

Wyrtki (1962) found that the theoretical profile (3.14), with $\alpha = 0$, agreed with the observed vertical profile of inorganic phosphorus in the Coral Sea, with a maximum concentration of P (phosphorus) occurring at a depth of about 2000 m (Fig. 3.8). More recently, Shaffer (1996) obtained excellent fits of theoretical profiles of NO_3 and PO_4 to data obtained from cruises in northern latitudes. In this case, different nonzero values of α were obtained for NO_3 and PO_4 profiles.

It seems unreasonable to assume that (3.12) also holds in the upper layer, especially at high latitudes, where seasonal variations, effects due to horizontal mixing, and wide fluctuations in the values of K_z and w exist, and where difficulties in the formulation of the nonconservative term become formidable. Nevertheless, Grill (1970) met with some success in applying (3.12) to the vertical distribution of inorganic silicate in the surface layer of the sea.

The vertical flux of nitrate across the pycnocline in the upper layer of the oligotrophic open ocean imposes a rigorous constraint on the rate of export of organic carbon from the surface layer of the sea ("new" production). Assuming horizontal fluxes to be small in relation to vertical fluxes,

$$\frac{\partial S}{\partial t} = \frac{\partial}{\partial z}\left(K_z \frac{dS}{dz}\right) - bS,$$

the rate of change of local nitrate concentration is equal to the turbulent diffusion of S and loss (biological sink) rate in the form of algal uptake. At steady state, and assuming constant K_z,

$$K_z \frac{d^2 S}{dz^2} - bS = 0.$$

Near-surface nitrate distributions in the open ocean, generally observed to increase exponentially with depth, are consistent with this first-order, concentration-dependent uptake by phytoplankton and a diffusional supply in the vertical direction. Even this simple model demonstrates that phytoplankton uptake exerts a control on both the vertical nitrate flux and gradient (Lewis et al., 1986). However, the specific rate of uptake, b, may itself vary with depth. If b follows the exponential profile of irradiance in the surface layer (Beer's law), then the steady-state relation becomes

$$K_z \frac{d^2 S}{dz^2} - aI_0 e^{-kz} S = 0.$$

Lewis et al. (1986) solved this equation subject to $dS/dz = 0$ (no nitrate flux at the sea surface) and $S = S_0$ at $z = 0$:

$$S(z) = S_0 \frac{\mathbf{K}_1(\beta)\mathbf{I}_0(\xi) + \mathbf{I}_1(\beta)\mathbf{K}_0(\xi)}{\mathbf{K}_1(\beta)\mathbf{I}_0(\beta) + \mathbf{I}_1(\beta)\mathbf{K}_0(\beta)}$$

$$= S_0 \beta (\mathbf{K}_1(\beta)\mathbf{I}_0(\xi) + \mathbf{I}_1(\beta)\mathbf{K}_0(\xi)),$$

where $\beta = (2/k)(aI_0/K_z)^{1/2}$, $\xi = \beta \exp(-1/2kz)$, and $\mathbf{I}_m$, $\mathbf{K}_m$ are modified Bessel functions of order m. The shape of this concentration profile increases approximately exponentially with depth, and a nonlinear fit of the parameters β and S_0 to the observed nitrate profile match the data well. The corresponding value of vertical nitrate flux (supply) agrees reasonably well with independent estimates of new production (i.e., production not fed by recycled NH_3) based on ^{15}N uptake, though the 95% confidence interval spans a relatively wide range.

Coastal areas are often characterized by marked upwelling of water caused by Ekman pumping (i.e., interactions of wind and the Coriolis force driving vertical advection; Mann and Lazier, 1996). The supply of nutrients by upwelling and their effect on primary productivity are among the central issues of marine ecology. Several international, multidisciplinary research efforts have addressed the interactions of physical and biological processes of coastal upwelling ecosystems (*Deep-Sea Research*, **24**(1), 1977; Dengler, 1985; MacIsaac et al., 1985). In principle, Lewis et al.'s (1986) steady-state equation could be extended to include the advective flux of nutrients due to upwelling (or downwelling). The addition of an advective term with velocity w (positive downward) to their equation yields

$$K_z \frac{d^2 S}{dz^2} - w \frac{dS}{dz} - aI_0 e^{-kz} S = 0,$$

which, using their boundary conditions, is satisfied by

$$S(z) = S_0 \xi^{-Pe} \beta^{1+Pe}(\mathbf{K}_{Pe-1}(\beta)\mathbf{I}_{Pe}(\xi) + \mathbf{I}_{Pe-1}(\beta)\mathbf{K}_{Pe}(\xi)),$$

where Pe = Peclet number ($= w/(kK_z)$), a dimensionless parameter expressing the relative importance of advection and turbulent diffusion over the length scale, k^{-1}). For no advection, $w = 0$, and the solution reduces to that of Lewis et al. (1986). Upwelling reduces the proportion of nutrient demand in the euphotic zone met by eddy diffusion, and the concentration profile becomes relatively flat (Swaney, unpublished).

While our understanding of the nutrient dynamics of the sea has been advanced by one-dimensional, linear models like those presented above, and their analytical solutions continue to give insight (e.g., Shaffer and Sarmiento, 1995; Shaffer, 1996), some caution should be used when using random walk models to simulate these processes, especially when the diffusivity varies spatially (see Visser, 1997). Importantly, knowledge of the biochemical reaction kinetics of oceanic nutrients has greatly progressed so that the non-conservative term, R, is often evaluated using more complex nonlinear forms. Anderson et al. (1978) constructed a diffusion-reaction model for the vertical distribution of anaerobic NO_2 in the sea that incorporates advective and diffusive processes with Michaelis–Menten reaction kinetics for NO_2 in the following equation:

$$\frac{\partial S}{\partial t} = K_z \frac{\partial^2 S}{\partial z^2} - w \frac{\partial S}{\partial z} + P(z)\left(1 - \lambda \frac{S}{S+M}\right),$$

where $P(z)$ represents the rate of oxidation of organic matter by denitrifying bacteria, λ is the mean ratio of the potential rate of NO_2 reduction to the potential rate of NO_3 reduction by denitrifying bacteria, and M is the Michaelis–Menten constant for NO_2 uptake. Anderson et al. (1978) successfully applied this model to interpret the observed pattern of anaerobic NO_2 in the eastern tropic North Pacific Ocean.

Michaelis–Menten kinetics have been used in many other studies to analyze, for example, the status of Si as a limiting nutrient in diatoms (Nelson and Treguer, 1992; Dugdale et al., 1995), the inhibition of NO_3 uptake by NH_4 and the effect on production (Harrison et al., 1996; Elskens et al., 1997), and effects of resource competition among multiple species (Fong et al., 1994). Mathematical modeling in recent years (e.g., Fasham et al., 1990; Sarmiento et al., 1993; Doney et al., 1996; Hurtt and Armstrong, 1996) frequently rely on such nonlinear nutrient kinetics. Increasingly, ecological relationships are being incorporated into these models, which make them analytically intractable but behaviorally rich [e.g., (1) allometrically scaled nutrient uptake and other organism size-based parameters (Armstrong, 1994) and (2) resource-ratio arguments that help explain the relationships between multiple nutrient profiles and biological production (Carpenter et al., 1992; Tilman et al., 1982)]. Much remains to be learned of the interplay between diffusion of nutrients and these biological processes (Owens, 1993; Jumars, 1993; Mann and Lazier, 1996).

Before ending this section, we note that Eq. (3.12), or the corresponding time-varying version,

$$\frac{\partial S}{\partial t} = K_z \frac{\partial^2 S}{\partial z^2} - w \frac{\partial S}{\partial z} + R,$$

can also be used as a model for the distribution of dissolved oxygen in water bodies. A boundary condition makes oxygen flux at the surface (reaeration) proportional to the difference between ambient concentration (O_2) and the atmospheric equilibrium value (O_{2s}):

$$F_{O_2} = v(O_{2s} - O_2)|_{z=\text{surface}}.$$

The constant of proportionality (mass transfer velocity) is generally assumed to be strongly dependent on wind speed at the surface (Liss and Merlivat, 1986; Wanninkhof, 1992).

In investigations of lake and river water quality, a solution of a simple form of Eq. (3.12) has a venerable history in water-quality engineering as the dissolved oxygen "sag" curve response to a point source of oxygen-consuming organic waste in which the water is being reoxygenated from the surface (Streeter and Phelps, 1925). This topic is covered thoroughly by Thomann and Mueller (1987).

In the sea and other bodies of water, phytoplankton in the photic zone produce oxygen, and predation and decomposition of organic matter by heterotrophs imply the regeneration of nutrients and the consumption of oxygen (Wyrtki, 1962; Tsunogai, 1972b). In addition, near the sea surface, an oxygen concentration in excess of the atmospheric equilibrium concentration by about 2% to 3% can result from bubble injection from wind-driven breaking waves in addition to biological production (Broecker and Peng, 1982; Liss and Merlivat, 1986; Craig and Hayward, 1987). The balance between these processes tends to result in a net production of oxygen during the daylight hours near the surface of the water column ($R > 0$) and a net consumption at depth ($R < 0$). Integrating the time-varying form of Eq. (3.12) over the water column results in a direct relationship between the change in the average oxygen content, the diffusion of oxygen at the boundaries, and the depth-integrated net production in the water column. This is the basis of the whole-ecosystem productivity estimation method introduced by Odum (1956) and since used by many others (e.g., Odum and Hoskin, 1958; Nixon and Oviatt, 1972; Kemp and Boynton, 1980; Emerson et al., 1993; DeGrandpre et al., 1997). Stigebrandt (1991) used the relationship between the oxygen flux boundary condition at the sea surface, oxygen concentration, and primary production in order to estimate depth-integrated production in the Baltic.

As far as the oceanic dispersion process is concerned, both oxygen and nutrients behave in the same manner, with the difference in boundary conditions, primarily at the sea surface, being largely responsible for the discrepancy between nutrient and oxygen profiles. Accordingly, it might be expected that there would be a close agreement in deep water between the amount of nutrients such as phosphate and the oxygen depletion. In reality, the matter is not so simple.

The layer of maximum PO_4 and other nutrients does not necessarily coincide with that of minimum oxygen (Fig. 3.8). A simple plot of the phosphate content against the "apparent oxygen utilization" may not reveal the expected relationship of chemical equivalence, as the effect of diffusion must be taken into consideration (Sugiura, 1964; Sugiura and Yoshimura, 1964). Also, "fractionation" effects occur that change the relative proportions of oxygen and nutrients with depth because of progressive change in the stoichiometric ratios of nutrients and oxygen in sinking organic matter as it is remineralized (Shaffer, 1996). Reactions closer to and within sediments are discussed below.

3.2.1 Subsurface Productivity and Chlorophyll Maximum

In the absence of other factors, the profile of chlorophyll production in the sea could be expected to be related directly to productivity, and therefore to the balance between corresponding light profiles and nutrient profiles (e.g., Wolf and Woods, 1988). For example, a nitrite maximum in the vertical is commonly associated with the chlorophyll maximum at the base of the euphotic zone (French et al., 1983). French et al. (1983) showed diel changes in nitrite concentration in which the nitrite maximum is produced mainly by phytoplankton during the day by reduction of nitrate. They used the vertical balance equation (3.12) for estimating K_z and w from temperature or salinity data ($R = 0$) combined with the rate (R) estimate of chlorophyll a or ATP. Gorfield et al. (1983) found that a subsurface maximum in electron transport system (ETS) activity (microbial activity) was associated with the (secondary) nitrite maximum and particle maximum.

Other factors also influence the chlorophyll depth profile. First, the ambient light conditions near the surface may actually inhibit production in shade- (depth-) adapted phytoplankton (Neale et al., 1991). Higher chlorophyll levels present in low-light-adapted species can result in a separation between the depth of the chlorophyll maximum and the productivity maximum (Venrick, 1982; Jumars, 1993).

Second, phytoplankton sinking through the water column transport chlorophyll to depths below that of chlorophyll production. Conversely, phytoplankton can be transported upward by advection (upwelling) or turbulent diffusion (Takahashi et al., 1985). These rates of transport may not be constant with depth as sinking rate can be an active response of phytoplankton to a reduction in light level with depth (Steele and Yentsch, 1960; Bienfang, 1993). Resistance to sinking by the pycnocline at the bottom of the mixed layer (cf. Roman et al., 1986; Vandevelde et al., 1987) and to variations in turbulent diffusion with depth (Jamart et al., 1977; 1979) can affect the position of the chlorophyll maximum.

Finally, predation by zooplankton and other organisms can affect the shape of the chlorophyll profile. Evidence exists for zooplankton density maxima that are coincident with either the productivity maximum (Roman et al., 1986) or the chlorophyll maximum. Tsuda et al. (1989) estimated that 56% to 100% of chlorophyll production at the chlorophyll maximum was consumed by microzooplankton (<95 µm) at a site in the subtropical North Pacific. Revelante and Gilmartin (1990) observed a maximum of ciliated protozoa coincident with the subsurface chlorophyll maximum and oxygen maximum in the Northern Adriatic. Herman (1989) observed maximum copepod densities coincident with the subsurface productivity maximum, and not at the chlorophyll maximum, in the Eastern subtropical Pacific. Townsend et al. (1984) observed highest copepod densities at the depth of the chlorophyll maximum in the Gulf of Maine. It may be true that both the chlorophyll and productivity maxima are preferred sites for grazing, depending on the season and the size of the grazer (Le Fevre and Frontier, 1988; Mann and Lazier, 1996).

3.2.2   Flocs, Aggregates, and Marine Snow

The nature of organic and inorganic matter in aquatic systems is directly related to the availability and diffusion of nutrients discussed above. As much of this material exists as particulate organic matter (POM) in aggregates >500 µm in diameter comprised of microorganisms, inorganic particles, transparent exopolymer particles (TEM), and detritus, it is prudent to examine the nature and formation of these aggregations, which are also known as *flocs* and *marine snow*. Aggregates lead to temporal and spatial patchiness in the environment and consequently are of importance for nutrient cycling (Posedel and Faganeli, 1991; Brzezinski et al., 1997), microbial activity (Silver et al., 1998), and the downward flux and potential retention of autotrophic production in the euphotic zone (Kiørboe et al., 1998). Ultimately, aggregates contribute to the downward flux of material in the water column and, therefore, remove nutrients from aquatic ecosystems (Fowler and Knauer, 1986).

There are historical accounts of bathyscaphe observations of relatively large particles in suspension, which were referred to as marine snow. Nishizawa et al. (1954) provided photographic evidence of such material a decade before Riley (1963) described 5-µm- to mm-sized organic aggregates in Long Island Sound. These aggregates were interpreted to have formed on surfaces, including those on air bubbles, from the agglutination of organic material from live and dead phytoplankton cells. In subsequent reports, Riley et al. (1964, 1965) noted variability in aggregate distribution both in space, which was strongly correlated to phytoplankton abundance in upwelling regions, and in depth, for which a decline was observed with distance from the surface, but concentrations remained consistent below depths of 3000 m.

The role of physical processes in the formation of aggregates as noted by Riley (1963) involved the accumulation of organic and inorganic materials. Sea-surface slicks, which are areas of accumulations of such materials (Posedel and Faganeli, 1991), may provide some of the material that eventually becomes incorporated into aggregates. Slick formation was directly related to periods when wind speeds were less than 7 m/s, which was quite frequent even in areas considered to be highly energetic (Romano and Marquet, 1991). Carlson (1987) noted that the microlayers associated with slicks can be extremely viscous. Issues related to the bulk properties of these conditions and those found in aggregations are presented in Jenkinson and Biddanda (1995). In addition to physical processes, there are important biological contributions to the formation and nature of aggregates, which were noted from careful in situ observation and collection (Silver et al., 1978).

The composition of marine aggregates may be primarily organic, diatomaceous due to the intrinsic stickiness of diatoms (Kiørboe et al., 1998), or mucous due to the castoff mucous feeding apparatus of pelagic zooplankters (e.g., larvaceans; Hansen et al., 1996; Kiørboe et al., 1996; also see Halloway and Cowen, 1997). In the cases of larvacean mucous, house-derived aggregates can be many dm in length and while spatially rare may contribute significantly to the POM distribution in marine waters (Silver et al., 1998). Early observations of POM in coastal inlets noted temporal and spatial variability in the composition and content of aggregates (Kranck, 1980). Similar diversity in lake aggregates was found by Grossart et al. (1997). They noted four types of aggregates in Lake Constance: (1) algal-derived; (2) zooplankton exoskeleton-derived; (3) cyanobacterial-derived; and (4) those of indeterminate derivation. Conversely, riverine aggregates appear to be both detrital and inorganic material packed within a bacterial-derived microfibril matrix (Droppo et al., 1997). In this case, the bacterial composition of extracellular polymers is of significance, as is the morphology (i.e., porosity and water content) of the aggregate. Similar observations have been made for marine aggregates (Alldredge and Silver, 1988). The shape of aggregates is fractal, especially with respect to length and mass relationships, which is important for aggregation formation as described by coagulation theory (Jackson and Burd, 1998). The fractal dimension (D) of marine snow is generally around 1.7, which is less than a value of 3 predicted for a sphere (Logan and Wilkinson, 1990). Aggregates smaller than 200 μm also appear to be fractal with values ranging from 1.7 to 2.3 depending on the sampling location and the technique used to estimate D (Jackson et al., 1997; Li et al., 1998). Importantly, Li et al. (1998) interpreted the low value of D for small aggregates to be a consequence of formation through coagulation, which would make them similar to marine snow.

Droppo et al. (1997) noted the complexity in the formation of riverine

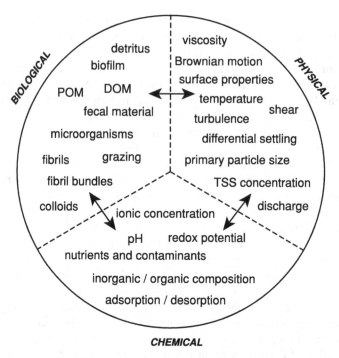

FIGURE 3.9. Biological, chemical, and physical factors influencing the aggregation and fragmentation of freshwater flocks (after Droppo et al., 1997).

aggregates involving biological, chemical, and physical factors (Fig. 3.9). Many of these factors also apply to marine aggregates. With respect to the physical processes, Jackson (1990) modeled the formation of algal flocs during algal blooms using a coagulation process, in which shear coagulation is thought to be of importance in the formation of aggregates in the upper ocean (McCave, 1984). Coagulation theory models collision rates via Brownian motion, fluid shear, and differential sedimentation of particles. Jackson (1990) predicted that there would be a two-stage process of aggregate formation, including low aggregate formation during the development and decline of the bloom, and high aggregate formation during the peak of the algal bloom. The rapid transition between the two stages, as indicated by a critical algal concentration, was predicted to be related inversely to the fluid shear, size of the cell, and stickiness of the cell surface. Importantly, aggregate size was predicted to influence the chemotaxis and attachment of bacteria to aggregates (Jackson, 1989). Mesocosm studies have supported the predictions of coagulation theory (e.g., Dam and Drapeau, 1995), and modeling theory has been refined (Jackson and Lochmann, 1993; see Jackson and Burd, 1998). Field studies by Riebesell (1991) confirmed the two-stage process of aggregate formation in his study of algal blooms in the

North Sea, where biological processes and shear were found to affect aggregate formation.

Shear stress can be estimated from wind speeds at 10 m (w_{10}) that create wind stress on the water surface and that lead to friction velocity (u_*) via

$$u_* = (C_d \rho_{\text{air}}/\rho_{\text{water}})^{1/2} w_{10},$$

where ρ_{air} and ρ_{water} are the density of the air and the water, respectively. The turbulent energy dissipation rate (ε) in the mixed layer in the upper ocean is a function of u_*:

$$\varepsilon = a u_*^2 f \, \exp[z/(b u_*/f)],$$

where a and b are constants (53.5 and 0.068, respectively) and f is the Coriolis parameter ($f = 2 \times 7.29 \ 10^{-5} \sin \phi$, where ϕ is the latitude). The shear rate (γ) is a function of ε and the kinematic viscosity (v):

$$\gamma = (\varepsilon/v)^{1/2}.$$

The smallest eddies associated with these conditions are determined via the Kolmogorov length scale (λ)

$$\lambda = (v^3/\varepsilon)^{1/4},$$

which is on the order of mm in the open ocean (Mitchell et al., 1985). Hill et al. (1992) examined the encounter rates of particles via eddies and determined that the relative velocity between particles was more important than viscous forces even though the spatial scales affecting encounters (e.g., λ) indicated viscous conditions.

While fluid shear contributes to the collision rate necessary for the coagulation process leading to the creation of aggregates, it may also lead to the disaggregation of established aggregates. The loss of aggregates is thought to occur due to sedimentation, consumption and degradation by organisms, dissolution, and fragmentation via surface erosion, pressure fluctuations, and filament fracture (Alldredge et al., 1990). Alldredge et al. (1990) examined the strength of four types of marine aggregates under different shears (γ). They discovered that, while fragmentation was possible via pressure fluctuations and filament fracture, the γ required to fragment aggregates were seldom found in nature (i.e., the turbulent energy dissipation rate, ε, was too low). They suggested that animal grazers are likely responsible for the vertical distribution of POM in the water column. Kiørboe's (2000) recent evaluation of the colonization of marine aggregates by invertebrate zooplankton supports this assertion.

High temporal resolution sampling of marine aggregates revealed seasonal as well as diel variation (day > night) in marine snow concentration measured at 270-m depth (Lampitt et al., 1993). This variation is thought to be driven by diel variation in (1) phytoplankton production due to solar inputs, (2) grazing by vertically migrating zooplankton, and/or (3) turbulence-induced fragmentation. Turbulence is generally reduced in the upper mixed layer during the day due to stratification caused by solar warming. The release of this potential energy at night leads to increased turbulent kinetic energy (i.e., shear). Ruiz (1997) modeled the growth dynamics of marine aggregate concentration (C) as

$$dC/dt = \text{aggregation} - \text{fragmentation} + \text{growth} - \text{grazing} - \text{sedimentation}$$

using parameter values measured or estimated from observations. Neither diel variation in phytoplankton growth nor diel variation in zooplankton grazing resulted in diel variation in C. Diel variation in turbulence, however, led to diel oscillations in the concentration of marine aggregates. Diel variation in the turbulent energy dissipation rate, where values of ε between 10^{-9} and 7×10^{-8} m²/s³, reduced C through fragmentation at a greater rate than was produced through coagulation. This is because fragmentation scales with $\varepsilon^{1/2}$, whereas coagulation scales with $\varepsilon^{1/3}$ for large particles (i.e., >0.01 cm) (see Hill et al., 1992). In other words, the nightly increase in ε results in more than eight times more fragmentation of large particles at night compared to the day, which is approximately twice the rate of coagulation of smaller particles over the same period. The magnitude of these ε-values are at least an order of magnitude less than the ε required to fragment marine aggregates in the laboratory as reported above (Alldredge et al., 1990).

Aggregates appear to accumulate at physical discontinuities in the water column associated with density stratification (MacIntyre et al., 1995). The likely cause of this phenomenon is water column shear caused by velocity gradients ($\partial u/\partial z$), horizontal intrusions, and turbulent mixing. This phenomenon was modeled via the accumulation of aggregates due to density-related sedimentation and accumulation due to interactions with turbulent flow. In the first model, the sinking speed (u_z) of an aggregate of radius r is given by

$$u_z = \sqrt{c\Delta\rho},$$

where $c = 2grC_d\rho_{\text{water}}$, g is gravity, and $\Delta\rho$ is the density difference (i.e., excess density). The density difference is a function of the solid (f_s), porosity (f_p), and mucous (f_m) content of the aggregate such that $f_s + f_p + f_m = 1$,

$$\Delta\rho = f_s\Delta\rho_s + f_p\Delta\rho_{af} + f_m\Delta\rho_{af},$$

where $\Delta\rho_s$ and $\Delta\rho_{af}$ is the density difference with respect to the solid and aggregate interstitial fluid (e.g., mucous), respectively. The model predicts that u_z would be reduced for highly porous aggregates with moderate $\Delta\rho$.

The duration of the reduction in u_z on approaching the density discontinuity was examined through a comparison of the time scales related to molecular diffusion ($t_D = r^2/2D$, where D is the diffusion coefficient), pressure-driven flow through the aggregate ($t_p = 2r/u_I$, where the interstitial flow $u_I = 3Pu/2r^2$ and P is the porosity), and shear-driven flow through the boundary layer ($t_\tau = \theta r/u_s$, where θ is angle subtending the boundary layer, and the maximum shear-driven flow $u_s = 3P^{1/2}u/2r$). For compact detrital marine snow, MacIntyre et al. (1995) determined that t_D for heat was more rapid than either t_p or t_τ, but it was also much slower for t_D for salt. Thus, the sinking velocity would be reduced near the halocline. For larger and more porous diatom aggregates, t_D for heat was more rapid than either t_p or t_τ, which were comparable to t_D for salt, indicating that the aggregate would have to be located within the halocline for a reduction in u_z. The higher the mucous content of the aggregate, the longer the duration of reduced u_z was predicted. In general, the salinity difference of the stratification was determined to be of greater importance than were the temperature differences.

In the second model of accumulation, MacIntyre et al. (1995) used random walk models to demonstrate that low-porosity aggregates accumulated due to the turbulent mixing at the discontinuity. High-turbulence intensities were predicted to disrupt the accumulation and fracture large porous aggregates. Regardless of the mechanism involved, the persistence of aggregates at discontinuities may be on the order of hours to days, which would have important ecological implications for trophic transfer.

3.2.3 Benthic–Pelagic Coupling

As noted above, marine snow and other detritus falling from the productive surface layer through the water column are acted upon by a variety of pelagic organisms, which regenerate mineral nutrients. Much of this regenerated material takes the form of fecal particles, which play an important role in transporting organic matter (i.e., nutrients) into deeper water. For example, Deibel (1990) provides settling velocities for fecal matter from neritic salps and dolioids, which eventually reaches the benthos, thereby coupling the benthic and pelagic zones.

The issue of benthic–pelagic coupling in the nearshore has received considerable attention by marine ecologists, especially with respect to bivalve ecosystems as presented below. This is due, in part, to the question of whether aquatic ecosystems are controlled by limits to primary production by phytoplankton or by consumption by grazers (e.g., bottom-up and top-down controls, respectively; Fretwell, 1987; Menge, 1992; Wildish and Kristmanson, 1997; Gili and Coma, 1998). In the former, a lack of nutrients limits phytoplankton growth, while in the latter, pelagic grazing is the limiting factor. Benthic suspension feeders may also contribute to the consumption of phytoplankton, depending on the water depth, water column mixing, and phytoplankton concentrations. These issues have been examined under unidirectional conditions in numerical, laboratory, and field studies (reviewed by Dame, 1996; Wildish and Kristmanson, 1997). The situation in lakes is complicated by the interaction of other forcing functions such as wind-generated waves, stratification, and surface-gravitational seiching (see Fischer et al., 1979; but see Ackerman et al., 2001).

Benthic bivalves are perhaps one of the most significant taxa of benthic suspension feeders, and consequently, benthic–pelagic coupling is best known in these systems (see Dame, 1996). Research on these and other suspension-feeding invertebrates has been recently reviewed by Wildish and Kristmanson (1997). The strongest case for benthic–pelagic coupling is provided by the detection of concentration boundary layers that have been observed over marine bivalves in estuaries (Dame, 1996). The modeling of these systems has confirmed that turbulent transport is the key physical process supplying phytoplankton to the bivalves (Fréchette et al., 1989). Evidence of suspension feeding-induced concentration boundary layers has also been observed in lakes (e.g., Ackerman et al., 2000; 2001). Similar reductions in phytoplankton biomass near coral reefs have also been reported (Yahel et al., 1998).

The mass transport of waterborne material (i.e., seston) to suspension-feeding bivalves is influenced by fluid dynamics (Cloern, 1991; Koseff et al., 1993; O'Riordan et al., 1995). For example, velocity has a positive effect on suspension-feeding rates measured at low velocities and a negative one at higher flow rates (Wildish and Kristmanson 1997; see review in Ackerman, 1999). While the positive effects at low velocity may be due to the local replenishment of resources, the mechanisms responsible for the decline in feeding at higher flows may be due to dynamic pressure differences in the flow and lift–drag effects that lead to behavioral responses in the bivalves (Ackerman, 1998a, 1999).

At larger scales, an increased velocity would replenish depleted resources with fresh seston as a function of the ambient flow and mixing within the water column. Physical modeling of siphonal flow in laboratory flow is consistent with this conclusion and suggests that there may also be significant refiltration in bivalve beds (Monismith et al., 1990; O'Riordan et al., 1995). The seston concentration (C_l) at bivalve beds has been modeled as

$$U\frac{\partial C_l}{\partial x} + w_s\frac{\partial C_l}{\partial z} - \frac{\partial}{\partial z}\left(K\frac{\partial C_l}{\partial z}\right) = \Psi(Q, C_\infty, E, H),$$

where U is velocity, w_s is the settling of seston, Q is the pumping rate, E is the filtering efficiency, H is the intake height, and Ψ is the filter feeding (O'Riordian et al., 1995). This model relates the physical mixing processes affecting the delivery of seston, to the physiology of filter feeding (Koseff et al., 1993; Butman et al., 1994; Wildish and Kristmanson, 1997). It requires information on the physical mixing processes and the ability of the bivalves to feed under these conditions. Neither the physical mixing of coastal waters nor the bivalve-filter-feeding physiology in them is well understood (Dame, 1996; Wildish and Kristmanson, 1997). This situation is similar for other taxonomic groups.

Benthic–pelagic coupling, depending on the context, may also refer to the relationships between the deposition of sediments from the water column to the sea floor and the benthic foodweb, the effect of benthic oxygen demand on pelagic primary production (Hargrave, 1973), the relationship between

diagenetic processes and nutrient availability, or the effect of benthic predators on pelagic populations. Graf (1992) reviewed many of these processes.
All of these phenomena are related to the coupling of the benthic to the
pelagic zones in a narrower sense—the magnitudes of advective and diffusive
terms in a mass-balance equation describing the transport of nutrients and
organisms across the interface. In a sense, the mixed layer of the benthic
zone ($\approx$ top 10 cm) plays a role analogous to the mixed layer at the surface
of the water column. It is a region of active biogeochemical processes and
may serve as a source or sink of nutrients to the water column above it. The
level of activity of this zone depends on the degree of advective transport
(sediment accumulation at the surface) and the degree of turbulent and
biodiffusion.

While the top 10 cm or so of the sediment profile is a mixed layer subject
to the mechanical mixing of sediments by organisms (bioturbation; Sect.
3.5.2), and the thin layer just below the surface (Brinkman layer) is subject
to the effects of shear, generated in the overlying waters, most transport in
the sediment profile is due to porewater diffusion, dispersion, and advection
(see Fig. 3.10; Boudreau, 1997). Immediately above the sediment surface
are boundary layers similar to those of the atmosphere (Sect. 2.6.1) in which
the presence of the benthic surface is manifested by the effects of diffusion
and shear on the composition and velocity profiles of the overlying waters
(Thibodeaux, 1996). In the diffusive sublayer, a linear concentration profile
of dissolved materials exists between the top of the layer and the benthic
surface. The lowest layers are frequently associated with a benthic nepheloid
(cloudy) zone in which benthic material is continually or episodically re-

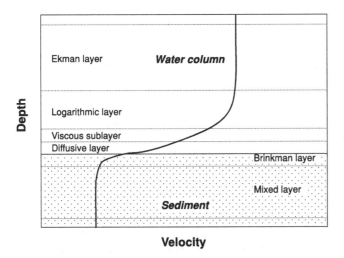

FIGURE 3.10. Velocity profile and various boundary layer features above and below
the seafloor (not to scale) (after Svensson and Rahm, 1991).

suspended by benthic organisms or flow events (cf. Richardson et al., 1993; Pudsey and King, 1997; Pilskaln et al., 1998).

From the standpoint of benthic diagenesis as described by advection–diffusion-reaction equations, the coupling between the benthic and pelagic zones occurs in the relationship between the deposition rate of sediments falling to the benthic surface (marine snow, etc.) and the rate at which the surface is increasing in elevation above a fixed datum. Neglecting time-dependent compaction of sediments, this allows the constituents of the benthic zone to be considered to be moving downward relative to the benthic surface, at a constant rate equivalent to the sedimentation velocity. At the benthic surface, in the absence of surface reactions and hydraulic gradients, the *diffusive* flux of dissolved nutrients from the boundary layer above the surface must equal the interstitial flux in the sediments relative to the surface (the sum of a diffusive and an advective term). Similarly, the *advective* flux of solid-phase nutrients (particles) in the water column is equal to the sum of a diffusive and advective flux in the sediments, where the diffusive term is typically due primarily to bioturbation (Sect. 3.5.2). A detailed treatment of the boundary conditions at the sediment–water interface is presented in Boudreau (1997), but also see Thibodeaux (1996) for detailed examples.

3.3   Diffusion of Spores

In antiquity people believed that winds carried diseases to man, animals, and crops: "when the wind is in the East 'tis good for neither man nor beast." This concept takes its modern form in terms of the problem of diffusion and transport of minute organisms in the air, which was the subject of an entire issue of the *Philosophical Transactions of the Royal Society Series B* in 1983 ("The Aerial Transmission of Disease," vol. **302**, pp. 437–604). Often, these microbes are *spores* such as the endospores of bacteria, the spores of Actinomycetes, fungi, ferns, and the pollen of flowering plants. We will also include the seeds and fruits of plants in our discussion of spores. In addition to settling under the influence of gravity, spores diffuse passively in the atmosphere.

The process of spore dispersal has four principal states: (1) spore liberation; (2) dispersion in the air or in water; (3) deposition; and (4) germination. Here we are concerned primarily with stage (2) in the air. Stages (1) and (3) are important only insofar as they specify the initial and boundary conditions.

A *dispersal spore* may consist of one or more cells. Its specific gravity ranges from 0.5 to 1.5; usually it is slightly heavier than water. Its size varies approximately from 1 to 100 μm with the exception of bacteria; for fungi, the range is 3 to 30 μm, and for pollen, it is 20 to 60 μm. Seeds and fruits, collectively known as diaspores, may be considerably larger than this range. Since spores are heavier than air, they tend to fall at a constant terminal velocity in still air.

3.3.1   Settling Velocity

According to Stokes' law, a small sphere of radius a moving with velocity v in still, viscous fluid with viscosity η is subject to a resistive force F, given by

$$F = 6\pi a\eta v.$$

When this force of resistance balances the immersed weight of the sphere (absolute weight minus the force of buoyancy), the sphere falls with a constant velocity. We call this terminal velocity the *settling velocity*, v_s. Constructing a force balance, we obtain

$$\underset{\text{resistance}}{6\pi a\eta v} = \underset{\text{weight}}{4/3\pi a^3\rho_1 g} - \underset{\text{force of buoyancy}}{4/3\pi a^3\rho g} = 4/3\pi a^3(\rho_1 - \rho)g, \qquad (3.16)$$

where ρ_1 is the density of the sphere, ρ is the density of the fluid, and g is the acceleration of gravity. From (3.16) we have

$$v_s = 2(\rho_1 - \rho)ga^2/9\eta. \qquad (3.17)$$

Since the density of spores is roughly 1 g/cm^3 and that of air is of the order of 10^{-3} g/cm^3, the influence of the buoyancy force may be ignored for spores falling in the atmosphere. This is not true for spores in water since $\rho_1 \cong \rho$; a spore may fall slowly in water or even float on the surface.

Stokes' law holds satisfactorily for spores of radius $a = 1$ to 100 μm falling in the atmosphere, for which the calculated settling velocity varies from 0.01 to 100 cm/s (see Fig. 3.11). In water, Stokes' law holds from microbes (plankton) of radius $a = 1$ to 250 μm, for which the calculated settling velocity varies from 10^{-6} to 0.1 cm/s. For an excellent review on the settling velocity of plankton in the sea, consult Smayda (1970). Stokes' law does not apply under more inertial conditions (i.e., Re > 0.5) due to turbulence, but v_s can be estimated before the onset of turbulence using the following relationship:

$$v_s = 2K_I a_0 \left(\frac{\rho_1 - \rho}{\rho}\right)^{2/3} \Big/ v^{1/3},$$

where K_I is a dimensional constant, and a_0 is the relative radius given by

$$a_0 = a - \zeta A,$$

where ζ is a constant (0.4 for spheres), and A is the largest radius that would satisfy Stokes' law (Dallavalle, 1948; also see Vogel, 1994). Sundby (1983) applied these models for v_s to the eggs of three pelagic fish species where Re < 5 and $K \approx 19$ (cgs). The terminal velocity was predicted to range from 0.2 to 2.5 mm/s corresponding to Re between 0.2 and 3, and these were presented in composite figure relating a and the density difference, $\Delta\rho$.

The relationship between particle shape and settling velocity has been an area of considerable interest. Komar et al. (1981) derived a semiempirical equation for the settling velocity of ellipsoidal and cylindrical particles sinking in water. This model compared favorably with settling velocity data

64   Akira Okubo, Josef Daniel Ackerman, and Dennis P. Swaney

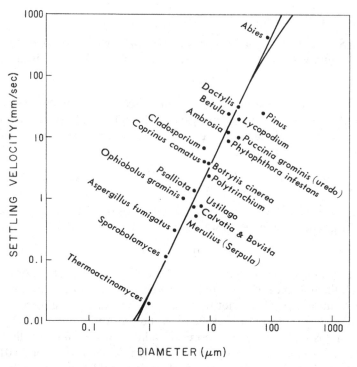

FIGURE 3.11. Observed settling velocities of spores and pollen related to diameter. The straight line represents calculated settling velocity of smooth sphere (density 1) from Stokes' law (copied with permission from Fig. 3 in Gregory, P. H.: *The Microbiology of the Atmosphere*, 2nd ed. J. Wiley & Sons, 1973).

obtained from the cylindrically shaped fecal pellets produced by copepods and euphausids. Takahashi and Honjo (1983) measured the sinking speeds of the skeletal remains of 55 species of marine protists known as radiolarians. v_s ranged from 13 to 416 m/day, i.e., 1.5×10^{-4} to 4.8×10^{-3} m/s. These speeds are generally lower than those predicted by Stokes' law. Nonsphericity can account for only part of this departure, but other factors, such as density, may also be of importance. Interestingly, the density of settling particles has been measured only directly, rather than inferred via settling rate experiments, in a number of species (e.g., Oliver et al., 1981; Ackerman, 1989, 1997b; see below and Sect. 3.2.2). Jackson (1989) revealed the importance of $\Delta\rho$ to settling velocity by reexamining the phytoplankton reviewed by Smayda (1970) and demonstrating the nonlinearity of $\Delta\rho$ with the cell diameter ($d = 2a$; $\Delta\rho \propto d^{-0.83}$). In this case, $v_s \propto d^{1.17}$ for phytoplankton, which was confirmed by Deibel (1990) for fecal pellets ($v_s \propto d^{1.14}$). Ross and Quentin (1985) measured the sinking rates of embryos of the Antarctic krill, *Euphausia superba*, at two hydrostatic pressures (10 and 1000 dbar). No significant differences in the sinking rates of embryos were found at any time during embryonic development. This result implies that while the water density increases at greater depths, the density of embryos also increases, thus maintaining the same density difference. Conversely, Cambalik et al. (1998) used a new motion analysis and experimental apparatus to examine the effects of salinity (i.e., density) on the ascending eggs of Atlantic menhaden. Ascending rates were found to be greatest at intermediate salinity and less at later stages of embryonic development.

When the concentration of particles is higher than a critical value, a new phenomenon occurs by which a group of particles settles as a whole to form a vertical density current. Bradley (1965, 1969) suggested the importance of this phenomenon in the natural environment. Both theoretical and experimental studies of the settling behavior of a group of particles has been examined (Adachi et al., 1978; Crowley, 1976, 1977; Thacker and Lavelle, 1978). Weiland et al. (1984) examined the sedimentation of two-component mixtures of solids and observed that (1) when the total suspension concentration is less than 10% by volume, both solid components settled at reduced rates, (2) for concentrations greater than 10%, the relative motion between particles leads to instabilities and "fingering settling" occurs. In this context, Batchelor and van Rensburg (1986) describe bulk vertical streaming motions that developed when two different species of small particles are dispersed uniformly in a fluid and are settling under gravity. They also present an instability model to explain the phenomenon.

Bienfang (1981) introduced a sinking-rate method that was suitable for the analysis of natural assemblages. Results from this method indicate (1) higher sinking rates in assemblages dominated by large cells, (2) decreased sinking rates after nutrient enrichment, and (3) buoyancy response to light levels. This would indicate that the settling velocity of phytoplankton depends largely on the conditions of the cells, with senescent cells sinking faster than the actively growing ones. The sedimentation of colonial cyanobacteria (*Microcystis aeruginosa*), which are buoyant, can be accounted for by their aggregation with colloidal precipitation (Oliver et al., 1985). Similar coagulation processes in lakes can be sufficiently rapid to significantly affect the downward flux of suspended material (Weilenmann et al., 1989). Results indicate that calcium ions act as destabilizing agents, while dissolved organic matter stabilizes particles and retard coagulation. Since the chemistry of lakes varies widely, natural coagulation rates should differ among lacustrine systems.

The coagulation of particulate matter into marine snow was discussed in Sect. 3.2.2, as was a model for the settling velocity of marine snow, which incorporated the porosity and mucous content of the aggregate. In situ settling velocities measured using stroboscopic photography have revealed size-related differences in v_s, with large aggregates settling slower than small particles (Asper, 1987). As described above, the dimensions (D) of marine aggregates are fractal between $D = 1.7$ and 2.3, which is less than 3, which

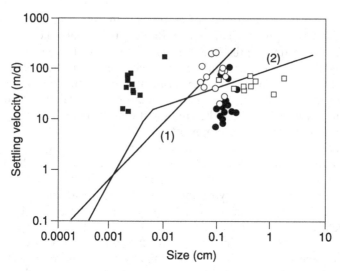

FIGURE 3.12. Size dependence of aggregates on settling velocity. Legend: ○ Bedford Basin, ● diatom aggregates, □ and ■ in situ measurements, line (1) from Jackson (1989), line (2) from Stokes' law (after Ruiz, 1997).

would be expected for a sphere or cube (see Jackson and Burd, 1998). Johnson et al. (1996) generated fractal aggregate models in the laboratory using latex microspheres. Their analysis revealed that the settling velocity of aggregates was on average 4 to 8.3 times higher than predicted by Stokes' law or the porosity models described in Sect. 3.2.2, which assume a homogeneous distribution of particles in the aggregate. The largest deviation from theoretical predictions occured when D was less than 2 and the porosity was overestimated. These phenomena are also evident in Ruiz's (1997) review of the relationship between settling velocity and aggregate size presented in Fig. 3.12.

The study of solid particles in turbulent flow is a very complex process. However, for spherical Stokesian particles whose size is smaller than the Kolmogorov microscale ($\lambda = 2$ to 7 mm) of turbulence, the average settling velocity equals the settling velocity in still water. Nonlinearities in the drag force may result in reduction of the average settling velocity (see Lande and Wood, 1987). Lande and Wood (1987) analyzed a physical model of particle motion utilizing information on sinking rates of particles and the rate of turbulent diffusion as a function of depth to estimate the average time that particles, starting at a given depth, remain in the surface mixed layer or the euphotic zone. The total duration of suspension at each depth before the particles leave the upper layers, whether the first time or permanently, was also estimated. They achieved this by projecting the stochastic trajectories of the individual particles instead of tracking the concentration of continuously distributed substances as the normal approach.

3.3.2   Diffusion Model

The analysis of experimental data on the dispersion of clouds of spores, pollen, and diaspores from the standpoint of atmospheric diffusion theory has provided insight into the dispersal of plants and other sessile organisms (Gregory et al., 1961; Gregory, 1973; Rombakis, 1947; Schrödter, 1960; Raynor et al., 1970, 1972, 1973; Okubo and Levin, 1989; Greene and Johnson, 1992, 1996).

It was in relation to epidemiology, i.e., the spread of disease-producing agents, that Rombakis (1947), probably for the first time, and Schrödter (1960) dealt with the theoretical aspects of spore dispersion in the atmosphere, using a simple model of turbulent diffusion. The problem is considered only in two-dimensional space, where the x-axis lies downwind along the ground and the z-axis is vertically upward. The origin of the coordinate system is placed at the source of spores (Fig. 3.13).

Assuming constant advective flow and diffusivity, the concentration of S of a steady-state distribution of spores satisfies

$$u\frac{\partial S}{\partial x} - w_s\frac{\partial S}{\partial z} = \frac{\partial}{\partial z}\left(K_z\frac{\partial S}{\partial z}\right), \tag{3.18}$$

where u is the wind velocity, w_s is the settling velocity of the spores, S is the concentration of spores in the air, and K_z is the vertical diffusivity (assumed constant here). We ignore diffusion in the x direction (see Sect. 2.6.1) as well as nonconservative terms.

Under the conditions that at $x = 0$ a cloud of N_0 spores is released and that the diffusive flux vanishes at the ground, we solve (3.18) to find

$$S = N_0\left(\frac{u}{\pi K_z x}\right)^{1/2}\exp\left\{-\frac{(z + w_s x/u)^2}{4K_z x/u}\right\}. \tag{3.19}$$

Inspection of (3.19) shows that the concentration distribution of spores over height z is a Gaussian plume, with mean and variance that vary linearly with distance x. The condition at the ground means that the diffusive flux would vanish at $z = 0$ if there were no settling, i.e., $w_s = 0$. A more reasonable treatment of boundary conditions is given in further discussion in this section. The number of spores N found above height z per unit horizontal area

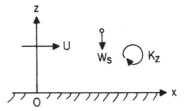

FIGURE 3.13. Model for spore diffusion in the air. U: wind velocity, w_s: settling velocity of spores, K_z: vertical diffusivity.

is given by integrating (3.19) over z:

$$N = \int_{z}^{\infty} S dz' = N_0 \left[1 - \Phi \left\{ \frac{(z + w_s x/u)}{2(K_z x/u)^{1/2}} \right\} \right],$$ (3.20)

where Φ is the error function, which is defined by

$$\Phi(a) \equiv \frac{2}{\sqrt{\pi}} \int_{0}^{a} e^{-b^2} \, db, \qquad a \geq 0,$$

and its numerical values lie between $\Phi(0) = 0$ and $\Phi(\infty) = 1$.

We shall now discuss some characteristics of spore dispersion on the basis of (3.20). We define the probable boundary of spore dispersal to be a curve both above and below of which 50% of the total number of spores released are found. We can obtain the equation of this curve by setting $N = N_0/2$ or by equating the argument of the error function (3.20) to 0.4768 since $\Phi(0.4769) = 1/2$. It follows that

$$\frac{(z + w_s x/u)}{2(K_z x/u)^{1/2}} = 0.4769.$$ (3.21)

The curve representing the boundary is thus seen to be a parabola passing through the origin (Fig. 3.14). Of course, we are free to choose other boundary curves characterized by different percentages, say $N = N_0/10$, which implies that $\Phi(1.1631) = 9/10$. However, this alters only the details of the argument, e.g., the numerical coefficients of the parameters in (3.22).

From (3.21) we can calculate (a) the range of dispersal X, i.e., the point at which the curve intersects the x-axis, (b) the height of dispersal, Z_m, i.e., the highest point of the curve, and (c) the duration of dispersal $T = X/u$, i.e., the effective flight duration of a spore. The results are found to be

$$X = 0.91 K_z u/w_s^2, \qquad Z_m = 0.2274 K_z/w_s, \qquad T = 0.91 K_z/w_s^2. \quad (3.22)$$

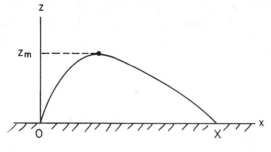

FIGURE 3.14. Curve representing the probable boundary of spore dispersal. X: range of dispersal, Z_m: height of dispersal.

TABLE 3.1 Calculated values of the range of dispersal (X), height of dispersal (Z_m), and duration of dispersal (T) for various values of diffusivity (K_z), wind speed (u), and settling velocity of spore (w_s)

(1)

K_z (m^2/s)	u (m/s)	w_s (cm/s)	X (km)	Z_m (m)	T
1	4	2	9.1	11.4	38 min
2	4	2	18.2	22.7	1 h 16 min
2	8	2	36.4	22.7	1 h 16 min
2	8	1	145.6	45.5	5 h

(2) $K_z = 1$ m^2/s, $u = 6$ m/s

size of spore (μ)	w_s (cm/s)	X (km)	Z_m (m)	T
small 5 × 3	0.035	44571	650	86 days
medium 14 × 6	0.138	2867	165	5 days 13 h
large 20 × 16	0.975	57.4	23.3	2.7 h
Phytophthora	1.3	32.3	17.5	1.5 h

(Remark) Reported range of dispersal for *Phytophthora* is between 200 m and 64 km.

(3) Various values of K_z (m^2/s)

size of spore (μ)	w_z (cm/s)	$K_z = 1$ Z_m (m)	T	$K_z = 2$ Z_m (m)	T	$K_z = 5$ Z_m (m)	T	$K_z = 10$ Z_m (m)	T
small 5 × 3	0.035	650	86 d	1300	172 d	3250	430 d	6500	860 d
medium 14 × 16	0.138	165	5.5 d	330	11.1 d	825	27.7 d	1650	55.4 d
large 20 × 16	0.975	23.3	2.7 h	46.6	5.4 h	116.5	13.5 h	233	27 h
Phytophthora	1.3	17.5	1.5 h	35.0	3.0 h	87.5	7.5 h	175	15 h

These quantities serve as parameters indicating the extent of spore dispersal. Note that both Z_m and T do not depend on wind velocity; only the range of dispersal is influenced by this parameter. Table 3.1 shows calculated values of X, Z_m, and T for various values of diffusivity, settling velocity, and wind velocity (Schrödter, 1960). The range of dispersal is inversely related to the settling velocity. Thus, the smaller the size of the organism, spore, or diaspore, the farther it will be blown if other parameters are unchanged. However, the other parameters are not likely to remain fixed. For example, female first-instar larvae of cochineal insects (*Dactylopius austrinus*) develop long, wax filaments on their dorsal surfaces that reduce their settling velocity and thereby enhance dispersal (Moran et al., 1982). The males have fewer, shorter filaments and are not dispersed as far. Augspurger and Franson (1987) found the deposition distance of samaras (winged fruit) of the tropical tree *Tachigalia versicolor* to be proportional to the average horizontal wind speed and inversely proportional to the square root of wing loading (weight of seed over projected cross-sectional area) over a range of sizes; they noted, however, that the effect of variation in wind speed is greater than that of variation in mass (see also Greene and Johnson, 1992).

Deviations of the flow-field parameters (velocity and diffusivity) from their steady-state values influence dispersal. Variance in deposition distance can be related to variance in wind direction, velocity, and turbulence (Matlack, 1992; Greene and Johnson, 1989, 1996). Matlack (1992) observed expected relationships between primary dispersal and seed size with *Betula lenta* seeds in a laboratory airstream, but noted only weak relationships in the field, presumably due to variations in the flow field.

Despite these caveats, the values of the parameters for spore dispersal given in Table 3.1 are not unreasonable, and Rombakis' (1947) model serves as a rough evaluation of spore dispersal. However, the fact that the model does not incorporate the structure of the atmospheric boundary layer renders it unsatisfactory. More advanced modeling relies on Sutton's equation diffusion (see Sect. 2.6.1) for atmospheric diffusion (Sutton, 1943; Schrödter, 1960).

The boundary condition at the ground creates some problems in obtaining an analytical solution. If there were no ground, the effect of spore settling could be taken into account simply by replacing the vertical coordinate z by $z + w_s t$ for an instantaneous source or by $z + w_s x/u$ for a continuous source, in the solution of atmospheric diffusion presented in Chap. 2. The settling effect becomes significant at places farther than a certain distance from the source. Spores settle by a vertical distance $w_s t$ during time t after release. At the same time vertical diffusion tends to spread spores by a vertical distance $(2K_z t)^{1/2}$. Hence, if we define a critical time t_c by $w_s t_c = (2K_z t_c)^{1/2}$ or $t_c = 2K_z/w_s^2$, the settling effect becomes important for times $t \geq t_c$. Likewise we may define a critical distance x_c as $x_c = u t_c = 2K_z u/w_s^2$, using the mean wind speed, u; the settling effect becomes important for distances $x \geq x_c$. For instance, when $u = 5$ m/s and $K_z = 100$ cm^2/s, the critical distance of settling for *Phytophthora infestans* ($w_s = 1.3$ cm/s) is $x_c = 600$ m. Microscopic spores of the order of 10 μm in size can travel several tens of kilometers from the source before settling comes into effect. Considering the fact that x_c is rather large, we sometimes use Sutton's formula (2.23) without correction due to settling. This practice is permissible for evaluation near the source area— say, within 1 km.

Godson (1957) applied Sutton's theory to the diffusion of particles that settle at the ground. Starting with Sutton's solution (2.23) in the absence of a ground level and adding the effect of the settling velocity,

$$S = \frac{Q}{\pi u c_y c_z x^{2-n}} \exp\left[-\left\{\frac{y^2}{c_y^2 x^{2-n}} + \frac{(z + w_s x/u)^2}{c_z^2 x^{2-n}}\right\}\right].$$

Note that the factor 2 in front of Q has been dropped. The boundary effect has not yet been taken into account. Godson's proposal for the reflective boundary condition at the ground is to add a mirror-image line source below the ground level. Thus,

$$S = \frac{Q}{\pi u c_y c_z x^{2-n}} \exp\left\{-\frac{y^2}{c_y^2 x^{2-n}}\right\} \left[\exp\left\{-\frac{(z+w_s x/u)^2}{c_z^2 x^{2-n}}\right\} \right.$$

$$\left. + \int_0^\infty \exp\left\{-\frac{(z+w_s x/u+\eta)^2}{c_z^2 x^{2-n}}\right\} f(\eta)\, d\eta\right], \qquad (3.23)$$

where the function f represents a relative intensity of the image source per unit vertical length, which must be determined in such a way that at $z = 0$, $\partial S/\partial z = 0$, i.e., such that the reflection condition is satisfied at the ground. This method is applied to a ground source; Godson also treats an elevated source.

More recently, Okubo and Levin (1989) used a solution of (3.18) to estimate the deposition rate of seeds at ground level from an elevated source ($x = 0$, $z = H$), incorporating the effect of atmospheric structure by considering $u = u(z) = u_0 z^\alpha$ and $K_z = K_z(z) = k u_* z$, resulting in the crosswind-integrated surface deposition rate $Q(x)$:

$$Q(x) = \frac{M w_s}{H \bar{u} \Gamma(1+\beta)} (f(x))^{1+\beta} \exp(-f(x)), \qquad (3.24)$$

where

$$f(x) = \frac{H^2 \bar{u}}{2(1+\alpha)\bar{K}_z x}, \qquad \bar{u} = \int_0^H u(z)\, dz/H, \qquad \bar{K}_z = \int_0^H K_z(z)\, dz/H,$$

where $\beta = w_s/(k u_*(1+\alpha))$, $k =$ von Karman's constant $(= 0.41)$, $u_* =$ friction velocity (see Sect. 2.6.1), and M is the rate of seeds released per unit length of a crosswind line source at height H.

Deposition rates estimated from (3.24) are similar to those resulting from a "tilted Gaussian plume" solution like (3.19) for small and large seed masses (see Fig. 3.15), but they differ somewhat at intermediate values; the modal value of the dispersal distance (i.e., the so-called seed rain) from the source is somewhat smaller than that of the Gaussian model for intermediate values of seed mass.

This seed rain is an important biological phenomenon for the ecology of plants (e.g., Harper, 1977; Willson, 1992). The aerial dispersal of diaspores has been reviewed by Van der Pijl (1972), Burrows (1975a, b, 1987), Harper (1977), Niklas (1992), and Raven et al. (1999). The small size of "dustlike" seeds of orchids allows them to be entrained into wind current, which affects their dispersal. A large number of seeds disperse ballistically, including *Impatiens* (touch-me-nots) and Mistletoes, which may disperse far from the

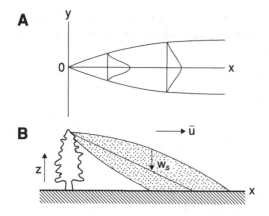

FIGURE 3.15. Gaussian plume model viewed from above in *A*, and tilted plume model viewed from the side in B. *H*: release height, *ū*: average wind speed, w_s: settling velocity, z: height (after Okubo and Levin, 1989).

source plant. Tumble weeds (*Salsola*) detach and are the ultimate wanderers, releasing seeds as they roll in the wind. Poplars (*Populus*) and Willows (*Salix*) have diaspores with woolly hairs that reduce settling velocities and render them as floating seeds. Plumed fruits of composites, including the dandelion (*Taraxacum*), are specialized fruits with a parachute-like pappus (plume) that affects their dispersal. These plants have evolved toward maximizing their dispersal by (1) decreasing the mass of the seed, (2) increasing the ratio of the pappus to the fruit, (3) increasing the aerodynamic drag of the pappus, and (4) elevating the height of release (Harper, 1977). Perhaps the most interesting aerial dispersers are the plane-winged and autogyroscopic-winged diaspores of conifers (seeds) and the samaras (fruit) of a wide variety of plants. These asymmetrical diaspores autogyrate once they reach terminal velocity (w_s), which is related to the square root of the wing loading (described above) and the conical section traced (area $= A_D$) as the diaspore autogyrates, given by

$$w_s = \sqrt{2}\left(\frac{W}{\rho A_D}\right)^{1/2},$$

where W is the weight and ρ is the density (see Ward-Smith, 1984; Niklas, 1992). Diaspore dispersal also occurs on the water surface via various modes of floatation (specialized hairs, disclike shapes) and rafting of vegetative tissues, and beneath the water surface via sedimentation (Van der Pijl, 1972; Cook, 1987; Ackerman, 1998b). The classic example of flotation involves coconuts (*Cocos*) that utilize Archimedes' principle, but many familiar plants like cattails (*Typha*) have hairs that decrease the surface tension to render themselves buoyant. Churchill et al. (1985) documented gas-bubble dispersal in the seagrass *Zostera marina*, in which gas bubbles attached to a small fraction of seeds can lead to dispersal distances orders of magnitude greater than without gas bubbles (e.g., maximum of 200 m versus 1 m). Underwater seed rains are likely to be analogous to those described above for aerial diaspores.

3.3.3   Experiments on Spore Dispersal

One of the earlier experiments on spore dispersal is credited to Stepanov (1935), who used artificial sources of spores; in one experiment he released 1.2×10^9 spores of *Tilletia caries* to study their diffusion. Gregory (1973) reexamined Stepanov's data, together with data obtained with *Lycopodium* spores, in light of Sutton's theory of atmospheric diffusion. Gregory found that the data agreed with Sutton's formula for the lateral variance

$$\sigma_y^2 = 1/2 c_y^2 x^{2-n}$$

if $c_y = 0.64$ m$^{1/8}$ and $n = 0.25$; the data were found to be incompatible with a Fickian diffusion model with constant diffusivity, variance $\sigma_y^2 = 2K_y x / u$.

Dispersion studies conducted by the Meteorology Group at Brookhaven National Laboratory (Long Island, New York) in the 1970s with the pollen of ragweed (*Ambrosia* spp.) and timothy, *Phleum pratense*, also supported Sutton's model (Raynor et al., 1970, 1972, 1973). Ragweed pollen grains are spherical and about 10 μm in radius. Their settling velocity is approximately 1.5 cm/s. Ragweed pollen was released continuously at the height of 1.5 m above ground level, from point sources and from circular area sources of various sizes. Samples were taken at four heights (0.5 to 4.6 m) and at four or five distances from the source, to a maximum of 69 m. The sample counts were converted to average concentration in grains per 1 m^3 and to deposition in grains per 1 m^2.

Figure 3.16 shows some horizontal distributions of pollen concentration at a height of 1.5 m and some cross-sectional distributions along the plume centerline. The mean wind speed at 1.5 m was 2.9 m/s. Figure 3.17 shows

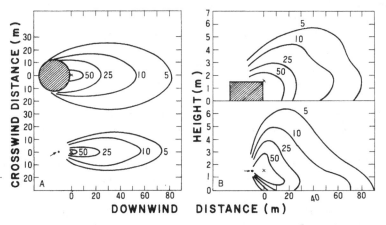

FIGURE 3.16. Diffusion of the pollen of ragweed from circular and point sources. A: Horizontal distributions at a height of 1.5 m. B: Vertical distributions along the plume centerline. The concentration at the point ($\times$) is taken as 100 units (from Raynor et al., 1970).

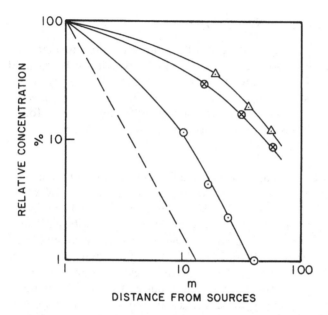

FIGURE 3.17. Variations of mean relative centerline concentration of ragweed pollen with distance. ——○—— data from a point source, ——⊗—— data from a circular source of 18.3 m in diameter, ——△—— data from a circular source of 27.4 m in diameter. (– – –) theoretical curve due to Sutton (from Raynor et al., 1970).

examples of the variations of mean relative centerline concentration with distance. Also shown is a point source calculated from Sutton's formula and corrected for deposition. The slope of Sutton's curve is similar to that of the experimental point source curve beyond 10 m. The area source curves also tend toward a similar slope at distances increasing with source size.

The ratio of the standard deviations of plume height and width, i.e., σ_z/σ_y, ranges from 0.10 to 0.25. These values are smaller than the value of 0.5 given by Sutton's theory (Sect. 2.6.1), indicating that lateral dispersion predominates over vertical dispersion close to the ground.

The mean velocity of deposition, calculated from the measured amounts of deposition on the ground, was found to be 5.05 cm/s, considerably larger than the computed settling velocity of 1.56 cm/s. Higher values of the velocity of deposition were found near the source.

It was estimated that about 1% of the pollen grains remain airborne at a distance of 1 km from the source. A single ragweed plant can release in excess of 10^6 grains per day during the height of the pollination season; even if only 1% of these remain airborne, the implied concentration of airborne ragweed pollen during the season is quite large. For an estimate we take an area inhabited by 10 ragweed plants per m^2. If the airborne pollen grains were distributed uniformly from ground level up to a height of 100 m, the air would contain 1000 grains per m^3.

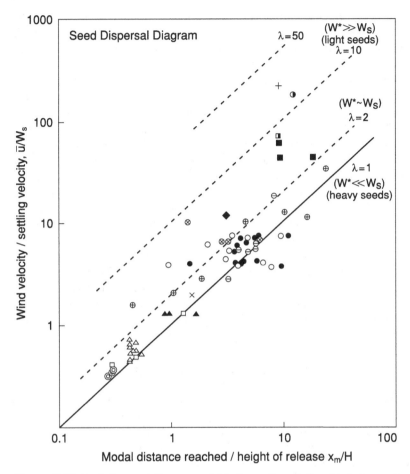

FIGURE 3.18. Seed dispersal diagram, which shows the relationship between the mode of dispersal distance and the mean wind speed (nondimensionalized form) for various values of λ ($\lambda = (\bar{u}/W_s)/(x_m/H)$). Legend: x_m: mode of seed dispersal distance, H: height of seed source, $\bar{u}$: mean wind speed, W_s: denotes settling velocity of seed. Here $\lambda = (\bar{u}/W_s)/(x_n/H)$. ⊚ Augspurger-Hogan, △ Augspurger-Franson, ⊗ McEvoy-Cox. ◆ Gross, □ Cremer, ⊕ Platt-Weis, ▲ Watkinson, ⊖ Isaac. ●○ Morse-Schmidt, × Colwell, ■ + Srceramula-Ramalingam, ⬛ Tonsor. ◐ Raynor. (From Okubo and Levin, 1989).

Okubo and Levin (1989) collected dispersal data from 15 field studies and tabulated the dispersal parameters to compare with their dispersal equation (3.24) (see Table 1 in Okubo and Levin, 1989). The resulting dispersal diagram (Fig. 3.18) plots average wind speed normalized by seed settling velocity against modal dispersal distance normalized by the height of seed release on a log scale. Heavy seeds match the predicted behavior especially well, with the modal dispersal distance $\approx Hu/w_s$, where H is height of release, u is average wind velocity, and w_s is settling velocity. Lighter seeds, with settling velocity smaller than the local turbulent mixing velocity w^* (speed at which the particle would "settle" via turbulence in the absence of gravity), tend toward a modal distance $\approx Hu/w^*$.

Spore distribution in higher atmospheric strata over longer distances is a very important problem from the viewpoint of epidemiology as well as of genetics and paleoecology (Moore, 1976; Mollison, 1977a; Aylor et al., 1982; Aylor, 1986). For example, the aerial transport of fungal spores has been known to be responsible for spreading plant diseases over distances of 500 km or more. A logical framework for estimating long-distance transport of viable spores involves (1) the reproduction of spores at the source area, (2) the escape of spores into the air above the canopy, (3) the transport and diffusion of spores in the atmosphere, (4) the survival of airborne spores, and (5) the deposition of spores onto a distant crop. The survival of spores and deposition or washout during rainfall are difficult to predict. On a much larger scale, the panspermia hypothesis is doubtful given Weber and Greenberg's (1985) experiments involving the inactivation of *Bacillus* by vacuum ultraviolet radiation to simulate interstellar conditions. Remarkably, the damage produced by low temperatures (e.g., 10 K) was less than the damage due to UV radiation in the range of 200 to 300 nm, but both were sufficient to kill the cells.

3.3.4   Vertical Distributions over Wide Regions

Thus far we have dealt with spore dispersion from given local sources. Now we shall discuss vertical distributions of dust, spores, and pollen grains over the globe as a whole or as a mean state over a wide area. In this case the effects of local source are smoothed out, and horizontal advection and diffusion may be ignored after averaging over a large horizontal area. We thus consider only the steady-state distribution in the vertical direction determined by

$$\frac{d}{dz}\left(K\frac{dS}{dz}\right) + \frac{d}{dz}(wS) = 0, \qquad (3.25)$$

where K is a virtual diffusivity in the vertical direction, which incorporates the effect of various processes arising from the averaging, such as small-scale vertical convection. In this sense K is not identical with K_z, which is appropriate for small-scale diffusion processes. By the same token, w should be interpreted as a virtual settling velocity. Both K and w depend in general on height.

Integrating (3.25) over z from the base height z_0 (which could be the top of a crop canopy, or the ground, $z = 0$) to a height z, we obtain

$$K\frac{dS}{dz} + wS = \left(K\frac{dS}{dz} + wS\right)_{z=z_0} \equiv P, \qquad (3.26)$$

where P is the upward vertical flux of S at the base. That is, under the assumption of steady state and no net horizontal flux, average vertical flux is constant $(= P)$. Integrating (3.26) once more over z, we find

$$S = \left[S_0 + \int_{z_0}^{z} P/K \exp\left\{ \int_{z_0}^{z'} w/K \, dz'' \right\} dz' \right] \exp\left\{ - \int_{z_0}^{z} w/K \, dz' \right\}, \qquad (3.27)$$

where S_0 denotes the concentration at $z = z_0$. The total number of airborne spores per unit surface area is given by

$$N = \int_{z_0}^{\infty} S \, dz. \qquad (3.28)$$

Some special models are considered below.

Model 1. K, w: constant, $z_0 = 0$, $P = 0$.
   From (3.27) and (3.28),

$$S = S_0 \exp(-wz/K),$$
$$N = S_0 K/w. \qquad (3.29)$$

Spore concentration is seen to decrease exponentially with height (Fig. 3.19).

Model 2. $K = a(z + z_1)$, $w = $ constant, $z_0 = 0$, $P = 0$ (a and z_1 are constants).

$$S = S_0 \left(1 + \frac{z}{z_1} \right)^{-\lambda}, \qquad \lambda \equiv w/a,$$
$$N = \begin{cases} S_0 z_1/(\lambda - 1), & \text{if } \lambda > 1, \\ \infty, & \text{if } \lambda \leq 1. \end{cases} \qquad (3.30)$$

The distribution represented by (3.30) corresponds to the vertical profile of insects due to Johnson (1969). See Sect. 6.2.

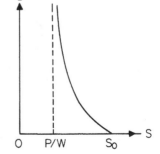

FIGURE 3.19. Theoretical distribution of spore concentration (S) in the vertical direction (z) ("normal" type).

Model 3. $K = az(1 - z/d)$, $w =$ constant, $z_0 > 0$ (a and d are constants).

$$S = \frac{P}{w} + \left(S_0 - \frac{P}{w}\right)\left(\frac{d - z_0}{d - z}\frac{z}{z_0}\right)^{-\lambda}, \qquad \lambda \equiv w/a,$$

$$N = \infty.$$

(3.31)

This distribution results from a parabolic profile of K that occurs at near-neutral conditions (Clarke, 1970; Foda, 1983). If $P = 0$, then $S = 0$ at $z = d$, corresponding to a diffusion ceiling or "inversion lid," above which the concentration is assumed to be negligible. In this case, N depends on λ. Foda (1983) discusses (3.31) with $P = 0$ in the context of dust transport from the desert to the sea, where the surface boundary condition changes from $S = S_0 > 0$ to $S = 0$ (a perfect sink).

Model 4. K, w: constant, $P \neq 0$, $z_0 = 0$.

$$S = P/w + (S_0 - P/w)\exp(-w/Kz),$$

$$N = \infty.$$

(3.32)

If $wS_0 > P > 0$, i.e., the flux at the ground is smaller than that of settling, S decreases with height and approaches a constant value, P/w (Fig. 3.19). If $P > wS_0 > 0$, i.e., the flux at the ground is larger than that of settling, S increases with height and approaches P/w (Fig. 3.20). In other words, depending on the value of the flux at the ground, the concentration does not necessarily decrease with height. Craigie (1945) observed vertical distributions of *Puccinia graminus* uredospores over Manitoba, Canada, which look similar to Fig. 3.20. However, spore distributions that exhibit decreasing concentration with height are more commonly observed.

If we consider the case $P = 0$ and choose $K = 10$ m²/s and $w = 3$ cm/s as typical values, the concentrations at heights of 1 km and 10 km are calculated from (3.29) to be, respectively, e^{-3} and e^{-30} of the concentration at the ground S_0. On the other hand, microscopic spores for which, say $w = 0.03$ cm/s, are expected to diffuse into the stratosphere, above 10 km.

Data from the Krakatao eruption of August 1882 (Flohn and Penndoff, 1950; Deirmendijan, 1973; Lamb, 1970), from the Mount St. Helens erup-

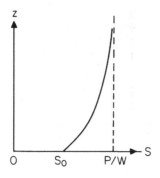

FIGURE 3.20. Theoretical distribution of spore concentration (S) in the vertical direction (z) ("abnormal" type).

tions of May and June 1980 (Danielsen, 1981; Pollack, 1981), from the El Chichón eruption of April 1982 (Robock and Matson, 1983), and from thermonuclear bomb tests (Machta, 1959) have shown that fine dust particles, once carried to heights above 10 km, are transported by the prevailing winds and spread over global dimensions on a time scale of weeks. Dust has been shown to be carried from terrestrial sources in Northern Africa to the Atlantic and from Asia to the Northern Pacific (Prospero, 1981, 1990; Parrington et al., 1983; Chen, 1985; Zhang et al., 1992). Smaller dust plumes originate in North America and Europe. This worldwide dispersion of dust is relevant to the problem of mineral nutrient fertilization of the open ocean (Martin et al., 1991; Duce et al., 1991; Prospero et al., 1996) and to long-distance dispersal of microbes (Close et al., 1978). As long as distantly dispersed microbes remain viable, they can reproduce after deposition and represent a potential threat to crops, livestock, and natural populations, as well as to humans.

An obvious limitation of the one-dimensional, steady-state equations described above is that they do not incorporate the details of known patterns of atmospheric circulation, scavenging and other atmospheric processes, and the distribution of sources that sophisticated numerical models include, e.g., GESAMP (Duce et al., 1991), the GFDL Global Chemical Transport Model (Levy et al., 1982), and others (Prospero et al., 1996). Variations in the surface cover within a region (e.g., open water, vegetative canopy, bare soil) affect the surface boundary condition as well as the average profiles of wind velocity and eddy diffusivity, so (3.26) is most appropriate over large regions of uniform cover. However, it can also be considered as a side-boundary condition for horizontal transport at the interface of two regions of different cover (cf. Foda, 1983).

3.3.5   Wind and Water Pollination

The last section on spore dispersal will focus on the transport and capture of pollen by receptive surfaces (e.g., pollination droplets and flowers), i.e., pollination, in aerial and aquatic environments. In these cases, pollination occurs via the fluid media, not an animal vector, and is therefore referred to as *abiotic*. While some early concepts considered abiotic pollination to be largely a stochastic event (e.g., Faegri and Van der Pijl, 1979), there is sufficient evidence to indicate that wind pollination (i.e., anemophily) and water pollination (i.e., hydrophily) represent sophisticated fluid mechanical solutions to the problem of pollen release, dispersal, and capture (see below and the recent review by Ackerman, 2000).

Anemophily is much more common than hydrophily, in that about 98% of the plants that pollinate abiotically are pollinated by wind (Faegri and Van der Pijl, 1979). Indeed, some of the earliest seed plants were pollinated by wind (Niklas, 1992), as are present-day gymnosperms (e.g., conifers), catkin-bearing trees (e.g., birches, oaks), and herbaceous plants, including many important cereals (e.g., grasses, sedges) (see reviews in Faegri and Van der Pijl, 1979; Regal, 1982; Whitehead, 1983; Niklas, 1985, 1992; Proctor et al., 1996; Ackerman, 2000). As mentioned, traditional views held that anemophily was a wasteful and inefficient process because pollen-to-ovule ratios can be as high as 10^6 : 1 and the aerodynamics of pollination were not understood (Faegri and Van der Pijl, 1979). The large pollen-to-ovule ratios are not unique or representative of all anemophilous plants, and some researchers have interpreted the large production of pollen to be related to the differences in metabolic costs relative to biotically pollinated plants (e.g., Niklas, 1992) and intramale competition (e.g., Midgley and Bond, 1991). Pollen are small relative to biotically pollinated plants (i.e., 20–60 μm versus ≤200 μm; Harder, 1998) and are unornamented without surface oils, etc. (the so-called pollenkitt), to eliminate the possibility of clumping (Crane, 1986). Experiments have shown that clumped pollen have higher settling rates than unclumped pollen (Niklas and Buchmann, 1988; see Sect. 3.3.1).

The receptive reproductive structures of wind-pollinated plants include strobuli (cones), catkins (flexible tassel-like inflorescences), and spikelets of grass flowers, which in the latter two cases have brush or featherlike stigmas to aid in pollen capture (Proctor et al., 1996). These reproductive structures are neither showy or colorful, nor do they contain the nectaries, odors, and other rewards that biotically pollinated plants use to attract pollinators. These generalizations, however, are not necessarily valid for the large numbers of families with species that are secondarily abiotically pollinated (cf. Renner and Ricklefs, 1995; Ackerman, 2000).

Ecologically, anemophily is associated with the spatial or temporal separation of male and female reproductive structures (i.e., monoecy and phenological allochrony, respectively), which may encourage outcrossing (Charlesworth, 1993; Renner and Ricklefs, 1995). Atmospheric conditions influence pollen release, which usually occurs under dry conditions and before leafing, which favor dispersal (Whitehead, 1983). There are important geographic associations of anemophilous plants with "stressful" environments, where species diversity is low, which leads to higher proportions of conspecifics, and conditions may be dry (e.g., high latitudes versus tropical rainforests; Regal, 1982; Whitehead, 1983). However, the interpretation of the reasons behind these associations has been called into question (Midgley and Bond, 1991).

The majority of the data related to the mechanisms of wind pollination in plants can be attributed to Niklas and co-workers (see reviews in Niklas, 1985, 1992), who used techniques to visualize the flow patterns (e.g., helium bubbles) and transport of pollen (stroboscopic photograph) around female reproductive structures in wind tunnels. Essentially, they showed that female organs and associated vegetative structures (e.g., bracts) modify the airflow in ways that favor pollination, as can be seen in Fig. 3.21. These modifications include changing flow patterns, flow patterns directed to receptive structures, downstream eddies that generate circulation, and alternating

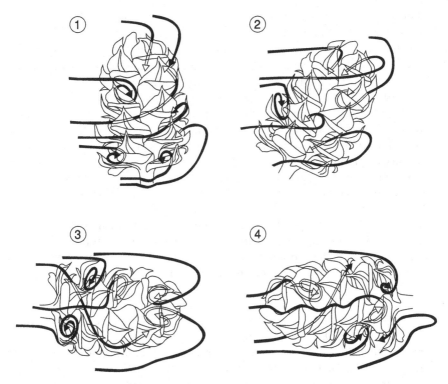

FIGURE 3.21. Characteristic airflow patterns around *Pinus* female cone obtained from statistical surveys at 5 m/s flow from left. Note the different orientations of the cone with respect to the direction of flow (from Niklas and Paw U, 1983).

Reynolds numbers (Niklas, 1985). Pollen capture, which is essentially an aerosol process, is favored in these fluid dynamic environments. The strongest support for anemophily as a highly canalized process comes from data that indicate there is aerodynamic segregation among species in the environment. For example, Niklas and Paw U (1983) demonstrated that pollination by the same pine species occurred at higher frequencies than pollination by congenerics in laboratory experiments. Niklas and Buchmann (1987) demonstrated striking pollen discrimination between pollen and ovules of two species of *Ephedra* in the laboratory, which were supported, in part, by field observations reported by Buchmann et al. (1989). More recently, Linden and Midgley (1996) demonstrated pollen discrimination among four species of anemophilous plants from four different families (!) in South Africa. Their field survey revealed that while stigmatic surfaces included alien pollen, a minimum of 40% of conspecific pollen was found on stigmas (>80% in one case), which was always greater than the background frequencies obtained on gel-coated slides. These observations confirm that wind pollination is a highly canalized process involving the interaction of aerodynamics and biology.

Although well known (see reviews in Faegri and Van der Pijl, 1979; Proctor et al., 1996), water pollination is also proving to be a highly canalized process although data are scant and only recently available (Ackerman, 1995, 1998b, 2000). Pollination in most aquatic vascular plants occurs in the air through either biotic pollination or anemophily. The apparently high proportion of anemophilous aquatic plants is due to systematic rather than functional affiliation (Cook, 1988). Hydrophily occurs in a limited number of aquatic plants and is categorized by the location of pollen transport. Ephydrophily is characterized by pollination on the water surface, in which surface and wind currents cause pollen and/or anthers "raft" to emergent or slightly submerged (i.e., meniscus) female flowers and inflorescences. Cox (review in 1988) has expended considerable effort in applying random search theory, which considers Brownian recurrence in two and three dimensions and predicts that the probability (P) of a surface ship moving along a path (L) in a given area (A) finding a submerged target is given by

$$P = 1 - e^{-(WL)/A},$$

where W is the width of the sonar beam (Koopman, 1956). Unfortunately, while the theory is valid, the application to pollination is flawed because, among other things, it assumes pollen transport to be random and therefore recurrent, which is not the case as wind-generated and other surface movements are directional (Ackerman, 1995, 1998b). Hyphydrophily, or underwater pollination, occurs in less than 20 genera of freshwater and marine plants (i.e., seagrasses; Ackerman, 2000). The most common mechanism of hyphydrophily in freshwater plants, which do not experience regular water currents, is for pollen, which is released higher in the water column, to "shower" down to enlarged stigmatic surfaces (Guo et al., 1990). Interestingly, a form of self-pollination can occur on the surfaces of air bubbles that may form underwater around flowers and inflorescences (Philbrick, 1988).

True hydrophily, i.e., submarine pollination that occurs through the action of water currents, occurs in the seagrasses, which are the only "higher plants" in the marine environment (den Hartog, 1970). Intertidal seagrasses may pollinate on the surface when they are exposed at low tides, but this is not the case when they or subtidal populations exist at considerable depths (see Ackerman, 1998b). Although this functional group of about 50 species is comprised of 12 or 13 genera in five families within three clades (Les et al., 1997), a number of consistent features are associated with hydrophily (Ackerman, 1995, 2000). For example, seagrass pollen has evolved convergently to filamentous shapes, either directly as pollen, as in the case of the Potomagetonales, or functionally, as in the case of the Hydrocharitales (Ackerman, 1995). In the case of the former, the genus *Amphibolus* is reported to have pollen that is 6 mm long, while in the latter, the genus *Thalassia* has spherical pollen that is released in monofilamentous aggregates or germinate precociously, and the genus *Halophila* releases monofilaments composed of linear tetrads of pollen. The strong association of filamentous pollen shapes and hydrophily is discussed below (also see Pettitt, 1984).

The female flowers of seagrasses are generally elongated stigmatic surfaces of highly reduced green-colored flowers, which, as in the case of anemophilous flowers, lack petals, nectaries, or other features characteristic of biotically pollinated plants (see drawings in den Hartog, 1970). As in the case of anemophily, flowers can be monoecious, can be dioecious, or have phenological separation between pollen release and pollen reception (Pettitt et al., 1981; Les et al., 1997). Pollen-to-ovule ratios have been estimated in only one species, and these were on the order of $10^4 : 1$ (Ackerman, 1993). The seagrasses are found in coastal areas throughout the world, including Hudson's Bay (den Hartog, 1970).

Field and laboratory observations have been made of the pollination mechanisms of seagrasses in a number of species; these are reviewed in Pettitt (1984). Unfortunately, these observations are limited taxonomically, and this is even more acute with respect to the documentation of the mechanics of submarine pollination mechanisms, which exists only for the north temperate seagrass, *Zostera marina*. Pollination in this species was examined in the laboratory using stroboscopic photography and in the field using pollen models and gel-coated surfaces (Ackerman 1989, 1997a, b). Hydrophily appears to be governed more by smooth and more viscous conditions (i.e., lower Re) than the turbulent eddies described for anemophily. The phenological emergence of female flowers from within the inflorescence results in an increase in the shear stress (τ) in the local flow (Ackerman, 1997a). When the filamentous pollen (2.7 mm × 7.5 μm) encounter this environment, they rotate and cross streamlines toward female flowers on account of τ and thereby increase the opportunity for pollination (Fig. 3.22; Ackerman, 1997b). Given two-dimensional flow, the axial force (F_A) that causes these movements is given by

$$F_A = \frac{\pi \mu L^2 \gamma}{8(\ln 2r - 1.75)} \cos 2\phi,$$

where μ is the viscosity, γ is the shear rate, L and r are the length and aspect ratio, respectively, of the pollen, and ϕ is the angle to the normal (Forgacs and Mason, 1958). Unlike spherical pollen, filamentous and functionally filamentous pollen need only be in the vicinity of female flowers to pollinate by (1) direct interception on stigmas, (2) rotating within half a pollen length of stigmas, and (3) by being redirected through streamlines toward flowers. Field observation confirmed the flow conditions necessary for these laboratory-based observations (Ackerman and Okubo, 1993), as did the differential recovery of filamentous versus spherical pollen models (Ackerman, 1989). Recent observations of the submarine pollination of *Amphibolis* in the field were consistent with these findings (Verduin et al., 1996). These obser-

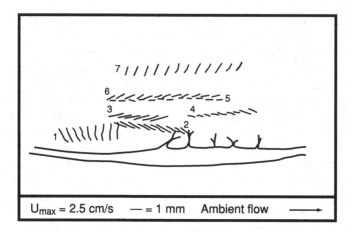

FIGURE 3.22. A composite tracing of seven selected pollen pathlines around *a Zostera marina* inflorescence in side view. Pollen rotate and cross through fluid streamlines (pathlines 1, 3, and 4) as they travel through the water with a maximum instantaneous pollen velocity (U_{max}) presented. This motion increases the potential for pollination. (From Ackerman, 1997b.)

vations and the strong convergence of filiform pollen in seagrasses indicate that filiform pollen is a functional adaptation for submarine pollination.

While data on the fluid dynamic segregation of pollen by sympatric species do not exist, genetic studies reveal that submarine pollination is successful in maintaining relatively high outcrossing rates in several species (Ruckleshaus, 1995; Waycott and Sampson, 1997). There are a large number of similarities between wind and water pollination; however, observations and characterization of the mechanics are lacking for a large number of taxa (Ackerman, 2000). This is an area rich in possibilities.

3.4   Dispersal of Gametes and Organisms

The evolution of reproductive mechanisms in animals originated with the dispersal of gametes and the results of their union in the external media, i.e., water. Subsequent time and evolution have led to the myriad of internal reproductive mechanisms that are more common to the reader. In this section, we concentrate on the relatively recent developments involving organisms that broadcast their gametes into the water column, where they are fertilized externally. Readers are directed below and to a review by Okubo (1994) for additional information on the diffusion of gamete dispersion.

While external fertilization can be found in most major taxonomic groups, much of our current understanding stems from a few well-researched model organisms such as echinoderms (e.g., starfish and sea urchins, especially the

TABLE 3.2 Factors influencing the fertilization of broadcast spawning invertebrates (adapted from Levitan, 1995)

A. Gamete	B. Individual	C. Population	D. Environmental
(1) *Sperm* Morphology Behavior Swimming speed Longevity	(1) *Spawning Behavior* Aggregation Synchrony Posture Spawning rate	(1) *Size & Spatial Aspects* Population size Population density Nearest-neighbor distance	(1) *Fluid Dynamics* Turbulence Mixing Water column discontinuities Fronts
(2) *Egg* Size Jelly coat Surface chemistry	(2) *Morphology* Size Fecundity	(2) *Demography* Sex ratio Age structure Size structure	(2) *Topography* Complexity Water depth Shelter
(3) *Condition* Age Compatibility	(3) *Condition* Age Energy reserves	(3) *Inter-Population* Other spawning species Predation	(3) *Water Quality* Temperature Salinity pH

genus *Strongylocentrotus*), corals (scleractinian and gorgonians), worms, and fish. The process of gamete dispersal has four principal states: (1) gamete release; (2) gamete dispersion in water; (3) gamete recognition; and (4) fertilization. We will examine aspects of each of these stages with the exception of (3), which is beyond the scope of this work (see Epel, 1991).

3.4.1 Broadcast Spawners and External Fertilization

One of the most essential aspects of reproduction is finding a mate, and this is of no less importance for broadcast spawners whose gametes (sperm and eggs) must be able to contact one another in the water column. Several patterns can be observed in nature in which mass spawners congregate and/or participate in synchronous mass spawning events. In addition, a number of factors affect gametes, individuals, populations, and environments that influence reproductive success in broadcast spawners (see Table 3.2).

Denny and Shibata (1989) used Gaussian plume models (see Sect. 2.6) to examine the physical conditions that favored successful reproduction in broadcast spawners under turbulent conditions. Their models predicted that the best opportunities for reproduction occur (1) when organisms aggregate in large numbers, (2) when they spawn synchronously, and (3) when there is weak to moderate energy in the environment. There is considerable empirical support for these predictions. Pennington (1985) provided empirical support for each of these predictions in experiments involving sea urchins in which fertilization success was examined in the field. Importantly, gamete concentrations are diluted rapidly, and fertilization success decreased with distance from the sperm source, which is similar to what was reported in

Sect. 3.3.3 for spore dispersal (Pennington, 1985; Levitan, 1991; Babcock et al., 1994; Benzie et al., 1994). Mead and Denny (1995) confirmed the last prediction by examining the fertilization success of sea urchin gametes exposed to different shear stress in a Couette cell. They found that while moderate shear led to increased rates of fertilization, high rates of shear led to lower fertilization success and developmental problems for embryos.

The size, motility, and viscosity of gametes may also influence reproductive success. Sperm are much smaller than eggs and are the motile of the two gametes. Vogel et al. (1982) modeled the fertilization success (φ) resulting in contact between sperm (S) and eggs (E) in the water using a Poisson distribution

$$\varphi = 1 - \exp\left(-\frac{\beta S}{\beta_0 E}(1 - e^{-\beta_0 E \tau})\right),$$

where β is the rate of fertilization, β_0 is the rate of collision ($\beta_0 = v\sigma$, where v is the sperm swimming velocity and σ is the cross-sectional area of the egg), and τ is the half-life of the sperm. It is evident from the model that egg size and aspects of sperm performance are important for fertilization success. Levitan (1993) noted that an increase in egg size (i.e., σ) among broadcast spawners was a mechanism by which females compete for sperm, as sperm limitation is considered to be common among broadcast spawners (Levitan, 1995; Lasker et al., 1996). Experiments with sea urchins confirmed this prediction in that the largest eggs required the least number of sperm for successful fertilization (Levitan, 1996). The one important stabilizing force is the fact that increased egg size also increases the probability of polyspermy (fertilization by more than one sperm), which results in reproductive failure (Styan, 1998).

Lifestyle can have a profound influence on broadcast spawners. Sessile broadcast spawners can be found at high densities, can be hermaphrodites, and can also reproduce asexually. Conversely, mobile organisms must meet at the correct time and place to ensure successful reproduction. We will examine these phenomena in sessile and mobile broadcast spawners.

Reef-building or scleractinian corals reproduce sexually either as broadcast spawners or as brooders (Fadlallah, 1983). Spawning can be synchronous over a period of several days or a period of a month, or seasonal with continuous spawning or monthly cycles (Richmond and Hunter, 1990). There are spectacular mass spawning events (i.e., 90% of species) that occur during the week following the full moon in the austral spring in the Great Barrier Reef of Australia (Babcock et al., 1986). Individual species appear to spawn at distinct times of the night, which may be a way to maximize fertilization success and limit hybridization (Harrison et al., 1984). The spawning species release sperm and eggs, which are both buoyant and float to the surface where they form large slicks (Harrison et al., 1984). The dispersal of these slicks and the interaction of island wakes and winds were reviewed by Pattiaratchi (1994). At least three different modes of gamete release occur in different species of corals: (1) extruded buoyant egg bundles that break at the surface; (2) streams of buoyant egg–sperm bundles; and (3) sperm release followed by sticky eggs (Harrison et al., 1984). Analogous variations in gamete release (i.e., plume versus strings versus clumps) have been observed in other invertebrates. The non-Newtonian nature of the gamete "packages" can lead to "shear thinning," where the form of gamete release is a function of the rate of shear (e.g., viscous packages under low shear; Thomas, 1994). This may account for some of the variation observed within species.

Sea urchins are important grazers in the benthic environments, which also participate in mass spawning events. In this case, there would need to be some type of environmental cue to lead the urchins to aggregate. Lamare and Stewart (1998) report such a mass spawning event in which *Evechinus* urchins aggregated along a 100-m-long line at densities up to 30 individuals per m^2 at a depth of 5 m to spawn (they are normally found at densities of 1.5 individuals per m^2 between 3 and 15 m depth). The buoyant eggs and sperm formed a 0.5-m-high cloud that was restricted by a halocline at 3-m depth. The low-salinity interface is believed to have increased the contact between sperm and eggs because it acts as a barrier to gamete dispersal.

The image of spawning salmon should be a familiar one to many readers. Salmon are anadromous, returning to their natal stream to spawn. Resident male and females excavate a redd in the gravel bottom of the stream and then release eggs and sperm in a cloudlike mass (Groot and Margolis, 1991). The fertilization of many marine fish can be more complex, involving stereotyped behaviors and territoriality in some cases. Such is the case in the tropical angelfish, *Pygophlites diacanthus*, in which the male and female courtship swim leads to the female's external release of eggs in the water column. The male flips his tail and swims downward, causing the egg mass to form a 50-cm-diameter vortex ring, which ascends upward at 27 cm/s (Gronell and Colin, 1985). The vortex confines the eggs and sperms and thereby increases gamete contact and fertilization. Okubo (1988) modeled this system by considering the impulse (I) of the force required to generate the vortex ring

$$I = \pi \rho R^2 X,$$

where ρ is the density, R is the radius of the vortex ring, and X is the circulation around the ring given by

$$X = 2\pi \int_0^a \xi \omega(\xi) \, d\xi,$$

where ξ is the radial coordinate in the vortex, and ω is the vorticity given by

$$\frac{\partial \omega}{\partial t} = \upsilon \left(\frac{\partial^2 \omega}{\partial \xi^2} + \frac{1}{\xi} \frac{\partial \omega}{\partial \xi} \right),$$

where v is the kinematic viscosity. The predicted ascent speed and maximum height obtained by the vortex agreed with observation. A diffusion model of this system confirmed that a higher fertilization would be expected through the concentration of gametes confined within the vortex ring.

3.4.2 The Dispersal of Fish Eggs and Larvae in the Sea

Marine scientists have long recognized that currents and turbulence in the sea may transport and disperse fish eggs and larvae over distances that are very important from the standpoint of regional ecology and gene flow (e.g., Scheltema, 1971c, 1975, 1986; Levin, 1983; Rothlisberg et al., 1983; Scheltema and Carlton, 1984; Norcross and Shaw, 1984; Emlet, 1986; McGurk, 1987; Roughgarden et al., 1987; Boicourt, 1988; Frank et al., 1993; Okubo, 1994). Eggs are considered to diffuse more or less passively. Larvae in the early stages of development may still be carried by water, although in the later stages they can perform their own motion, in particular vertical migration. Aggregation effects (i.e., patches) may be observed in the distribution of fish eggs and larvae for the same reasons they are observed in other plankton. As outlined by Okubo (1984), these include (1) behavioral reactions of the larvae to environmental factors (Sinclair, 1988; Houde, 1997), (2) food-chain associations in predator–prey interactions (McGurk, 1987; Sinclair and Iles, 1989), (3) aggregative feeding behaviors of larvae, and (4) mechanical retention in "structures" of the flow field (e.g., convective cells, gyres, and fronts between different water masses (Okubo, 1994; Bailey et al., 1997; Dickey-Collas et al., 1997; Munk et al., 1999).

Well-controlled diffusion experiments using eggs or larvae are difficult to perform. Ordinarily the distributions of eggs and larvae in the sea are measured and interpreted in the light of oceanic diffusion theory (Strathmann, 1974; Talbot, 1974, 1977; McGurk, 1988; Kim and Bang, 1990; Hill, 1990, 1991; Possingham and Roughgarden, 1990). Often the available data of larval distribution are not sufficiently detailed to permit a direct and accurate comparison with the theory. Despite these problems, estimates of the dispersal of eggs and larvae have been made using dye tracers, such as Rhodamine B (Hirano and Fujimoto, 1970; Talbot, 1977; Becker, 1978), fluorescein (Koehl et al., 1993; Koehl and Powell, 1994), and small particles (Koehl and Powell, 1994). At larger scales, satellite-tracked drifters, drogued to match the depths of larval concentrations, have been used for several years to estimate larval dispersal patterns (Hinckley et al., 1991; Bailey et al., 1997).

The mortality of eggs or larvae must be taken into account in a discussion of their dispersal patterns; however, the variance of the spatial distribution from an instantaneous source would not be altered by mortality, provided the process is density-independent, i.e., behaves according to first-order kinetics. This can be seen easily; the effect of the density-independent mortality process on the spatial distribution is simply to bring in a multiplicative factor e^{-vt}, with v the death rate and t the time after release, to the distribution without mortality, i.e., the distribution of a conservative quantity (Hill, 1990, 1991). Since the factor does not depend on spatial location, the variance—which is the second moment of the radial coordinate from the center of mass of a patch—is invariant. The variance will doubtlessly be influenced by mortality if it is density-dependent or if the source is continuous (McGurk, 1988).

Whether or not the mortality of an egg or larva is density-dependent, it is likely to be organism size-dependent because of both the effect of size on its average residence time in the water column and its desirability as food (McGurk, 1986). Mackenzie et al. (1994) suggested a dome-shaped relationship between the level of turbulence of the flow and ingestion rates of larval fish.

Some studies of larval dispersal conducted in estuaries in conjunction with dye tracer experiments suggest that larvae cannot diffuse as rapidly as might have been expected on purely physical grounds (Talbot, 1974). It seems probable that the larvae tend to aggregate near the bottom and thus do not partake in the full movement and mixing of the water. If the shear effect in the vertical direction is responsible for longitudinal dispersion (see Sect. 2.6.2), a strong tendency for larvae to concentrate near the bottom gives rise to a large decrease in the effective thickness on which the shear can operate, thus suppressing the horizontal dispersion. If this conclusion is correct, it is important to know the vertical distribution of larvae in order to evaluate their horizontal dispersal (Strathmann, 1974). Talbot (1977) described an extensive program of investigation of the distribution of plaice eggs and larvae in the North Sea. In this area the vertical shear effect was much more important than horizontal dispersion characterized by a constant diffusivity (Bowden, 1965). The effects of the vertical distribution and migration of larvae on their horizontal transport remain an active area of research (Hill, 1995).

As a means of genetic exchange and transportation, the advection of eggs and larvae by tides and prevailing currents plays a crucial role, as recognized over a century ago by Wallace (1876). Several studies have since suggested that major oceanic currents, such as the Kuroshio, are much more important in the movement of fish eggs and larvae than diffusion (Fig. 3.23), although large-scale eddies may play an equally important role in the dispersal (Hirano and Fujimoto, 1970; Fujimoto and Hirano, 1972; Hill, 1991; Scheltema, 1995; Bailey et al., 1997; Dickey-Collas et al., 1997). Sinclair (1988) examined how currents can break up and redistribute marine larval populations. Possingham and Roughgarden (1990) used one- and two-dimensional advection–diffusion models to examine the relative importance of eddy diffusion, longshore currents, and available habitat in organisms with a coastal adult phase and a pelagic larval phase.

An interesting example of dispersal on the ocean surface is provided by Ikawa et al. (1998), who examined the dispersal of ocean skaters (*Halobates*). This genus includes the only species (five) of truly oceanic insects, which live

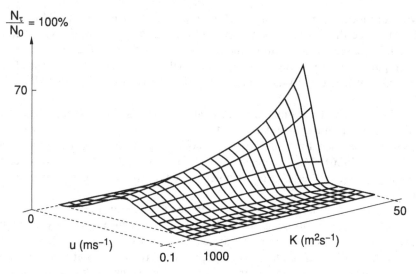

FIGURE 3.23. Relative contribution of advection (u) and diffusion (K) in larval survival at recruitment site 300 km downstream from the center of release (from Hill, 1991).

on the surface film of the major oceans between $\pm 40°$ N. Their dispersal is governed by oceanic diffusion (i.e., turbulence), which is greater than the Kolmogorov scale (see Sect. 2.6.2). In this case, information from Okubo diagrams (e.g., Okubo, 1971) was used to determine that (1) dispersal could be as great as 1250 km in 60 days, (2) encounter rates between males and females could be on the order of 11 per day, and (3) slow growth rates could be offset by the relatively long life span of the insects. Importantly, the estimated encounter rates between the sexes was more than an order of magnitude greater than what was predicted by random encounters.

On the basis of extensive surveys of larvae throughout the North and Equatorial Atlantic Ocean, Scheltema (1971a, b; 1995) summarized the evidence for long-distance larval dispersal. According to Scheltema (1971a), the probability of successful transoceanic larval dispersal, p, is equal to the product of the drifting coefficient, p_d, and survival coefficient, p_s:

$$p = p_d p_s. \tag{3.33}$$

An approximate value for the drifting coefficient may be derived from a knowledge of drift-bottle and drift-card recoveries. Based on 156,276 bottles released along the North American coast, Dean F. Bumpus of the Woods Hole Oceanographic Institution evaluated the percentage of recovery in the Eastern Atlantic Ocean to be 0.2%. Also, Stander et al. (1969), using 1800 drift-cards released along the Northwest African Coast and subsequently recovered in the Western Atlantic Ocean, evaluated the recovery to be 0.18%. The percent recovery in both directions is remarkably similar. Consequently, an average value of 1.9×10^{-3} may be assigned to p_d. As probably not all bottles or drift-cards that actually made the crossing were recovered, the value of p_d is considered conservative.

The value of the survival coefficient, on the other hand, is much more difficult to estimate. Scheltema assumed

$$p_s = 2/N_f,$$

where N_f is the fecundity or total number of eggs produced per female. The numerical factor 2 is derived on the basis of an equal number of males and females. For the studied species of gastropod veligers, a reasonable estimate of N_f is 5×10^6 eggs. Then the probability that a particular larva from an egg mass will survive and ultimately reach the opposite coast of the Atlantic Ocean is estimated from (3.33) to be

$$p = 1.9 \times 10^{-3} \times 2/5 \times 10^6 = 7.6 \times 10^{-10},$$

or approximately one in eight billion. This probability is extremely small, but if we estimate the number of successful larvae, N_s, by multiplying this p by the number of larvae produced by a population, we obtain approximately $N_s = 2$, 2×10^3, 2×10^6, and 2×10^9, respectively, for populations of 10^3, 10^6, 10^9, and 10^{12} individuals.

Strathmann (1974) examined short-term advantages of dispersing the sibling larvae of sedentary marine invertebrates. Although the rate of diffusion in the sea increases with time, the advantage in adding an extra day to the pelagic life of larvae probably decreases as the duration of the pelagic phase increases, because the added increment of spread, i.e., $\Delta\sigma_r^2(t)$, is smaller relative to the spread already achieved, i.e., $\sigma_r^2(t)$, the horizontal variance.

One difficulty with the above estimates of dispersal probability is the assumption that p_d and p_s are independent. In fact, Bailey et al. (1995, 1997) suggested that the conditions associated with high larval transport rates (e.g., high wind and currents and depressed primary production) are likely to be associated with depressed larval feeding rates and elevated larval mortality. Nevertheless, the significance of large-scale passive larval transport to the global distribution of even sedentary marine organisms is indisputable (Scheltema, 1995).

3.5 Transport Across the Solid Interface

As discussed above, the interface between a fluid medium (air or water) and a saturated porous medium (the soil or benthic sediments) represents an abrupt transition in flow properties, but for simple physical situations (e.g., one-dimensional flows at a "flat" interface, assumed to be at depth $z = 0$), continuity of solute fluxes and concentrations are assumed at the boundary. If the flux is considered to be the sum of advective and (Fickian) diffusive

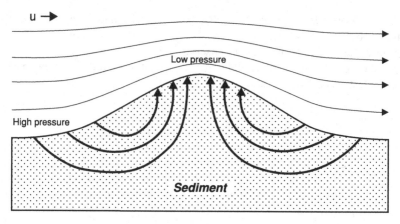

u →

Low pressure

High pressure

Sediment

FIGURE 3.24. Effect of "wavy" topography on the interstitial flow within sediments. u: velocity. (After Boudreau, 1997.)

flux, then this is equivalent to the conditions $C_f = C_p|_{z=0}$ and $dC_p/dz = D_f/(\phi D_p)\,dC_f/dz|_{z=0}$, where the subscripts f and p refer to the fluid and porous media, respectively, D is the coefficient of diffusion or dispersion for the media, and ϕ is the porosity of the soil or sediment.

In the case of solid-phase material falling to the sediment surface (e.g., marine snow) and accumulating, the interfacial boundary moves upward at a rate equal to the sediment accumulation rate. From the frame of reference of the interfacial boundary, this is equivalent to an advective flux downward for both dissolved and solid-phase materials. Boudreau (1997) provides a comprehensive summary of these and other issues of the benthic interface.

For more complex geometries, such as boundaries of variable ("wavy") elevation (e.g., a field furrowed by plowing, or a seafloor rippled by wave action) in which flow occurs in the fluid medium along the interface, Bernoulli's principle dictates that differences in pressure will exist between high and low points on the boundary (Vogel, 1994). Assuming that the solid medium is permeable, the result will be pressure-induced advection in the porous medium (Shum, 1992, 1993; Boudreau, 1997) (Fig. 3.24). This "ventilation" may be beneficial to resident organisms in the sediment or soil (see next section). Organisms living at the interface may also directly modify its exchange properties. This is discussed in greater detail in Sect. 3.5.2.

3.5.1 Fluid Exchange in Animal Burrows

We end this chapter with the investigation of a fluid-mechanical problem relating to diffusion and ventilation in animal bores and burrows. Our discussion begins with terrestrial organisms, principally small mammals, and then we examine analogous systems involving invertebrates in aquatic envi-

WIND

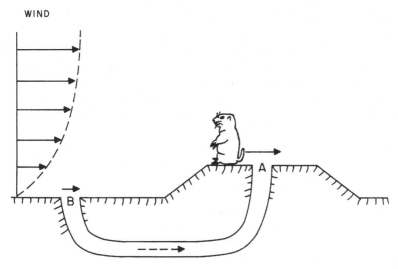

FIGURE 3.25. Schematic picture of a prairie-dog burrow, with wind profile and burrow ventilation.

ronments. Readers are also referred to an excellent symposium volume edited by Meadows and Meadows (1991), which examines the environmental impacts of burrowing animals in terrestrial and aquatic environments.

Figure 3.25 is a schematic of a prairie-dog (*Cynomys ludovicianus*) burrow. The main burrow opening A is located at a mound, while the other opening B, which presumably serves as an emergency exit, is located at the ground level. Vogel and Bretz (1972) suggested that such a structure in burrows may play an important role in ventilation and thus in determining habitat suitability (see review in Vogel, 1994). Importantly, burrowing activities may lead to significant environmental effects such as changes in plant abundance and distribution, and alteration in soil stability (see Meadows and Meadows, 1991).

At least two physical mechanisms may create wind-induced unidirectional bulk flow or "passive ventilation" (Vogel et al., 1973; Vogel, 1977a). The first is due to pressure differences generated by variations in velocity along a streamline, i.e., Bernoulli's principle. The second, "viscous sucking" or "viscous entrainment," depends on the viscosity of the fluid, and as a result stagnant fluid in a burrow may be pulled out of an aperture by adjacent rapidly moving fluid.

Either mechanism for inducing flow in a burrow requires that the openings to the burrow must be dissimilar. Differences may be in height above the ground, location on unlevel ground, or merely variations in the shape or form.

Let us study this problem more quantitatively from a fluid-mechanical standpoint. We apply Bernoulli's theorem (Tani, 1951; Batchelor, 1967) to a stream filament along the ground (Fig. 3.25):

$$(1/2u^2 + p/\rho + gz)_{\text{at A}} = (1/2u^2 + p/\rho + gz)_{\text{at B}}, \quad (3.34)$$

where u, p, ρ, g, and z, are, respectively, the wind speed, (total) pressure, density, acceleration of gravity, and height of the point considered above a fixed horizontal plane. In (3.34) we assume a steady flow of an incompressible inviscid fluid of uniform density; the atmosphere approximately satisfies these conditions.

The total pressure is the sum of the static pressure, p_0, and the dynamic pressure, p_1. Since the static pressure balances the force of gravity, we have

$$(p_0/\rho + gz)_{\text{at A}} = (p_0/\rho + gz)_{\text{at B}}. \quad (3.35)$$

Subtracting (3.35) from (3.34), we obtain

$$(1/2u^2 + p_1/\rho)_{\text{at A}} = (1/2u^2 + p_1/\rho)_{\text{at B}}.$$

Hence,

$$(p_1/\rho)_{\text{at B}} - (p_1/\rho)_{\text{at A}} = 1/2(u_A^2 - u_B^2). \quad (3.36)$$

Ordinarily, the vertical profile of the wind near the ground looks like that illustrated in Fig. 3.25. Thus, the wind speed at A is greater than at B, and the right-hand side of (3.36) becomes positive. Accordingly, the dynamic pressure at B is higher than that at A. The pressure difference sets the air in the burrow in motion, in the direction of the *dashed line* shown in Fig. 3.25. In other words, the principle is the same as that of the Venturi tube; the dynamic pressure at a point where the fluid moves swiftly is lower than the dynamic pressure at a point where the fluid moves slowly, and the burrow corresponds to a U-tube manometer attached to the Venture tube.

We shall now make some rough calculations of the flow in the burrow. We may regard the burrow as a cylindrical tube of radius a and assume that the air flow inside is laminar. The pressure gradient between A and B is $(P_B - P_A)/\ell$, where ℓ is the distance between A and B along the burrow. Applying Poiseulle's law (Tani, 1951; Prandtl and Tietjens, 1957; Batchelor, 1967) leads to the result that the flux of fluid in the burrow, Q, due to the pressure gradient is given by

$$Q = \pi a^4 (P_B - P_A)/8\nu\ell$$

or

$$Q = \pi a^4 (u_A^2 - u_B^2)/16\nu\ell, \quad (3.37)$$

where ν is the kinematic viscosity of air and (3.36) is used. If V denotes the mean velocity of ventilation in the burrow, $Q = \pi a^2 V$. Hence, from (3.37),

$$V = \pi a^2 (u_A^2 - u_B^2)/16\nu\ell. \quad (3.38)$$

Furthermore, the time of ventilation, i.e., the time required for the air inside the burrow to be refreshed, T, is given by

$$T = \pi a^2 \ell / Q = \ell / V$$

or

$$T = 16 v \ell^2 / a^2 (u_A^2 - u_B^2). \tag{3.39}$$

As an example characterizing prairie-dog burrows, we take $\ell = 20$ m, $a = 6$ cm (Sheets et al., 1971), $U_A = 20$ cm/s, $U_B = 10$ cm/s (Vogel and Bretz, 1972), and $v = 0.15$ cm^2/s. Thus, from (3.38) and (3.39), we calculate

$$V = 2.25 \text{ cm/s} \quad \text{and} \quad T = 14.8 \text{ min.}$$

The effectiveness of this ventilation is noteworthy. However, the actual burrows often do not meet all the conditions that Bernoulli's principle requires.

The Reynolds number (the ratio of inertial to viscous forces) for this flow is calculated to be Re $= 2aV/v = 180$, confirming the assumption of laminar flow. Turbulence can occur in a circular tube at a Reynolds number Re $\cong 2000$ or larger.

The second mechanism, however, viscous sucking, produces in general the same directional flow in a burrow as the previous mechanism, that is, fluid will leave an aperture exposed to higher external velocities and enter an aperture at the other end that is exposed to lower external flow. Viscous sucking, however, results in some additional complications of practical importance. For example, a large hole generally provides a better exit than a small hole; thus, fluid will enter a small aperture and leave through a larger one even if there is no difference in external velocity between the ends of the burrow. Also, location near the upwind edge or atop a sharp-edged crater improves the performance of an aperture as an exit for fluid.

Because of the considerable complexity of the mechanism of viscous entrainment as well as of the dual mechanisms that may operate simultaneously, Vogel (1977a) attempted to develop a semiempirical approach for predicting the internal velocity. The driving force, F, for the induced flow can be represented by

$$F = 0.15 \rho A u^2, \tag{3.40}$$

where A is the cross-sectional area of the apertures, u is the free-stream velocity, and the numerical coefficient has been determined using experimental models of burrows. Equation (3.40) may be understood in terms of an analogy to the frictional law for plants (see Sect. 3.1). Combining this expression for the driving force with Poiseulle's law, we obtain

$$V = 6 \times 10^{-3} A u^2 / v \ell, \tag{3.41}$$

which states that for a given fluid and a given aperture geometry, the internal flow velocity is proportional to the square of the external flow velocity.

However, this relationship must break down as u is increased, insofar as internal flow relieves the induced pressure. The breakdown point depends on the burrow system. For example, measurements on models of prairie-dog burrows (Vogel et al., 1973) show that relation (3.41) holds up to a V/u ratio of about 0.1; above this value the measured curve indicates a linear relation,

$$V = ku + C$$

where k and C are constants.

Animal burrows are generally curved rather than straight. Secondary burrow flow may occur due to the curvature of the burrows. Perhaps more importantly, burrows are usually highly branched and thus differ greatly from a simple straight tube. This was demonstrated by Izumi (1973), who excavated occupied rat dens and revealed a complex network structure in his study of Norway rats in a natural environment. Davies and Jarvis (1986) confirmed this complexity and noted temporal variation in burrow construction and maintenance in two mole rat species in South Africa. They reported an average burrow length of 132 m, or 3.9 g/m in terms of biomass per unit length, in their review of more than 20 burrowing mammal species. Burrow length appears to vary indirectly with food supply, while burrow diameter appears to match closely the diameter of the animal. In terms of metabolic output, the rate of oxygen consumption while burrowing bears a linear relationship with the rate of burrowing. For example, the total energetic cost of constructing the burrow of a mole rat amounts to 79% of the estimated digestible energy available from its food source (geophyte corms) in the area (Du Toit et al., 1985).

Vogel et al. (1973) estimated that a prairie dog in a burrow has an oxygen supply sufficient for about 10 hours. Without the bulk movement of air (passive ventilation) and without diffusion through the soil, molecular diffusion through the burrow alone would require time $T \sim \ell^2/2D \sim (2000)^2/2 \times 0.2 = 10^7$ s to refresh the prairie-dog burrow; in other words, the air inside the burrow would stagnate over several months.

In fact, many animals live in burrows where they are often completely separated from the free atmosphere by a medium that does not allow bulk gas transfer; yet ultimately they must exchange carbon dioxide and oxygen with the free atmosphere. How do they manage to survive?

Wilson and Kilgore (1978) examined the exchange of respiratory gases in these animal burrows with the free atmosphere and concluded that molecular diffusion through the soil is of paramount importance. Wilson and Kilgore base their discussion on simple mathematical models for the steady-state diffusion of respiratory gases in sterile burrow-soil systems (Fig. 3.26).

The flux of respiratory gases between an element of the burrow and the free atmosphere through the soil atmosphere, F, is given by

$$F = -2/3\kappa\varepsilon D(S_\alpha - S), \tag{3.42}$$

where S_a and S are, respectively, the concentrations of gas in the free atmosphere and in the burrow element, D is molecular diffusivity in the free atmosphere, ε is soil porosity, and κ is a shape factor describing the geometry

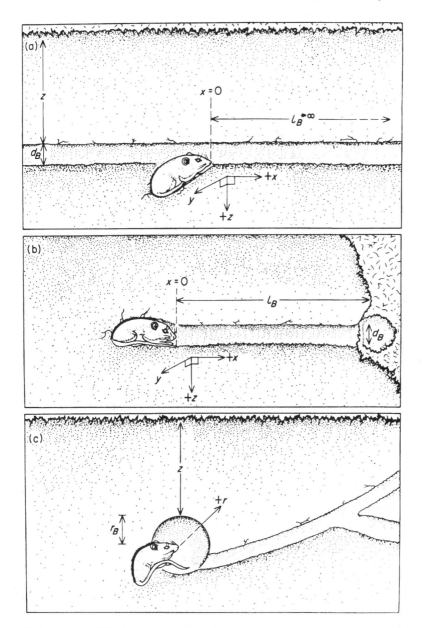

FIGURE 3.26. Diffusion models of respiratory gases in mammal burrows. (a) Model 1 (infinitely long burrow); (b) Model 1 (finite burrow open to the ground); (c) Model 2 (spherical chamber). (From Wilson and Kilgore, 1978; this figure was provided by K. J. Wilson and D. L. Kilgore.)

of the flux for a particular spatial configuration of the burrow element and the groundfree atmosphere interface; thus, κ depends on the burrow depth and the burrow diameter. Expression (3.42) is, in fact, Fick's law of diffusion, stating that the flux of gas between these two surfaces is proportional to the difference of the concentrations at the surfaces.

Two different mathematical models are constructed.

Model 1. The mammal rests in a long, cylindrical burrow. Assuming that the concentration of gases across the burrow (y–z direction) is uniform, we consider the flux balance of respiratory gas in an element of burrow with length Δx at distance x from the origin (Fig. 3.26). In the case of steady-state diffusion, the flux of gas in the burrow at x equals the sum of the flux at the distance $x + \Delta x$ and of the flux between the element and the free air through soil surrounding the burrow element. Thus, we have

$$-DA(dS/dx)_x = -DA(dS/dx)_{x+\Delta x} - 2/3\kappa\varepsilon D\ell\Delta x(S_\alpha - S)_x, \quad (3.43)$$

where A is the cross-sectional area of the burrow and ℓ is the burrow circumference. Expanding, i.e., $(dS/dx)_{x+\Delta x} = (dS/dx)_x + \Delta x(d^2S/dx^2)_x + \cdots$, substituting this into (3.43), and taking the limit as Δx vanishes, we obtain

$$\frac{d^2S}{dx^2} + \frac{2\kappa\varepsilon\ell}{3A}(S_\alpha - S) = 0. \quad (3.44)$$

Equation (3.44) has the general solution

$$S(x) = S_\alpha + C_1 e^{\beta x} + C_2 e^{-\beta x}, \quad (3.45)$$

where C_1 and C_2 are constants determined from the appropriate boundary conditions, and $\beta = (2\kappa\varepsilon\ell/3A)^{1/2}$.

i. For an infinitely long burrow (Fig. 3.26a), the boundary conditions are $-DA(dS/dx)_x = \pm q/2$ at $x = 0$, i.e., animal location, and $S \neq \infty$ at $x \pm \infty$, where q denotes the source (or sink) strength of respiratory gases. The plus sign in front of q is applied to the right of the point $x = 0$, and the minus sign is applied to the left of this point.

The solution for this case is then given by

$$S(x) = S_\alpha + q/2AD\beta e^{-\beta x}, \qquad x \geq 0, \quad (3.46)$$

$$S(x) = S_\alpha + q/2AD\beta e^{\beta x}, \qquad x \leq 0. \quad (3.47)$$

We thus find the concentration of respiratory gases at the location of the animal ($x = 0$), S^*, to be

$$S^* = S(0) = S_\alpha + q/2AD\beta. \quad (3.48)$$

ii. For a finite burrow with one end that opens to the ground (Fig. 3.26b), the boundary conditions are $-DA(dS/dx)_x = q$ at $x = 0$, and $S = S_\alpha$ at $x = \ell_B$. The solution for this case is then given by

$$S(x) = S_\alpha + \frac{q}{AD\beta}\frac{\sinh\beta(\ell_B - x)}{\cosh\beta\ell_B} \quad (3.49)$$

and

$$S^* = S(0) = S_\alpha + \frac{q}{AD\beta}\tanh\beta\ell_B. \qquad (3.50)$$

Model 2. The mammal rests in a spherical chamber remote from the ground surface (Fig. 3.26c).

Assuming that the mammal's body interferes negligibly with gas transfer and that the concentration of gases in the burrow chamber is spherically symmetric, we obtain the steady-state diffusion equation in a sphere (Carslaw and Jaeger, 1959):

$$\frac{d}{dr}\left(r^2\frac{dS}{dr}\right) = 0, \qquad (3.51)$$

where r is the radial coordinate. Equation (3.51) has the general solution

$$S(r) = C_1 + C_2/r,$$

where the constants C_1 and C_2 are determined from the following boundary conditions:

$$-4\pi r^2 D \, dS/dr = q \qquad \text{(source strength) at any } r$$

and

$$D \, dS/dr = 2/3\kappa\varepsilon D(S_\alpha - S) \qquad \text{at } r = r_B.$$

We then obtain the solution

$$S(r) = S_\alpha + \frac{q}{4\pi Dr} - \frac{q}{4\pi D}\left(\frac{1}{r_B} - \frac{3}{2\kappa\varepsilon r_B^2}\right), \qquad (3.52)$$

and the concentration of respiratory gases at the location of the animal $(r = r_s)$ is

$$S^* = S(r_s) = S_\alpha + \frac{q}{4\pi D}\left(\frac{1}{r_s} - \frac{1}{r_B} + \frac{3}{2\kappa\varepsilon r_B^2}\right). \qquad (3.53)$$

If the gas exchange is to be adequate for a mammal, the following inequalities must be satisfied:

$$(S_\ell - S^*)_{O_2} < 0 < (S_\ell - S^*)_{CO_2} \qquad (3.54)$$

where S_ℓ denotes the lethal concentration of a respiratory gas.

Based on these mathematical models, Wilson and Kilgore (1978) calculated the concentration gradient of respiratory gases along long, narrow burrows for standard mammals of various masses at different soil porosity. It was found that the distance from the mammal at which gas concentrations are negligibly different from the free atmosphere does not exceed 7.5 body lengths; for soils of high porosity, the distance is less than about 3 body lengths. On the other hand, even in highly porous soils, the greatest mass permissible by the inequalities of (3.54) is about 0.4 kg, which is noticeably toward the lower end of the range of mammalian mass.

Wilson and Kilgore (1978) also calculated the source concentration of CO_2, S^*, in a burrow chamber under various conditions. It was found that (i) soil porosity has a strong effect on the respiratory environment, (ii) the depth of the chamber, Z, has only a weak effect, and (iii) soil temperature within the *thermoneutral zone* of the resident mammal has a weak effect on this transfer. However, below the thermoneutral zone, where temperature modifies the rate of production or consumption of respiratory gas, the effect is strong. Large mammals are more restricted in the design and setting of burrows than small mammals. The mathematical models indicate that normothermic eutherian mammals with masses much in excess of 0.5 kg are precluded from an indefinite occupation of deep burrows in most field conditions.

Gupta and Deheri (1990) extended Wilson and Kilgore's (1978) inequality (3.54) for fluxes of gas in spherical burrows (i.e., Model 2). They show that (3.53) can be simplified to provide the following result:

$$S^* \approx S_a + 2\beta r_B^{3/2} r_s \left[\frac{1}{r_s r_B^3 - 1} + \frac{1}{r_B(r_s - r_B^3)} \right].$$

Given that $\beta = (2\kappa\varepsilon\ell/3A)^{1/2}$, this implies that soil porosity (κ) and the geometry of the burrow (ℓ and A) have a strong effect on the ventilation of the burrow.

Weir (1973) measured air flow through the mounds of termites, *Macrotermes subhyalinus* (Rambur), and showed that air passed into the mound at certain openings. The number of such openings varied: Small mounds 0.5 to 1.0 m high might have two or three openings, and larger mounds 2.0 to 3.0 m high might have ten or more openings.

For small mounds, air was found to enter from basal and peripheral openings and to escape from central openings on top of the mound. This is consistent with the Venturi effect. For large mounds, the circulation pattern was less clearly defined. Some peripheral and basal openings did not always function as well-defined entrances for air but could function as exits under some wind conditions. Weir (1973) suggested that vortices caused by the mound structure probably reverse the air flow. Central openings consistently function as air exits, however.

Airflow measurements made with a small-scale mound subject to environmental air speed of 2.5 to 2.7 m/s result in a mean volume of air passing through the mound of 18.5 l/s (i.e., >1 m^3/min). If we assume a mean radius of the tubes of 10 cm, the above flux corresponds to a quite swift mean speed of ventilation of 59 cm/s. The air circulation in these mounds provides the colony of termites with a degree of homeostatic regulation.

Vogel (1977b) measured flow velocities through and immediately adjacent to the excurrent openings (oscula) of living sponges in their natural environment and found that the flow through the oscula was positively correlated to the ambient flow. The result indicates that the passage of water through sponges is, at least in part, induced by ambient currents.

Diffusion models similar to those of Wilson and Kilgore (1978) have been considered by Withers (1978), who examined both subaquatic and subterranean animal burrows. It was concluded that due to very small molecular diffusivity in water only the smallest aquatic burrowing animals, such as boring sponges and burrowing barnacles, might rely on diffusion to sustain their metabolic demands, but also see below. On the other hand, subterranean ectotherms may be able to rely on gaseous diffusion, except when they aggregate together in large numbers, and subterranean endotherms, because of their high metabolic rates, are likely to encounter low concentrations of O_2 and high concentrations of CO_2 in their burrows.

An interesting example of ventilation was reported by Korhonen (1980), who provided information on the winter burrows of voles, which are constructed in snow. Temperature differences within the burrows led to an outflow of air provided that the external temperature was not too low. In that case, there was an inflow of air that cooled the burrows. Measured CO_2 levels were never too low to be of concern metabolically, and this is likely due to exchange within the burrow systems, through the snow, and with the water in the snow.

Burrows constructed by infaunal animals are a conspicuous feature of marine sediments. Of particular interest in the study of burrowing organisms are the population dynamics of the species, the bioturbation activity of the animal, the bioturbation effect on the sediments, and the alterations of the pore water chemistry.

Dworschak (1983) investigated the burrow structures of the thalassinid shrimp, *Upogebia pusilla*, on a tidal flat using resin casts, which revealed burrow structures that ranged from simple U-shaped tubes to U-shaped tubes with branches. One of the external openings is raised above the surface, forming a conelike mound. Allanson et al. (1992) investigated the flow in burrow structures of a related species (*Upogebia africana*) using flow chamber and field study. The induced flow through the burrow was analogous to the flow induced through the prairie-dog burrow, the mean circulation velocity of which is given by (3.38); note the need to include π in the relationship. They estimated that the mean discharge velocity would range between 1.2 and 1.7 cm/s for various cone heights, bed roughness, and boundary layers, for a free surface velocity of 15 cm/s. The induced flow was found to be proportional to the Reynolds number and the aspect ratio of the burrow.

Animals in the field showed a significant orientation toward the induced current, but there was no preferred orientation of the burrow openings, i.e., the flow induction mechanisms work in all directions as long as there is a height difference between burrow openings. Importantly, animals showed reduced pumping activity via their pleopods during tidal exchanges compared to periods of no flow. Recent investigations of the morphology and pumping activities of another thalassinid shrimp (*Callianassa*) indicate that the active burrow flow is dominated by a coordinated series of metachronally oscillating motions of the pleopods, which can be energetically expensive (Stamhuis and Videler 1998a, b, c). These results suggest that the induced flow in burrows may play an important role in the feeding and respiratory behavior of these animals.

A number of fish species are also known to burrow and undertake a number of important functions within burrows (Atkinson and Taylor, 1991). For example, tilefish seek shelter and construct burrows at depths of 100 to 300 m on the continental shelf. Grimes et al. (1986) reported that vertical burrows are the primary habitat of these fish, and these burrows occur between depths of 80 and 305 m. It is believed that these burrows are formed by a combination of oral excavation by tilefish, secondary bioerosion by associated fauna, and the swimming motion of the tilefish. The mean depth of the burrows is 1.7 m, while the upper cone has an average diameter of 1.6 m. By injecting a dye marker from the *Johnson-Sea-Link* submersible, Grimes et al. (1986) found that the larger secondary burrows located at the burrow margin were interconnected to the main burrow shaft. The interconnected burrow structures provide a means of passive ventilation of the tilefish burrow systems at velocities approaching 50 cm/s! However, active burrow ventilation by the fish swimming motion may be equally or more important than passive ventilation.

Passive ventilation of burrows and other biological structures (e.g., tubes) also takes place in the absence of animals. Ray and Aller (1985) suggested that this phenomenon would have important implications for the chemistry of sediments. Essentially, the induced flow out of the biological structure generates a downward flux in the sediments surrounding the tube that then supplies the induced flow (see Fig. 3.27). Meyers et al. (1988) demonstrated that the spatial distribution of meiofauna responds quickly to subtle changes in sediment chemistry, principally oxygen and sulfide levels, caused by adding or removing model tubes of infaunal organisms. The cycling of water via a passive ventilation of relict biological structures such as the proteinaceous or parchmentlike tubes of polychaete worms was examined by Libelo et al. (1994). They provided a theoretical model for the induced outward and sediment flow, which was then examined using computational techniques. Results indicate that the water flux in the sediments varies with the boundary layer velocity, the hydraulic conductivity of the sediments, and the spatial distribution of tubes. They showed that the water flux rate through sediments is on the order of 10^{-3} cm/s, which is comparable to the pumping rates due to the active pumping of animals.

The exchange of water in swamps is different from that of the systems discussed above, principally due to the high clay content of the sediments and the surface-water slopes. Few animals in these systems create mounds around their burrows. Water exchange in animal burrows appears to be driven by pressure differences rather than induced flows due to Bernoulli's

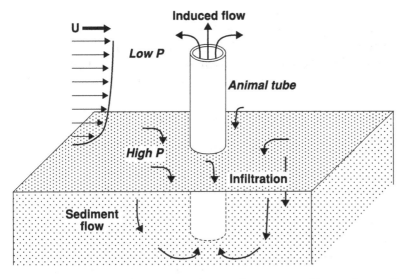

FIGURE 3.27. Cycling of water through the sediments created by passive ventilation of relict biological structures (after Libelo et al., 1994).

principle (Ridd, 1996). A surface-water slope of 1/1000 is predicted to create a pressure head of 0.5 mm for openings separated by 0.5 m, which would lead to a significant burrow flow. Ridd (1996) used the fact that burrow flow is dominated by friction, the Darcy–Wiesbach equation,

$$\frac{\Delta p}{L} = \frac{f \rho V^2}{2D},$$

where f is the friction factor, Δp is the pressure difference, L and D are the length and diameter of the burrow, and ρ and V are the density and velocity of the water, and the oscillation of water level, to predict theoretically the flow in the burrow. A rate of 0.1 to 1 cm/s was predicted for a 1-m-long burrow with a 1-cm diameter. Field measurements of burrow velocity were less than 3 cm/s, which corresponds well with the theoretical estimate. Given the density of burrows and these flow estimates, 10^3 to 10^4 m^3 of water (i.e., 0.3% to 3% of the swamp water volume) is estimated to flow through the burrows per km^2 of mangrove swamp per tidal cycle.

While many burrowing organisms benefit from passive ventilation of their burrows, a number of air-breathing organisms (e.g., crustaceans, insects, spiders) also live within burrows in marine sediments (Wyatt and Foster, 1991). Unlike the situations discussed above, these organisms must avoid flooding their burrows, which are typically less than 3 mm in diameter. Interestingly, air bubbles trapped in the neck of a staphylinid beetle burrow act as a temporary plug, which enables the beetles to plaster over the burrow neck to prevent flooding. Maitland and Maitland (1994) developed a theoretical model to explain these observations by considering the surface tension forces between the trapped air bubble, the burrow sediments, and the advancing tide. Their theoretical prediction of 3 mm for the maximum burrow diameter that can trap a bubble is consistent with observations from nature.

3.5.2   Bioturbation and Related Effects

The mixing of sediments by organism activities is called *bioturbation*. It also takes place in soil on land by burrowing creatures. In fact, Darwin (1881) examined the stirring of soils by the burrowing earthworms, and his detailed observation of these organisms and the way that they affect ecological and geological processes still deserves careful attention (cf. Cook and Linden, 1996; Bouche and Al-Addan, 1997).

Bioturbation occurs widely in aquatic sediments—lakes, estuaries, coastal waters, and the deep sea. The benthic organisms responsible for bioturbation include clams, worms, and crustacea, among others (Aller and Aller, 1992; McCall and Soster, 1990; Madsen et al., 1997). Many burrows formed by marine animals are lined with thin layers of organic material. The permeability of these linings to solute diffusion plays an important role in determining the chemical composition of surrounding sediment and the burrow habitat (Aller, 1980, 1988; Fenchel, 1996a, b). In addition, the activities of these organisms constitute mixing processes in their own right (e.g., "burrow and fill mixing"; Gardner et al., 1987; Wang and Matisoff, 1997; Boudreau, 1997, 2000).

In addition to the mixing of sediments, biogenic "nonlocal" transport of nutrients (i.e., not dependent on local concentration gradients) may result from the movement of organisms between different sediment layers during the period between feeding and defecation (Boudreau and Imboden, 1987; Boudreau, 1997) by "conveyor-belt" transport through the digestive tract of the organism (e.g., the deposit feeding behavior of some benthic worms; Boudreau, 1986b, 2000; Soster et al., 1992; Blair et al., 1996) or by porewater irrigation (exchange of water in benthic burrows for water at the sediment surface by organismal pumping activity; Emerson et al., 1984; Christensen et al., 1987; Martin and Banta, 1992; Boudreau and Marinelli, 1994).

Recently, Boudreau (1998) proposed a general model for the depth variability of bioturbation in the benthic mixed layer limited by the availability of labile organic matter consumed by bioturbating organisms. The effective bioturbation coefficient, $D_b(z)$, is found to drop from its value at the surface $(z = 0)$ to zero at the mixed layer depth $(z = L)$ (approximately 10 cm) as $(1 - z/L)^2$. The corresponding steady-state concentration profiles of substances that obey an advection–diffusion-reaction equation subject to this depth-variable mixing can be expressed in terms of modified Bessel functions (Swaney, 1999).

3.5.3   Notes on Solute Transport in Soils and Sediments

As in the sea, nutrients within terrestrial ecosystems may limit the activity of primary producers, but in this case the milieu is soil, often considered an ecosystem in its own right. In recent decades, our understanding of the physical transport of nutrients and contaminants (e.g., pesticides) in soil and bedrock has grown enormously. Transport processes in soils are radically different from those of the sea or the air. Flows are characterized by low Reynolds numbers instead of the turbulent conditions frequently found in marine and terrestrial systems and constrained to occur in the interstices between soil particles (de Marsily, 1986; Marshall and Holmes, 1988). The flow of water through unsaturated porous media can itself be described as a diffusive process in which the diffusion coefficient (diffusivity) is a nonlinear function of moisture content (Childs and Collis-George, 1950). The complexities of such processes are beyond the scope of this volume. In saturated soils and sediments (e.g., in aquifers or at the bottom of the sea), the flux of water is proportional to the gradient of hydraulic head or pressure (Darcy's law) and frequently can be considered to be in steady state (de Marsily, 1986). In such cases, nutrients and other dissolved chemicals (solutes) follow a solute-transport mass-balance equation similar to (2.16) (written here for the one-dimensional case):

$$\varphi \frac{\partial S}{\partial t} = \frac{\partial}{\partial z}\left(\varphi D_z \frac{\partial S}{\partial z}\right) - \frac{\partial(\varphi w_z S)}{\partial z} + R,$$

where φ = porosity of the medium, D_z = the hydrodynamic dispersion of the medium (mixing due to microscale motions through the porous medium), w_z = the steady-state flow velocity of water through the medium, S is solute concentration, and R is the local net gain or loss of material due to chemical transformations. In aquatic sediments, this equation is often referred to as the diagenetic equation. Boudreau (1997, 2000) provides an excellent discussion of this equation and its variations.

Complications can arise from several factors, including variable water content in the vadose (unsaturated) zone of soils (Hayashi et al., 1998), which affects both transport and biogeochemical transformations; degree of adsorption of the solute to sediment and other particles (Charbeneau and Daniel, 1993; Wu et al., 1997; Bengtsson et al., 1993); gas-phase transport (Washington, 1996, and in the case of pollutant spills, transport in other liquid phases); and time-varying boundary conditions (e.g., the water table variation in response to evapotranspiration or tides). Charbeneau and Daniel (1993) and Mercer and Waddell (1993) reviewed some of these issues. Spatiotemporal patterns can result from spatial variability of physical characteristics of soils and bedrock (Wierenga et al., 1991; Arocena and Ackerman, 1998) such as soil layers and macropores and fractures of both biogenic and physical origin (cf. Stone, 1993; Li and Ghodrati, 1994, 1995; Boll et al., 1997), as well as inherent instabilities of the flows (Glass et al., 1989a, b; Liu et al., 1994). However, as in the sea and atmosphere, relatively simple analytical solutions to advection–diffusion-reaction equations (e.g., Ogata and Banks, 1961; Lindstrom et al., 1967; Aller, 1980; Enfield et al., 1982; van Genuchten and Alves, 1982; Boudreau, 1986a, b, 1987; Jury et al., 1987; Toride et al., 1993; Angelakis et al., 1993) have been used in some cases to estimate the steady-state and transient distributions of nutrients and contaminants in soils, benthic sediments, and groundwater.

Largely in response to concerns about environmental pollution, a variety of numerical advection–diffusion-reaction models now exist with the goal of assessing groundwater quality and the spread and distribution of contaminant plumes (e.g., Carsel et al., 1984; Pacenka and Steenhuis, 1984; van der Heijde et al., 1985; Wagenet and Hutson, 1989; Wagenet and Rao, 1990; Pennell et al., 1990; Follet et al., 1991). In the last decade, a more general class of models, called *transfer-function* models, have extended the idea of the advection–diffusion model beyond simple Fickian diffusion by considering the probability distributions of the travel times of solutes moving through soils of various configurations (cf. Jury et al., 1986, 1990; Jury and Roth, 1990).

4
Diffusion of "Smell" and "Taste": Chemical Communication

Akira Okubo, Robert A. Armstrong, and Jeannette Yen

The natural world is filled with a vast variety of "invisible," "inaudible" smells upon which animals depend for their lives through chemical communication. For the behavior of many mammal species, in particular carnivora and nocturnal animals, the role of the olfactory sense is more important than that of the auditory sense. Animals living in underground burrows, deep-sea fish, and many species of insects make use of smell. In the deep sea, with its low turbulence and perpetual darkness, scent may perhaps be even more important for aquatic animals than for terrestrial organisms (Hamner and Hamner, 1977).

Chemical communication between animals occurs upon the completion of three essential steps: (a) the release of a chemical (olfactory or gustatory) signal; (b) the transmission of that signal through the environment; and (c) its reception by another individual.

Diffusion apparently controls process (b). However, an understanding of the entire process of olfactory or gustatory response in animals requires a knowledge of chemotaxis as well as the mechanics of diffusion of chemical signals. In essence, the environmental transmission of chemicals can be attributed to a passive diffusion process; thus, the diffusion theory presented in Chap. 2 may well be applicable. On the other hand, certain aspects peculiar to chemical communication in animals deserve special discussion in this chapter.

4.1 Diffusion of Insect Pheromones

Pheromones are "odor" chemicals released by animals and utilized for chemical communication between members of the same species. They thus contrast with allomones and kairomones used for communication between members of different species. The diffusion theory of insect pheromones was founded by Wilson (1958) and Bossert and Wilson (1963). Although their theoretical treatment was still quite elementary, significance should be attached to their pioneering role in this field.

Some important practical aspects of the diffusion problem concern the following:

i) the amount of pheromone released, in terms of initial concentration or source intensity;
ii) the rate of transmission in terms of components due to, for example, wind velocity and diffusion;
iii) the effective duration of the signal and its fade-out rate; and
iv) the most efficient mode of transmission.

The last of these items may have some significance from an evolutionary standpoint.

A discussion of a few models of pheromone diffusion under various conditions or release follows.

4.1.1   Instantaneous Emission in Still Air

Ignoring the effects of wind and turbulence and assuming an isotropic diffusion of Fickian type with a constant diffusivity D, we can write the three-dimensional diffusion equation as

$$\frac{\partial S}{\partial t} = D\left(\frac{\partial^2 S}{\partial x^2} + \frac{\partial^2 S}{\partial y^2} + \frac{\partial^2 S}{\partial z^2}\right), \tag{4.1}$$

where S is the concentration of pheromone, t is time, and x, y, and z are Cartesian spatial coordinates.

It is supposed that M molecules of pheromone are released instantaneously at the origin ($x = y = z = 0$). The solution of (4.1) in an infinite domain is (Carslaw and Jaeger, 1959)

$$S(x, y, z, t) = M/(4\pi Dt)^{3/2} \exp(-r^2/4Dt), \tag{4.2}$$

where $r^2 = x^2 + y^2 + z^2$. Isoconcentration contours are thus seen to form spherical surfaces about the origin.

The ground is assumed to be a reflecting plane at $z = 0$, implying the boundary condition:

$$\partial S/\partial z = 0 \quad \text{at } z = 0. \tag{4.3}$$

Equation (4.3) states that no flux of substance occurs across the surface.

The method of images (Carslaw and Jaeger, 1959) can now be used to show that instantaneously, if the point source is located at the ground, the solution to the diffusion problem on a semiinfinite domain ($z > 0$) embodied in (4.1) and the boundary condition (4.3) is simply twice the solution in infinite space given by (4.2) (see Fig. 4.1). Thus, for a ground source,

$$S(x, y, z, t) = 2M/(4\pi Dt)^{3/2} \exp(-r^2/4Dt). \tag{4.4}$$

Let C (mole/cm^3) be a minimum or threshold concentration of substance, such that the receiving animal can respond only to concentrations C and

FIGURE 4.1. Diffusion of pheromone in still air from an instantaneous source on the ground.

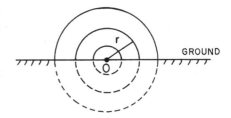

greater. Here (mole/cm^3) denotes the number of molecules per cm^3. As in Sect. 3.3.2, we shall investigate the way in which the isoconcentration surface represented by the threshold concentration changes with time. Clearly, that surface will initially remain in the vicinity of the source, gradually spread out, reach a certain maximum distance from the origin, and then shrink back to zero, i.e., the origin. The radius of the sphere representing the threshold concentration is determined from the value of r at which $S = C$ (constant) in (4.2) or (4.4): $r = R(t)$. The time variation of R is illustrated in Fig. 4.2. As can be seen, the effective radius of odor chemicals attains a maximum, R_m, at time t_m, and becomes zero at t_f, at which time chemical communication ceases.

For an infinite space:

$$R_m = 0.527(M/C)^{1/3}, \qquad (4.5)$$

$$t_m = (0.0464/D)(M/C)^{2/3}, \qquad (4.6)$$

$$t_f = et_m = (0.126/D)(M/C)^{2/3}. \qquad (4.7)$$

One simply replaces M in the above equations by $2M$ for ground sources. Note that R_m depends only on the ratio M/C and is independent of D.

Actually we often know little about the values of M, C, and D; however, these parameters may be estimated using the theoretical relations obtained. Thus, knowledge of R_m enables evaluation of M/C, and a combined knowledge with t_m determines D from (4.6). Wilson (1958) studied alarm commu-

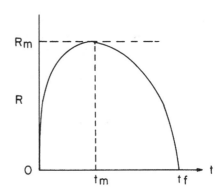

FIGURE 4.2. Time variation of the effective radius (R) representing a threshold concentration of pheromone.

nication with the harvest ant, *Pogonomyrmex badius*. The experiments were designed to determine the above parameters. It was found that the maximum alarm radius of the alarm pheromone was 6 cm, the arrival time was 32 sec, and after the arrival the signal ceased within 35 sec. From the measured volume of the mandibular glands of several minor workers, Wilson estimated an average value of M of 6.26×10^{16} moles. It was then calculated that $C = 4.5 \times 10^{13}$ (mole/cm^3) and $D = 0.43$ cm^2/s.

The short range and duration of the pheromone should be desirable for the alarm system, avoiding unnecessary persistence of the signal after the period of danger has passed. If necessary, the signal may be relayed to farther distances by a release of pheromones from one of the ants initially receiving the alarm.

4.1.2 *Continuous Emission in Still Air*

As in the previous case, a ground source is assumed to be located at the origin, $x = y = z = 0$. Let us suppose that a pheromone is released at the constant rate Q (mole/s). Since the solution from an instantaneous source is known, i.e., (4.4), the principle of superposition may be invoked to obtain the solution for a continuous source. This principle is often used to obtain the solution of a continuous source by integrating over time, i.e., "superposing" the solution of diffusion from an instantaneous source. It is based on the simple concept that a continuous source may be regarded as a continuous release of instantaneous sources. The principle may also be used for a spatially extended source by integrating the solution for a point source over the space in question.

The amount of release during an infinitesimal time dt is denoted as Qdt. Replacing M in (4.4) by Qdt and integrating with respect to time, we obtain

$$S(r, t) = \int_0^t 2Q/(4\pi Dt')^{3/2} \exp(-r^2/4Dt') \, dt'$$

$$= Q/2\pi Dr[1 - \Phi\{r/(4Dt)^{1/2}\}], \tag{4.8}$$

in which Φ is the error function.*

As the release of pheromone continues for a long time ($t \rightarrow \infty$), the concentration approaches the steady-state limit,

$$S(r) = Q/2\pi Dr. \tag{4.9}$$

---

*The error function is defined by

$$\Phi(a) = (2/\pi^{1/2}) \int_0^a e^{-b^2} \, db, \quad a \geq 0,$$

and its numerical values lie between $\Phi(0) = 0$ and $\Phi(\infty) = 1$.

In other words, the concentration decreases linearly with distance from the source. The radius of the sphere containing concentrations above the threshold value C surrounding the source, R_m, can be deduced from (4.9) to be

$$R_m = Q/2\pi CD. \qquad (4.10)$$

This corresponds to the maximum effective distance for chemical communication associated with long-term release from the source.

Let us return to the *P. badius* alarm system and apply the above results. Supposing that worker ants arrive at the alarm zone at a constant rate of one individual each five seconds, it is seen that $Q/C = M/C \div 5 = 1400/5 = 280$ cm^3/s. Taking $D = 0.43$ cm^2/s as before, the sphere surrounding the alarm zone, which contains concentrations above the attractive threshold of the pheromone, is found to possess a radius $R_m = 104$ cm. Thus, the maximum effective radius is about 1 m; the time required for the alarm to propagate to a distance of half this radius can be found to be 6400 s $= 1$ h 47 min.

4.1.3 Continuous Emission from a Moving Source

Herein the diffusion of substances associated with chemical trails is considered. Such insects as ants leave a trail of pheromone by releasing the substance as they move across the ground. It is assumed that the release of pheromone lasts for a sufficiently long time. The odor substance evaporates, and a diffuse pheromone cloud extends along the ground and into the air. The trail serves to lead other workers to food sources.

We define u to be the velocity of the ant and define the origin to coincide with the location of the insect. With reference to this coordinate system attached to the moving ant (Fig. 4.3), the steady-state distribution of substance in the pheromone trail obeys the following advection–diffusion equation:

$$u\frac{\partial S}{\partial x} = D\left(\frac{\partial^2 S}{\partial y^2} + \frac{\partial^2 S}{\partial z^2}\right), \qquad (4.11)$$

where diffusion in the direction of the ant's locomotion (x-axis) is ignored (an assumption that is usually accurate).

If Q is the rate at which the ant emits its pheromone, the solution of (4.11) is given by

$$S(x, y, z) = (Q/2\pi Dx)\exp\{-u(y^2 + z^2)/4Dx\}. \qquad (4.12)$$

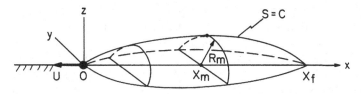

FIGURE 4.3. Boundary surface of a pheromone trail emitted by a moving ant (C: threshold concentration; U: ant's walking speed).

It is noted that, if we set $x/u = t$ in (4.11), the resultant equation becomes a two-dimenional diffusion equation, the solution of which, for an instantaneous source, is the two-dimensional version of (4.4). The substitutions $M \to Q/u$ and $t \to x/u$ in the solution of (4.4) thus yield (4.12).

As before, the boundary of the trail can be defined by some threshold concentration C; thus, the equation representing the boundary becomes the surface for which $S(x, y, z) = C$ in (4.12) (Fig. 4.3). The point X_f at which the surface intersects the x-axis determines the maximum length of the trail. From (4.12) we have

$$X_f = Q/2\pi CD. \tag{4.13}$$

The maximum trail radius in a plane perpendicular to the x-axis, R_m, is

$$R_m = (2Q/e\pi Cu)^{1/2}, \tag{4.14}$$

and it occurs at $x = X_m = X_f/e$.

From experiments on fire ant trails, Wilson (1962) obtained $X_f = 42$ cm, $R_m = 1$ cm. Taking the velocity of the trail-laying ant to be $u = 0.4$ cm/s, it can be seen from (4.14) that $Q/C = 1.71$ cm^2/s, and from (4.13) that $D = 0.00649$ cm^2/s. This value for D seems very small for the molecular diffusivity of the substance in air; this may be due to the fact that not all of the substance is initially in the gaseous state. Since the substance is quite volatile, however, the model is still acceptable with a value of D reduced by an appropriate factor to account for the evaporation time. The fade-out time of the trail due to diffusion is about 100 s. In other words, 100 s after the last ant returns from an exhausted food source, the trail disappears so that the use of unproductive old trails is avoided. For more details about odor trails in ants, see Wilson (1971).

4.1.4   Continuous Emission of Pheromone in a Wind

In comparison with the transmission of ant pheromones, the effective range of the sex attractants of larger insects such as moths is far greater. As may be expected from the results of Sects. 4.1.1 and 4.1.2, such long-range communication is possible only by taking advantage of the wind, i.e., the movement of the environmental medium itself.

Since the wind not only carries the pheromone molecules but also creates its own turbulence, pheromone diffusion cannot be treated by the simple methods of Sect. 4.1.3. One must enter the realm of turbulent diffusion, the characteristics of which depend on the wind structure and properties of boundary surfaces. In effect, we must use the theory of diffusion in atmospheric boundary layers described in Chap. 2.

Bossert and Wilson (1963) employed Sutton's diffusion Eq. (2.23). Analogously to Sect. 4.1.3, they were able to obtain the characteristics of sex attractant pheromone plumes. Thus, the maximum length, X, maximum width, Y, and maximum height, Z, of the pheromone plume determined by a

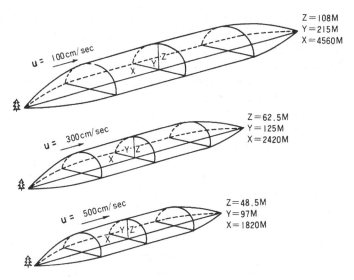

FIGURE 4.4. Threshold volume of sex attractant pheromone downwind from a single female gypsy moth at various wind velocities. The maximum values of the dimensions are given (from Bossert and Wilson, 1963).

certain threshold concentration, C, are given by

$$X = (2Q/C\pi c_y c_z u)^{1/2-n}, \tag{4.15}$$

$$Y = c_y (2Q/C\pi c_y c_z ue)^{1/2}, \quad \text{and} \tag{4.16}$$

$$Z = c_z (2Q/C\pi c_y c_z ue)^{1/2}. \tag{4.17}$$

From experiments with the gypsy moth, *Porthetria dispar*, it was found that Q/C for a single female is in the range from 1.87×10^{10} to 3.03×10^{11} cm^3/s. Taking 10^{11} cm^3/s to be a typical value for Q/C and taking the diffusion parameters as $n = 0.25$, $c_y = 0.4$ cm$^{1/8}$, $c_z = 0.2$ cm$^{1/8}$, the maximum values X, Y, and Z can be calculated from (4.15)–(4.17) for various wind speeds (Fig. 4.4).

Note that for the range of winds shown in Fig. 4.4, stronger winds reduce the effective range of attraction. This is due to the increased turbulent diffusion associated with higher wind velocities, which spreads the attractant more uniformly through space. The result is thus seen to be fundamentally different from the case of constant diffusivity in still air outlined in Sect. 4.1.3. It is seen from the calculated result (Fig. 4.4) that male moths can detect females located at a distance of as much as 1 km.

The experiments of Kaae and Shorey (1972), which demonstrated that the pheromone-releasing behavior of noctuid moth (*Trichoplusia ni*) females is greatly influenced by wind velocity, are worthy of mention. Females were found to spend more time releasing pheromones when exposed to air velocities ranging from 0.3 to 1.0 m/s than higher or lower velocities. It appears that evolution has endowed the females with the ability to recognize wind conditions that are favorable to chemical communication.

Using diffusion models similar to those given herein, Bossert (1968) analyzed the transmission of chemical information associated with time-varying pheromone release. It is found in this case as well that a moderate wind can make communication practical over a longer range.

Thus far, we have discussed the diffusion of pheromones. However, the mechanism by which insects actually follow the pheromone cloud presents an entirely different problem. It is found that the concentration gradients of pheromone as calculated directly from the diffusion theory are certainly too small to be detected by source-seeking insects.

Schwink (1954), Wright (1964), and Farkas and Shorey (1972) are among those who consider anemotaxis to play an important role in the orientation process. Thus, male moths respond to the attractant pheromone of females by initiating an upwind flight; if they by chance miss the pheromone plume, they engage in a zigzag movement until the plume is rediscovered, and then they continue to fly upwind. By repeating this chemo-anemotactic process, the male moth can finally locate the female. This kind of zigzag movement is not only used by flying insects in chemotaxis, but also by ants in following odor trails laid on the ground (Hangartner, 1967). According to Rust and Bell (1976), nonflying insects such as the cockroach also use chemo-anemotaxis in response to air that is laden with sex pheromone.

Gillies and Wilkes (1974) discovered in experiments with host-seeking mosquitoes that their search flight before the detection of host stimuli is generally in the downwind direction; upon encountering the odor plume of a host, they turn around and track back upwind toward the host. This constitutes a very efficient host-seeking strategy, as, for a given output of energy, the insect is able to cover a much larger area than if it were restricted to upwind movements.

Almost all the models of pheromone diffusion, including those of Bossert and Wilson, are based on the assumption that the substance spreads in a continuous, regular plume when released continuously, as shown in Fig. 4.5(a). Actually, the diffusion pattern of the substance consists of numerous

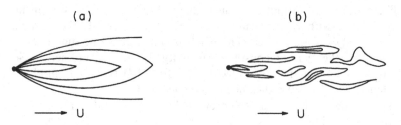

FIGURE 4.5. Diffusion plumes in the environmental fluid of velocity U. (a) idealized plume; (b) realistic plume (filamentary plume).

meandering filaments generated by turbulence in the environmental fluid (Fig. 4.5(b)). Thus, the mean concentration of pheromone does not vary regularly in space, but rather the pheromone concentration fluctuates at random and on the average tends to increase toward the source. Such a temporal and spatial pattern of concentration fluctuation might provide new information to insects which the imaginary plume shown in Fig. 4.5(a) lacks, and it is possible that this random pattern enhances the chemotactic behavior of insects. However, appropriate modeling of the pattern of Fig. 4.5(b) would be extremely difficult due to the need to know not only $\bar{S}$, but also the statistical properties of S', and furthermore, the probability density $p(S, \mathbf{x}, t)$ mentioned in Sect. 1.3. Csanady's book (Csanady, 1973) provides a discussion of the variance of concentration fluctuations in the environment. For another development, consult Meyers et al. (1978).

4.2   A Diffusion Problem Concerning the Migration of Green Turtles

Determination of the basis for animal orientation and navigation during migration has proved to be a vexedly difficult problem. The various theories that have been proposed differ from species to species (Orr, 1970). In this section, we shall examine the study by Koch et al. (1969) of the migration of the green turtle, *Chelonia mydas*, which provides an example of directional information that is intimately associated with diffusion.

A tagging experiment (Carr, 1967) showed that green turtles hatched on Ascension Island in the South Atlantic traveled over a distance of 1200 nautical miles to the Brazilian coast for feeding; females then returned to the same island, and even to the same beach of birth, to lay eggs (Fig. 4.6).

We should like to determine the mechanism of their homing migration. How is it possible for green turtles to return without fail to a solitary island only 10 km wide and removed from the South American coast by over 1200 nautical miles? Celestial navigation seems to be unacceptable. Visual and auditory cues are most unlikely because of the distance. The only remaining sensory cue is "smelling" or "tasting."

According to Sverdrup et al. (1942), the South Atlantic Equatorial Current flows approximately from east to west in the latitudes between 0° and 20° S, and it forms a prevailing current more than 1000 miles wide. It extends from the surface to a depth of several tens of meters with a mean speed of 1 kt (50 cm/s). If one rode on the current, it would take about 50 days to travel from Ascension Island to the coast of Brazil.

Koch et al. (1969) speculated that an unknown chemical substance originating at Ascension Island is carried westward by the equatorial current, and that this substance provides a chemical stimulus to the migrating green turtle. The substance would tend to mix uniformly within a surface layer,

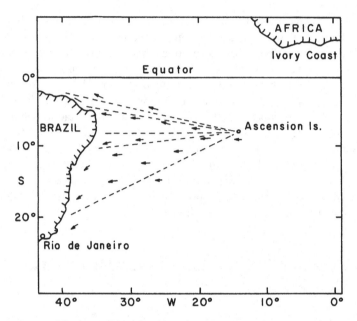

FIGURE 4.6. Ascension Island in the South Atlantic Ocean. Arrows represent surface current trends (from Koch et al., 1969).

50 m deep on the average, and flow toward the coastal region of Brazil subject to horizontal diffusion.

To facilitate calculation of the concentration distribution, we assume that the substance is chemically stable for the period of travel. The x-axis is taken to be in the east–west direction, and the y-axis, in the north–south direction; the substance is assumed to mix uniformly within a depth of h in the vertical direction.

The diffusion equation for a Fickian model reads

$$u\frac{\partial S}{\partial x} = D\frac{\partial^2 S}{\partial y^2}, \tag{4.18}$$

where u is the velocity of the current, S is the substance concentration, and D is a constant diffusivity. The diffusion in the x direction is ignored. When Q is the rate of release of the substance at the source ($x = y = 0$), the solution of (4.18) is given by

$$S(x, y) = \{Q/h(4\pi Dux)^{1/2}\}\exp\{-uy^2/4Dx\}. \tag{4.19}$$

A model of oceanic diffusion that is more realistic than the Fickian model is provided by the solution of Joseph-Sendner (1958) as given by (2.26). However, (2.26) is the solution for an instantaneous source; to modify it for our case of continuous release, we could invoke the principle of superposition directly. Koch et al. (1969) chose rather to use an approximate method of the superposition (Frenkiel, 1953; Gifford, 1959), corresponding to the oft-used approximation that diffusion in the direction of the mean flow may be ignored in evaluating concentration from a continuous source. In effect, the Joseph-Sendner solution (2.26) representing diffusion in the x and y directions is replaced by a one-dimensional version of its solution representing diffusion only in the y direction; that is,

$$S(x, y) = (m/2hPt) \exp(-y/Pt). \qquad (4.20)$$

The transformation $m \rightarrow Q/u$, $t \rightarrow x/u$ applied to this formula [see the similar method used in Eq. (4.12)] leads to the solution for a continuous source:

$$S(x, y) = (Q/2hPx) \exp(-uy/Px). \qquad (4.21)$$

For calculations of the concentration distribution from (4.19) and (4.21), Koch et al. assumed $Q = 1$ mol/s, $u = 24$ N.M./day (nautical miles per day), $h = 50$ m, (Fickian) $D = 5 \times 10^7$ cm²/s (Montgomery, 1939; Montgomery and Palmén, 1940), and (Joseph-Sendner) $P = 1$ cm/s. Figures 4.7 and 4.8 illustrate the result.

Though the Joseph-Sendner model provides a more accurate description for oceanic diffusion than the Fickian model, there is little difference between the two models at large distances from Ascension Island. Koch et al. conclude that either model suggests that the dilution factor is not so great as to exclude chemical perception, by an aquatic animal in the coastal water of Brazil, of a substance released from Ascension Island.

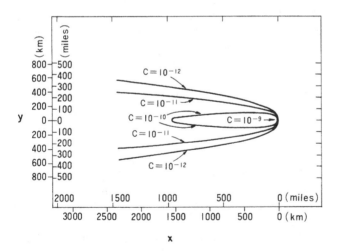

FIGURE 4.7. Steady-state concentration based on the Fickian diffusion model; x: distance from Ascension, y: distance from stream-axis, C: concentration in moles/ℓ. It is assumed that the release rate is 1 mol/s, mean current velocity is 24 nautical miles/day, thickness of mixed layer is 50 m, and diffusivity is 5×10^7 cm²/s (from Koch et al., 1969).

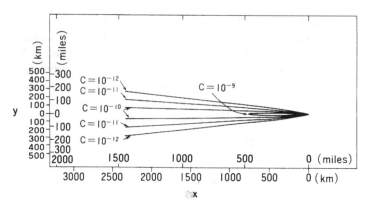

FIGURE 4.8. Steady-state concentration based on the Joseph-Sendner diffusion model. Most of the parameters are used as in Fig. 4.7. The value of P (diffusion velocity) is taken to be 1 cm/s (from Koch et al., 1969).

But how could turtles be able to follow the chemical trail? The precise mechanism remains unknown. According to calculations, the concentration gradient is extremely small. Thus, at a point 1600 km west of Ascension Island, a turtle would have to swim approximately 80 km north or south to experience a tenfold change in concentration. The streamwise gradient is even smaller; in fact, much more so.

As northwest trade winds prevail over the oceanic region concerned, the direction of the current may approximately coincide with the direction of incident waves and swells. It seems conceivable that a turtle might use the waves as a directional clue to its orientation. Of course, the identity of the chemical substance, if it exists, should be investigated in the future. Moreover, a precise estimate of the amount of release will be a prerequisite for a critical evaluation as to whether or not the concentration in the water is sufficient to allow green turtles to detect the substance by smell or taste.

The migratory behavior of the green turtle in other areas should also be compared with that of the Ascension Island population. In fact, Balazs (1976) conducted tagging studies in the Hawaiian Archipelago. He found that, unlike the *Chelonia* population migrating to Ascension Island, Hawaiian green turtles migrating from the southeast to French Frigate Shoals were moving *with* prevailing currents. This would prohibit the direct use of chemical cues originating from the breeding site for possible navigational purposes, as Koch et al. hypothesize.

Finally, Carr and Coleman (1974) have proposed the hypothesis of sea-floor spreading to explain open-sea migratory adaptations in the Ascension Island *Chelonia* population in an evolutionary framework; the ancestors of *Chelonia mydas* subpopulation, which now migrates 1200 nautical miles from Brazil to Ascension Island for breeding, were induced to swim ocean-ward for increasing distances during the gradual separation of South America and Africa in the earliest Tertiary. Perhaps a small plume of chemicals arising from the island might have provided the turtle with a pathway to the final landfall.

4.3   Chemical Communication in Aquatic Organisms

In the example of the previous section, we used large-scale eddy diffusivity to calculate the average concentration of a chemical attractant downstream from a point source. For the case of green turtles, this assumption is probably appropriate, since the "point source," Ascension Island (area: 88 km^2), is large enough to interact with the mesoscale eddy field in the surface ocean. However, if we consider parcels of attractant (or any other chemical) as they leave a much smaller source, we see that in the short term they will expand only slowly due to molecular diffusion and small-scale eddy diffusion. The succession of such parcels describes a filament (Fig. 4.5), which may become tortuously folded by larger-scale eddy motions as it proceeds downstream. Over time, the position of the filament changes; over long enough time, the ensemble average of the filaments produces a large-scale diffusive plume.

The problem for this section is to explore the situations in which each description—diffusion plume or filamentary plume—is appropriate. For the green turtles of Section 4.2, the source is large, minimizing the filamentary nature of the plume; and the turtles themselves swim relatively slowly, so that they can get repeated whiffs of attractant as they proceed upstream, averaging over any filamentary structure that is present. For smaller sources and faster-moving receptors, however, the filamentary structure will dominate.

In this section we consider two applications. First we consider the problem of chemical communication in aquatic organisms whose body sizes span several orders of magnitude. We then discuss an "inverse" problem from chemical oceanography, where the sensor (a sediment trap) is fixed, and the problem is to relate the signal sensed by the trap to photosynthetic production in the surface layer of the ocean.

4.3.1   Temporal and Spatial Scales for Chemical Communication

Physical Scales.   The effects of the physical environment on the structure and transmission of odors in fluids differ significantly at small versus large temporal and spatial scales. Over time intervals of seconds, there is little effect of microbial decay or turbulent eddy diffusion. At these scales, flow has been attenuated by viscous forces, and the low Reynolds number of this regime creates a laminar flow field. Within this field, the major force affecting the odor structure is molecular diffusion. If, at these time scales, flow is created biologically at small spatial scales—for example, in the feeding current or wake of a copepod—then advection would dominate over diffusion in the distribution of the odorant. Turbulent distribution of odorants into filaments forming a plume occurs at large spatial and temporal scales where currents and tides impose directionality on the distribution of the odors. Current research focuses on how such physical variation in odorant structure affects the chemosensory guidance systems of aquatic organisms.

Biological Scales. Given these scaling differences in the physical distributions of odorants, organisms ranging in size from bacteria to lobsters show temporal and spatial variations in their response to chemical signals. Furthermore, each organism, as a consequence of individual size and speed, lives in its own biologically created Reynolds number regime that will interact with the Re regime of the physical environment. Small organisms like bacteria and protozoans live in a low Re realm where signals are transported in a predictable manner along a diffusional gradient. Copepods swim slowly and live in an intermediate (1–100) Re realm where signals are still transported in a predictable manner along a diffusional gradient or along orderly streamlines. In contrast, larger organisms like crabs and lobsters swim faster and live in a high Re realm. The flow is turbulent and is only predictable in a statistical sense. The problem with this statistical structure is that the animals respond in a near-instantaneous manner modified by specific reaction times. Recent advances in chemical ecology have focused on how aquatic organisms use chemicals for communication and the impact of physically derived turbulence.

Bacteria. Bacteria (0.84 to 73 μm for motile genera; 0.14 to 12 μm for nonmotile genera; Dusenbery, 1997) respond to the diffusive component of the odor field and exhibit kinesis (stimulus-induced movement, without directional orientation) but not taxis (directed motion). A chemical gradient is detected by comparing differences in the chemical concentration using either spatial or temporal sampling (Dusenbery, 1998). If the bacteria seek the signal source and experience higher chemical concentrations, they continue to move in a straight line; if they do not experience increased concentration, then they tumble in a random direction. If the uptake rate is greater than the rate of diffusion to the cell, there will be a deficit in the area surrounding the cell. To improve conditions, a cell can move either by its own form of motility or by utilizing turbulence. However, at the scale of bacteria, viscous forces minimize the effectiveness of turbulence to cause relative movement between a cell and its local fluid environment (Lazier and Mann, 1989). Since physical mixing cannot enhance nutrient flux, large bacteria rely on self-generated movement (Dusenbery, 1997).

Protozoans. Two main marine protozoan (20–200 μm) groups are flagellates and ciliates (Fenchel, 1987), swimming from 200 μm/s (flagellates) to 1000 μm/s (ciliates). Many protists show both taxis and kinesis. Some flagellates exhibit helical chemotaxis to help in orientation (Crenshaw, 1996). At the scales of protists, molecular diffusion rates limit rapid signal transport to less than hundreds of micrometers. To enhance nutrient flux, protozoans in the size range of 63 to 100 μm benefit from commonly observed levels of small-scale turbulence (Karp-Boss et al., 1996). Organisms less than 200 μm are considered too small to communicate by pheromones (Dusenbery and Snell, 1995), so Wolfe (2000) proposed that protozoans could still signal cells in close proximity with a non–steady-state excretion.

Copepods. Copepods (1–10 mm) can detect chemical molecules transported to their receptors by molecular diffusion. However, this is a slow process, so the copepod can use its feeding current to increase the rate of transport by advective flow. The feeding current generated by copepods is a low Re feature (Strickler, 1982). Hence, when the active space of an algal cell is entrained in the flow field, it is deformed in a predictable way and the particle path is predetermined by the position of entry into the flow field (Alcarez et al., 1980). When the odors activate antennulary sensors, the copepod is alerted to the approach of a cell from a specific location in the three-dimensional space of the flow field. Within the feeding current, the Peclet number is greater than 1, and advection distorts the distribution of odorant on a time scale too fast for diffusion to erase (Andrews, 1983). Here, advection enhances aquatic communication by accelerating the process of chemical transport beyond the limitations imposed by molecular diffusion.

However, many copepods (1 mm in length, 1 mm/s swimming speed, $\text{Re} \approx 1$) do live at or below the Kolmogorov scale, the size of the smallest eddies formed by physical mixing where molecular diffusive forces dominate, and flow is attenuated by viscous forces and is isotropic. Within these Kolmogorov eddies, copepods exhibit a remarkable behavior of precisely following the three-dimensional trails left by their mates (Doall et al., 1998; Weissburg et al., 1998; Yen et al., 1998). Female copepods leave a pheromone in their laminar wake at a concentration and location that remain detectable by male copepods. Here, instead of limiting chemical communication, the slow rate of diffusion enhances chemical communication by allowing the trail to persist long enough for the male to find it. Observations that certain species of copepod reside in layers of high Richardson's number (Gallager et al., 1997) or low energy dissipation rate ε (Incze, 1996) also provide evidence of the importance of water stability to preserve these communication signals for successful mating and subsequent recruitment into the population. Shrimp (Hamner and Hamner, 1977) follow chemical trails by combining chemotaxis with geotaxis to travel down the scent trail to scavenge on the sinking foodfall. This can occur in the deep sea, where water is stable.

Decapods. In contrast to bacteria and protists that rely on molecular diffusion for mass flux, crabs and lobsters (1 cm to greater than 10 cm) respond to the advective transport of the chemical (Basil and Atema, 1994; Atema,

1996; Weissburg, 1997). These decapods live in a relatively high Reynolds realm where flow has a strong effect on mixing the odorant. According to current ongoing research, signals emitted into a turbulent flow are deformed by the turbulent features that pick up the odorant from the point of emission; these deformations in the odor field appear to persist despite differences in the ambient velocity field at least under some conditions (Weissburg, 2000). These organisms respond to the immediate structure of the plume with reaction times of less than a few seconds. The combination of edge detection and the detection of odor filaments with the detection of fluid flow guides their search to the odor source. Snails and starfish that may respond at longer time scales, integrating the rate of contact and/or concentration of the individual filaments, could indeed be responding to the time-averaged structure of the odor plume (Weissburg, 2000). Analyses of the population response of crabs and lobsters also may show that, as a group, the response to a stationary source (not live, moving prey) represents the time-averaged structure.

Summary. For bacteria and protozoans, the final collection of approaches and paths taken to reach the source most closely matches the statistically time-averaged diffusing odor. Similarly, but at the large scale, the tracks of turtles reconstruct the statistically time-averaged plume structure. In between, we see some fantastically intriguing chemosensory behavior where at intermediate Re, copepods follow a trail that is slowly expanding spatially due to molecular diffusive forces, but temporally for times less than those where turbulent eddy diffusion begins to erase the trail. The copepod's behavior matches the instantaneous structure of the signal. At higher Re, the element of chaos leads to unpredictable spatial distributions of the filaments within odor plumes. Due to this unpredictability, crabs and lobsters rely on two modalities to find the source: odor and flow. When the odor is detected, the response is to move upstream. Without flow, the crabs are not able to find the source. The paths taken to the odor source define neither the time-averaged nor the instantaneous filament structure. (See Table 4.1.)

TABLE 4.1. Scaling of the fluid physical regime with biological responses of aquatic organisms

Size	Organism	Physical Regime	Biological Response
10 μ	Bacteria	Molecular diffusional gradient	Kinesis
100 μ	Protozoans	Molecular diffusional gradient	Kinesis and taxis
1 mm	Copepods	Molecular diffusional gradient and laminar flow	Orient to trails and predictable flow
10 cm	Decapods	Turbulent flow	Orient by whiffs and directional currents
1 m	Turtles	Eddy diffusional gradient	Track gradients?

4.3.2   Models of Chemical Communication in Aquatic Organisms

Use of laser-induced fluorometry and other flow-visualization techniques (Weissburg, 2000) confirms that the structure of a turbulent odor plume is comprised of many individual filaments of odorants, as depicted in Fig. 4.5. Current research in the aquatic realm seeks to couple an instantaneous visualization of an odor plume's *physical* structure with *behavioral* observations of aquatic organisms orienting to the odor. Reactions to the filaments instead of the time-averaged edge of the plume would confirm the importance to the organism of the fine-scale structure of the odorant. The behavioral responses of moths (Vickers and Baker, 1994) to an odor plume created in a wind show that both the increment in the odorant concentration in the direction of the odor source and the temporal and spatial patterns of concentration fluctuation do provide information to guide their chemotactic behavior. In contrast, orientation in turbulent aquatic odor plumes occurs largely via simple mechanisms of edge detection (that is, by detecting spatial patterns in odorant levels: Webster et al., 2001), with a generalized up-current response. Recent research (Webster and Weissburg, 2001) suggests that the instantaneous structure is too unpredictable to provide accurate information on source location. Hence, the complexity of the internal instantaneous structure of an odor plume still limits our present ability to model odor tracking in a turbulent environment.

In contrast, the simplicity of the laminar viscosity-dominated environment of small organisms like copepods and bacteria makes them more amenable to modeling efforts. For instance, Andrews' (1983) model of the transport of phytoplankton odor entrained in a sheared feeding current of a copepod shows a separation of the leading edge of the active space of the odorant. Activated sensors along the copepod antennule would detect the odor prior to the arrival of the odor source, the algal cell. With this advance warning, the copepod could have time to redirect the streamline or reorient to the approach of the alga. This indeed was the behavior observed when *Eucalanus* perceived the algal cell 1250 μm away and 450 msec prior to ingestion (Koehl and Strickler, 1981). These dynamics of odor deformation were recently examined electrochemically at the time and space scales of copepod flow fields (Moore et al., 1999).

Likewise, we can apply equations for Fickian diffusion [Eqs. (4.18) and (4.19)] to model the use of the three-dimensional mating trails for chemical communication between copepods. If λ is the ratio of the initial concentration over the threshold concentration for detection and A is the source area, we can calculate the length of the detectable trail x^* and its fade-out time t^* for a small (600 μm) zooplankter ($A = 10^{-4}$ cm^2, $U = 10^{-1}$ cm/s; $D = 10^{-5} - 10^{-8}$ cm^2/s for small to large molecules, respectively) relying on a large molecular weight pheromone. (See Table 4.2.). For trail lengths over 10 cm, turbulent processes become increasingly effective in reducing the fade-out time.

TABLE 4.2. Length of the detectable copepod mating trail x^* and its fade-out time t^* for different values of λ, the ratio of the initial pheromone concentration over the threshold concentration for detection

λ	x^*	t^* (s)
1	0.1 mm	0.1
10	1 cm	10
100	100 cm	1000
1000	100 m	100,000

If the threshold for detection is on the order of 10% of the original concentration ($= 10$), then the predicted lifetime of a three-dimensional trail diffusing by molecular process is on the order of tens of seconds for lengths of 10 cm or less. And indeed, the longest documented trail followed by a mate-tracking copepod (Doall et al., 1998; Weissburg et al., 1998; Yen et al., 1998) was 13.7 cm, and the oldest trail 10.3 s old. The male also precisely follows the trail of the female, indicating that the trail structure is changing only very slowly. This confirms that eddy diffusion is minimal at these short (<10 s), small (<10 cm) scales, and zooplankton of about 1 mm would be most effective in using pheromonal trails.

4.3.3   An Inverse Problem: Estimating the "Statistical Funnel" of Sediment Traps

In the previous section, the problem was for an individual to follow a plume to its source. In the present section, we consider a case where the "organism" (here a sediment trap) is fixed, and the problem is to relate the mass of sediment particles collected to their source in the surface layer of the ocean.

In the ocean, carbon dioxide is converted by phytoplankton photosynthesis into organic carbon. Through processes of aggregation and grazing, these organic compounds are formed into larger particles. Mineral materials such as calcium carbonate and silica are also produced by some groups of phytoplankton; these materials are denser than seawater, allowing the particles to sink. This sinking flux acts as a "biological pump" for carbon from the surface ocean to the deep ocean; the strength of this pump (along with the "solubility pump" driven by temperature) determines the rate at which the oceans can take up carbon dioxide from the atmosphere.

Chemical oceanographers deploy sediment traps at various depths in the ocean to capture this flux. A major goal has been to relate the fluxes caught in traps to surface production by phytoplankton, which can be estimated using satellite imagery. A major problem with making this linkage is that particles sink rather slowly (the canonical value is 100 m/d) when compared to currents, so that the source area for particles, even for a trap at 3-km depth, may be hundreds of km away. Around this mean flow field, there are also strong eddy motions, and in some places the flow field meanders considerably. For example, Deuser et al. (1988) noted that in the Sargasso Sea off Bermuda, the dominant pattern was seasonal and was easily captured by sediment trap collections; but off the northeast coast of South America, low-salinity filaments from the Amazon and Orinoco rivers meandered over the trap site, creating a signal in the traps that looked random despite the fact that satellite images showed coherent structures.

The statistical connection between surface productivity and sediment trap records was explored in a series of papers by Deuser et al. (1988, 1990), Siegel et al. (1990), and Siegel and Deuser (1997). Deuser et al. (1988) coined the term "statistical funnel" to describe the notion that particles collected by sediment traps originate at the surface intersection of a cone-shaped statistical structure whose apex is the sediment trap. By correlating the seasonal pattern of satellite chlorophyll observations near Bermuda to the seasonal pattern of deposition in sediment traps at 3200 m, Deuser et al. (1990) and Siegel and Deuser (1997) showed that the center of the surface opening of the statistical funnel for these traps was approximately 200 km northeast of the trap mooring, implying that sinking particles move almost horizontally, with a very small vertical vector component (Siegel and Deuser, 1997).

The size of the statistical funnel was explored in a series of simulations by Siegel and Deuser (1997). Releasing particles at the surface to see where they ended up would not work, since a vanishingly small proportion of particles would reach the trap. Instead, they released particles at the trap and allowed them to float to the surface with characteristic velocities of 50 m/d, 100 m/d, and 200 m/d. (Alternatively, they allowed the particles to sink backward in time.) As the particles rise from one depth to the next, their horizontal velocities may change slightly. The statistics of this change are described by a Lagrangian autocorrelation structure with time scale $\tau = 10$ d, which means that the correlation of velocity after 10 d with the original velocity is only $1/e$. This correlation structure is simulated using a recursion relationship; for the x direction (east), the eddy-induced velocity u'_i of particle i is given by

$$u'_i(z, t + \Delta t) = (1 - \Delta t/\tau)u'_i(z, t) + \sqrt{(2\Delta t/\tau)}\,\sigma_u(z)r,$$

where z is depth, Δt is the time interval used in the simulation, r is a random number drawn from a standard normal distribution $N(0,1)$, and $\sigma_u(z)$ is the square root of the turbulence-induced variance $\sigma_u^2(z)$ at depth z. This variance is directly proportional to the eddy kinetic energy at any depth z, which is higher near the surface; its mathematical form near Bermuda was estimated from data to be

$$\sigma_u^2(z) = 30 \exp\{3.7 \exp(-z/600)\}.$$

The simulation was run using several thousand particles released at each of several depths, and with several sinking velocities. Each particle follows a

TABLE 4.3.

| | | \multicolumn{3}{c}{H (m)} | | |
		500	1500	3200
	50	151	224	264
V (m/d)	100	81	130	168
	200	43	71	99

discrete trajectory that could be considered the center of a filament perhaps 100 m in diameter (Siegel and Deuser, 1997). Particles intersect the surface at time H/V, where H is the depth of the trap and V is the sinking velocity. Remarkably, even with the assumptions of autocorrelated motion and variation of σ_u with depth, the resulting statistical funnels are circular normal (Armstrong and Siegel, 2001); the sizes of the statistical funnels can therefore be characterized uniquely by their standard deviations s (km); see Table 4.3.

Table 4.3 contains values of 2.80 s (in km) from the simulations of Siegel and Deuser (1997). (Because the distributions are circular normal, 98% of particles originate at distances $\leq 2.80\ s$.) As the table shows, the area from which particles are drawn increases with trap depth H and decreases with sinking velocity V. Even in the case of very fast (200 m/d) particles sinking to a very shallow (500 m) trap, the size of the statistical funnel is still very large, with a 98% origination contour of 43 km. For slow-sinking (50 m/d) particles and deep (3200 m) traps, the 98% collection distance becomes enormous (264 km).

However, large distances do not imply that surface patterns of productivity cannot be correlated with patterns from subsurface particle collectors, at least if one is interested in averages over long time periods. For example, Deuser et al. (1990) found that the coherence between time series of trap values and of satellite chlorophyll values had a correlation coefficient $r = 0.96$ when (i) the values within each two-week period were averages over several years, (ii) the surface box studied was aligned 200 km northeast of Bermuda, and (iii) the transit time from the surface to the trap was assumed to be 1 month. So, in analogy to previous examples, a diffusion model seems to be quite useful for describing suitably averaged time series of trap data.

However, for shorter deployments, Siegel and Deuser (1997) pointed out that the "funnel" at any one time is only a filament of perhaps 100-m diameter at the surface, and that this filament can move quite a distance (due to the passage of mesoscale eddies) during the course of a short (30 d) trap deployment. At these short time scales, heterogeneity in sinking rates may also be important. At short time periods, the filament model is appropriate, and predicting the surface origin of fluxes at depth would require exquisite knowledge of fluid motions for hundreds of kilometers around the trap.

5
Mathematical Treatment of Biological Diffusion

Akira Okubo and Daniel Grünbaum

In the lifetime of most animals there occurs a time when the site of inhabitation is abandoned in favor of migration. Thus, in an environment changing through space and time, the most probable strategy for a new individual to adapt to survive and reproduce may not necessarily consist of remaining to compete with its parents or congeners, but may rather consist of migrating elsewhere to find an empty niche to inhabit (Taylor and Taylor, 1977; see also Lidicker and Caldwell, 1982). As a result the spread of population, i.e., dispersal, takes place. Such animal movement includes nomadism, whereby animals wander with no particular direction in search of sustenance, in a manner that resembles the random walk; and migration, which may be either periodic as animals move from one habitat to another in a repetitive cycle, or nonperiodic, implying a certain degree of permanence to the move. In addition, animals may display a restricted movement as they carry on their daily activities within a given domain of their habitat (home range).

The migration and dispersal of animals, while containing subjective elements that may not be totally controlled by animals, by and large constitute a ceaseless, active effort on the part of the animal to put itself in advantageous circumstances. However, the movement of two individuals placed in the same environment is not identical. It is necessary to consider animal motion as a random variable (see Sect. 2.1). Nevertheless, the random motion of animals in general cannot be considered to be that of a "simple diffuser" such as the random walker.

A degree of success has been achieved in the analysis of dispersal of animal populations by starting with a direct analogy to the random walk or physical diffusion, with an additional consideration of intra- and interspecific population interaction. The work of Skellam (1951) has provided a profound and lasting contribution to this approach. However, a more realistic model of biological diffusion must be built by properly combining the following concepts: correlated random walks; diffusion incorporating space–time variation of parameters and nonuniformities; treatment of individual interaction after the fashion of the many-body problem; and statistical treatment using computer simulation. The formulation of such models alone necessitates a better grasp of the natural occurrence of movements of animal individuals and populations.

5.1   Animal Motion and the Balance of Acting Forces

We may assume that Newton's three laws of motion also apply to animal motion:

1. The forces acting on an organism at rest or moving with constant velocity are in balance.
2. When the force balance is broken, a temporal change of the momentum of an organism occurs in the direction of the net force. The rate of change of momentum is equal to the net force. If we consider times that are small compared with those required for noticeable loss or gain in weight of the organism, the change in mass can be ignored, and the net force may be equated to the product of mass and acceleration.
3. When organism A applies a force to the environment or object B, B reacts by applying a force of the same magnitude in the opposite direction on A. However, we must stipulate that when B is also an organism, this action–reaction law may not necessarily apply.

Among the forces acting on organisms are such external forces as gravity and pressure from environmental fluids; but the forces that ultimately control animal motion are, rather, internal in nature. They include animal reaction to changes in elements of the environment, the instinct to search for food or escape from enemies, and sex drive.

To clarify the balance of forces acting on an organism, we consider an example of a squirrel clinging to a tree (Fig. 5.1). Gravity acts vertically downward on the center of gravity of the animal. If m is the squirrel's mass and g is the acceleration of gravity, then the magnitude is $G = mg$. This force tends to propel the animal downward; to stop this occurrence, a force acting in the opposite direction is required. To obtain this opposing force, the squirrel pushes itself downward by clinging to the tree. The tree reacts by providing an equal force on the squirrel in an upward direction. If we denote the magnitude of this force by P, then $P = G = mg$. A squirrel reacting to danger and climbing the tree must apply a downward force (push) greater than G to the tree. Thus, the net force of propulsion is $P - G$. If the squirrel cannot obtain the necessary reaction from the tree, its greatest efforts will not allow it to climb. For example, if the tree is perfectly smooth, the squirrel cannot push down on the tree, and without the necessary reaction it must suffer the fate of falling to the ground under the influence of gravity. In effect, animals propel themselves forward by exerting a somewhat greater

FIGURE 5.1. Balance of forces in the vertical direction acting on a squirrel clinging to a tree.

backward force on their surroundings. Forward propulsion is obtained as a reaction to a backward force applied to the ground in the case of terrestrial animals, to water in the case of aquatic animals, and to air in the case of airborne animals.

Once motion has commenced and a velocity difference developed between the organism and the surrounding fluid, the animal meets with resistance. This resistive force (drag) may be expressed, in analogy to (3.2), as

$$R = \frac{1}{2} C_D \rho A u^2. \tag{5.1}$$

Here C_D is the drag coefficient, ρ is the density of the surrounding fluid, u is the velocity difference, and A is a characteristic area of the animal (for a general reference, see Vogel, 1994). Unfortunately no single definition of A exists that is suitable for all body shapes; the frontal area, i.e., the projected area of the body in the direction of flow, the greatest projected area, and the total surface area are conveniently used for A (Alexander, 1968).

At Reynolds numbers between about 10^3 and 10^5, the drag coefficient is nearly constant for a body of a given shape; thus, the drag is approximately proportional to the square of the speed. At low Reynolds numbers below about 100, C_D increases rapidly with decreasing Reynolds number. At extremely low Reynolds numbers, say below 1, C_D becomes inversely proportional to the Reynolds number; thus, the drag is linearly proportional to both the speed and the viscosity of the surrounding fluid [see Stokes' law, (3.15)]. The reader may find a paper by Purcell (1977) on organism locomotion at low Reynolds numbers interesting.

The implication of this Reynolds number dependence of drag is worthy of attention in animal locomotion. For example, each newly hatched gypsy moth larva is equipped with a long thread of silk, which serves to increase the drag force on the larva by the additional viscous drag force on the silk. As a result, the larvae can be carried in suspension by relatively weak winds (McManus, 1973). The method of transport is analogous to the early instar dispersal of the Douglas-fir tussock moth (Mitchell, 1979) and to the gossamer flight of orb-weaving spiders (Nishiki, 1966; Vugts and Van Wingerden, 1976; Tolbert, 1977; Miller, 1984; Plagens, 1986). Aerodynamical aspects of gossamer flight are discussed by Humphery (1987) and Greenstone (1990). A laboratory test with tussock moth instars showed that the settling rate for unfed first instars with no silk threads was 87.1 cm/s compared to 14.5 cm/s for larvae with threads 2.4 m long. Thus, larvae with very long threads dispersing from trees 30 m tall into very high winds could travel downwind as far as 500 m.

The same mechanism was studied in aquatic environments by Sigurdsson et al. (1976) with young post-larval bivalve mollusks, which secrete byssus threads for drifting (see also Lane et al., 1985). Prezant and Chalermwat (1984) reported that adults of a freshwater bivalve, *Corbicule fluminea*, secrete long mucous threads through their exhalent siphons that act as draglines to float the animal into a water column. Prezant and Chalermwat speculate that this mechanism may help in the downstream or interstream dispersal of this rapidly spreading Asiatic clam. The spread of this exotic across North America since its accidental introduction about 50 years ago has led to controversy over the mode of transport that could account for its invasiveness. Although the dispersal of post-metamorphic molluscs by means of mucous threads has been studied in some detail, factors triggering post-metamorphic drifting are only recently being explored. Martel and Chai (1991a, b) and Martel and Diefenbach (1993) studied these factors in the laboratory as well as in the field.

The faster an animal moves, the greater the resistance; thus, there exists a limit to the speed an animal can obtain for a given propulsion, where a steady (acceleration-free) motion is attained. In this case the propulsion P and resistive force must be of the same magnitude and directed oppositely to balance each other (Fig. 5.2). The velocity is given by

$$u = (2P/C_D\rho A)^{1/2}. \qquad (5.2)$$

For many terrestrial animals this balance is determined by the greatest propulsion that the animal can possibly obtain. In other words, the animal

FIGURE 5.2. Balance of forces in the horizontal direction acting on a fish swimming with a constant velocity.

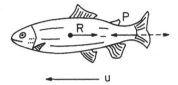

may move at a constant maximum velocity such that this propulsion just equals resistance (see Sect. 8.5).

Especially in marine animals, resistance to motion may be modified by changes in shape or posture at different speeds. Cowles et al. (1986) measured the drag and swimming speed of a bathypelagic mysid (*Gnathophausia ingens*) using a wind tunnel and a high-resolution transducer. The relationship between velocity and drag was obtained for sizes ranging from 6 to 12 cm (total length) and speeds from 0 to 15 cm/s. Drag on the bodies of dead mysids increased *linearly* with water velocity, a relationship not predicted for high Reynolds numbers by standard fluid dynamic equations. This linear increase is due to a change with increasing speed in the angle of the negatively buoyant mysid's body relative to the oncoming current. Such a linear relation would markedly increase the cost of locomotion in this mysid *at low speeds* compared to that of a neutrally buoyant organism that tends to have the relationship predicted by hydrodynamic equations.

5.2   Taxis and the Equation of Motion

We express Newton's second law of motion analytically as

$$m\frac{d\mathbf{v}}{dt} = \mathbf{F},$$ (5.3)

where m is the animal's mass, assumed to be invariant during motion, $\mathbf{v}$ is the animal's velocity (vector), and $\mathbf{F}$ is the (vector) sum of the acting forces. Furthermore, velocity is the time rate of change of position (vector) $\mathbf{x}$

$$\mathbf{v} = \frac{d\mathbf{x}}{dt}$$

so that (5.3) may be rewritten as

$$m\frac{d^2\mathbf{x}}{dt^2} = \mathbf{F}.$$ (5.4)

According to (5.3) and (5.4), if we specify $\mathbf{F}$, the motion of an individual animal is uniquely determined given initial position and velocity.

A mode of animal movement in response to stimuli is *taxis*, by means of which an animal is able to choose its environment and habitat in the midst of the processes of migration and dispersal. (The meaning of taxis used herein includes not only directional movement but also nondirectional movement, kinesis.) In response to a gradient of chemical attractant, organisms may exhibit taxis (i.e., directionally biased random motion) or kinesis (i.e., spatially biased random motion). See Lapidus (1980), Lapidus and Levandowsky (1981), and Tranquillo and Alt (1990) for terminology concerning various forms of biased random walks and behavioral responses to sensory stimuli.

Investigation of animal motion in response to stimuli is still largely at the qualitative stage (Fraenkel and Gunn, 1961). This contrasts with physics, which established itself with classical mechanics based on Newton's laws of motion. Animal motion in response to stimuli does not necessarily follow set rules. There are comparatively predictable responses such as *topotaxis*, in which an animal responds to a stimulus from a set direction by moving in another specific direction, but even in this case individual reaction varies somewhat. On the other hand, there are movements so irregular as to be regarded as a sort of random motion; examples are *klinokinesis*, which involves change in the frequency and direction of movement in response to the stimulus, and *orthokinesis*, in which the speed of movement is altered. However, even kinesis is not entirely random; the degree of response depends on the strength of the stimulus, so that the motion appears to be deterministic in a statistical sense. For example, animal populations that display positive kinesis in response to light tend to concentrate in dark regions where their movement is reduced.

Thus, when considering the equation of motion of organisms, it is necessary to divide the forces acting on an individual into random components and nonrandom components,

$$m\frac{d\mathbf{v}}{dt} = m\frac{d^2\mathbf{x}}{dt^2} = \mathbf{R} + \mathbf{K} + \mathbf{A}, \tag{5.5}$$

where $\mathbf{R}$ is the resistance force, $\mathbf{K}$ is the nonrandom force, and $\mathbf{A}$ is the random force. The force of gravity is included in $\mathbf{K}$. Insofar as the force contains a random component, the resulting motion and resistance are of course random.

As a special case, we consider a "quasi-equilibrium" case of a dispersing animal population, in which the average rate of change of momentum is small and in which the total applied forces, averaged over the population, are approximately balanced at each instant. Then (5.5) becomes

$$O = \bar{\mathbf{R}} + \bar{\mathbf{K}} + \bar{\mathbf{A}} = \bar{\mathbf{R}} + \bar{\mathbf{K}}. \tag{5.6}$$

$\mathbf{A}$ is a random force with an average of zero, so that the nonrandom force and the resistance force are balanced in the average. Now we assume that the resistance is simply proportional to velocity:

$$\bar{\mathbf{R}} = -mk\bar{\mathbf{v}}, \tag{5.7}$$

where k is a resistance coefficient. Note that (5.7) is Stokes' law (refer to Chap. 3), which strictly applies only to minute organisms. However, some experimental estimates suggest that flow is laminar around swimming zooplankton, even at high velocities up to 10 cm/s (Re $\approx$ 100). Olgivy and Dubois (1981) calculated drag for swimming bluefish. The fishes' drag is linearly proportional to swimming speed in a range of 0.1 to 1.4 m/s, while the drag of a wooden model of similar size and shape is proportional to speed squared. These results suggest that, at least in some organisms, novel biofluid dynamical mechanisms may reduce drag in such a way as to make the velocity more similar to (5.7) than expected on simple physical grounds.

From (5.6) and (5.7), $\bar{\mathbf{R}}$ can be eliminated, and since $\bar{\mathbf{K}}$ is equal to $\mathbf{K}$ itself,

$$\bar{\mathbf{v}} = \bar{\mathbf{K}}/mk = \mathbf{K}/mk. \tag{5.8}$$

According to (5.8), if the space–time variation of $\mathbf{K}$ is known, the average velocity of the population can be obtained by dividing the nonrandom force by the resistance coefficient.

For the quasi-equilibrium hypothesis to be valid, it is required that the characteristic time in (5.5) for accelerative equilibrium of the resistance and propulsive forces be less than a gross time scale of the motion. In particular, when the resistance force is linearly proportional to velocity, the characteristic time is the relaxation time, k^{-1}.

5.3 Extension of the Random Walk Model and the Equation of Biodiffusion

Let us return to the random walk model of Sect. 2.2 and consider the equation governing the statistics of a particle. We define the probability that a particle released from the origin at $t = 0$ reaches point x by time t to be $p(x,t)$. At one time interval earlier, i.e., at time $t - \tau$, the particle was at either point $x - \lambda$ or $x + \lambda$ (Fig. 2.1). If we call α the probability that a particle will move to the right in time unit τ, and β the probability that the particle will move to the left ($\alpha + \beta = 1$), then

$$p(x,t) = \alpha p(x - \lambda, t - \tau) + \beta p(x + \lambda, t - \tau). \tag{5.9}$$

In Sect. 2.2, we considered the case that $\alpha = \beta = 1/2$. A random walk in which the probabilities of movement to the right and left are equal is called a *simple*, or *isotropic*, random walk (Spitzer, 1976).

To obtain a diffusion equation from (5.9), it is assumed that λ and τ are small compared to, respectively, x and t, and that each term on the right-hand side of the equation can be expanded in a Taylor series in x and t,

$$p(x - \lambda, t - \tau) = p(x,t) - \lambda \partial p/\partial x - \tau \partial p/\partial t + \lambda^2/2\, \partial^2 p/\partial x^2 + \lambda\tau\, \partial^2 p/\partial x \partial t$$
$$+ \tau^2/2\, \partial^2 p/\partial t^2 + \cdots,$$

$$p(x + \lambda, t - \tau) = p(x,t) + \lambda\, \partial p/\partial x - \tau\, \partial p/\partial t + \lambda^2/2\, \partial^2 p/\partial x^2 - \lambda\tau\, \partial^2 p/\partial x \partial t$$
$$+ \tau^2/2\, \partial^2 p/\partial t^2 + \cdots. \tag{5.10}$$

All of the right-hand derivatives are evaluated at (x, t). If (5.10) is substituted into (5.9) and the relations $\alpha + \beta = 1$, $\alpha - \beta \equiv \varepsilon$ are used,

$$\frac{\partial p}{\partial t} = -\frac{\lambda \varepsilon}{\tau} \frac{\partial p}{\partial t} + \frac{\lambda^2}{2\tau} \frac{\partial^2 p}{\partial x^2} + \lambda \varepsilon \frac{\partial^2 p}{\partial x \partial t} + \frac{\tau}{2} \frac{\partial^2 p}{\partial t^2} + \cdots, \qquad (5.11)$$

where the parameters λ, τ, and ε are assumed to be constant.

Now let us consider the limit as these parameters go to zero. We shall not do this indiscriminately; rather we shall suppose that as τ becomes small, λ and ε decrease so as to be of the same order of magnitude of $\tau^{1/2}$. In other words, in the first and second terms on the right-hand side of (5.11),

$$\lim_{\lambda, \varepsilon, \tau \to 0} \lambda \varepsilon / \tau = u, \qquad \lim_{\lambda, \tau \to 0} \lambda^2 / 2\tau = D. \qquad (5.12)$$

Since the other right-hand terms converge to zero, the following equation is obtained:

$$\frac{\partial p}{\partial t} = -u \frac{\partial p}{\partial x} + D \frac{\partial^2 p}{\partial x^2}. \qquad (5.13)$$

This is the equation of diffusion for the random walk that results from the limiting process. If p is multiplied by the total number of released particles, the particle concentration S is obtained, so that

$$\frac{\partial S}{\partial t} = -u \frac{\partial S}{\partial x} + D \frac{\partial^2 S}{\partial x^2}. \qquad (5.14)$$

Equation (2.4) corresponds to (5.14) when advection u is absent ($\alpha = \beta = 1/2$).

Skellam, the founder of biodiffusion theory, has emphasized the necessity of formulating a more realistic model that takes into account interaction between individuals, stimuli, and animal response to environment (Skellam, 1972, 1973). As a first step, let us examine some aspects that the random walk model lacks.

(i) Statistical Correlation Between Steps. An animal's motion during a small time period has a tendency to proceed in the same direction as it did in the immediate period before. In other words, successive movements are not mutually independent. The random walk model does not allow for the possibility of correlation between successive steps.

A random walk model with correlated steps was developed by Goldstein (1951), who showed that in the limit the model leads to the so-called telegraph equation. The details may be found in the original paper; the resulting equation for probability is

$$\frac{\partial^2 p}{\partial t^2} + \frac{1 - \gamma}{\tau} \frac{\partial p}{\partial t} = \frac{\lambda^2}{\tau^2} \frac{\partial^2 p}{\partial x^2}. \qquad (5.15)$$

Here γ is a parameter representing the correlation between successive steps, and the other symbols are the same as those used for the simple random walk. Also, terms of vanishing order have been neglected. If limits are taken such that

$$\lim_{\lambda,\tau\to 0} \lambda^2/\tau^2 = v^2, \qquad \lim_{\substack{\gamma\to 1 \\ \tau\to 0}}(1-\gamma)/\tau = 2/T, \qquad (5.16)$$

Eq. (5.15) becomes the telegraph equation

$$\frac{\partial^2 p}{\partial t^2} + \frac{2}{T}\frac{\partial p}{\partial t} = v^2\frac{\partial^2 p}{\partial x^2}, \qquad (5.17)$$

where T is a characteristic time of step correlation.

The telegraph equation possesses elements of both diffusion and wave motion; probability p or concentration S is transmitted with wave speed v while being dispersed. In particular, when $t \gg T$, (5.17) can be approximated as a form of the diffusion equation

$$\frac{\partial p}{\partial t} = \frac{v^2 T}{2}\frac{\partial^2 p}{\partial x^2} \qquad (5.18)$$

in which the diffusivity is expressed as $v^2 T/2$. The physical difference between the telegraph equation and the diffusion equation, (5.13), is that for the former the velocity of dispersion is finite, whereas for the latter it is infinite. This must not be confused with the velocity of a wave front of a diffusing population to be discussed in Sect. 10.2. The meaning herein is that instantaneously after release from a point source, an infinitesimal concentration reaches infinitely far; e.g., the point-source solution to (5.14), $S(x,t)$ has the property that if $t > 0$ then $S(x,t) > 0$ for all values of $|x|$, no matter how large.

The difference between the two equations can further be understood in terms of limits. To derive the telegraph equation, limits were taken [see (5.16)] in such a way that $\lambda/\tau = v =$ finite, i.e., a finite walk speed is implied. On the other hand, to derive the diffusion equation, limits were taken [see (5.12)] in such a way that $\lambda^2/\tau =$ finite, i.e., an infinite walk speed is implied; $\lambda/\tau = ((\lambda^2/\tau)1/\tau)^{1/2}$, which becomes infinite in the limit as $\lambda, \tau \to 0$. Thus, the diffusion equation is characterized by a type of infinite dispersion speed. Since no organism can spread or "propagate" with an infinite speed, the telegraph equation is basically more realistic than the diffusion equation when applied to animal dispersal problems. However, in the limit of $t \gg T$, the difference between the two equations is, for all practical purposes, very small. In other words, as long as we consider times that are much longer than the duration of walk correlation, T, the results of the diffusion equation obtained from the random walk can be used.

On the other hand, for $t \ll T$, (5.17) takes the form of the wave equation

$$\frac{\partial^2 p}{\partial t^2} = v^2\frac{\partial^2 p}{\partial x^2}. \qquad (5.19)$$

Thus, organisms starting their motion from an origin first spread as a wave in all directions with velocity v. This is, in fact, more realistic in animal dispersal than the movement predicted by the diffusion equation.

A nonlinear equation for population movement and reproduction was derived by Dunbar and Othmer (1986). Spatial distribution of populations was described by

$$\frac{\partial^2 p}{\partial t^2} + \frac{2}{T}\frac{\partial S}{\partial t} = v^2 \frac{\partial^2 S}{\partial x^2} + \left(\frac{\partial}{\partial t} + \frac{2}{T}\right)F(S), \qquad (5.20)$$

where S is concentration, T is a characteristic time constant, v is velocity, and $F(S)$ represents the population dynamics. Holmes (1993) also derived the same equation from a discrete correlated random walk (developed by Goldstein, 1951) with the inclusion of population dynamics (see also Holmes et al., 1993). Their studies compare the results of telegraph-type dispersal models with diffusion-type models for a logistic population by calculating travelling wave speeds. For all the real data analyzed, both models give nearly the same speeds. Nevertheless, for organisms with high population growth and low movement rates, the disparity in range expansion rates could be serious (see also the section on travelling wave problems in Chapter 10).

Skellam (1973) considered the random movement of an organism that is correlated in the direction of movement. Calculating the square of the distance of dispersal, i.e., the variance, he was able to show that the variance is asymptotically proportional to time, a result that agrees with the diffusion equation and lends support to the above theory. Earlier, Taylor (1921) treated the same problem in connection with turbulent diffusion.

Let's explain this in terms of a one-dimensional example. We let x_j $(j = 1, 2, \ldots, n)$ be the displacement of an organism after a succession of time intervals τ. We assume that the length of each step is constant; $|x_j| = \lambda$. The square of the distance from the origin after $n\tau$ time has elapsed is given by

$$R_n^2 = (x_1 + x_2 + \cdots + x_n)^2.$$

Taking the average of this quantity, we obtain

$$\bar{R}_n^2 = \overline{(x_1 + x_2 + \cdots + x_n)^2} = \bar{x}_1^2 + \bar{x}_2^2 + \cdots + \bar{x}_n^2 + 2(\overline{x_1 x_2} + \overline{x_1 x_2} + \cdots)$$

$$= n\lambda^2 + 2(\overline{x_1 x_2} + \overline{x_1 x_2} + \cdots + \overline{x_r x_s}). \qquad (5.21)$$

We postulate that there is correlation only between successive steps. If we define $\overline{x_j x_{j+1}}/\lambda^2 = \gamma$, then

$$\overline{x_j x_{j+2}}/\lambda^2 = \gamma^2, \qquad \overline{x_j x_{j+s}}/\lambda^2 = \gamma^s,$$

and (5.21) becomes

$$R_n^2 = n\lambda^2 + 2\lambda^2\{n\gamma + (n-1)\gamma^2 + (n-2)\gamma^3 + \cdots + \gamma^n\}$$

$$= \lambda^2\left\{n + \frac{2n\gamma}{1-\gamma} - \frac{2\gamma^2(1-\gamma^n)}{(1-\gamma)^2}\right\},$$

where we exclude the case $\gamma = 1$, so that $|\gamma|$ is less than unity. Thus,

$$R_n^2 = \lambda^2 \frac{1+\gamma}{1-\gamma} n - \frac{2\lambda^2\gamma^2(1-\gamma^n)}{(1-\gamma)^2}. \tag{5.22}$$

As n becomes large, γ^n becomes small, and the second term on the right-hand side becomes negligible; $R_n^2 \to \lambda^2(1+\gamma)/(1-\gamma)n$. In other words, in the limit the variance depends linearly on time, i.e., Fickian diffusion. Obviously, if the walk is completely uncorrelated ($\gamma \equiv 0$), $R_n^2 = \lambda^2 n$ for any n.

The derivation of the telegraph and diffusion presented earlier in this section assumed that individuals employ a particular type of random walk, in which individuals take discrete instantaneous steps. Othmer et al. (1988) contrasted this behavior, which they termed a *position jump process*, to a behavior in which an individual's velocity undergoes discrete instantaneous changes, which they termed a *velocity jump process*. In a velocity jump process, the individual's position remains continuous in time. Distributions of individuals performing both position jump and velocity jump behaviors can be approximated by telegraph- and diffusion-type equations. However, the exact and complete distributions (from which simpler forms can be deduced) are given by equations that include integrals over time, position, and/or velocity. For the position jump process, Othmer et al. (1988) derived an expression for the distribution of a population,

$$p(\mathbf{x}, t) = \delta(\mathbf{x})\delta(t) + \int_0^t \int_X \phi(t - \tau)T(\mathbf{x}, \mathbf{y})p(\mathbf{y}, t)\,d\mathbf{y}\,dt, \qquad \mathbf{x} \in \mathbf{X}, \tag{5.23}$$

for a population that is initially concentrated at the origin ($p(\mathbf{x}, 0) = \delta(\mathbf{x})$). In (5.23), $\phi(t)$ is the *waiting-time density*, so that the probability of an individual not jumping by time τ after the previous jump is $1 - \int_0^\tau \phi(s)\,ds$. $T(\mathbf{x}, \mathbf{y})$ is the probability density for a jump from position $\mathbf{y}$ to $\mathbf{x}$.

For the velocity jump process, the redistribution of population density is described by

$$\frac{\partial p}{\partial t} + \nabla_{\mathbf{x}} \cdot p = -\lambda p + \lambda \int_V T(\mathbf{v}, \mathbf{v}')p(\mathbf{x}, \mathbf{v}', t)\,d\mathbf{v}', \tag{5.24}$$

where λ is the rate of velocity changes and $T(\mathbf{v}, \mathbf{v}')$ is the transition probability from velocity $\mathbf{v}'$ to $\mathbf{v}$. Othmer et al. (1988) obtained expressions for spatial moments of population distribution that may be useful in determining the type of behavior best describing real animal trajectories and estimating their parameters.

5.3.1 Correlated Random Walks

Kareiva and Shigesada (1983) derived expressions for mean squared displacement in correlated random walks in which the angles between the directions and lengths of successive movements are chosen from prespecified random distributions. They used their model to describe ovipositing cabbage white butterflies (*Pieris rapae*). Cain (1989) analyzed patterns of clonal growth in goldenrod plants (*Solidago altissima*) in fields, relating the statistics of rhizome length, branching angles, and rhizome numbers to the temporal process of vegetative spread. He found that *S. altissima* can be modeled as a correlated random walk. Also see Hall (1977) for an analysis of correlated random walks in the slime mold, *Dictyostelium discoideum*.

Marsh and Jones (1988) considered, in addition to the behavior assumed by Kareiva and Shigesada, three alternative types of random walk behaviors. These four models assume two-dimensional space and movement trajectories that consist of sequences of straight-line paths or steps: Model I, in which step length η and direction of movement θ are independent; and Model II, in which η depends on the direction of movement relative to some fixed compass direction. For each of Models I and II, behaviors can be specified as (a) *oriented*, in which θ is chosen relative to some fixed compass direction, or (b) *unoriented*, in which θ is chosen relative to the direction of the previous step. Model I(b) corresponds to the case considered by Kareiva and Shigesada. Marsh and Jones developed summary statistics to discriminate between paths resulting from these random walk types. For Model I(a), i.e., oriented movement with no length–direction correlation, the mean squared displacement after the nth step is

$$\langle R_n^2 \rangle = n \langle \eta^2 \rangle + n(n-1)\langle \eta \rangle^2 \alpha^2, \tag{5.25}$$

where $\langle . \rangle$ denotes expected value and $\alpha = \langle \cos(\theta_j) \rangle$. Marsh and Jones defined the quantity

$$\Delta = \frac{1}{n^2}\left\{ \left(\sum \cos(\theta_j)\right)^2 + \left(\sum \sin(\theta_j)\right)^2 \right\}$$
$$- \frac{1}{(n-1)^2}\left\{ \left(\sum \cos(\omega_j)\right)^2 + \left(\sum \sin(\omega_j)\right)^2 \right\}, \tag{5.26}$$

where ω_j is the angle between the directions of steps j and $j+1$. For Model I(a), they found

$$\Delta = -\frac{1}{n(n-1)} + \left(1 - \frac{1}{n}\right)\alpha^2 - \frac{2(n-2)}{(n-1)^2}\alpha^2\alpha_2 - \frac{(n-2)(n-3)}{(n-1)^2}\alpha^4. \tag{5.27}$$

In (5.27), $\alpha_2 = \langle \cos(2\theta_j) \rangle$. For Model I(b), i.e., unoriented movement with no length–direction correlation, they found

$$\langle R_n^2 \rangle = n\langle \eta^2 \rangle + 2\langle \eta \rangle^2 \frac{c}{1-c}\left(n - \frac{1-c^n}{1-c}\right), \tag{5.28}$$

with $c = \langle \cos(\omega_j) \rangle$, and

$$\Delta = -\frac{1}{n(n-1)} + \frac{2c}{n(1-c)^2}\left(\frac{c^n}{n} - c + 1 - \frac{1}{n}\right) - \left(1 - \frac{1}{n-1}\right)c^2. \tag{5.29}$$

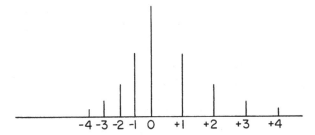

FIGURE 5.3. Random walk with spatially nonuniform step lengths.

For Model II(a), i.e., oriented movement with length–direction correlation,

$$\langle R_n^2 \rangle = n\langle \eta^2 \rangle + n(n-1)\langle \eta \cos(\theta) \rangle^2 + \langle \eta \sin(\theta) \rangle^2. \qquad (5.30)$$

Marsh and Jones showed that for some biologically reasonable movement distributions, Δ is positive for Model I(a) but negative for Model I(b), suggesting that this statistic is useful in discriminating between paths generated by the two types of random walks.

(ii) Nonhomogeneity of the Parameters of the Walk. The characteristic parameters of animal motion vary in space and time according to such factors as nonhomogeneity of stimuli and environment, choice of habitat, and variation in animal behavior. Thus, it is unrealistic to suppose that such parameters as λ and τ are constant. In addition, the step parameters may vary from individual to individual, so that in a model of population dispersal, these parameters themselves must be regarded as random variables.

As a simple example of diffusion in which the length of each step is not spatially constant, we consider the case illustrated in Fig. 5.3 with a skewing to the left.

A diffusion equation that accounts for the effect of spatial nonuniformity can be written as (Chapman, 1928)

$$\frac{\partial S}{\partial t} = \frac{\partial^2}{\partial x^2}(DS), \qquad (5.31)$$

where the diffusivity D depends on x. When nonhomogeneities of stimuli and environment produce an oriented response in animals (taxis), the probability of movement in a specific direction increases, and the associated random walk is accompanied by advection.

(iii) The Effect of Nonrandom Forces Acting on an Individual. A good example of an external, nonrandom force acting on an individual undergoing a random walk is provided by the diffusion of spores that fall in the atmosphere under the influence of a gravitational field. Gravity accelerates the spore, but this is quickly balanced by resistance due to the viscosity of the air; the particle then descends with a constant settling velocity (Sect. 3.3.1).

In other words, gravity causes advection of a population in the direction of the force. Likewise, topotaxis or escape behavior due to stimuli can also be interpreted as advection caused by external forces.

(iv) Interference Between Individuals.   Random walk models that take some account of interference between individuals (attraction, repulsion) are nearly nonexistent; for a rare exception, see Schwarz and Poland (1975). The problem falls into the category of many-body problems with random motion and is extremely difficult. Interaction between individuals must likely be involved in both the advection and diffusion terms. An intuitive approach suggests that u and D should be functions of population density S as well as of time and space. Gurney and Nisbet (1975), Gurtin and MacCamy (1977), and Shigesada and Teramoto (1978) have developed this approach in the problem of dispersing populations. The details will be seen in Sects. 5.6 and 10.5. Thus, the diffusion equation becomes nonlinear in S, and one must expect to encounter considerable mathematical difficulty in solving it. (By a "nonlinear equation" we mean that at least one term contains the dependent variable more than once as a product or division.)

(v) Nonuniformity of the Use of Space and Time.   Animals utilize spatial selectivity in their movement. Dogs cannot climb trees. Moles dig around rocks as they progress. The restriction of "territory" also exists, such that one individual may not cross the marked path of another. In terms of a random walk, not all of the lattice points of space can be utilized. Thus, we should perhaps deal with the problem in terms of a random walk with "forbidden" lattice points or of restricted walks (Barber and Ninham, 1970).

A similar restriction can also be applied to time. The utilization of time by animals is not uniform. They do not always move around, but sometimes rest; we should regard such periods as "time standing still."

It is possible to consider other factors ad infinitum; however, let us discuss a useful model of biodiffusion that takes into account at least some aspects of the above points.

5.3.2   Patlak's Model

Patlak (1953a) developed an extended version of the random walk model, accounting for correlation between successive steps, nonisotropic environment (nonhomogeneity), and external forces. Furthermore, the model assumes that the speed c and time interval τ of the step of each particle are not constant; rather, each possesses some probability distribution.

Let $S(x, t)$ be the particle concentration. We will calculate the flux $J_x(x_0, t)$ at point x_0. This flux is given by the difference between the number of particles that pass x_0 from left to right and the numbers that pass from right to left in a unit time interval. In other words, it is the difference between the number of particles starting from points $x' < x_0$ that move in the positive direction and pass x_0, and the number of particles starting from points

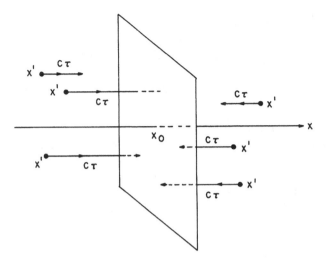

FIGURE 5.4. Patlak's model of a generalized random walk.

$x' > x_0$ that move in a negative direction and pass x_0, in a unit of time (Fig. 5.4). No matter where the particles come from, only those that satisfy the condition $|x_0 - x'| \equiv r \leq c\tau$ contribute to the flux (Fig. 5.4), which is thus given by

$$J_x(x_0, t) = \int_0^\infty dc \int_0^\infty d\tau \int_0^{c\tau} dr \frac{q^+(c, \tau; x', t)}{\tau} S(x', t)$$

$$- \int_0^\infty dc \int_0^\infty d\tau \int_0^{c\tau} dr \frac{q^-(c, \tau; x', t)}{\tau} S(x', t). \qquad (5.32)$$

In (5.32), q^+/τ is the probability density that a particle located at x' at time t moves to the right with step speed c and step time τ in a unit time, and likewise q^-/τ expresses the probability density of movement to the left.

If the integrands of (5.32) are expanded about x_0 in a Taylor series and only first-order terms are retained, the result is

$$q^+(c, \tau; x', t)S(x', t) = q^+(c, \tau; x_0, t)S(x_0, t) - r\frac{\partial}{\partial x}(q^+ S)_0,$$

$$q^-(c, \tau; x', t)S(x', t) = q^-(c, \tau; x_0, t)S(x_0, t) - r\frac{\partial}{\partial x}(q^- S)_0, \qquad (5.33)$$

where $(.)_0$ denotes evaluation at point x_0.

Instead of q^+ and q^-, let us introduce the probability densities ω^+, ω^- that a particle will change its direction at x_0 and t and move, respectively, to the right or left with speed c and step time τ; according to Patlak, the following relations can be established:

$$q^+(c, \tau; x_0, t) = \frac{\tau\omega^+(c, \tau; x_0, t)}{\int_0^\infty dc \int_0^\infty d\tau\, \tau\omega^+(c, \tau; x_0, t)} = \frac{\tau\omega^+}{\bar{\tau}}, \tag{5.34}$$

$$q^-(c, \tau; x_0, t) = \frac{\tau\omega^-(c, \tau; x_0, t)}{\int_0^\infty dc \int_0^\infty d\tau\, \tau\omega^-(c, \tau; x_0, t)} = \frac{\tau\omega^-}{\bar{\tau}}, \tag{5.35}$$

where the overbar denotes averaging.

Rewriting (5.32) with the use of (5.33), (5.34), and (5.35), we obtain

$$J_x(x_0, t) = \left\{ \int_0^\infty dc \int_0^\infty d\tau \frac{c\tau}{\bar{\tau}}(\omega^+ - \omega^-) \right\} S(x_0, t)$$

$$- \frac{\partial}{\partial x} \left\{ \int_0^\infty dc \int_0^\infty d\tau \frac{c^2\tau^2}{2\bar{\tau}}(\omega^+ + \omega^-) S(x, t) \right\}_0 \tag{5.36}$$

or

$$J(x_0, t) = \frac{\overline{(c\tau)}_d}{\bar{\tau}_0} S(x_0, t) - \frac{\partial}{\partial x} \left\{ \frac{1}{2} \frac{\overline{c^2\tau^2}}{\bar{\tau}} S(x, t) \right\}_0. \tag{5.37}$$

In (5.37), averaged quantities are in general functions of time and space. The quantity $\overline{(c\tau)}_d$ represents the difference between the value of $\overline{c\tau}$ for a particle moving to the right and that to the left; Patlak further decomposes this to

$$\overline{(c\tau)}_d = \overline{(c\tau)}_1 + \beta(x, t)\overline{(c\tau)}. \tag{5.38}$$

The term $\overline{(c\tau)}_1$ is the difference of the average step length due to movement in the positive direction $\overline{(c\tau)}^+$ and the step length averaged over both directions $\overline{c\tau}$. The term $\beta(x, t)$ represents the effect of increased step length in a specific direction due to, for example, external forces; it represents the degree of nonsymmetrical motion due to such external forces as stimuli or persistence of motion of the organism in a specific direction, resulting in an animal changing to one direction more often than another. In the derivation of the equation, it should be noted that the restrictions $\overline{(c\tau)}_1 \ll \overline{(c\tau)}$ and $|\beta| \le 1$ have been postulated.

Using (5.38) to rewrite the flux equation (5.37), we can obtain the diffusion equation

$$\frac{\partial S}{\partial t} = -\frac{\partial}{\partial x}\left[\left\{ \frac{\overline{(c\tau)}_1}{\bar{\tau}} + \frac{\beta(x, t)\overline{(c\tau)}}{\bar{\tau}} \right\} S \right] + \frac{\partial^2}{\partial x^2}\left(\frac{1}{2}\frac{\overline{c^2\tau^2}}{\bar{\tau}} S \right), \tag{5.39}$$

where all of the terms $\overline{(c\tau)}_1$, $\bar{\tau}$, $\overline{(c\tau)}$, $\overline{(c^2\tau^2)}_1$ are, in general, functions of x, t, and S (and the concentration of other species if interference occurs).

Patlak furthermore decomposed $\beta(x, t)$ into a term representing directional persistence and a term representing the effect of external forces; however, the resulting expressions are extremely complicated and apparently have little application without further reduction.

At this point, if the forms

$$\frac{(\overline{c\tau})_1}{\overline{\tau}} + \frac{\beta(x,t)(\overline{c\tau})}{\overline{\tau}} \equiv u_e(x,t),$$

$$\frac{1}{2}\frac{\overline{c^2\tau^2}}{\overline{\tau}} \equiv D(x,t)$$

(5.40)

are defined, then (5.39) takes the form

$$\frac{\partial S}{\partial t} = -\frac{\partial}{\partial x}(u_e S) + \frac{\partial^2}{\partial x^2}(DS).$$

(5.41)

If this is compared with the equation of advection and diffusion of Chap. 2,

$$\frac{\partial S}{\partial t} = -\frac{\partial}{\partial x}(uS) + \frac{\partial}{\partial x}\left(D\frac{\partial S}{\partial x}\right),$$

(5.42)

it is noticed that the expression of the "diffusion term" is somewhat different. A bit of manipulation of (5.41) leads to

$$\frac{\partial S}{\partial t} = -\frac{\partial}{\partial x}\left\{\left(u_e - \frac{\partial D}{\partial x}\right)S\right\} + \frac{\partial}{\partial x}\left(D\frac{\partial S}{\partial x}\right),$$

(5.43)

which, with the definition

$$u_e - \frac{\partial D}{\partial x} \equiv u,$$

(5.44)

has the same form as the equation of advection and diffusion. Of course, if the diffusivity D is not dependent on space, (5.41) and (5.42) take the same form with $u_e = u$. As a model for interpretation of animal taxis and kinesis, (5.41) seems more logical than (5.42) (Patlak, 1953a, b; Skellam, 1973). See also the Note of Sect. 5.4.

Turchin (1991) applied Patlak's framework to dispersal in foraging insects. He defined a "residence index" that reflects equilibrium distributions of population density, given observations of the spatial variation of movement characteristics such as duration and speed of movements. Observed distributions were well predicted by this index in case studies using four different insect species. Grünbaum (1998) developed analogous statistics representing the travel time between low-resource and high-resource parts of the habitat and the relative densities of foragers in the two condition types.

5.3.3 The Fokker–Planck Equations

Patlak's method may be considered an extension of the diffusion theory of gas kinetics to organisms. A more formal but elegant method is the derivation of the Fokker–Planck equation used in the analysis of Brownian motion.

Brownian motion, the endless irregular motion of pollen particles in liquid observed with the aid of a microscope and described by the English botanist Brown (1828), attracted the interest of many eminent physicists such as Einstein (1905) and Smoluchowski (1916) and has furthermore enjoyed considerable favor with mathematicians. During the course of research on Brownian motion, not only random walk theory, but also such important fields as random processes, random noise, spectral analysis, and stochastic equations were developed. Certainly Brown could not have dreamed that the motion of pollen in a single drop of water provided the stimulus for such achievements. For the mathematical treatment of Brownian motion, consult Wax (1954), Nelson (1967), and Hida (1975).

Let $S(x', t)$ be the density of a biotic population at point x' and time t. The concentration at x a short time Δt later, i.e., $S(x, t + \Delta t)$, can be calculated by multiplying the transitional probability that a population at x' at time t will move to x in time Δt by the concentration at x' and t and summing over all values of x'.

Thus, if $\phi(x', t; x, t + \Delta t)\, dx'$ is the probability that a population between points x' and $x' + dx'$ at time t will move to x by time $t + \Delta t$ (Fig. 5.5), we can write

$$S(x, t + \Delta t) = \int_{-\infty}^{\infty} S(x', t)\phi(x', t; x, t + \Delta t)\, dx'. \qquad (5.45)$$

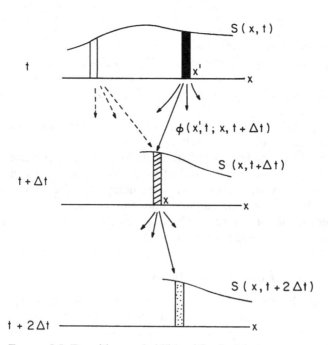

FIGURE 5.5. Transition probabilities (ϕ) of a Markov process. $S(x, t)$: concentration distribution at point x and time t.

Clearly, an individual that was at x' at time t will move to some point x by time $t + \Delta t$, so that

$$\int_{-\infty}^{\infty} \phi(x', t; x, t + \Delta t)\, dx = 1. \tag{5.46}$$

Note that in (5.45) the probability that a population should move from its state at t to its state at $t + \Delta t$ has been assumed to be independent of the states previous to time t. Such a process determined only by a "one-generational dependency" is known as a *Markov process*.

To obtain a partial differential equation for $S(x, t)$ from (5.45), the method of Srinivasan and Vasudevan (1971) and Skellam (1973) may be employed. [Readers who are not interested in the derivation itself may skip to Eq. (5.58).] First, the characteristic function $\Pi(x', t; 0, t + \Delta t)$ of $\phi(x', t; x, t + \Delta t)$ is defined:

$$\Pi(x', t; 0, t + \Delta t) \equiv \int_{-\infty}^{\infty} \phi(x', t; x, t + \Delta t) e^{i\theta(x - x')}\, dx, \tag{5.47}$$

$$\phi(x', t; t + \Delta t) \equiv \frac{1}{2}\pi \int_{-\infty}^{\infty} \Pi(x', t; 0, t + \Delta t) e^{-i\theta(x - x')}\, d\theta. \tag{5.48}$$

Equations (5.47) and (5.48) state that ϕ and Π are the Fourier transforms of each other. Now we shall expand $e^{i\theta(x - x')}$ in terms of its argument;

$$e^{i\theta(x - x')} = 1 + i\theta(x - x') + \frac{(i\theta)^2}{2!}(x - x')^2 + \cdots. \tag{5.49}$$

The substitution of (5.49) into (5.47) provides

$$\Pi(x', t; 0, t + \Delta t) = \int_{-\infty}^{\infty} \phi(x', t; x, t + \Delta t)\, dx$$

$$+ i\theta \int_{-\infty}^{\infty} (x - x')\phi\, dx + \frac{(i\theta)^2}{2!} \int_{-\infty}^{\infty} (x - x')^2 \phi\, dx + \cdots. \tag{5.50}$$

We then define the following:

$$\int_{-\infty}^{\infty} (x - x')^j \phi(x', t; x, t + \Delta t)\, dx \equiv m_j(x', t, \Delta t). \tag{5.51}$$

Equation (5.50) is rewritten with the aid of (5.46) and (5.51) to give

$$\Pi = 1 + i\theta m_1 + \frac{(i\theta)^2}{2!} m_2 + \cdots, \tag{5.52}$$

146   Akira Okubo and Daniel Grünbaum

which is then inserted into the integrand of (5.48):

$$\phi(x',t;x,t+\Delta t) = \frac{1}{2}\pi \int\limits_{-\infty}^{\infty} \left(1 + i\theta m_1 + \frac{(i\theta)^2}{2!}m_2 + \cdots\right) e^{-i\theta(x-x')}\, d\theta. \quad (5.53)$$

Equation (5.53) is substituted into (5.45), resulting in

$$S(x,t+\Delta t) = \frac{1}{2}\pi \int\limits_{-\infty}^{\infty}\int\limits_{-\infty}^{\infty} dx'\, d\theta \left(1 + i\theta m_1 + \frac{(i\theta)^2}{2!}m_2 + \cdots\right) e^{-i\theta(x-x')} S(x',t)$$

$$= \int\limits_{-\infty}^{\infty} dx'\, S(x',t) \left\{ \frac{1}{2}\pi \int\limits_{-\infty}^{\infty} e^{-i\theta(x-x')}\, d\theta + \frac{m_1}{2\pi} \int\limits_{-\infty}^{\infty} i\theta e^{-i\theta(x-x')}\, d\theta \right.$$

$$\left. + \frac{m_2}{2\pi} \int\limits_{-\infty}^{\infty} \frac{(i\theta)^2}{2!} e^{-i\theta(x-x')}\, d\theta + \cdots \right\}$$

$$= \int\limits_{-\infty}^{\infty} dx'\, S(x',t) \left\{ \delta(x-x') - \frac{\partial}{\partial x}\delta(x-x')m_1(x',t,\Delta t) \right.$$

$$\left. + \frac{1}{2}\frac{\partial^2}{\partial x^2}\delta(x-x')m_2(x',t,\Delta t) + \cdots \right\}$$

$$= S(x,t) - \frac{\partial}{\partial x}\{m_1(x,t,\Delta t)S(x,t)\}$$

$$+ \frac{\partial^2}{\partial x^2}\left\{\frac{1}{2}m_2(x,t,\Delta t)S(x,t)\right\} + \cdots. \quad (5.54)$$

In the preceding calculations, the relation

$$\frac{1}{2}\pi \int\limits_{-\infty}^{\infty} e^{-i\theta(x-x')}(i\theta)^s\, d\theta = -\left(\frac{\partial}{\partial x}\right)\delta(x-x') \quad (5.55)$$

was used. Here $\delta(x-x')$ is the Dirac delta function: It is defined such that for $x \neq x'$, δ vanishes; for $x = x'$, δ becomes infinitely large; and such that $\int_{-\infty}^{\infty} \delta(x-x')S(x',t)\, dx' = S(x,t)$. Furthermore, we expand

$$S(x,t+\Delta t) = S(x,t) + \Delta t\left(\frac{\partial S}{\partial t}\right) + \cdots. \quad (5.56)$$

By the use of this expansion, we rewrite (5.54) and take the limit as $\Delta t \to 0$; we obtain

$$\frac{\partial S}{\partial t} = -\frac{\partial}{\partial x}\{M_1(x,t)S\} + \frac{\partial^2}{\partial x^2}\left\{\frac{1}{2}M_2(x,t)S\right\}, \quad (5.57)$$

where

$$M_1(x, t) = \lim_{\Delta t \to 0} \frac{m_1(x, t, \Delta t)}{\Delta t}$$

$$= \lim_{\Delta t \to 0} \frac{1}{\Delta t} \int_{-\infty}^{\infty} (x'' - x)\phi(x, t; x'', t + \Delta t)\, dx'', \qquad (5.58)$$

$$M_2(x, t) = \lim_{\Delta t \to 0} \frac{m_2(x, t, \Delta t)}{\Delta t}$$

$$= \lim_{\Delta t \to 0} \frac{1}{\Delta t} \int_{-\infty}^{\infty} (x'' - x)^2 \phi(x, t; x'', t + \Delta t)\, dx'', \qquad (5.59)$$

and it assumed that higher-order terms

$$M_j(x, t) = \lim_{\Delta t \to 0} \frac{m_j(x, t, \Delta t)}{\Delta t} \quad (j \geq 3) \qquad (5.60)$$

go to zero in the limit. (This implies that in a small time interval Δt, factors that change the location of an organism must also be small.) Equation (5.57) is the *Fokker–Planck equation* or *forward Kolmogorov equation*.

Skellam (1973) defined $M_1/M_2 = (1/2)\partial/\partial x(\ln \gamma)$ and modified (5.57) to

$$\frac{\partial S}{\partial t} = \frac{1}{2}\frac{\partial}{\partial x}\left\{\gamma \frac{\partial}{\partial x}\left(\frac{M_2}{\gamma}S\right)\right\}, \qquad (5.61)$$

in which he called $M_2 S \gamma^{-1} \equiv \Gamma$ the "dynamic level." Γ is a more meaningful quantity than the concentration S. The reason for this is that the biological flux can be expressed in terms of Γ as $J_x = -(1/2)\gamma\partial\Gamma/\partial x$, implying that diffusion takes place from high values of density Γ to low values.

In Eq. (5.57), M_1 is the first moment, and M_2 is the second moment of displacement. They correspond, respectively, to the terms $(\overline{c\tau})_1/\overline{\tau} + \beta(x, t)\overline{c\tau}/\overline{\tau}$ and $\overline{c^2\tau^2}/\overline{\tau}$ in (5.39) of Patlak. We depend on the equation of motion of an organism to provide an independent evaluation of these moments. In one dimension,

$$m\frac{d^2 x}{dt^2} = F. \qquad (5.62)$$

If (5.62) is integrated twice over time, the following is obtained:

$$x(t + \Delta t) - x(t) = u_0 \Delta t + \int_t^{t+\Delta t} dt' \int_0^{t'} (F/m)\, dt'', \qquad (5.63)$$

where u_0 is the velocity at time $t = 0$. As discussed in Sect. 5.2, F includes random components, so that velocity and displacement are also random quantities. By dividing the average net displacement over a time interval Δt

by Δt and taking the limit as $\Delta t \to 0$, we obtain M_1:

$$M_1(x, t) = \lim_{\Delta t \to 0} \overline{\frac{x(t + \Delta t) - x(t)}{\Delta t}} = \bar{u}_0 + \lim_{\Delta t \to 0} \frac{1}{\Delta t} \int_t^{t+\Delta t} dt' \int_0^{t'} \overline{(F/m)} \, dt''. \quad (5.64)$$

Likewise,

$$M_2(x, t) = \lim_{\Delta t \to 0} \overline{\frac{\{x(t + \Delta t) - x(t)\}^2}{\Delta t}}$$

$$= \lim_{\Delta t \to 0} \frac{1}{\Delta t} \overline{\left\{ \int_t^{t+\Delta t} dt' \int_0^{t'} (F/m) \, dt'' \right\}^2}. \quad (5.65)$$

However, F in general depends not only on t but also on x and dx/dt, so that (5.64) and (5.65) provide nothing more than formal expressions for M_1 and M_2. To evaluate these parameters, it is necessary to determine the form of F in the problem of concern and to integrate the equation of motion. Thus far, limited success has been obtained only for certain special cases (Nelson, 1967; Soong, 1973; Sture et al., 1974).

A well-known example is provided by the Langevin equation (Langevin, 1908; Hori, 1977) describing Brownian motion under the influence of a random external force:

$$m \frac{d^2 x}{dt^2} = -mk \frac{dx}{dt} + A(t), \quad (5.66)$$

where k is a resistance coefficient, and $A(t)$ is a random force. Equation (5.66) can easily be integrated for x, and thus M_1 and M_2 can be directly calculated (Uhlenbeck and Ornstein, 1930); however, the use of a quasi-equilibrium assumption, as discussed in Sect. 5.2, simplifies the evaluation. Thus, we have

$$0 = -mk \frac{dx}{dt} + A(t). \quad (5.67)$$

Integration of this equation over a small interval of time yields

$$x(t + \Delta t) = x(t) + \int_t^{t+\Delta t} A(t')/mk \, dt'. \quad (5.68)$$

Accordingly,

$$M_1(x, t) = \lim_{\Delta t \to 0} \overline{\frac{x(t + \Delta t) - x(t)}{\Delta t}} = \lim_{\Delta t \to 0} \frac{1}{\Delta t} \int_t^{t+\Delta t} \bar{A}(t')/mk \, dt' = 0 \quad (5.69)$$

since the average of a random force is zero. Similarly,

$$M_2(x, t) = \lim_{\Delta t \to 0} \frac{\overline{\{x(t + \Delta t) - x(t)\}^2}}{\Delta t}$$

$$= \lim_{\Delta t \to 0} \frac{1}{\Delta t} \iint_{t}^{t+\Delta t} \frac{\overline{(A(t')A(t''))}}{m^2 k^2} \, dt' \, dt''. \tag{5.70}$$

We now make the assumption that the random force is self-correlated only for times t', t'' whose interval is very small; this provides essentially the same model as that of the random walk. It is found that

$$M_2 = 2\overline{A^2}T/m^2k^2, \tag{5.71}$$

where $\overline{A^2}$ is the power of the random force and T is a characteristic time of correlation. As a result of (5.69) and (5.71), the Fokker–Planck equation (5.57) reduces to the Fickian diffusion equation:

$$\frac{\partial S}{\partial t} = D \frac{\partial^2 S}{\partial x^2} \quad (D \equiv \overline{A^2}Tm^2k^2). \tag{5.72}$$

So far we have dealt with the density function of a biotic population at a point x and time t. We can extend the treatment to the density function in phase space, i.e., space of position x and velocity v of a species, by considering $P(x, v, t)$ to be the density of population at point x with velocity v at time t.[1]

For this phase-space process, the Fokker–Planck equation has been generalized to (Chandrasekhar, 1943; Wang and Uhlenbeck, 1945; Soong, 1973)

$$\frac{\partial P}{\partial t} = -v \frac{\partial P}{\partial x} - \frac{\partial}{\partial v}(A_v P) + \frac{1}{2} \frac{\partial^2}{\partial x^2}(A_{xx} P)$$

$$+ \frac{1}{2} \frac{\partial^2}{\partial x \partial v}\{(A_{xv} + A_{vx})P\} + \frac{1}{2} \frac{\partial^2}{\partial v^2}(A_{vv} P), \tag{5.73}$$

where

$$A_v \equiv \lim_{\Delta t \to 0} \frac{\overline{v(t + \Delta t) - v(t)}}{\Delta t},$$

$$A_{xx} \equiv \lim_{\Delta t \to 0} \frac{\overline{\{x(t + \Delta t) - x(t)\}^2}}{\Delta t},$$

$$A_{xv} \equiv A_{vx} = \lim_{\Delta t \to 0} \frac{\overline{\{(x(t + \Delta t) - x(t))(v(t + \Delta t) - v(t))\}}}{\Delta t},$$

$$A_{vv} \equiv \lim_{\Delta t \to 0} \frac{\overline{\{v(t + \Delta t) - v(t)\}^2}}{\Delta t},$$

---

[1] A Fokker–Planck equation with two variables (position and velocity) is often called *Kramer's equation* (Riskin, 1984).

and A_v, A_{xx}, A_{xv}, and A_{vv} can be evaluated from the following set of equations:

$$\frac{dx}{dt} = v,$$

$$\frac{dv}{dt} = \frac{F}{m}.$$

For cases of practical interest, we divide F/m into two parts:

$$\frac{dv}{dt} = \frac{F}{m} = a(x, v, t) + b(x, v, t)\lambda(t), \tag{5.74}$$

where a and b are deterministic (nonrandom) functions and λ is a random function of the so-called white-noise-type such that

$$\overline{\lambda(t)} = 0,$$

$$\overline{\lambda(t)\lambda(t + \Delta t)} = \delta(t).$$

We then obtain (Arnold, 1974; Soong, 1973)

$$A_v = a(x, v, t),$$

$$A_{xx} = 0,$$

$$A_{xv} = A_{vx} = 0,$$

$$A_{vv} = b^2(x, v, t),$$

where Ito's integral (Ito, 1944, 1946, 1951) is used to evaluate the statistical characteristics. The Fokker–Planck equation (5.73) is then written as

$$\frac{\partial P}{\partial t} = -v\frac{\partial P}{\partial x} - \frac{\partial}{\partial v}\{a(x, v, t)P\} + \frac{1}{2}\frac{\partial^2}{\partial v^2}\{b^2(x, v, t)P\}. \tag{5.75}$$

The spatial density function of a biotic population, $S(x, t)$, can be calculated by integrating $P(x, u, t)$ over u:

$$S(x, t) = \int_{-\infty}^{\infty} P(x, u, t)\, du. \tag{5.76}$$

A controversy appeared concerning the integration of the stochastic differential equation of the type of (5.74). The so-called Stratonovich calculus (Stratonovich, 1966), in which the usual rules in calculus continue to apply, and the so-called Ito calculus, in which the rules are changed, have been considered for the integration procedure. If we use Stratonovich's integral, A_v appears as $a(x, v, t) + (1/4)\partial b^2(x, v, t)/\partial v$ (Goel and Richter-Dyn, 1974). Mathematically speaking, the Ito calculus is more fundamental and general than the Stratonovich calculus (van Kampen, 1981). The divergence arises only when the white noise is multiplied by a function of the solution of the equation such as $b(x, v, t)$ in (5.74). Otherwise, the Ito and Stratonovich interpretations of the solution of the differential equation coincide; this is the case where b is a function of time only. Smith (1983) shows that a digital simulation of a noise-induced phase transition using an algorithm consistent with the Ito calculus is in agreement with the predictions of that theory, whereas experiments with an analogue simulator yielded results in agreement with predictions of the Stratonovich theory. The reader may find Mortensen's (1969) and Gray and Coughey's (1965) papers helpful.

Note.   From Chap. 6 on, only a few of the biodiffusion equations developed in this chapter will be used, and we will rely on a most simple form of the diffusion equation. Some explanation is in order lest the reader feel swindled. The problem of biodiffusion in ecological systems is sufficiently complex, especially when intra- and interspecific relations are included, such that the complicated biodiffusion equations of this chapter impede an analytical development and tend to mask basic insight into the problem. When one is faced with a new problem, the appropriate method is to investigate the results obtained from the simplest possible model and then gradually develop a more realistic model. On the one hand, an immediate saunter into a jungle of messy equations in which one loses sight of the goal is a waste of energy. On the other hand, we shall surely progress beyond the simplest models some day, and it is at that time that the material of this chapter will prove its value.

5.4   Application to Taxis

The application of biodiffusion equations to problems involving taxis has met with some success as regards lower organisms.

Keller and Segel (1971a) have constructed a diffusion model of chemotaxis in bacteria (*Escherichia coli*). The rate of random turning of bacteria depends on the concentrations of chemical attractants: klinokinesis. Herein we suppose that the step length of the bacterial random walk is ℓ and that the frequency of changing step direction f depends on the concentration $C(x)$ of chemical attractant; $f = 1/T = f\{C(x)\}$. Since C varies with position, so does f. We also assume that the size of the bacterial receptive organ is much smaller than ℓ, but Keller and Segel have analyzed a more general case.

For this model of klinokinesis, we can put $(\overline{c\tau})_1 = 0$, $\beta = 0$, $\overline{c^2\tau^2}/\overline{\tau} = \ell^2/\tau = \ell^2 f(C)$ in Patlak's equation (5.39), so that (5.39) takes the form

$$\frac{\partial S}{\partial t} = \frac{\partial^2}{\partial x^2}\left\{\frac{1}{2}\ell^2 f(C(x))S\right\} = -\frac{\partial}{\partial x}\left\{\chi\frac{C}{dx}S\right\} + \frac{\partial}{\partial x}\left(\mu\frac{\partial S}{\partial x}\right), \qquad (5.77)$$

where we define

$$\chi \equiv -\frac{1}{2}\ell^2\, df/dC, \qquad (5.78)$$

$$\mu \equiv \frac{1}{2}\ell^2 f(C). \qquad (5.79)$$

In the above, χ is called the chemotactic coefficient, and μ is called the motility (μ is in fact the diffusivity).

The chemotaxis of bacteria is thus determined by diffusion, delineated by the motility (diffusivity), which depends on the concentration of chemical substance, and advection, which depends on the gradient of the concentration of chemical substance. When the frequency of direction change of the bacteria increases with increasing concentration, advection occurs in the direction of decreasing concentration; and when the frequency decreases with increasing concentration, advection occurs in the direction of increasing concentration.

Keller and Segel (1971b) were able to use this model of chemotaxis to explain the phenomenon of wavelike propagation of bands of certain species of bacteria under the influence of a chemical substrate. For these travelling bands to develop, the chemotactic coefficient per unit concentration must be greater than the motility.

More complex models for collective motions of chemotactic cells have been developed (Segel and Stoeckly, 1972; Raman, 1977; among others). In these models the advection–diffusion equation for the cell density is coupled with the diffusion equation of the chemical attractant concentration. Also, Oosawa and Nakaoka (1977) have presented a model to describe the tactic behavior of microorganisms. A unique aspect of this model is that individual cells are considered to be particles having *internal state variables*. The tactic behavior is determined by the internal state of those cells, and environmental conditions such as temperature gradients influence the behavior of a cell through its internal state. On the basis of this concept, Oosawa and Nakaoka obtained a method for relating the tactic behavior of microorganisms to changes of the environment with time. The presence of internal state dynamics that respond to the environment, and that subsequently modulate movement behavior, is of fundamental importance in understanding how organisms select their environments. Analytical and numerical methods are only recently being developed (see Sect. 5.7), and the experimental basis for characterizing internal state dynamics remains sparse for all but a few organisms. Internal states represent a historical component of behavior that can complicate the classification of behaviors and the determination of whether aggregation will result (Doucet and Drost, 1985).

Kareiva and Odell (1987) developed a detailed model of ladybird beetles (*Coccinella septempunctata*) foraging on aphids. They based their model on the observation that beetles change their walking direction more frequently when their stomachs are full after consuming prey, but hold their walking speed relatively constant. Kareiva and Odell formulated a model in which local prey density determines consumption rates of prey and gut fullness of predators. Predator behavior is assumed to depend on gut fullness. A detailed set of experiments provided parameters for internal state dynamics and movement behaviors. This model gave accurate predictions of predator and prey distributions in a one-dimensional experimental array of goldenrod plants. Random walks in which movements of cells are mediated by internal states play important roles in human health and have been extensively studied both experimentally and theoretically. For example, detailed analyses of cytomechanics and receptor mediated behavior are presented by Tranquillo and Lauffenburger (1987) and their colleagues. They present a unified model of persistent and biased random walks, in which leukocytes are represented as integrated systems that sense and respond to noisy receptor signals. See also Tranquillo et al. (1988) and Tranquillo and Alt (1996) for more recent developments.

The fact that animal population distributions exhibit aggregation in certain places in the presence of kinesis alone, without directional taxis, can be demonstrated with the equation of Patlak (5.39) and the Fokker–Planck equation (5.57). For example, if in (5.39) we set $(\overline{c\tau})_1 = 0$ and $\beta = 0$,

$$\frac{\partial S}{\partial t} = \frac{\partial^2}{\partial x^2}\left(\frac{1}{2}\frac{\overline{c^2\tau^2}}{\overline{\tau}}S\right). \tag{5.80}$$

Let us place an animal population in a finite space and supply a nonuniform stimulus. The animals react by beginning a random movement under the influence of kinesis. After a aufficiently large time, we may suppose that the concentration reaches some steady state. Setting the left-hand side of Eq. (5.80) to zero, it is found that

$$\frac{d}{dx}\left(\frac{1}{2}\frac{\overline{c^2\tau^2}}{\overline{\tau}}S\right) = A \quad \text{(constant)}. \tag{5.81}$$

Equation (5.81) is an expression for flux. Since in a closed space the flux at the boundaries must vanish, we obtain $A = 0$. Integrating (5.81) once more,

$$\left(\frac{1}{2}\overline{c^2\tau^2}/\overline{\tau}\right)S = B \quad \text{(constant)} \tag{5.82}$$

or

$$S = 2B\overline{\tau}/\overline{c^2\tau^2}. \tag{5.83}$$

Thus, the steady-state distribution is proportional to $\overline{\tau}$ and inversely proportional to $\overline{c^2\tau^2} = \ell^2$. In the case of orthokinesis, $\overline{\tau}$ is constant; (5.83) can be written as $S = 2B/(\overline{c^2}\overline{\tau})$, from which one can see that animals concentrate in regions where their step speed is small, i.e., where their motion becomes slow. In the case of klinokinesis, $\overline{c}$ becomes constant, so that concentration occurs where $\overline{\tau}$ is small, i.e., where the frequency of direction change is large. In the diffusion equation with zero advection, $\partial S/\partial t = \partial/\partial x(D\partial S/\partial x)$, the steady-state solution implies that S must be constant everywhere, even if D is allowed to vary in time and space, in contrast to (5.80).

The above discussion applies to the steady-state distribution. The nature of the unsteady distribution prior to this is more complicated; unless the effective advection due to kinesis, i.e., $-\partial/\partial x(\overline{c^2\tau^2}/2\overline{\tau})$, is somewhat greater

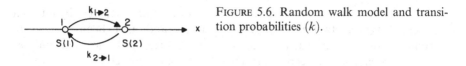

than the diffusivity $\overline{c^2\tau^2}/2\bar{\tau}$, the development of the above-mentioned steady-state distribution of a population released from a point source may require a considerable amount of time (see Sect. 5.5).

Note. Let us return to the random walk model in order to reexamine the difference between Eqs. (5.41) and (5.42) (Fig. 5.6). We consider densities at the points at time t to be $S(1, t)$ and $S(2, t)$. Defining the transition probability that an individual will move from point 1 to point 2 in a small time τ to be $k_{1\to2}$, and likewise denoting the transitional probability from 2 to 1 as $k_{2\to1}$, it is seen that in time τ the population concentration $k_{1\to2}S(1, t) - k_{2\to1}S(2, t)$ moves in the positive x direction from 1 to 2. If the distance between the two points is λ, then the population flux between them is given by

$$\mathbf{J} = \frac{\lambda^2}{\tau}\{k_{1\to2}S(1, t) - k_{2\to1}S(2, t)\}/\lambda. \tag{5.84}$$

The transition probabilities in general depend on the conditions at each of the points and in between; according to the nature of the dependency, the following three cases can be considered (Skellam, 1955, 1973).

1. $k_{1\to2} = k_{2\to1} = k(1; 2)$. The transition probabilities are assumed to be equal, both depending on the average conditions between the two points;

$$\mathbf{J} = \frac{\lambda^2}{\tau}k(1; 2)\frac{S(1, t) - S(2, t)}{\lambda}. \tag{5.85}$$

In the limit as τ and $\lambda \to 0$,

$$\mathbf{J} = -D(x)\frac{\partial S(x, t)}{\partial x}, \tag{5.86}$$

in which $D(x) = \lim_{\lambda, \tau\to0} \lambda^2/\tau k(1; 2)$.

This is essentially the Fickian expression for flux for the case of a spatially varying diffusivity; it leads to (5.42) (apart from advection). As the transition probabilities depend on the average state between the two points ("neutral transition"), this formulation should be appropriate for the diffusion of physical properties and inorganic material and, in some cases, for biological diffusion (taxis), when the flux is always directed from high concentration to low concentration.

2. $k_{1\to2} = k(1), k_{2\to1} = k(2)$. In this case, the transition probabilities depend only on conditions at the point of departure ("repulsive transition");

$$\mathbf{J} = \frac{\lambda^2}{\tau}\frac{k(1)S(1, t) - k(2)S(2, t)}{\lambda}. \tag{5.87}$$

In the limit as τ and $\lambda \to 0$,

$$\mathbf{J} = -\frac{\partial}{\partial x}\{D(x)S(x, t)\}, \tag{5.88}$$

in which $D(x) = \lim_{\lambda, \tau \to 0} \lambda^2/\tau k(1)$.

This also provides an expression for flux in the case of spatially varied diffusivity, but one based on transition probabilities dependent only on the state at the point of departure. If we inspect the contents of the Patlak or Fokker–Planck model leading to (5.41), it should become clear that the model is based on transition probabilities that depend only on the departure point. Thus, the above expression and that of (5.41) are seen to correspond. Certain kinds of animal taxis and dispersal are best handled in terms of "repulsive transition" by which the transition probabilities depend only on departure conditions. In this case the flux may also be written as

$$\mathbf{J} = -\frac{\partial D}{\partial x}S - D\frac{\partial S}{\partial x} \equiv u_d S - D\frac{\partial S}{\partial x}, \tag{5.89}$$

which can be regarded as the Fickian diffusion with an additional effective advection in the direction of decreasing D (the direction of decreasing repulsion).

3. $k_{1 \to 2} = k(2), k_{2 \to 1} = k(1)$.   In this case the transition probabilities depend on conditions at the point of arrival ("attractive transition"):

$$\mathbf{J} = \frac{\lambda^2}{\tau} \frac{k(2)S(1, t) - k(1)S(2, t)}{\lambda}. \tag{5.90}$$

In the limit as τ and $\lambda \to 0$, we can write

$$\mathbf{J} = -D^2(x)\frac{\partial}{\partial x}\left\{\frac{S(x, t)}{D(x)}\right\}, \tag{5.91}$$

in which $D(x) = \lim_{\lambda, \tau \to 0} \lambda^2/\tau k(1)$.

Again an expression for flux with a spatially varying diffusivity is obtained, but one that is applicable to the case where the transition probabilities depend only on the state at the point of arrival. It is thought that some types of animal taxis may be approached in terms of attractive transition. The flux expression applicable to this case may be written as

$$\mathbf{J} = \frac{\partial D}{\partial x}S - D\frac{\partial S}{\partial x} \equiv u_d S - D\frac{\partial S}{\partial x}, \tag{5.92}$$

which can be regarded as Fickian diffusion with an additional effective advection in the direction of increasing D (increasing attraction).

Note that in the cases of repulsive and attractive transition, the flux is not necessarily directed down the concentration gradient. Thus, certain characteristics of animal diffusion and taxis can be explained with the use of these models. Figure 5.7 shows schematically the differences among these three models.

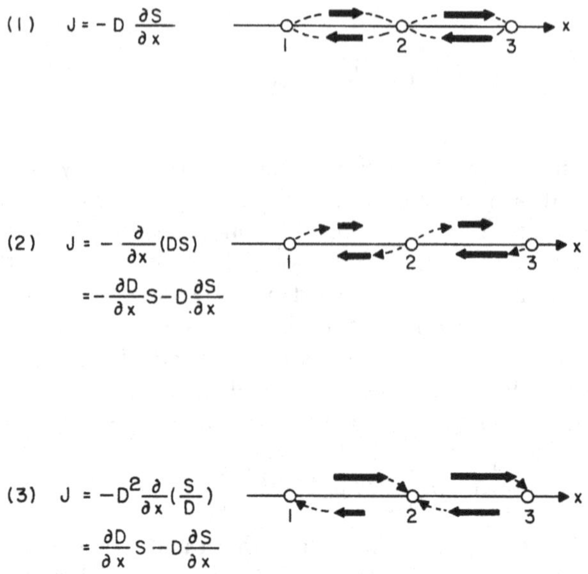

(1) $J = -D \dfrac{\partial S}{\partial x}$

(2) $J = -\dfrac{\partial}{\partial x}(DS)$

$= -\dfrac{\partial D}{\partial x}S - D\dfrac{\partial S}{\partial x}$

(3) $J = -D^2\dfrac{\partial}{\partial x}\left(\dfrac{S}{D}\right)$

$= \dfrac{\partial D}{\partial x}S - D\dfrac{\partial S}{\partial x}$

FIGURE 5.7. Population fluxes associated with spatially varying diffusivity. According to the nature of the dependence of the transition probability on the conditions at each of the points and in between, the expression for the flux differs: (1) neutral transition (Fickian type); (2) repulsive transition; (3) attractive transition.

The tactic behaviors discussed so far have involved responses to environmental variables. An interesting elaboration that relates these behaviors to evolution and population dynamics was developed by Grindrod (1988). He proposed that organisms may directly assess and climb *gradients in fitness* (i.e., instantaneous reproduction rate). Specifically, Grindrod assumes that the population distribution is governed by

$$\frac{\partial S}{\partial t} = \delta \Delta S - (1-\delta)\nabla(\underline{w}S) + SE(S,x,t) - \varepsilon\Delta\underline{w} + \underline{w}$$

$$= \lambda\nabla E(S,x,t). \tag{5.93}$$

In (5.93), δ represents the probability of dispersing randomly, $1-\delta$ the probability of dispersing deterministically in response to spatial variation in reproduction rate, and $\underline{w}$ is the average velocity of these deterministic dispersers. λ and ε are constants chosen such that $\underline{w}$ is a local average of the ideal dispersal velocity, with sharp local variations smoothed out. With a reproduction rate E that depends in a nonlinear fashion on the local population, Grindrod shows a variety of population trajectories depending on initial distributions for single species and for two species in competition.

5.5   Simulation of Taxis

Computer simulation provides an approach for stochastic investigation of animal movements. The method requires an abstraction of actual animal motion into certain elements, e.g., speed, direction, activity, and rest periods, and an evaluation of the statistical distribution of each of these processes. For this purpose, the assumed relations may be based on actual data (see Sect. 8.3) or on theoretical considerations. The required distributions and algorithms can then be programmed into a computer to simulate animal motion. The result is then compared with data to test the applicability of the model; if necessary, alterations are made. A well-verified model can then be used for further simulations under various different conditions. Progressing one step beyond, one might, with the addition of other processes associated with population ecology (growth, death, predation, competition, etc.), be able to develop a fairly complete model of an ecosystem.

We will briefly discuss research on animal taxis by means of simulation. Rohlf and Davenport (1969) have used computer simulation to investigate the effects of simple kinesis, orthokinesis, and topotaxis. An individual is placed at the origin of an (x, y)-coordinate system and is moved according to the probabilities associated with the various forms of taxis.

(1) Simple kinesis: The probability of moving in the same direction as previously is $1/2$; the probability of turning to the right or left is $1/4$ in either case; and the probability of moving backward is 0.

(2) Klinokinesis: The probability of changing direction increases (decreases) in proportion to the organism's position along the x-axis.

(3) Orthokinesis: The length of a step increases (decreases) in proportion to the position of the organism along the x-axis.

(4) Topotaxis: A gradient of stimulus is taken in the x direction; when the individual is directed parallel to the y-axis, the probability that the next step will be in the positive x direction is $2/3$, and in the negative direction, $1/3$; when the individual is faced in the x direction, the probabilitiy of a turn is $1/2$.

(5) A model including all of the above is considered. In this fashion a simulation was performed using 500 individuals, supplying each with 100 basic moves. The following results were found:

   (i) On the average, animals show no tendency for displacement in the direction of a stimulus gradient due to klinokinesis alone (see the previous section).

   (ii) Animal density tends to increase in regions where activity is reduced, under the influence of orthokinesis alone.

   (iii) When positive orthokinesis, whereby activity increases in regions of strong stimulus, is combined with negative klinokinesis, whereby the frequency of changing direction increases in regions of weak stimulus, a pronounced aggregation of animals occurs in regions of weak stimulus.

(iv) If positive orthokinesis, negative klinokinesis, and positive topotaxis are combined, animals show a strong tendency to aggregate toward the source of the stimulus. Rohlf and Davenport extended this model to include animal sensory adaptation and performed simulations with the combination of taxis and kinesis.

(v) If sensory adaptation is included together with positive orthokinesis and negative klinokinesis, animals concentrate in regions of strong stimulus even in the absence of topotaxis.

(vi) Positive orthokinesis and positive klinokinesis work together to negate the effect of sensory adaptation.

5.6   Advection–Diffusion Models for Biodiffusion

Neither Patlak's model nor the Fokker–Planck equation for biodiffusion takes into account interference between individual organisms, although the parameters pertinent to the model may be regarded as being dependent on the population density in general. Shigesada and Teramoto (1978) presented a mathematical model of advection and diffusion to explain the spatial distribution of animal populations that are principally controlled by interference between individuals and other environmental conditions. The formulation is based on the assumption that animals move under the influence of the following fundamental forces: (1) a dispersive force associated with random movement of animals; (2) an attractive force, which induces directed movement of animals toward favorable environments; and (3) population pressure due to interference between individual animals. The force associated with population pressure was originally discussed in a quantitative fashion by Morisita (1952, 1971; also see Sect. 6.1).

Thus, Shigesada and Teramoto present their model equation for the one-dimensional case as

$$\frac{\partial S}{\partial t} = \frac{\partial^2}{\partial x^2}[\{\alpha(x) + \beta(x)S\}S] + \frac{\partial}{\partial x}\left\{\frac{\partial \Phi}{\partial x}S\right\}. \tag{5.94}$$

Here $S(x,t)$ is the density of animals; $\alpha(x) + \beta(x)S$ represents the virtual diffusivity, dependent on spatial inhomogeneity and population pressure; and $\Phi(x)$ denotes the potential of the environmental attraction, which induces the advection velocity $u(x) = -\partial \Phi/\partial x$ toward favorable regions. The form of the virtual diffusivity is consistent with Morisita's semiempirical formulation. Shigesada and Teramoto derived Eq. (5.94) from a microscopic model in which individuals perform a biased random walk. The transition probabilities depend on the conditions at the point of departure, i.e., repulsive transition (see Sect. 5.4), so that the diffusive flux can be expressed as (5.88).

The steady-state distribution of animals is obtained by equating the right-hand side of (5.94) to zero and integrating with respect to x:

$$\frac{d}{dx}\{(\alpha + \beta S)S\} + \frac{d\Phi}{dx}S = -J \quad \text{(constant)}, \tag{5.95}$$

where J represents the flux of organisms. When an animal population is placed in a closed region, this flux must vanish. If it is furthermore assumed that α and β are constants, (5.95) can be integrated once more to yield

$$\alpha \ln\{S(x)/S_0\} + 2\beta(S(x) - S_0)(\Phi(x) - \Phi_0) = 0, \tag{5.96}$$

where S_0 and Φ_0 are the corresponding values at $x = 0$.

We shall examine the density distribution given by (5.96). When animals disperse independently of each other ($\beta \equiv 0$), they are distributed according to the relation

$$S(x) = S_0 \exp\{-(\phi - \phi_0)/\alpha\}. \tag{5.97}$$

In other words, organisms tend to aggregate in regions where the environmental potential is low. When interference occurs ($\beta \neq 0$), the aggregative tendency toward favorable environments is counteracted by the force of population pressure, which encourages dispersal; the more animals that are present, the more noticeable this effect may be.

Even though the environmental potential is uniform, spatial variation in the diffusivity and the population pressure effect can produce inhomogeneity in the steady-state density distribution; this situation somewhat resembles that already discussed through Eqs. (5.81)–(5.83). Shigesada and Teramoto proposed experimental procedures to estimate the kinetic parameters $\alpha(x)$, $\beta(x)$, and $\Phi(x)$.

Shigesada et al. (1979) extended their model to the populations of two animal species that have almost the same affinity for the environment and are under the influence of population pressure due to intra- and interspecific interference. The time changes of the population densities $S_1(x, t)$ and $S_2(x, t)$ are given by

$$\frac{\partial S_1}{\partial t} = \frac{\partial^2}{\partial x^2}\{(\alpha_1 + \beta_{11}S_1 + \beta_{12}S_2)S_1\} + \frac{\partial}{\partial x}\left(\gamma_1 \frac{\partial \phi}{\partial x}S_1\right), \tag{5.98}$$

$$\frac{\partial S_2}{\partial t} = \frac{\partial^2}{\partial x^2}\{(\alpha_2 + \beta_{21}S_1 + \beta_{22}S_2)S_2\} + \frac{\partial}{\partial x}\left(\gamma_2 \frac{\partial \phi}{\partial x}S_2\right), \tag{5.99}$$

where $\alpha_i + \beta_{ij}S_j$ ($i, j = 1, 2$) are the virtual diffusivities of the ith species, dependent on spatial inhomogeneity and intra- and interspecific population pressure, and γ_i ($i = 1, 2$) are the coefficients of affinity for the environment.

The steady-state distributions $S_1^*(x)$ and $S_2^*(x)$ are obtained by equating the right-hand sides of (5.98) and (5.99) to zero and integrating with respect to x. When we assume that the fluxes of organisms vanish at the boundaries and α_i, β_{ij}, and γ_i are all constants, the following equations are obtained:

$$\frac{dS_1^*}{dt} = -\gamma_1 \frac{d\phi}{dx} S_1^* \{\alpha_2 + \beta_{21} S_1^* + (2\beta_{22} - \gamma_1^{-1}\gamma_2\beta_{12})S_2^*\}A^{-1}, \qquad (5.100)$$

$$\frac{dS_2^*}{dt} = -\gamma_2 \frac{d\phi}{dx} S_2^* \{\alpha_1 + \beta_{12} S_2^* + (2\beta_{11} - \gamma_1\gamma_2^{-1}\beta_{21})S_1^*\}A^{-1}, \qquad (5.101)$$

where $A = (\alpha_1 + 2\beta_{11} S_1^* + \beta_{21} S_2^*)(\alpha_2 + \beta_{21} S_1^* + 2\beta_{22} S_2^*) - \beta_{12}\beta_{21} S_1^* S_2^* > 0$.

The qualitative features of the solution of (5.100) and (5.101) are best understood by plotting the isoclines given by $dS_1^*/dx = 0$ and $dS_2^*/dx = 0$ in the (S_1^*, S_2^*) plane as is often done in the dynamical equations of two species populations. Thus, Shigesada et al. (1979) showed that (1) at the point of intersection of the phase trajectory with the isocline $dS_1^*/dx = 0$, $S_1^*(x)$ becomes maximum; likewise, at the point of intersection with the isocline $dS_2^*/dx = 0$, $S_2^*(x)$ becomes maximum, and (2) at the point where the trajectory crosses the line $S_1^* = S_2^*$, the population density curves $S_1^*(x)$ and S_2^* as functions of x cross each other. Typical patterns of S_1^* and S_2^* are shown in Fig. 5.8. (It should be remarked that in Fig. 5.8, $d\phi/dx > 0$ is assumed. If $d\phi/dx < 0$, the patterns become those of Fig. 5.8 with the direction of the x-axis reversed.)

An interesting case is seen in Fig. 5.8(b), in which the gradients of two population density curves have opposite signs at the intersecting point. This type of pattern suggests the possibility of "habitat segregation," i.e., two similar species segregate each other in their habitat. As will be shown in Sect. 10.5, this spatial segregation acts to stabilize a system of competing species by relaxing the interspecific competition and hence gives rise to the coexistence of two similar species.

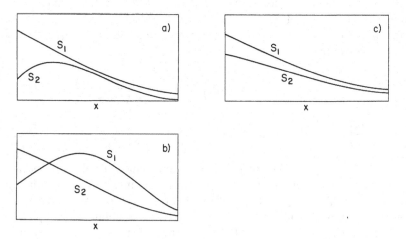

FIGURE 5.8. Typical patterns of habitat segregation of two populations undergoing advection and density-dependent diffusion (from Shigesada et al., 1979).

5.7   Internal State-mediated Taxis

Starting with pioneering works by Keller and Segel (1971a), Oosawa (1977), Alt (1980), Dickinson and Tranquillo (1995), and many others, a large literature has developed that is coming to fruition with modeling methods for describing complex behaviors of individual cells or organisms, statistical techniques for estimating parameters for these models, and mathematical analyses that predict the population distributions resulting from these behaviors. We defer to these original sources for the technical details and historical development. We close this chapter by giving a brief demonstration of how models of individuals undergoing internal state-mediated random may be projected to distributions of populations, over time and space scales that reflect these behaviors' important ecological, evolutionary, and physiological effects.

We consider a one-dimensional velocity jump process (Othmer et al., 1988) in which densities of right- and left-moving individuals are given by

$$\frac{\partial p^+}{\partial t} + V\frac{\partial p^+}{\partial x} + \frac{\partial}{\partial y}(f(y,S)p^+) = \lambda(y)(p^- - p^+),$$

$$\frac{\partial p^-}{\partial t} - V\frac{\partial p^-}{\partial x} + \frac{\partial}{\partial y}(f(y,S)p^-) = \lambda(y)(p^+ - p^-).$$

(5.102)

In (5.102), $p^+(t,x,y)$ represents the density of individuals moving to the right at constant speed V at position x with internal state y at time t. $p^-(t,x,y)$ is the corresponding density for individuals moving to the left. $\lambda(y)$ is the probability per unit time of an individual's turning from one direction to the other, which is assumed to be a function only of the internal state, y. The operator f is the rate of change of an individual's internal state when it is exposed to an environmental variable $S(x)$. For instance, in the example below, f represents the stochastic changes in the fraction of bound receptors, and S is the concentration of an attractant binding to those receptors (see Dickinson and Tranquillo, 1995; and Grünbaum, 2000, for details and additional examples).

The total population density of individuals at a point in time and space is

$$P(t,x) = \int_Y (p^+(t,x,y) + p^-(t,x,y))\,dy.$$

(5.103)

A *taxis equation* approximating the diffusion and advection of this population has the form

$$\frac{\partial P}{\partial t} = \frac{\partial}{\partial x}\left(D(S)\frac{\partial P}{\partial x} - \chi(S)P\right).$$

(5.104)

Equation (5.104) implicitly contains the details of individuals' internal responses to the environmental variable, S, and their behavioral responses to internal state, and how individuals are distributed over the various internal states and directions. However, these details no longer appear in (5.104). Instead, we derive expressions for the diffusion coefficient, $D(S)$, and taxis coefficient, $\chi(S)$, that approximate the complete dynamics but retain the concise form of (5.104).

We assume that spatial gradients are smooth, i.e., that an individual's environment changes only slightly between turns. Specifically, we assume that this smoothness is characterized by a scaling parameter, $\varepsilon = V/\lambda_0 L$. Here, L is the length scale over which the attractant changes significantly (e.g., $L = S/\dfrac{\partial S}{\partial x}$), and λ_0 is a "typical" turning rate. It is shown in Grünbaum (2000) that (5.104) gives good approximations for population distributions resulting from (5.102) when $\varepsilon \ll 1$ and when D and χ are taken to be

$$D(S) = \frac{V^2}{\lambda_0} \int_Y g(y, S)\, dy,$$

$$\chi(S) = \frac{V^2}{\lambda_0} \int_Y h(y, S)\, dy,$$

(5.105)

where $g(y, S)$ and $h(y, S)$ are solutions to a set of partial differential equations,

$$2\lambda(y)g + \frac{\partial}{\partial y}(f(y, S)g) = \lambda_0 \xi_0(y, S),$$

$$2\lambda(y)h + \frac{\partial}{\partial y}(f(y, S)h) = \lambda_0 \frac{\partial \xi_0}{\partial S}.$$

(5.106)

In (5.104), $\xi(y, S)$ is the equilibrium state distribution of individuals in a fixed attractant level S, given by the solution to

$$\frac{\partial}{\partial y}(f(y, S)\xi_0) = 0; \qquad \int_Y \xi_0(y, S)\, dy = 1.$$

(5.107)

If the scaling parameter ε is small, then ξ_0 is a good estimate of the state distribution in the full equations, (5.102), when calculated for the local attractant concentration, $S(x)$. In this case, the product $P(t, x)\, \xi_0(y, S(x))$ is an estimate of the entire state-space distribution—virtually the complete information that would have resulted from solving the full equations. However, since the space and state distributions are calculated separately, this estimate requires only a small fraction of the computational effort of solving the full equations for the same distribution.

5.7.1   An Example: Receptor Kinetics-based Taxis

A simple caricature of "running" and "tumbling" bacterial taxis (Segel, 1977) will serve to illustrate how (5.104)–(5.107) translate models of individual

Receptor Dynamics

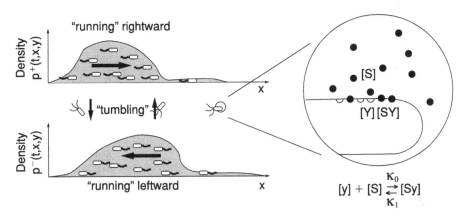

FIGURE 5.9. Schematic of rightward- and leftward-moving individuals. Inset shows schematic of surface receptor binding and unbinding kinetics, as used in the example of simplified bacterial taxis.

state dynamics and movement into population-level, advection–diffusion equations. In this example, the signal transduction mechanism consists of a single receptor type that reversibly binds attractant molecules according to the kinetics.

$$[y] + [S] \underset{\kappa_1}{\overset{\kappa_0}{\rightleftarrows}} [Sy] \tag{5.108}$$

In (5.108), $[y]$ is the fraction of unbound receptors, $[S]$ is the attractant concentration, and $[Sy] = 1 - [y]$ is the fraction of bound receptors (Fig. 5.9). κ_0 and κ_1 are the rate constants of binding and unbinding, respectively. Using the Fokker–Planck method of Sect. 5.3, we show the internal state operator f to be

$$f = \left(\kappa_1(1-y) - \kappa_0 Sy + \frac{1}{2N}(\kappa_1 - \kappa_0 S)\right) - \frac{1}{2N}(\kappa_1(1-y) + \kappa_0 Sy)\frac{\partial}{\partial y}, \tag{5.109}$$

where N is the number of receptors per cell. To complete the example, we assume that the turning rate has a strong response to receptor state when at high attractant concentration. This is achieved by the nonlinear functional form,

$$\lambda(y) = \lambda_0 + \lambda_1 \sqrt{y}. \tag{5.110}$$

Numerical Results.   As already stated, the approximation technique is based on the assumption that $\varepsilon \ll 1$. From the analysis, it is only possible to say

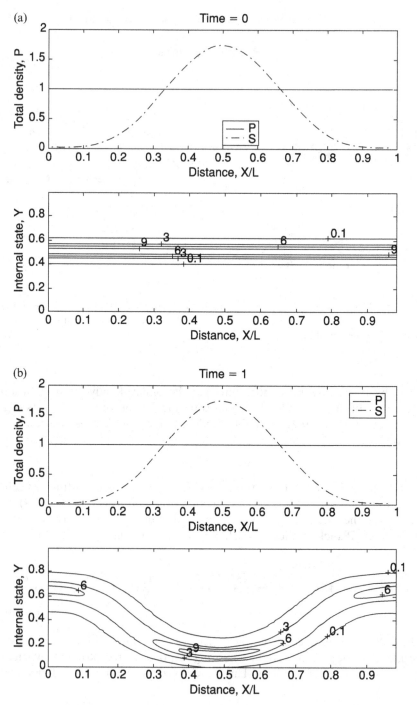

FIGURE 5.10. Sequence of state-space distributions in numerical solutions of the full hyperbolic system in (5.102). Parameters are $\kappa_0 = 2$, $\kappa_1 = 0.4$, $\lambda_0 = 4$, $\lambda_1 = -3.75$, $N = 100$, $V = 1$, and $X_{\max} = 10$. In each frame, the lower plot shows contours of the state-space density distribution, $p^+(t, x, y) + p^-(t, x, y)$, and the upper plot shows the spatial distribution of the total population, $P(t, x)$. The distribution of the attractant, $S(x)$, also appears in the upper plot ($S_0 = 0.025$, $S_1 = 1.725$).

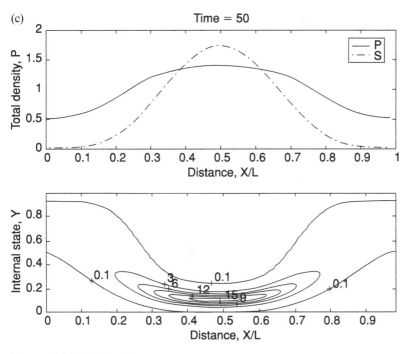

FIGURE 5.10 (*continued*)

that the approximation gets relatively better as ε gets smaller—assessing how accurate the approximation is for a particular value of ε requires actually calculating and comparing the solutions to each. This section shows comparisons of numerical solutions of the full equations in (5.102) to the approximation, (5.104)–(5.107), to evaluate accuracy for a range of ε's. For a standard set of turning and internal state parameters, we calculate transient and equilibrium distributions for a progression of domain sizes, $x \in [0, X_{\max}]$. The attractant distribution for these computations is

$$S(x) = S_0 + \frac{1}{4}S_1\left(1 - \cos\left(\frac{2\pi x}{X_{\max}}\right)\right)^2. \tag{5.111}$$

Thus, as the domain size increases, the length scale L characterizing the attractant distribution increases proportionately, and ε decreases proportionately.

We first examine the behavior of the full equations, (5.102), on a periodic domain with $X_{\max} = 10$, with an initial population distribution that is uniform in x and a tight Gaussian distribution in y. This initial distribution is shown in Fig. 5.10(a), followed by snapshots of the population at two subsequent times (b and c). In this simulation, the temporal development of the population shows two distinct phases (parameters are shown in the figure legend). In the first phase, the state distribution rapidly adjusts to near local equilibrium. Figure 5.10(b), the state-space contour plot at $t = 1$, shows that this near-equilibration in state is already largely complete. In contrast, the

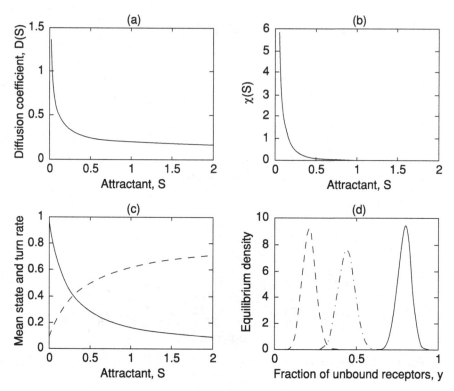

FIGURE 5.11. Results of the approximation procedure in (5.104)–(5.107): (a) diffusion coefficient, $D(S)$; (b) taxis coefficient, $\chi(S)$; and (c) average internal state, $\bar{y}(S)$ (solid line) and normalized turning rate, $\bar{\lambda}(S)$ (dot-dashed line). Also shown, (d), are equilibrium internal state distributions, $\xi_0(y, S)$, for selected attractant levels ($S = 0.05$, solid line; $S = 0.25$, dot-dashed line; $S = 0.75$, dashed line). Behavior parameters are as in Fig. 5.10.

distribution in space, shown in the upper plot, has barely changed over this short time interval. By $t = 100$, however, the space distribution has changed markedly (Fig. 5.10(c)), as the population collects in the part of the domain with largest S. Note that though overall population density in this part of the domain is higher at $t = 100$ than at $t = 1$, the state distribution of this population is nearly unchanged.

The results of the approximation procedure, applied to the same parameters, are summarized in Fig. 5.11. In addition to showing how the diffusion and taxis coefficients, D and χ, vary with attractant concentration, this figure shows two additional dynamically meaningful statistics that are readily derived from the approximation. The mean value of the internal state variable is

$$\bar{y}(S) = \int_Y y \xi_0(y, S) \, dy. \tag{5.112}$$

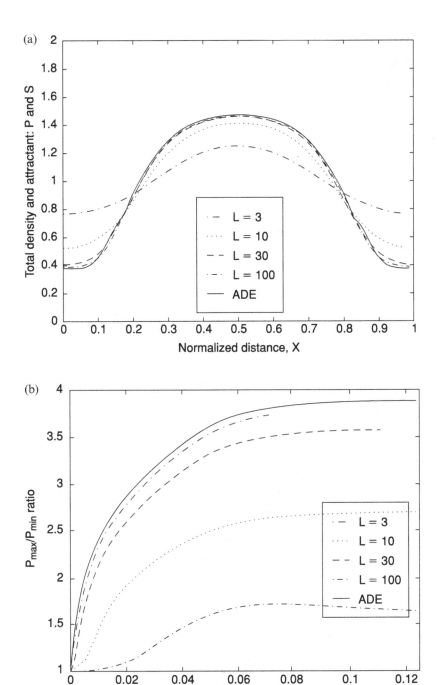

FIGURE 5.12. Comparison of exact to approximate total population distributions for a progression of length scales: (a) equilibrium population distribution; and (b) time series of an aggregation statistic, P_{max}/P_{min}. Shown are results for domain sizes $X_{max} = 3$, $X_{max} = 10$, $X_{max} = 30$, and $X_{max} = 100$, spanning a range in which the scaling parameter decreases from roughly $O(1)$ to $O(10^{-2})$. Time and space variables are normalized as described in the text. Attractant parameters are as in Fig. 5.10.

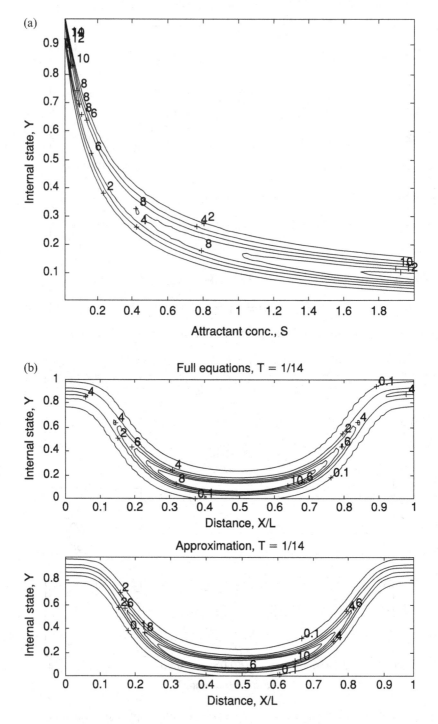

FIGURE 5.13. A comparison of the state-space distribution in the full equations (top frame) with an estimated distribution from the approximate equations (bottom frame). The estimate is derived by weighting the local state distribution, ξ_0, by the local total population density, $P(t, x)$. Parameters are as in Fig. 5.10, with $X_{\max} = 100$. The computational effort required to solve the approximate equation was lower by a factor of roughly 3000 compared to the full equation.

The normalized average turning rate is

$$\bar{\lambda}(S) = \frac{1}{\lambda_0} \int_Y \lambda(y)\xi_0(y, S)\, dy. \tag{5.113}$$

Finally, Fig. 5.11(d) shows the equilibrium state distributions, ξ_0, over a range of attractant concentrations, S.

How well does the approximation predict the full model? Figure 5.12 shows a series of direct comparisons between the two, using the same parameters as in Fig. 5.10, but varying the domain size, X_{max}. Again, increasing X_{max} corresponds to "stretching" the attractant distribution and thus decreasing ε. According to the small-ε theory, the advection–diffusion equation should predict the dynamics of (5.102) more and more accurately as X_{max} becomes large. To assess how large X_{max} must be for a good approximation, we first compare the equilibrium distribution of the total population in space (Fig. 5.12(a)). We then compare the temporal development of a statistic for the "clumpiness" of the population, the ratio P_{max}/P_{min} (Fig. 5.12(b)). In these plots, time and space variables are scaled, $X = x/X_{max}$ and $T = V^2 t / \lambda_0 X_{max}^2$, so that the results of the approximations (5.104)–(5.107) are invariant with changes in X_{max}, λ_0, and V.

In the simulation results, the approximation is poor for the smallest domain size, $X_{max} = 3$, which corresponds to $\varepsilon \sim O(1)$. For such a small domain size, the time scale of lateral transport is of a similar magnitude to the time scales of internal adjustment and directional exchange. This overlap in time scales is reflected by the "overshoot" in the P_{max}/P_{min} ratio: the population aggregates in space before the state distribution has equilibrated. However, when the domain size is increased to $X_{max} = 30$ ($\varepsilon \sim O(10^{-1})$), the approximation is much better for both the equilibrium state distribution and the temporal development of the aggregated population. For $X_{max} = 100$ and higher, the approximation is very accurate. Figure 5.13 shows that estimate of the state-space distribution, ξ_0, for one of these runs ($X_{max} = 30$) also agrees well with the true distribution. The agreement would be even better for larger domain sizes.

In the example, we deliberately chose parameters for which the turning rate spans at least an order of magnitude over the range of attractant concentration in the domain. Furthermore, our statistic of clumpiness, P_{max}/P_{min}, exaggerates errors at both the maximum and minimum of the resource distribution. Of these, the resource minimum is more significant, because in that part of the domain the turning rate is also at a minimum, and so ε is at a maximum. Therefore, we expect that the last type of error to converge with increasing X_{max} will be the population minimum, and this expectation is borne out in the simulations.

6
Some Examples of Animal Diffusion

Akira Okubo and Peter Kareiva

The previous chapter described the fundamentals of the mathematical theory of animal dispersal. The present chapter is concerned with some examples of animal dispersal and migration viewed in the light of biodiffusion.

6.1 Population Pressure and Dispersal

In his book *Animal Ecology*, Ito (1975) classifies animal dispersal into random dispersal and density-dependent dispersal and emphasizes the importance of the latter from the standpoint of population dynamics.

The relationship between animal dispersal and population density has been studied extensively with insects. Morisita (1950) ascertained a relation between population density and dispersal in natural populations of water striders. Later, similar relations were recognized in experiments with aphids (Ito, 1952) and with rice weevils (Kono, 1952), from which it was concluded that for each species there exists an associated population pressure that enhances population dispersal.

Morisita (1954) attempted to quantify this population pressure experimentally. He released ant lion larvae (*Glenuroides japonicus*) from a point and observed their dispersal. The movement pattern of individuals was classified as one of two types: one that dug holes in the vicinity of the release point ("normal individual"), and the other that dug holes after having traveled large distances from the release point ("abnormal individual").

Considering only the dispersal of normal individuals, Morisita found that the number of individuals, N, that settled inside a circle of area A centered at the release point could be expressed by

$$N = M\{1 - e^{-cA}\}, \tag{6.1}$$

where M is the total number of normal individuals and c is a parameter associated with dispersal. Equation (6.1) agrees with the result obtained from a two-dimensional normal distribution, i.e., a two-dimensional simple random walk. Also, the relation $\sigma_r^2 = (\pi c)^{-1}$ where c is the parameter in (6.1) and σ_r^2 is the horizontal variance (see Sect. 2.6.2), is found to hold.[1]

Analyzing data obtained with rice weevils (Kono, 1952) and with adzuki-bean weevils (Watanabe et al., 1952) as well as those obtained with ant lions, Morisita (1954) deduced the regression

$$\sigma_r^2(t) = \sigma_\infty^2 t/(t + T), \tag{6.2}$$

where t is time, and σ_∞^2 and T are parameters dependent on M. Morisita gave the following relationships:

$$\sigma_\infty^2 = M/(\alpha + \beta M),$$
$$T = \lambda/(\alpha + \beta M), \tag{6.3}$$

where α, β, and λ are constants. From (6.2), for $t \ll T$, i.e., for initial dispersal,

$$\sigma_r^2 \to M\lambda^{-1}t. \tag{6.4}$$

The dispersal increases proportionally with both population density and time. On the other hand, for $t \gg T$, i.e., for final dispersal,

$$\sigma_r^2 \to \sigma_\infty^2 = M/(\alpha + \beta M). \tag{6.5}$$

The dispersal becomes constant independent of time, and the dependence on M weakens as M increases. For high values of initial population density (M), the dispersion approaches the value $1/\beta$ independent of M.

---

[1] Number density S for the two-dimensional normal distribution is given by

$$S(x, y, t) = (M/\pi\sigma_r^2) \exp\{-(x^2 + y^2)/\sigma_r^2\} = (M/\pi\sigma_r^2) \exp(-r^2/\sigma_r^2).$$

Integrating the above over a circular area A around $r = 0$, we obtain

$$N = (M/\pi\sigma_r^2) \int_0^{(A/\pi)^{1/2}} \exp(-r^2/\sigma_r^2)2\pi r \, dr = M\{1 - \exp(-A/\pi\sigma_r^2)\}.$$

Thus, we find $\sigma_r^2 = (\pi c)^{-1}$. Inoue (1972, 1978) comments that the empirical law (6.1) does not necessarily mean that the dispersal process is a type of simple random walk.

Analyzing data from insect and mammal dispersal experiments, Inoue (1978) revealed that the regression line of $\ln(1 - N/M)$ to $r^2(= A/\pi)$ often does not pass through the origin but intersects the ordinate, i.e., the axis of $\ln(1 - N/M)$, at a negative value, say $-\lambda^2$. The interpretation of this result is that the intercept of the regression line with the ordinate represents the portion of individuals that remain at the release point throughout the experimental period; $-\lambda^2 = \ln(1 - m/M)$, where m is the number of individuals sedentary at the point. Along this line of thought, Inoue developed a new regression method for analyzing animal movement patterns.

The behavior of the initial variance is the same as that for Fickian diffusion so that the relation for diffusivity $D = M/4\lambda$ may be deduced; note that D is thus proportional to the population density. Morisita's empirical formulas (6.2) and (6.3) appear to be of general applicability in describing the time variation of the variance for insect dispersal from a point source (see Sect. 6.2.1).

The connection of population density with dispersal behavior has significance when viewed from the standpoint of social processes in communities (Ito, 1961). Also, Andrewartha and Birch (1954) attached great importance to dispersal as a reaction to crowding. Yet overpopulation does not necessarily lead to dispersal. As will be discussed in Chap. 7, many species sometimes form groups or swarms of high density.

In lieu of fitting density-dependent models to insect dispersal, several experiments have directly manipulated the number of animals that were released, and then used the diffusion coefficient as a response variable to test whether diffusion coefficients varied with density (Kareiva, 1983; Rosenberg et al., 1997). These experiments have typically found an increase in diffusion coefficients as density is increased, perhaps because the densities used as treatments are typically very high, and hence crowding effects might be expected.

6.2 Horizontal and Vertical Distributions of Insects in the Atmosphere: Insect Dispersal

The air is a plenum of insects. Table 6.1 shows an example of the altitudinal distributions of insects as sampled in the daytime for a period of 10,000 minutes (about 7 days) over the State of Louisiana (Glick, 1939).

The upper limit of the insect distribution is amazingly high. Thus, Glick caught ten individuals of diptera, hymenoptera, and homoptera by sampling

TABLE 6.1. Altitudinal distributions of insect numbers as sampled in the daytime for a period of 10,000 minutes over the State of Louisiana (Glick, 1939)

	Height (m)				
Order	61	305	610	914	1,524
Diptera	5,175	1,979	1,024	586	279
Coleoptera	2,225	519	161	98	51
Hymenoptera	1,646	508	235	113	62
Homoptera	1,493	571	361	284	100
Heteroptera	550	214	106	57	22
Thysanoptera	44	21	10	8	3
Lepidoptera	36	12	12	6	1
(Others)	81	55	19	20	15

for two hours at a height of 14,000 feet (4267 meters). Most insects found in the atmosphere above about 200 m are of small size and have large relative surface areas, rendering them suitable for suspension.

In the summer of 1967 a great number of planthoppers were collected on board a ship on duty at a weather station 500 km south of Sionomisaki, Japan. Since then, much attention has been paid to the long-distance migration of planthoppers (Asahina and Tsuruoka, 1970; Kishimoto, 1971; Inoue, 1974). Inoue (1971) speculated that planthoppers might concentrate in the southern part of a cold front associated with low atmospheric pressures and that the dimension of the aggregation might be about 1 km in length, 150 m in width, and 40 m in thickness.

Investigators often try to describe the vertical distribution of insects by empirical formulas such as that of Wolfenbarger (1946, 1959):

$$S = a + b \log Z + c/Z, \tag{6.6}$$

and of Johnson (1969):

$$S = c_1(Z + Z_e)^{-\lambda}, \tag{6.7}$$

where S is the number density of insects, Z is the height from the ground, and a, b, c, c_1, Z_e, and λ are constant parameters.

6.2.1 Dispersal of Insects

Because of their abundance and their high mobility as compared to animals of similar size, insects are suitable for the study of dispersal. A dispersal experiment consists of releasing a great number of insects from a given source and of observing the spatial distribution at various subsequent times.

Since three-dimensional dispersal is difficult to observe, we usually attempt to measure the two-dimensional distribution of dispersal by the use of traps placed near the ground. Sometimes the released insects are marked with radioactive tracers, etc., to distinguish them from others (Hawkes, 1972; Lamb et al., 1970).

The study of insect dispersal plays an essential role in estimating the areal spread of damage caused by a newly invaded pest or the spatial distribution of insects during the active period in spring subsequent to emergence from hibernation spots. Quantitative information concerning dispersal plays an essential role in the evaluation of pest control (Joyce, 1976; Stephens and Aylor, 1978).

The work of Japanese entomologists on insect dispersal has enjoyed a good international reputation. Let's begin by first looking at a laboratory study by Watanabe et al. (1952). Under nearly constant temperature, humidity, and illumination, they investigated the pattern of dispersal of adzuki-bean weevils (*Callosobruchus chinensis*) on a piece of paper. They found the pattern of dispersal about the release point to be approximated by the two-dimensional normal distribution. In other words, the dispersal pattern,

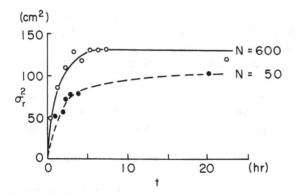

FIGURE 6.1. Time variations of the variance of the dispersal of adzuki-bean weevils: N denotes the number of individuals released (from Watanabe et al., 1952).

assumed to be isotropic, is given by

$$S(r, \theta, t) = (M/\pi\sigma_r^2) \exp(-r^2/\sigma_r^2), \tag{6.8}$$

in polar coordinates (r, θ), where the origin is taken to be the release point, and where M is the total number of insects released and σ_r^2 is the horizontal variance.

Figure 6.1 shows the time variation of the variance. It can be seen that the dispersal of adzuki-bean weevils is very rapid in the beginning but gradually slows down so that after several hours the insects tend to remain in a limited region (see Sect. 6.1). Such a pattern is one of the characteristic features of biodiffusion. Not only adzuki-bean weevils and rice weevils but also other insects (see Fig. 6.2) and other animal species commonly demonstrate this pattern of dispersal.

Since Dobzhansky and Wright (1943, 1947) carried out field experiments with *Drosophila* (*Drosophila pseudoobscura*), a number of important works have appeared: by Oda (1963) with catalpa scales; by Ito and Miyashita (1961) with green rice leafhoppers; by Aikman and Hewitt (1972) with grasshoppers; by Clark (1962) with grasshoppers; by Finch and Skinner (1975) with cabbage root flies; by Gillies (1961) with mosquitoes; by Jackson (in Johnson, 1969) with tsetse flies; and by Lamb et al. (1970) with weevils, to mention a few. An exhaustive list of examples of dispersal may be found in the papers by Wolfenbarger (1946, 1959) and a book by Turchin (1998). In addition, Kareiva (1983) reported that in eight different field experiments in which the movements of insects were tracked (spanning twelve different species), the data were well described by a simple diffusion model (i.e., a Gaussian distribution spreading linearly through. time). Although heterogeneous environments, environmental conditions, and interactions among individuals often add complications to simple diffusion models, the diffusion

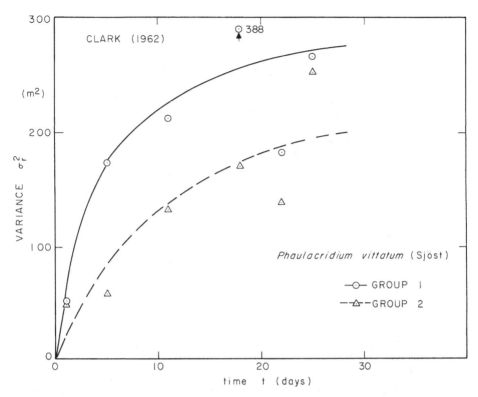

FIGURE 6.2. Time variations of the variance of the dispersal of grasshoppers; the total number of individuals released is about 200.

coefficient has become a standard tool for reporting rates of movement. In fact, diffusion coefficients are now often reported as a means of quantifying the effectiveness of predators at finding prey (Kareiva and Odell, 1987), the ability of threatened or endangered species to move among habitat fragments (Schultz, 1998), the impacts of vegetation structure on dispersal rate (Wetzler and Risch, 1984) and rate at which pests spread into new areas (Sharov and Leibhold, 1997).

Wolfenbarger (1946, 1959) made an ambitious attempt to synthesize a vast amount of dispersal data ranging from plant spores to small animals such as insects and fit them into Eq. (6.6) by taking Z to be the distance of dispersal. However, not every set of data fits the formula best. We often find that any kind of data can be made to fit an empirical formula that contains three parameters. It is an important task to model dispersal on the basis of a knowledge of factors influencing dispersion (Shigesada and Teramoto, 1978).

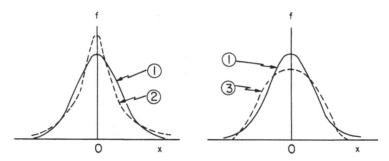

FIGURE 6.3. Three types of frequency distribution (f vs. x) with the same mean and variance: (1) normal (Gaussian), (2) leptokurtic, (3) platykurtic.

Summarizing data on insect dispersal, we point out some features:

i) The kurtosis[2] of a spatial distribution of individual insects is usually larger than 3.0—the distribution is thus said to be *leptokurtic* (Sokal and Rohlf, 1969). In other words, the distribution has higher amplitudes near its center and toward its tails than the normal distribution with the same mean and variance[3] (Fig. 6.3). This feature is noticeable especially in the early stages of dispersal (Dobzhansky and Wright, 1943; Aikman and Hewitt, 1972).

ii) The variance of insect displacements increases most rapidly in the beginning of dispersal, slows its rate of increase as time progresses, and eventually approaches a fixed value (Figs. 6.1 and 6.2). The variance curve appears to vary with the total number of individuals released even though environmental conditions are identical, and the asymptotic value of the variance is controlled by environmental factors.

iii) Environmental factors such as light, temperature, humidity, and wind speed affect the dispersal (Morisita, 1971). Generally speaking, high temperature, low humidity, and weak winds encourage dispersal.

Dobzhansky and Wright (1943) reason that leptokurtic distributions would result from heterogeneous populations of insects, some dispersing rapidly and others dispersing slowly. The superposition of two normal distributions with different values of variance produces a leptokurtic pattern shown in Fig. 6.4. By the same token, a mixture of data from various experiments with widely different rates of dispersion gives a greater kurtosis

---

[2] The kurtosis, β_2, is defined by $\beta_2 \equiv \mu_4/(\sigma^2)^2$, where σ^2 is the variance and μ_4 is the 4th-order central moment.

[3] If we express a density distribution function by $f = f_0 \exp(-cr^k)$, $\beta_2 = \Gamma(1/k)\Gamma(5/k)/\{\Gamma(3/k)\}^2$, where Γ is a gamma function. For a normal distribution, $k = 2$ and $\beta_2 = 3$. The distribution function for insects (and many other organisms) may be approximated by taking $k = 1 \sim 1/2$ (Richardson, 1970). Thus, the spatial pattern of biodiffusion is generally leptokurtic.

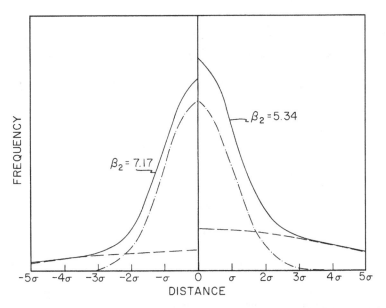

FIGURE 6.4. Superposition of two normal distributions with different values of variance and the same mean value. (Right): One part of normal distribution with the variance σ ($-\cdot-$) is mixed with one part of normal distribution with the variance 4σ ($- - -$). The compound distribution has the kurtosis of 5.34. (Left): One part of normal distribution with the variance σ ($-\cdot-$) is mixed with a half part of normal distribution with the variance 4σ ($- - -$). The compound distribution has the kurtosis of 7.17.

than the separate components (Wright, 1968). Also, James (1978) has studied an inverse problem, i.e., the estimation of the mixing proportion of two normal distributions by means of simple, rapid measurements.

Let us clarify the above matter in a more quantitative fashion. Take two populations that diffuse according to Fick's law with different diffusivities or variances. That is, the dispersal pattern of each population is Gaussian and given by

$$f_1 = \frac{n_0}{(2\pi)^{1/2}\sigma_1} \exp\left(-\frac{x^2}{2\sigma_1^2}\right), \tag{6.10}$$

$$f_2 = \frac{n_0}{(2\pi)^{1/2}\sigma_2} \exp\left(-\frac{x^2}{2\sigma_2^2}\right), \tag{6.11}$$

where the total number of individuals belonging to both populations is assumed to be equal and is denoted by n_0, and σ_1^2 and σ_2^2 are the variances for both populations. Furthermore, suppose that $\sigma_1^2 \geq \sigma_2^2$.

When individuals from two such populations are pooled, the compound distribution is given by the mean of (6.10) and (6.11).

That is,

$$f = \frac{n_0}{2(2\pi)^{1/2}} \left\{ \frac{1}{\sigma_1} \exp\left(-\frac{x^2}{2\sigma_1^2}\right) + \frac{1}{\sigma_2} \exp\left(-\frac{x^2}{2\sigma_2^2}\right) \right\}. \qquad (6.12)$$

The variance and the fourth-order central moments of (6.12) are given by

$$\sigma^2 = \frac{1}{2}(\sigma_1^2 + \sigma_2^2), \qquad (6.13)$$

$$\mu_4 = \frac{3}{2}(\sigma_1^4 + \sigma_2^4), \qquad (6.14)$$

respectively. Hence, we obtain the kurtosis

$$\beta_2 = \mu_4/(\sigma^2)^2 = 6(\sigma_1^4 + \sigma_1^4)/(\sigma_1^2 + \sigma_2^2)^2. \qquad (6.15)$$

As $\sigma_1^2 \geq \sigma_2^2$, we put $\sigma_1^2 = m\sigma_2^2$ $(m \geq 1)$ and substitute into (6.15). The result is

$$\beta_2 = 6(1 + m^2)/(1 + m)^2. \qquad (6.16)$$

For $m = 1$, i.e., homogeneous populations, $\beta_2 = 3$ (normal distribution). For $m > 1$, $\beta_2 > 3$, and the value of β_2 increases with m. As $m \to \infty$, $\beta_2 \to 6$. Accordingly,

$$6 \geq \beta_2 \geq 3.$$

We are thus able to show that a compound distribution taken from heterogeneous populations, even though individual distributions are Gaussian with different variances, becomes leptokurtic. [Skellam (1951a) demonstrates this more elegantly using a moment-generating function.] Fig. 6.4 illustrates this schematically.

As long as the ratio m remains constant, the kurtosis will be invariant even if both σ_1^2 and σ_2^2 vary with time. On the other hand, the kurtosis decreases with time if m decreases with time; such a case will occur when the population having a larger variance, σ_1^2, thus dispersing faster, tends to settle sooner than the population having a smaller variance, σ_2^2 (Dobzhansky and Wright, 1943).

Thirty years later Dobzhansky and Powell (1974) repeated the release experiment of *Drosophila* at the original site. This time, instead of laboratory-raised mutant flies as used previously, they released wild flies collected on the same day or one day before the experiment. It was discovered that: (i) the spatial distributions of the flies were *not* leptokurtic but close to Gaussian even one day after the release; (ii) the dispersal rate was appreciably greater than that of the previous experiment; and (iii) the variance two days after the release was usually less than double that one day after the release.

It is reasonable to suppose that some of the laboratory-raised flies were deficient in vigor, while the wild flies constituted a more or less homogeneous population as far as dispersal was concerned, and that probably the crowding excited the flies and stimulated them to disperse at a higher rate than they would otherwise.

The rapid initial dispersal might as well be due to the release procedure itself (Clark, 1962). If, on the other hand, it is due to population pressure associated with crowding, then the spatial distribution should be *platykurtic* (Sokal and Rohlf, 1969) as shown in Fig. 6.3, rather than leptokurtic; this has been noted by Aikman and Hewitt (1972). This invites further discussion with a mathematical model.

As an example of a field experiment, let us take a look at the dispersal of flea beetles in long linear strips of collards (one of the favored food plants of flea beetles). In these experiments, Kareiva (1982) marked beetles with micronized fluorescent powders, chilled the beetles so they would not be disturbed by the handling, and released them in central patches. One, two, or three days later beetles were collected with a gas-powered vacuum sampler and returned to the lab where they could be inspected for the presence of a mark. In experiments where the patches of food plants were relatively uniform in quality, the beetles spread out among the patches according to a Gaussian distribution whose variance increased linearly through time. However, when the patches of food plants had been experimentally manipulated to differ dramatically in quality, then the simple diffusion model had to be converted into a random walk model in which probabilities of movement depended on the local quality of food patches. Whereas Kareiva (1982) relied on mass-marking large collections of beetles for these experiments, other methods include individually marking insects or actually tracking and mapping the paths of individuals while they are moving through the environment (Kareiva and Shigesada, 1983; Schultz, 1998; Turchin, 1998). A particularly interesting innovation to insect diffusion experiments involves the boundary-flux approach developed by Fagan (1997). Fagan takes advantage of the fact that diffusion models predict a flux of individuals across any boundary, and then uses sticky strips that trap walking insects to estimate this flux. Fagan's method allows estimation of diffusion coefficients without having to mark insects or painstakingly record the positions of recaptured insects.

One assumption of purely diffusive movement is that the organisms are moving with infinite velocity and frequently reversing their direction. The fact that this sounds like an unreasonable assumption does not mean that diffusion models are entirely inappropriate—rather it merely means that these models cannot be applied to short timeframe predictions. In fact, the simplest diffusion models often describe well the dispersal of animal populations. One of the more realistic descriptions of animal movement proposed as an alternative to simple diffusion is a correlated random walk (Kareiva and Shigesada, 1983). According to this model, animals move at a particular finite velocity, change directions according to some frequency distribution, and tend to keep moving in the same direction. The telegraph equation that results from this process (see Sect. 5.3) includes both a diffusive term and a "wave term" (second time derivative of population density), with the importance of the wave term declining as organisms reverse directions more frequently (i.e., move in a "less correlated" manner). Surprisingly, the telegraph equation has almost never been applied to actual data describing animal movement (but see Holmes, 1993), although its discrete Markovian jump process analogue is commonly used to describe observations of movement paths (Kareiva and Shigesda, 1983). One reason the telegraph equation has not been used is that, even if we consider a population of dispersing organisms that adheres to a telegraph equation, as time progresses the diffusion term will come to dominate and eventually the population can be viewed as redistributing itself in a diffusive fashion. In the short run, the fact that the telegraph equation predicts a platykurtic distribution of dispersers about their release point may explain some of the departures from Gaussian distributions observed in field studies such as those reviewed by Kareiva (1983).

Biologists have long noted that even though animals may appear to move randomly locally, they also can alter their movement to move toward favored sites and toward or away from conspecifics. In Chaps. 4 and 7, detailed models of animal aggregation and grouping are explored. In general, the addition of density-dependent and advective forces is often warranted when describing insect movement. Many of the cases in which simple diffusion does not adequately describe insect movement are probably due to heterogeneous environments in which the organisms alter their pattern of movement in response to local conditions (Kareiva, 1983; Turchin, 1998). While it is straightforward to write extensions of diffusion models to represent these more complicated patterns of movement, it is much harder to estimate the key parameters or rate constants in these models. The two contrasting approaches that have been successfully used are as follows:

(i) Develop a mechanistic link between exact individual behavior and a resulting mass-action equation that then allows one to parameterize the mass-action model in terms of the behavioral attributes of individuals whose movement is tracked (Kareiva and Odell, 1987; Turchin, 1998), and

(ii) Use numerical methods that find the parameters in mass-action population models, which minimize the deviation between data and the solution to the partial differential equation (Banks et al., 1985, 1988).

Both approaches have their advantages and disadvantages. The translation of individual behavior into a mass-action model is the most intellectually appealing but is often impractical simply because the exact movements of individuals can only be followed under the most contrived situations. On the other hand, the "best fit" to a mass-action model often includes more than one "solution" because different sets of model coefficients can generate the same distribution of organisms in space (Banks et al., 1988).

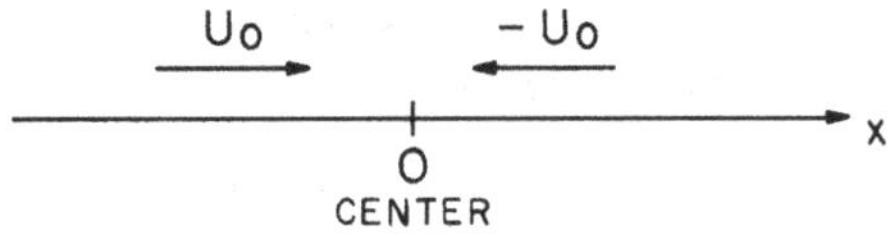

FIGURE 6.5. Diffusion model with the center of attraction. The speed of attractive flow is U_0.

6.2.2   Mathematical Models for Insect Dispersal

Dispersal due to population pressure may be appropriately modeled by expressing the advection and diffusion terms as functions of population density, as mentioned in Sects. 5.3 and 5.6. Let us consider a one-dimensional case. The model equation can be written as (neutral transition)

$$\frac{\partial S}{\partial t} = -\frac{\partial}{\partial x}(uS) + \frac{\partial}{\partial x}\left(D\frac{\partial S}{\partial x}\right) \tag{6.17}$$

where the advection and diffusivity generally depend on x, t, and S. The advection in (6.17) represents the effect of attraction of insects to a particular region. For instance, any tendency for concentration of insects around a point is interpreted to be the result of an attractive flow toward this point.

As a simple model, we take the center of attraction at the origin, $x = 0$, and assume that the speed of attractive flow is constant,

$$u = -u_0 \, \text{sgn}(x) \tag{6.18}$$

where $\text{sgn}(x)$ is a function defined to be 1 if $x > 0$ and -1 if $x < 0$ (Fig. 6.5). Assume the diffusivity to depend not on x and t, but on S alone. We thus take

$$D = D_0(S/S_0)^m, \quad m > 0, \tag{6.19}$$

where D_0 is the diffusivity for $S = S_0$ (a reference concentration). As $m > 0$, the diffusivity increases with S. The effect of population pressure is thus incorporated into D. Substituting (6.18) and (6.19) into (6.17), we find

$$\frac{\partial S}{\partial t} = u_0\frac{\partial}{\partial x}\{\text{sgn}(x)S\} + D_0\frac{\partial}{\partial x}\left\{\left(\frac{S}{S_0}\right)^m\frac{\partial S}{\partial x}\right\}. \tag{6.20}$$

The problem is to solve (6.20) under appropriate initial conditions. However, the equation is nonlinear with respect to S, and in general an analytical solution is difficult to obtain.

For such nonlinear problems, it may prove useful to pay attention to two limiting cases, i.e., small values of t (initial dispersal) and large values of t (final dispersal).

When the advection term is compared with the diffusion term, we realize that initially the concentration gradient, $\partial S/\partial x$, is very large near the origin from which individuals are released, and also diffusivity is high due to high density of individuals. Thus, we may ignore the advection term in the initial period of dispersal.

$$\frac{\partial S}{\partial t} = D_0 \frac{\partial}{\partial x} \left\{ \left(\frac{S}{S_0}\right)^m \frac{\partial S}{\partial x} \right\}$$

(6.21)

Pattle (1959) has given the solution to (6.21);

$$S = \begin{cases} S_0(t_0/t)^{1/m+2}(1 - x^2/x_1^2)^{1/m}, & |x| \leq x_1, \\ 0, & |x| > x_1, \end{cases}$$

(6.22)

with $x_1 \equiv r_0(t/t_0)^{1/m+2}$, $r_0 \equiv Q\Gamma(1/m + (3/2))/\pi^{1/2}S_0\Gamma(1/m + 1)$, and $t_0 \equiv r_0^2 m/2D_0(m + 2)$, Q being the initial flux of individuals from the origin and Γ being the gamma function. The population disperses only to a finite range $x = x_1(t)$, and the spatial pattern is platykurtic, i.e., flatter than the Gaussian distribution. This dispersal pattern is consistent with the predictions of Aikman and Hewitt (1972).

For the final period of dispersal, advection and diffusion play equally important roles. In fact, these two processes balance each other. Diffusion tends to spread the population from the center, while advection acts to attract it to the center. Eventually a *steady state* is established (Shigesada and Teramoto, 1978; Shigesada et al., 1979), so that $\partial S/\partial t = 0$. Under this condition we integrate (6.20) once with respect to x:

$$u_0 \operatorname{sgn}(x)S + D_0(S/S_0)^m \, dS/dx = 0,$$

(6.23)

where the integration constant becomes zero insofar as $S = 0$ at $|x| = \infty$. Integrating (6.23) once more over x and seeking a symmetric solution around $x = 0$, we obtain

$$S = \begin{cases} S_0(1 - mu_0/D_0|x|)^{1/m}, & |x| \leq x_b \equiv D_0/mu_0, \\ 0, & |x| > x_b \equiv D_0/mu_0 \end{cases}$$

(6.24)

where we take S_0 to be the population density at the center.

The population disperses only within a finite region. Figure 6.6 shows a plot of S/S_0 against $|x|/x_b$ for various values of m. The spatial distribution is not necessarily platykurtic as we now include the effect of advection. Depending on the density dependence of D, we obtain a leptokurtic distribution (see $m = 1/2$). The distribution of insects in a *kabashira* (a plumelike swarm of mosquitoes) observed by Okubo and Chiang (1974) looks very much like that of $m = 1/2$ (see Sect. 7.3).

Since x_b is finite, the variance of the distribution approaches a constant value at $t \to \infty$. The variance is zero at $t = 0$ for a point release, so that the time change of the variance of insect dispersal from a point source takes the form of Figs. 6.1 and 6.2.

Okubo (1972) demonstrates that a temporal variance behavior similar to those shown in Figs. 6.1 and 6.2 can be obtained for populations under a

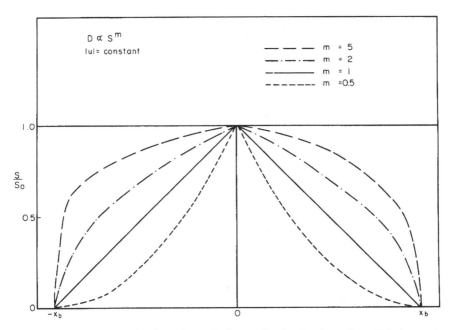

FIGURE 6.6. Spatial distribution of population under density-dependent diffusion and advection toward the center.

simple diffusion with a "harmonically bounded" attraction toward the center without invoking the population pressure effect on diffusion (Fig. 6.7). In this model, however, the spatial pattern of dispersion is simply Gaussian, and thus it extends its tails toward infinity.

Summarizing these considerations, we must admit that the development of mathematical models for insect dispersal is still immature. We recognize the necessity of introducing the population density effect to the advection–diffusion equation, (6.17), but its proper formulation must await future investigation. For the moment we may content ourselves with a simple model that has a constant diffusivity and a uniform flow of attraction toward the center of dispersal. Also, a simple random walk model might serve our purpose if the environment of dispersal is assumed to be homogeneous and if there is no interference between individuals. In this context, Rogers (1977) showed that the observed distributions of tsetse flies (*Glossina fuscipes* and *Glossina morsitans*) within their natural habitat conformed to the expectation of a simple diffusion model. For heterogeneous populations, we are required to divide them into subpopulations, each being regarded as uniform, and to treat each dispersal separately.

Another interesting problem is the spatial distribution of insects resulting from a combination of dispersal and intraspecific interaction. As a simple example, consider an insect population that is dispersing from a point ac-

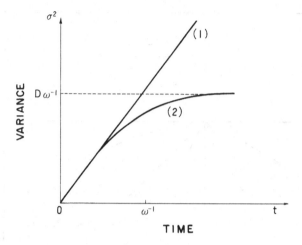

FIGURE 6.7. Time variation of the variance of population dispersal subject to a simple diffusion with a harmonically bounded attraction toward the center: D is the diffusivity and ω is the frequency of the harmonically bounded attraction. (1) without attraction, (2) with attraction.

cording to Fick's law, i.e., a random walk, and is reproducing at a constant rate. The mortality rate of parents is assumed to be constant. We may query as to what the distribution of eggs in space might be.

In one dimension the parent population obeys the following diffusion equation:

$$\frac{\partial S}{\partial t} = D\frac{\partial^2 S}{\partial x^2} - \mu S, \tag{6.25}$$

where S is the population density of parents, D is the diffusivity, and μ is the mortality rate. The equation for eggs is

$$\frac{\partial E}{\partial t} = \lambda S, \tag{6.26}$$

where E is the population density of eggs and λ is the rate of egg deposition.

The solution of (6.25) subject to the initial condition that, at $t = 0$, $\int_{-\infty}^{\infty} S\, dx =$ the initial total number of individuals $\equiv N_0$ is given by

$$S(x, t) = N_0/2(\pi Dt)^{1/2} \exp(-x^2/4Dt - \mu t). \tag{6.27}$$

Substitution of (6.27) into (6.26) and integration over t yields

$$E(x, t) = \lambda N_0/(\pi D)^{1/2} \int\limits_{0}^{t^{1/2}} \exp(-x^2/4D\eta^2 - \mu\eta^2)\, d\eta, \tag{6.28}$$

which satisfies the initial condition that at $t = 0$, $E = 0$. Horenstein (1945) provides a closed form of the integral of (6.28). The asymptotic form $(t \to \infty)$ of (6.28) is given by

$$E(x, \infty) = \lambda N_0/2(D\mu)^{1/2} \exp\{-\mu/D)^{1/2}|x|\}. \tag{6.29}$$

That is, the asymptotic distribution of eggs takes the form e^{-ax}, thus having a kurtosis of 6, characteristic of a leptokurtic distribution. This is one example which illustrates that in general a compound distribution differs from its parent distribution and often exhibits a leptokurtic pattern even if the parent distribution is Gaussian.

The problem can easily be extended to two- or three-dimensional dispersal, where the asymptotic distribution is expressed in general by a modified Bessel function of the second kind, i.e., the so-called K-distribution. In this context, Yasuda (1975) developed a mathematical model for a random walker who may stop his movement at any time. The probability of stopping time of the walker is distributed according to a gamma law. The resultant distribution of dispersal distance is proven to be the K-distribution.

The present model, Eqs. (6.25) and (6.26), is essentially the same as that of Broadbent and Kendall (1953), who discussed the dispersal of larvae of the helminth *Trichostrongylus retortaeformis*. The larvae are hatched from eggs in the excreta of sheep or rabbits and wander apparently at random until they climb and remain on blades of grass. There they may be eaten by another animal, in the intestines of which the cycle recommences.

The combination of (6.25) and (6.26) can be applied to the distribution of the larvae thus isolated on blades of grass. To this end, we regard S as the number density of larvae that are still free to perform a random walk, and E as the number density of the larvae that are settled upon blades of grass. Furthermore, we take $\mu = \lambda$. This means that the process of larvae climbing up and settling on blades of grass is considered as a loss of S with a constant rate μ, so that the population of free larvae lost becomes the population of larvae settled, E.

Williams (1961) studied a similar problem, that of the distribution of larvae of randomly moving insects. Spatial distributions of eggs and larvae of insects constitute a subject of practical importance (Jones, 1977), and the study of mathematical modeling for them should be encouraged (e.g., a study due to Kuno (1968) in Sect. 10.2.3).

Other examples of dispersal problems are the propagation of plant or animal diseases carried by insects, and pollination by bees (Bateman, 1947).

Morris' (1993) study of biased random movement in honeybees is noteworthy because it translated observations of individual bees into estimates of the advection velocity and diffusion coefficient in a standard advection–diffusion model [Eq. 6.17]. By obtaining a numerical solution to this model, Morris (1993) was able to predict the spread of pollen that had been marked with a dominant heavy anthocyanin gene (and that hence showed up in progeny as "purplish" seedlings).

6.3 Diffusion Models for Homing and Migration of Animals

The homing and migration of birds and fish, often with incredible accuracy over enormous distances, constitute one of the wonders of nature. Random search has been proposed as a possible mechanism for homing. In this section we will examine the random search hypothesis.

At first sight it seems most unlikely that a completely random search should be a factor in homing, but this possibility must be investigated more closely before we discard it. Wilkinson (1952) first demonstrated that random search could indeed explain some observed phenomena associated with bird homing.

Since then Jones (1959) and Saila and Shappy (1963) have promoted the idea that random search combined with a small amount of directional orientation (possibly of an olfactory nature) can theoretically provide reasonable homing results in fish. Also, Wilson and Findley (1972) showed that experimental data on bat homing could be interpreted in terms of the random search hypothesis. All of these studies suggest that random search may not be totally nonsensical as a homing mechanism.

The mathematical model of random search depends on the diffusion equation based on a random walk. The assumptions underlying the discussion by Wilkinson (1952) are as follows:

i) The animals search for home independently.
ii) The search is completely random and the animals have no memory; this assumption ensures that the results of the calculation are pessimistic.
iii) The random search is a diffusion process characterized by a constant diffusivity, D.
iv) All the animals search for a time not greater than t_0; the search effectively lasts only for a finite time either because the animal's incentive for homing disappears or because the experimenter gives up observation at the return point. As a matter of fact, the animals do not search all the time. Only a fraction of a calendar day is spent in search. We need a factor to determine the time for which the animal is actually in motion.

Let's consider one-dimensional diffusion along the x-axis (Fig. 6.8). The origin, $x = 0$, is taken to be at the home for which the animal aims, and

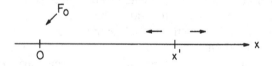

FIGURE 6.8. Model of random search in one-dimensional space. x': point of release of homing animals, 0: homesite.

$x = x'$ is taken to be the point of release of homing animals. This model is applicable, say, to the homing of sea birds searching along a coast. The diffusion equation is given by

$$\frac{\partial S}{\partial t} = D \frac{\partial^2 S}{\partial x^2},$$ (6.30)

where S is the number density of animals. The initial condition is that all of the individuals, numbering N_0 in total, are at $x = x'$ at $t = 0$.

As individuals arrive at the homesite, they "settle" there. This implies that the home, $x = 0$, is considered to be a "sticky" wall. Thus, the boundary condition at $x = 0$ is $S = 0$. In other words, there are no birds flying at the origin. This is not strictly the correct boundary condition; a more appropriate one would be that of radiation (Carslaw and Jaeger, 1959), but the error introduced is small in our case.

Equation (6.30) is solved subject to these initial and boundary conditions (see Carslaw and Jaeger, 1959, for example). The result is

$$S(x,t) = \frac{N_0}{2(\pi D t)^{1/2}} \left[\exp\left\{ -\frac{(x-x')^2}{4Dt} \right\} - \exp\left\{ -\frac{(x+x')^2}{4Dt} \right\} \right].$$ (6.31)

Note that (6.31) is applied only to the region $x \geq 0$.

Having obtained S, we then calculate the homing flux of birds at $x = 0$, F_0. The flux at the origin is defined to be the number of birds passing through a plane perpendicular to the x-axis at $x = 0$ from the positive side of x to the negative side of x per unit time. For this diffusive process the flux is given by

$$F_0 = D\partial S/\partial x|_{x=0}.$$ (6.32)

Substituting (6.31) into (6.32), we have

$$F_0 = N_0 x'/2t(\pi D t)^{1/2} \exp(-x'^2/4Dt).$$ (6.33)

Thus, we can calculate the number of birds that return before time t_0 after the release, N, as

$$N = \int_0^{t_0} F_0 \, dt = N_0[1 - \Phi\{x'/2(Dt_0)^{1/2}\}],$$ (6.34)

where Φ denotes the error function (see Sect. 3.3.2). The probability of return before t_0 is then given by

$$p^* = 1 - \Phi\{x'/2(Dt_0)^{1/2}\},$$ (6.35)

and the probability of not returning before t_0 is likewise given by

$$q^* = 1 - p^* = \Phi\{x'/2(Dt_0)^{1/2}\}.$$ (6.36)

The error function has the following properties; its value is close to zero when the argument is very small compared with one, and its value is essentially one when the argument is more than 2. In other words, the homing probability is nearly one when the distance x' between the release point and home is very small compared with a typical diffusion distance due to random search, $2(Dt_0)^{1/2}$, while the homing probability is nearly zero when $x' > 4(Dt_0)^{1/2}$.

We can calculate the average speed of homing, V_h, from F_0 and N:

$$V_h(t_0, x') = \int_0^{t_0} x'/tF_0(t, x')/N(t, x')\, dt$$

$$= \frac{2D}{x'} \left[1 + \frac{x'}{(\pi Dt_0)^{1/2}} \exp\left(-\frac{x'^2}{4Dt_0}\right)\left\{ 1 - \Phi\left(\frac{x'}{2(Dt_0)^{1/2}}\right)\right\}^{-1}\right].$$

$$(6.37)$$

Random search has the unique feature that the averaging homing speed is nearly independent of the distance between the release point and home. In fact, experimental data actually show no characteristic variation of the average homing speed with distance to release.

For a test of the random search hypothesis, Wilkinson (1952) compared (6.33) with experimental results for starlings. The values of D and t_0 for the starlings are taken to be $D = 8000$ (miles)^2/day and $t_0 = 12.5$ days (Wilkinson, 1952). Figure 6.9 shows the comparison of the percentage of return F_0/N_0 at time t versus release distance x'. The cut-off at high values of the abscissa is due to the fact that birds have a finite speed of flight; they cannot reach home in less than a certain time, while Fickian diffusion is characterized by an infinite speed of flight (see Sect. 5.3), so that we cannot apply (6.33) for very small values of t. Considering the difficulties in obtaining reliable data for homing, we may conclude that the theoretical result agrees fairly well with experiment.

Wilkinson (1952) also discussed a two-dimensional random search appropriate for the homing of land birds and certain types of migration; that is,

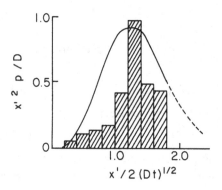

FIGURE 6.9. Comparison of the theoretical values of the return probability p at time t vs. release distance x' with those of observation with starlings. The probability and release distance are properly nondimensionalized. The solid curve is based on the random search model, and the histogram is based on the observation (from Wilkinson, 1952).

random search with a slight degree of orientation. Wilkinson's idea of random search has the advantage of explaining many observed facts of homing without assuming that the birds possess specialized sensory capacities of a still mysterious nature. At the same time, however, Wilkinson noted, "it is not suggested that this investigation proves that random search is indeed the mechanism by which the homing of wild birds is accomplished." There are clear cases where homing birds such as domestic pigeons demonstrate homing orientation.

In order to reveal the true mechanism of homing, it is essential at least to trace the actual movement paths, or trajectories of the animals involved. Telemetry will certainly play an important role to this end (Marshall, 1965; Southern, 1965).

Griffin and Hock (1949) attempted to follow homing gannets (*Morus basanus*) visually from slow-flying light aircraft. The tracking of nine gannets for periods from 68 minutes to over 9 hours was an outstanding achievement. Within 24 to 75 hours after release, five birds returned to a homesite 340 km (213 miles) distant. Figure 6.10 illustrates the trajectories of the nine birds. The data show that the gannets definitely did not fly directly home, but rather performed an irregular search.

Wilkinson (1952), using Griffin and Hock's data, estimated a mean flight speed v of 56 km/h and a mean free path ℓ of 76.8 km. These figures gave a value for diffusivity, $D = 25,800$ km^2/day, using $D = 1/4v\ell$ for two-dimensional diffusion. The experiment lasted five days. If we take 8 hours to

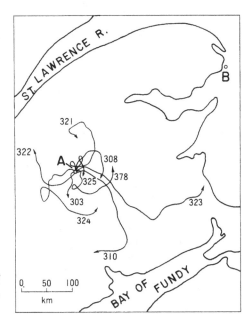

FIGURE 6.10. Homing trajectories of gannets. A: release point; B: homesite (from Griffin and Hock, 1949).

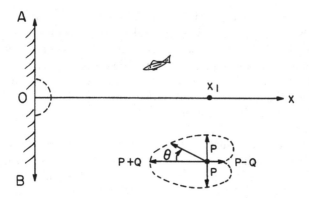

FIGURE 6.11. Model of random search with a low degree of orientation. x_1: release point; 0: homesite.

be an effective daily flight time of gannets, $t_0 = 5/3$ days. These values give a 60% return in theory compared with an observed return of 59%, and a mean speed of homing $V_h = 186$ km/day in theory compared with 163 km/day in observation. We note, however, that the effect of wind on bird migration and homing should be taken into consideration, as Rainey (1960) has suggested.

Saila and Shappy (1963) applied a random movement model to salmon migration. Their model is based on random searching combined with a low degree of orientation to an outside stimulus. The model is two-dimensional in a semiinfinite space with the origin taken at the mouth of the river to which the fish return. A radius of 74 km (40 nautical miles) from the stream mouth is arbitrarily chosen as the area within which a successful search at sea is ended (Fig. 6.11). The fish start homing at a point x_1 taken on the x-axis. The pattern of swim lengths, i.e., the mean free path, ℓ, at sea is chosen as

$$\ell = P + Q \cos \theta = P(1 + a \cos \theta), \qquad a = Q/P.$$

The mathematical expression (above) means that P represents a mean free path (step) of a purely random walk and Q represents a mean free path in the direction perpendicular to the coast, which is modified by homing orientation. Thus, the model is characteristic of a biased random search.

Saila and Shappy conducted a computer study for various combinations of P and Q, given $x_1 = 2224$ km (1200 nautical miles). The maximum endurance for migration was taken as 175 days and the mean swimming speed ranged from 4.6 to 9.3 km/h. It was shown that return probability increases significantly with a relatively small degree of orientation toward the shore. The precise mechanism of the orientation remains unknown.

Randomness in homing seems to exist also in animals other than birds and fish. Thus, Wilson and Findley (1972) obtained data on homing of bats (*Myotis nigricans*) released at various distances and compared the results with the random search model of Wilkinson (Table 6.2). The theoretical

TABLE 6.2. Comparison of homing data for *Myotis nigricans* with expected return frequencies calculated from Wilkinson's (1952) model

Release distance (km)	Homing probability (%)		No. of bats released
	Observed	Calculated	
3.5	94	100	17
5	92	100	12
10	70	100	17
16	33	31	12
38	25	11	16
50	5	8	39
104	0	3	21

return percentage was calculated on the basis of a familiar area of 13 km around the home roost. A test for goodness of fit shows no significant difference between the theoretical and observed return probabilities.

Also, Murie (1963), in homing experiments with deer mice, holds the view that random dispersal away from the release site is sufficient to produce the homing performances observed.

6.4   Model for Muskrat Dispersal and Biological Invasions in General

We are now at the stage where a classic work by Skellam (1951) deserves mention. Skellam succeeded for the first time in applying the diffusion model to mammal dispersal.

In 1905 a landowner in Bohemia inadvertently introduced muskrats (*Ondatra zibethica*) to Europe by allowing five of them to escape. They started to spread and repeated the process of reproduction and dispersal until they finally came to inhabit Europe in many millions (Elton, 1958). Figure 6.12 shows the apparent boundaries of dispersal for various years. If we plot the square root of the area enclosed by each boundary against time, as shown by the circles in Fig. 6.12, we obtain a quite linear relation between the two variables.

Skellam (1951) derived the linear relationship using a model of two-dimensional random dispersal combined with an exponential growth of population. If we take polar coordinates centered at the release point and assume that the dispersal is isotropic and has constant diffusivity, and that the growth is proportional to the population density, S, then muskrat dispersal is modeled by

$$\frac{\partial S}{\partial t} = \frac{D}{r} \frac{\partial}{\partial r} \left(r \frac{\partial S}{\partial r} \right) + \alpha S, \tag{6.38}$$

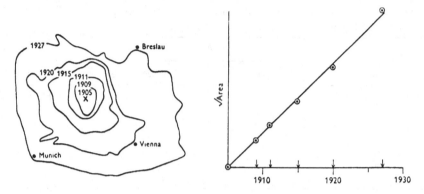

FIGURE 6.12. Spread of muskrats in Europe. (Left): Apparent boundaries of muskrat dispersal for various years. (Right): Relation between the effective radius of inhabitation and time. The circles are observed values, and the line is based on the theory (from Skellam, 1951).

where D is diffusivity and α is the net growth rate. [When a species invades and spreads in a habitat where competitors or natural enemies appear to be entirely lacking, we may assume that exponential growth can occur (see Chap. 10).] Solving (6.38) under the initial condition that at $t = 0$, m individuals are concentrated at $r = 0$, we obtain

$$S = (m/4\pi Dt) \exp(\alpha t - r^2/4Dt). \tag{6.39}$$

Integrating (6.39) over the area from $r = R(t)$, i.e., the radius of an approximate boundary of habitat occupied by the species at time t, to $r = \infty$, we find

$$m' = \int_R^\infty 2\pi r S\, dr = m \exp(\alpha t - R^2/4Dt). \tag{6.40}$$

In the above, m' is the total number of animals outside the boundary.

Now m' can be said to have escaped notice by a survey, so that m' must be quite small compared with the total number of animals reproduced during time t, i.e., $m \exp(\alpha t)$. Hence, Eq. (6.40) approximates as

$$R^2 = 4\alpha Dt^2. \tag{6.41}$$

Since the area inside the boundary is πR^2, its square root becomes $2(\pi \alpha D)^{1/2} t$, i.e., proportional to t. The linear relation is thus proved. The slope of the straight line shown in Fig. 6.12 is given by $2(\pi \alpha D)^{1/2}$. Detailed data on muskrat ecology are required in order to estimate α and D independently. This linear relationship seems valid also in cases of the invasion of the nine-banded armadillo (*Dasypus novemcinctus*) into the United States (Humphrey, 1974) and of the spread of larch casebearer (*Coleophora laricella*) in northern Idaho (Long, 1977).

Because ecologists have increasingly documented the environmental degradation due to invasions by exotic species (U.S. Congress, 1993), much recent research has aimed at modeling the range expansions of invading organisms, often using diffusion models (Shigesada and Kawasaki, 1997). Andow et al. (1990) provide an exemplary study in which they estimated rates of population growth and diffusion coefficients over short time frames (weeks to years) and then asked if a simple diffusion model could predict observed range expansions over time scales of decades or longer. For several species the simple diffusion models worked well; those species that failed to match up with the simple predictions exhibited range expansions much larger than predicted, suggesting some form of long-distance transport that was not detected in the abbreviated dispersal experiments. Subsequent to the Andow et al. (1990) study, several researchers have developed modifications of the simple diffusion models that afford better descriptions of more complicated invasion processes. The three complications that are especially noteworthy are: the consideration of population growth with an Allee effect (or reduced per capita reproduction when animals are scarce (Lewis and Kareiva, 1993), spread into a heterogeneous environment (Shigesada et al., 1986), and spread by a mix of standard diffusion and longrange "jumps" (Shigesada et al., 1995). Although these more complicated models are difficult to parameterize from field data, observations of invasions have prompted researchers to take some of these models very seriously. For example, one feature of many invasions that demands an explanation is the very long time lag that is often noticed between when a species first establishes a population in a novel region, and when that populations starts to spread rapidly though its new habitat (Kareiva et al., 1996). Veit and Lewis (1996) suggest that one explanation for such a pattern is reduced rates of population growth at low densities (the abovementioned Allee effect), which seems to explain well the pattern of the house finch invasion in Noah America.

In addition to predicting the speed of an invasion after the invasion has occurred, diffusion models of ecological invasions can be used in a practical manner. For example, using diffusion models of invasion, one can ask what type of hostile barrier might be needed to halt the spread of gypsy moths (Sharov and Liebhold, 1998), and outbreak of rabies (Murray et al., 1986), or the spread of genetically engineered microbes (Lewis et al., 1996). Although such models cannot be exact, they are especially useful as indicators of how much effort would need to be expended to halt an undesirable biological invasion. The analysis of Sharov and Liebhold (1998) is particularly ingenuous, because it redefines the invasion process to consist of the multiplication and dispersal of gypsy moth colonies as opposed to the movement of individual gypsy moths. In this way, a model is produced that is more in accord with what can be observed regarding gypsy moth dynamics, and that gets around the underestimation of spread that arises from observations of individual movements (which typically fail to detect long jumps). Whereas it is practical to detect the appearance of a new gypsy moth infestation far ahead of the advancing population, it is virtually impossible to detect a few individual gypsy moths drifting far away from their source populations.

6.5   The Dispersal of Animal-Borne Plants

This example provides an application of the principles of animal dispersal. We will look at Skellam's calculation (Skellam, 1951) of the rate of post-glacial spread of oak trees in North Britain. According to Reid (1899), in 20,000 years after the melting of the last Pleistocene ice sheet, the oaks advanced about 1000 km. Even if we overestimate the rate of spread result-ing solely from daughter oaks about their parent to be 1 m per year, it would have taken about a million years for the oak forests to be reestablished. Are we thus driven to accept the view that the last glaciation was not as extensive as previously believed, and to suppose that the oak population regenerated from scattered pockets which survived in favorable valleys?

Skellam attacked this problem by reversing the train of reasoning. Ap-plying the same model as that of muskrat spread to the problem, Skellam estimated a "diffusivity" for the advancement of the oaks. The oak does not produce acorns until it is sixty or seventy years old. We thus take the number of generations after the last glaciation to be about 300. It seems safe to assume that the average number of mature daughter oaks produced by a single parent oak was at most 9 million. (Such a figure might be considered outrageously high, but it may be reasonable when taken to be the rate of population growth at the periphery of the advancing front of the oak forest.) We thus have $e^{\alpha} \leq 9,000,000/1$, or $\alpha \leq \ln 9,000,000/\text{generation}$. Substituting into (6.41) this value of α and $R = 1000$ km at $t = 300$ generations, we obtain $D \geq 0.1735$ km^2/generation.

As the mean square distance of daugher oaks about their parent may be defined to be $4D$ per one generation, the mean dispersal distance is defined by the square root of the mean square distance, and hence $2\sqrt{D} \geq 0.83$ km. This distance is much larger than the distance to which the acorns from a parent oak are disseminated. It was thus concluded that animals such as rooks or small mammals (rodents?) must have played a major role as agents of seed dispersal.

In a similar analysis, Cain et al. (1998) used a diffusion plus population growth model to ask whether ant dispersal of seeds could explain the range expansion of woodland herbs following the Holocene retreat of glaciars in North America. Because Cain et al. (1998) were able to directly follow how far ants carried seeds as well as estimate the rate of plant population growth, parameters were experimentally obtained (as opposed to being fit to patterns of spread). By performing this analysis, Cain et al. were able to show that even with ant dispersal of seeds, the realized rate of range expansion following glacial retreat must have involved occasional long-distance colonization events (such as tornados or hurricanes, or seeds clinging to the feet of birds that flew long distances).

6.6   Diffusion Models as a Standard Tool in Animal Ecology

Since the first edition of this book, the application of diffusion models to dispersal has become commonplace. Perhaps the most notable contribution has been the monograph by Turchin (1998), in which practical methods for studying dispersal are discussed, with an emphasis on how to link field data to movement models. Studies of insect dispersal are especially prominent because it is relatively easy to handle and mark thousands of individuals and then recapture them at modest expense. In reviewing studies of animal movement it is important to heed the three "wrongs": wrong but useful, not necessarily that wrong, and woefully wrong.

First, let us consider the "wrong but useful" character of diffusion models. Clearly, the simplest diffusion models cannot possibly be exactly right for any organism in the real world—animals have too much behavior and the environment is far too complicated to be described by elegantly simple diffusion models. On the other hand, these models provide a concise standardized framework for estimating one of population biology's most important and most neglected parameters: the diffusion coefficient. Ecologists routinely estimate and report the intrinsic rate of population growth or the "r" of a logistic equation as representing some ideal rate of multiplication inherent to a species (even though no one expects simple exponential growth to hold, and no one expects this intrinsic multiplication rate to be a constant). Nonetheless, intrinsic rates of population growth are widely appreciated as one of the most fundamental rate constants in ecology.

The same approach should be extended to diffusion coefficients—they should be routinely measured and reported even though the simplest possible diffusion model from which they derive generally does not hold. In fact, the ratio of intrinsic rates of population growth relative to diffusion coefficients tells us a great deal about spatial population dynamics, and our understanding of ecology would be greatly advanced by making diffusion coefficients as common a measured biological attribute as intrinsic rates of population growth.

Second, let us consider the "not necessarily so wrong." Diffusion models are approximations of much more complicated processes. Even though no animals move with infinite velocity following amazingly tortuous paths, the fact is that net displacements of organisms are often described by Gaussian distributions or sums of Gaussian distributions. Specifically, the errors associated with diffusion approximations may be minimal for the range of parameter space in which many organisms operate. For example, Holmes

(1993) found that although the diffusion model could in theory be a terrible approximation of invasion velocity under the assumption of a correlated random walk, the diffusion approximation was expected to work well for the mixture of parameters that seemed to apply to most animals. There is no disputing that diffusion approximations can effectively describe the redistribution of populations that arises as a result of animal movement.

Last, let us consider those circumstances in which a diffusion model is woefully wrong. Such a case might be when animals interact socially so that their movements cannot be understood without attention to social groupings, or when animals are navigating according to some external cue with the goal of moving toward a particular place. But even in these cases, diffusion models provide a concise null hypothesis against which to evaluate complications, and a good first step in understanding animal movement is to quantify exactly how the data disagree with the predictions of a simple diffusion process.

In summary, diffusion models are gaining acceptance as the standard theoretical tool with which to examine patterns of organism dispersal and the consequences of that dispersal. When diffusion models take their place next to Leslie matrices, and Ricker equations or logistic models of population growth, ecology will have finally integrated spatial and temporal dynamics —an integration necessary for theoretical and pragmatic reasons.

7
The Dynamics of Animal Grouping

Akira Okubo, Daniel Grünbaum, and Leah Edelstein-Keshet

Note to the reader: *The study of animal grouping, both theoretical and observational, has undergone an explosive development in the two decades since the First Edition of this book appeared. To maintain the spirit of this book, and to keep our contribution reasonably concise, we have divided our chapter into two sections. The first includes Okubo's original chapter, modified slightly by us and according to revisions he noted before his death. The second includes a brief description of some of the ideas and approaches that are new or that have developed significantly in the recent past, especially those that we believe were most influenced by Okubo and his book. We do not attempt in this small space to be comprehensive, but instead, we refer the reader to many other existing reviews. Among these are Okubo's own excellent review of animal aggregations (Okubo, 1986) and a brief survey of mathematical approaches co-authored with one of us (Grünbaum and Okubo, 1994). Also see Pitcher and Parrish (1993).*

Though the term "grouping" is used very ambiguously in this chapter, we mean it to be a phenomenon such as insect swarming or fish schooling in which a number of animals are involved in movement as a group. In this chapter an attempt is made to describe the motion of swarming individuals, and to comprehend the grouping from the standpoint of advection–diffusion processes.

7.1 Physical Distinction Between Diffusion and Grouping

The frontispiece plate shows two aerial photographs, one of a locust swarm, and the other of a dye patch released at sea. Though size and appearance are similar, the two entities are fundamentally different in that the dye patch will soon diffuse and disperse, whereas the locust swarm will maintain its cohesion while travelling long distances for hours or days, despite considerable randomness in individual movement.

Observing a mosquito swarm, a rather familiar phenomenon on summer evenings, one may get the impression that individual mosquitoes fly completely at random. In fact, they do not exhibit a simple random flight. In so doing, a mosquito swarm would diffuse and soon cease to exist. In reality, the swarm persists for long times with little change in its dimensions ("quasi-stationary"). We must conclude, therefore, that an unknown factor operates against the power of diffusion. Before relating this factor to the behavior and ecology of animal grouping, we shall discuss the physical distinction between diffusion and swarming.

Let's consider a simple one-dimensional case (x-axis). Let $v_i(t)$ be the velocity and x_i the position of the ith individual in a swarm at time t. Starting at the origin ($x = 0$) at $t = 0$, its displacement at time t is given by

$$x_i(t) = \int_0^t v_i(t')\, dt'.$$

We are concerned not with individual values of x_i and v_i but with the statistical characteristics of swarming. As a statistical measure determining the swarm size, i.e., dimension, the variance of the displacements is calculated by averaging the square of x_i over the ensemble. To secure meaningful statistics, it is required that the number of individuals in an ensemble be sufficiently large. Also, for simplicity, we assume that individual movement is symmetrical with respect to the origin so that the group centroid (the average of x_i) remains at the origin. After some manipulation, we obtain

$$\overline{x^2} = 2t \int_0^t R_1(\tau)\, d\tau - 2 \int_0^t \tau R_1(\tau)\, d\tau \tag{7.1}$$

(Hinze, 1959), where

$$R_1(\tau) \equiv \overline{v(t')v(t'+\tau)} \tag{7.2}$$

represents the *velocity correlation function*, an ensemble average of the product of the velocity of an individual at one moment t' and the velocity of the same individual at a time τ later. If the random velocity is statistically stationary, the correlation function depends only on the time lag τ.

For ordinary diffusion, one anticipates an initially strong velocity correlation that weakens as the time lag is increased; the particle velocity loses its statistical dependence on past velocities as the random walk continues. $R_1(\tau)$ might depend on τ as depicted in either of the two fairly typical curves in Fig. 7.1, i.e., R_1 approaches zero at large values of τ. Thus, $\int_0^t R_1\, d\tau$, the area under the curve R_1 from $\tau = 0$ to $\tau = t$, approaches a constant value as t becomes large. We denote this limiting value as D. Similarly, the second integral on the right-hand side of (7.1) also approaches a constant value at large t. However, for large time the first term, which is proportional to t, will

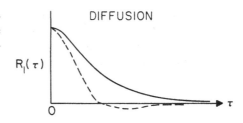

FIGURE 7.1. Relations between velocity correlation function (R_1) and time lag (τ) for particles undergoing turbulent diffusion.

exceed the essentially constant second term, so that asymptotically

$$\overline{x^2} = 2Dt. \tag{7.3}$$

Clearly, Eq. (7.3) is the relation for variance found for purely random dispersal, which is equivalent to Fickian diffusion.

If (7.1) is to describe swarming, then $\overline{x^2}$ must approach a constant for a steady-state swarm to be maintained. This is satisfied if the first integral vanishes asymptotically; therefore, the correlation function, which takes positive values for small t, must take negative values for at least some large values of τ. The $R_1(\tau)$ curve most likely oscillates about zero, as shown in Fig. 7.2. Furthermore, the oscillation must be such that $\int_0^t R_1 \, d\tau = 0$ and the positive areas of the $R_1 - \tau$ curve cancel the negative areas for sufficiently large values of t. In physical terms, the individual motions appear to resemble a periodic oscillation about $x = 0$. For this form of $R_1(\tau)$, the second integral on the right-hand side of (7.1) approaches a negative constant value, which we shall call $-L^2/2$. We thus obtain, from (7.1),

$$\overline{x^2} = L^2. \tag{7.4}$$

The variance attains a constant value; i.e., the size of the swarm becomes stationary.

The above discussion is somewhat formal and abstract. No explanation is given as to how the requisite shape or the correlation function emerges from more fundamental mechanisms or processes. We now calculate R_1 itself using a simple dynamical model. To this end we start with the equation of motion for individual animals (see Sect. 5.2):

$$m\frac{d^2x}{dt^2} = R + K + A. \tag{7.5}$$

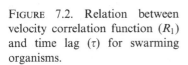

FIGURE 7.2. Relation between velocity correlation function (R_1) and time lag (τ) for swarming organisms.

We assume that (i) the frictional force, R, is proportional to the velocity, (ii) the nonrandom force, K, is attractive toward the center of the swarm and proportional to the distance from the center, and (iii) the random force, A, is a function of time only. Hence, (7.5) reads

$$m\frac{d^2x}{dt^2} = -mk\frac{dx}{dt} - m\omega^2 x + A(t), \tag{7.6}$$

where k is a frictional coefficient and ω is the frequency.

Solving (7.6) for x [see Uhlenbeck and Ornstein (1930) for details] and calculating $v = dx/dt$, we may also obtain the correlation function from (7.2), which is found to be

$$R_1(\tau) = \frac{\varepsilon^2}{2k}e^{-k\tau/2}\left(\cos\omega_1\tau - \frac{k}{2}\omega_1\sin\omega_1\tau\right), \tag{7.7}$$

where ε^2 is the spectrum[1] of the random force $A(t)$ per unit mass and $\omega_1 \equiv (1/2)(4\omega^2 - k^2)^{1/2}$. Using (7.7), the variance can be evaluated from (7.1) and is found to approach

$$\overline{x^2} = \frac{\varepsilon^2}{k\omega^2} = \text{constant} \tag{7.8}$$

as $t \to \infty$. That is, this dynamical model enables us to obtain the variance in terms of parameters that characterize the motion of individuals. The standard deviation, $\overline{(x^2)}^{1/2}$, scales the spatial extent of the swarm.

7.2 Formulation of Swarming by a Generalized Diffusion Equation

We see that mathematical models for the spatial distribution of animals, such as the density pattern within a swarm, cannot be based on the simple random walk. They must include a mechanism that opposes the action of diffusion, i.e., the flux of organisms through a plane perpendicular to the x-axis must consist of at least two components, one random and the other nonrandom.

If we assume a diffusion process for the random component and an advection process for the nonrandom component, these fluxes may be formulated as $-D\partial S/\partial x$ and uS, respectively (Fig. 7.3), where D denotes

---

[1] The spectrum of $A(t)$ is defined to be the Fourier transform of the correlation function of A, i.e., $\psi(f) = \int_{-\infty}^{\infty} \overline{A(t')A(t'+\tau)}e^{if\tau}\,d\tau$, where f is the frequency. When the values of $A(t)$ at two different times are not correlated at all, except for an infinitesimally small value of the difference, the spectrum attains a constant value that is not dependent on f, i.e., white noise. We set this value equal to $m^2\varepsilon^2$, i.e., $m^2\varepsilon^2 = \psi(f)$.

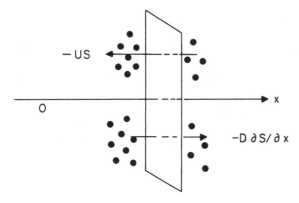

FIGURE 7.3. Organism fluxes due to advection (uS) and diffusion ($-D\partial S/\partial x$).

diffusivity and u denotes the mean drift of individual organisms passing through the plane.

Note that the random flow of organisms occurs from high concentration to low concentration, whereas the nonrandom flow of organisms occurs in the direction of the mean drift. In general, both D and u may vary within a swarm and may also depend on the concentration of organisms.

The total flux is given by

$$J = uS - D\frac{\partial S}{\partial x}.$$ (7.9)

As the swarming is quasi-stationary, the value of J must be the same through any plane. Furthermore, one finds no organisms ($S \equiv 0$) beyond the "swarm boundary," so the flux must vanish; $J = \text{constant} = 0$. From (7.9),

$$\frac{1}{S}\frac{dS}{dx} = \frac{u}{D}.$$ (7.10)

One can solve (7.10) for S as a function of x, given the ratio u/D; conversely, one can estimate u/D from the actual distribution of S as a function of x.

We can also evaluate the advective velocity and diffusivity from the equation of motion for individuals as described in Sect. 5.3 and then construct the advection–diffusion equation. In particular, the advection velocity can be calculated from the dynamical model [e.g., Eq. (7.6)] under the assumption of quasi-equilibrium. Neglecting averaged acceleration, $\overline{dx/dt} = \omega^2 x/k - \overline{A/mk}$. The average of dx/dt over individuals passing through a plane at a fixed value of x, $\overline{dx/dt}$, is the advection velocity, u. Since A is a random force, its average vanishes. We thus find $u = -\omega^2 x/k$. Also, $D = \overline{A^2}T/m^2k^2$ from (5.71). Hence, the diffusion equation for random populations subject to an attractive force proportional to the distance from the center can be expressed as

$$\frac{\partial S}{\partial t} = \frac{\omega^2}{k} \frac{\partial (xS)}{\partial x} + \frac{\overline{A^2}T}{m^2 k^2} \frac{\partial^2 S}{\partial x^2}. \tag{7.11}$$

Equation (7.11) is a Fokker–Plank equation for the Ornstein–Uhlenbeck process (Uhlenbeck and Ornstein, 1930; Wang and Uhlenbeck, 1945). Recent work by Grünbaum (1994) leads to a more detailed derivation of an Eulerian model (one-dimensional PIDE) for density-dependent swarming derived directly from a Lagrangian model incorporating individuals that seek to establish a favored "target" density in the group (see Sect. 7.9.3 for details).

7.3   Insect Swarming[2]

7.3.1   Locust Swarms

Although insect swarming is one of the most extensively studied cases of animal grouping, only a few works have concentrated on the motion of individuals in a swarm. Among these few, studies of the migration and swarming of the desert locust, *Schistocerca greraria*, by researchers at England's Centre of Overseas Pest Research (previously the Anti Locust Centre) have done much to reveal the orientation and flight behavior of swarming individuals. Figure 7.4 is taken from a paper by Waloff (1972). It demonstrates that the orientation of individual insects is somewhat irregular despite the fact that the swarm as a whole drifts downwind. [For mechanisms of directional flight in wind, see David (1986).] It is argued that the behavior at the edges provides a mechanism to keep the swarm together, allowing the swarm to migrate over long distances for many hours without disrupting its coherence. Recent theoretical work (Edelstein-Keshet et al., 1998) has demonstrated that the mechanism of turning back at the edge of the swarm cannot account for swarm cohesion since there is a loss of individuals at the rear edge of the moving swarm.

It is generally accepted that desert locust swarms migrate downwind (Rainey, 1989). Contrary to this general theory, Baker (1978) presented evidence in support of his hypothesis that swarms migrate toward "preferred" directions, but this was later refuted by Draper (1980). See also Baker et al. (1984) for observations of locust swarming with a 16-mm camera. High-speed films of swarms of *Locusta migratoria* were analyzed to measure the locust's body orientation, flight track, height, and speed of flight. In all swarms, the

---

[2] According to T. Ikawa, Morioka College, Japan, since Okubo's book there have been many advances on insect swarming from the perspectives of evolutionary ecology but few on the details of the individual motion within the swarm. The editors would like to thank her for input in this section.

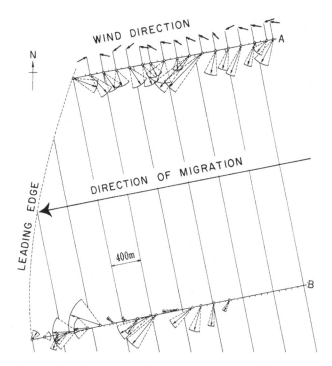

FIGURE 7.4. Orientation of insects in a swarm of Desert Locust. Mean orientations of flying locust with their angular deviations are shown. Mean surface Wind is 5.5 km/h. A: Transect at or near the northern edge of the swarm. B: Transect through a more central part of the swarm (from Waloff, 1972).

mean course angle and mean track angle relative to wind direction were significantly different from zero as the swarm as a whole drifted predominantly downwind. According to Rainey's calculation (Rainey, 1958, 1989), turbulent diffusion associated with a mean wind speed of 5 m/s would spread a swarm of an initial diameter of 8 km into one of 16, 40, and 80 km after 6, 30, and 70 hours, respectively. Aside from the effect of atmospheric turbulence, the random motion of individuals alone would spread a swarm of an initial diameter of 8 km into 16, 32, and 52 km after 6, 30, and 70 hours, respectively.

7.3.2 Experimental Techniques

Radar has provided a useful technique for the study of insect flight (Hardy et al., 1966; Atlas et al., 1970; Schaefer, 1976). An article by Schaefer (1976) contains an outline of the elements of radar entomology along with impressive photographs of radar echoes from flying grasshoppers, crickets, moths, dragonflies, butterflies, and desert locusts. Radar observation enables us to study such factors as day and night airborne insect populations (numbers and densities), species and sex identification, insect air speed and orientation, and wing-beat rates. When a single insect echo is observed, we can track the individual and study its movement pattern. The potential of radar in the study of insect swarms is clearly demonstrated by Schaefer's work. A research program involving the coordinated use of radar, specially instrumented aircraft, and observations from platforms extending above the forest canopy has established the massive scale and regularity of evening take-off flights of spruce budworm moths, *Christoneura fumiferana*, and of subsequent nocturnal dispersal at levels usually high above the ground (Greenbank et al., 1980). This integrated research program has provided new data for moth dispersal and spread of infestations, which are indispensable for insect control (Irwin and Thresh, 1988).

Video techniques have become inexpensive and have thus gained popularity. Images can be transferred directly to the computer. However, low resolution limits their usefulness for tracking movements, analyzing three-dimensional trajectories of flying insects, let alone swarming behavior.[3] Shinn and Long (1986) developed a technique for three-dimensional analysis of insect swarming using a video/digitizer/microcomputer system for use in the field. However, nonvideo-based techniques are presently more precise. See Okubo et al. (1981) on the three-dimensional structure of *Anarete pritchardi* swarms in the field in which they used the shadows of the midges in flight over a white swarm marker to reconstruct the full three-dimensional trajectories. (Other papers on this subject include Dahmen and Zeil, 1984; Riley et al., 1990; Zeil, 1986; and O'Brien, et al., 1986.)

Riley (1994) provides a good review of the application of video techniques for research on insect behavior with many examples of two-dimensional measurements of swarming. Examples of video-based studies of mosquito motion include Peloquin and Olson (1985) and Gibson (1985). The former investigated the events leading to swarm formation, the effect on swarm shape of other insects flying near the swarm, and the way that wind changes the shape and location of the swarm relative to a marker. Gibson (1985) compared flight patterns of male and female mosquitoes, *Culex pipiens quinquefasciatus*. Individual trajectories are elliptical loops whose foci gradually drift with respect to the swarm marker. Individuals do not seem to have preferred positions within the swarm, but appear to drift at random.

7.3.3   *Mosquitoes, Flies, and Midges*

Swarming of dipterous insects is also of interest in the context of population ecology (Syrjamaki, 1964; Downes, 1969). A review of insect swarming from the perspective of behavioral ecology and mating is given by Sullivan (1981). Characteristically, diurnal swarming is performed by a single sex, usually

---

[3] According to T. Ikawa, this is one reason why very few papers about insect movements in swarms have been published since the First Edition of Okubo's book.

male, with activity peaks at dusk and dawn; females are attracted by the swarm (possibly by sound or pheromones produced by the males) and mating results. Swarms consisting mostly or entirely of females have been reported. Female swarms of *Acropygia sp.* (formicidae) are of special interest not only for their number but also for the frequent occurrence of homosexual pairing by females (Eberhard, 1978). Even sexually segregated swarms occur in which both sexes form separate swarms in close proximity (Sullivan, 1981).

Conspicuous objects (stones, trees, water surfaces), animals, human beings, and special ground conditions can serve as swarm markers, i.e., locations about which swarming is initiated and maintained. Allan and Flecker (1989) presented a detailed investigation of a species of mayfly, *Epeorus longimanus* (ephemeroptera), in order to achieve a better understanding of "landmark" swarming. Mayflies usually swarm over water or within 100 m of a shoreline and execute a distinct vertical nuptial dance. The pattern of flight of swarming individuals varies widely from very regular movement, hovering and circling around a definite position in the air, to very irregular, almost random movement (Downes, 1969).[4] The relationship between the number of swarming individuals and the size of the swarm marker has been studied, for example, by Chiang (1961) and Koyama (1962). There is controversy about whether individuals respond independently to a swarm marker (Downes, 1955) or also to one another; Koyama (1962, 1974) suggests that the number of individuals of *Fannia scalaris* Fabricius in a swarm is limited through interaction among swarming flies. He observed that swarming males frequently chased and became entangled with each other. Details of chasing behavior of houseflies, *Fannia canicularis*, have been studied by Land and Collett (1974) and Zeil (1986) by means of high-speed filming. Also, Okubo and Chiang (1974) and Okubo et al. (1977), analyzing a high-speed filming of *Anarete pritchardi* swarming, showed interference between individuals and an inward force at the periphery of a swarm.

Anarete pritchardi is a midge with a length of 2 mm and a wingspan of 3 mm. The peak of swarming activity occurs around noon, and the midges are most active in bright, hot environments. A white swarm marker on the ground induces the insects to swarm in nearly horizontal flight about 5 cm above the ground. Photographic studies reveal that

(i) The swarm moves as a unit to and fro above the swarm marker, with swarm dimension pulsating around a mean size.

(ii) Trajectories of individuals appear to combine two extremes types: a "wide" pattern (relatively straight to-and-fro movement akin to that of a pendulum) and a "tight" pattern (short, zigzag motion similar to random flight).

---

[4] See Richter (1985) and Richter et al. (1985) for similar studies using acoustic scattering techniques in aquatic environments.

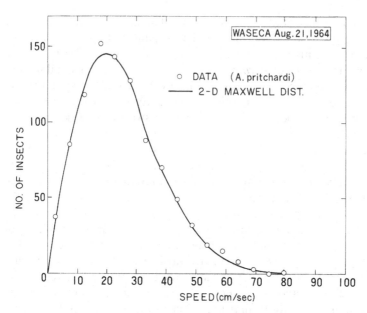

FIGURE 7.5. Speed frequency distribution of swarming midges. Data are compared favorably with the Maxwell speed distribution of gas kinetics (from Okubo and Chiang, 1974).

(iii) The velocity distribution of individuals is Gaussian with a maximum speed of 80 to 90 cm/s. The speed distribution can formally be represented by the celebrated Maxwell distribution of gas kinetics (Jeans, 1952) (Fig. 7.5).

(iv) The concentration distribution of swarming insects is slightly flatter than that of the Gaussian distribution with the same mean and variance: The swarm boundary is distinct (Fig. 7.6).

(v) The velocity autocorrelation coefficient[5] oscillates up and down about the zero line, as one might expect for swarming (see Sect. 7.1) (Fig. 7.7).

(vi) The midge moves with high accelerations, occasionally twice the acceleration of gravity, and it is subject to an inward force that is felt strongly at the swarm edge (Fig. 7.8).

(vii) The energy expenditure of swarming midges is estimated to be 3.5 kcal per kg of body weight per hour.

The ratio u/D can be evaluated from the concentration distribution S according to (7.10). To obtain a tentative value of the diffusivity of an insect near the edge, we take u to be 15 cm/s. This leads to an estimation of

---

[5] The correlation coefficient is a dimensionless form of the covariance in which the correlation function is divided by the average of the square of the variable under consideration. A correlation coefficient may take values only between −1 and +1.

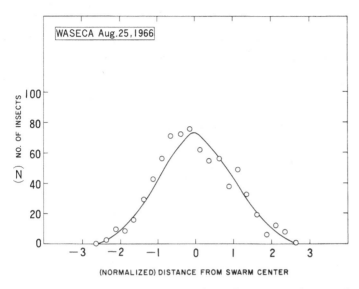

FIGURE 7.6. Concentration distribution of swarming insects (from Okubo and Chiang, 1974).

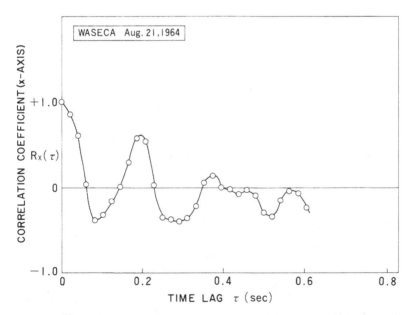

FIGURE 7.7. Velocity autocorrelation coefficient of swarming insects versus time lag (from Okubo and Chiang, 1974). The correlation coefficient is a dimensionless form of the covariance in which the correlation function is divided by the average of the square of the variable under consideration. A correlation coefficient may take only values between −1 and +1.

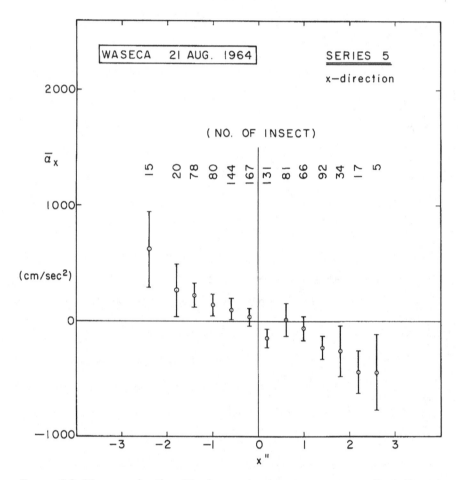

FIGURE 7.8. Mean acceleration ($\bar{a}$) of swarming insects versus normalized distance from the swarm center x'') (from Okubo et al., 1977).

D of 10 cm^2/s. Without a component of inward flow, the swarm would spread at the rate of $2(2D)^{1/2} = 9$ cm/s.

7.3.4   Marine Zooplankton Swarming

As of the First Edition, recent work on the swarming of zooplankton included Hebert et al. (1980) on a predatory copepod, Hamner and Carleton (1979) on copepods in coral reefs, and others. Okubo and Anderson (1984) discussed mathematical models for zooplankton swarms, using a Newtonian equation of motion with white-noise stochastic random force, a nonrandom attractive force toward the swarm center, and friction proportional to the organism velocity. Ambler et al. (1996) and Buskey et al. (1996) have studied a marine copepod *Dioithona oculata* that forms dense swarms in the mangroves of the Caribbean. Using photoreception, these 1-mm crustaceans locate a shaft of sunlight that serves as a swarm marker. Swarming is maintained for hours, despite currents of up to 2 cm/s (necessitating swimming at 20 body lengths per second) at a high energetic cost. The advantage is that the swarm keeps to a safe habitat during the day: Planktivorous fish stay outside the prop root habitat during the day, to avoid ambush by predators. Swarms disperse at night and reform at dawn.

Lobel and Randall (1986) studied swarms of the hyperiid amphipod underwater and noted formation of tornado-shape swirls. Swarms are predominantly females. The variations in size of Antarctic krill were investigated by Watkins (1986). O'Brien (1988, 1989) discusses direct observations of swarms of mysids and the internal structure of these swarms. A review of some of the advances in this area is given by Yen and Bundock (1997).

The advantage of studying zooplankton aggregation is that the phenomenon can often be seen in a small tank in the laboratory, where it can be filmed and investigated thoroughly. Work on plankton aggregation that results from turbulence in the ocean has also been carried out, e.g., by H. Yamazaki, whose research includes mixing and diffusion problems and the coupling of microscale physics and plankton in the ocean. See also Kils (1993) for turbulence-related patchiness in plankton.

7.4   Fish Schooling

Fish schools are much more highly organized than insect swarms. Neighboring fish in a school orient themselves similarly; the distance between individuals is uniform and the motion of individual fish is synchronized (Hunter, 1966). For a review of fish schools, see Pitcher (1998) and Pitcher and Parrish (1993). The reader may find Radakov's (1973) book on the ecology of fish schooling useful. Van Olst and Hunter (1970) studied the organization of schools of Pacific mackerel (*Somber japonicus*), northern anchovy (*Engraulis mordax*), and others, with the aid of photography. Symons (1971) also used photography to study spacing and density in schooling three-spine sticklebacks (*Gasterostens aculeatus*) and mummichog (*Fundulus heteroclitus*). Symons showed that fish were usually spaced regularly when the average distance between nearest neighbors was less than one fish length. At greater distances, i.e., lower density, spacing between fish was found to be either random or in a state of loose aggregations. This work suggests that some critical density is necessary for the occurrence of schooling.

Interfish distances estimated from two-dimensional measurements tend to be underestimated, and the nearest neighbor may even be incorrectly identified (Symons, 1971). Several techniques for the three-dimensional measurement of school structure including stereo and shadow have been developed. Pitcher (1973) determined the three-dimensional structure of fish schooling in an experimental tank by a photographic technique using a mirror. Dill et al. (1981) developed a new stereophotographic technique for analyzing the three-dimensional structure of fish schools.

In his pioneer paper, Parr (1927) discussed the formation of fish schooling in terms of the balance of two counteracting forces, i.e., attraction and repulsion between individual fish. Parr did not, however, formulate these attractive and repulsive forces. Breder (1954), making an analogy to intermolecular forces, assumed that the attraction between fish is accomplished primarily by means of vision.[6] Krause and Treger (1994) and Treger and Krause (1995) devised an experimental setting in which the responses (e.g., speed of motion) of individual fish to other fish and to small groups of fish could be measured. Breder assumed that attraction is inversely proportional to the mth power of the distance between fish, and repulsion (being associated with water movement, sound, odor, or taste in addition to vision) is inversely proportional to the nth power of the distance. Attraction can be expected to have a longer range than repulsion, so that $m < n$. The resultant force is

$$F = \underbrace{-a/r^m}_{\text{attraction}} + \underbrace{b/r^n}_{\text{repulsion}}, \qquad (7.12)$$

where a and b are constants. At some equilibrium distance r_e, given by

$$r_e = (b/a)^{1/(n-m)}, \qquad (7.13)$$

both forces balance exactly and the net force vanishes. (At shorter distances, repulsion dominates, and at longer distances, attraction dominates; see Fig. 7.9.) As a special case, Breder (1954) assumed that attraction (due to vision)

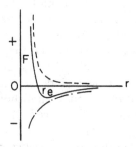

FIGURE 7.9. Attraction ($-\cdot-\cdot-\cdot-$) and repulsion ($----$) between two individuals as a function of separation r. F: resultant force; r_e: equilibrium distance.

---

[6] According to Hunter and Nicholl (1985), vision plays a primary role in the maintenance of most fish schools. Other sense organs, particularly the lateral line, are also important in coordinating movements and spacing of fish within the school (Pitcher et al., 1996), but it is unlikely that the lateral line alone is sufficient for maintaining the integrity of schools at night. (For example, the visual threshold for schooling northern anchovy adults, *Engraulis mordax*, was determined in the lab to be $6 \times 10^{-11} \, W/cm^2$, and the maximum depth of nightly schooling of northern anchovies in the sea is estimated to range from 15 to 40 meters.)

is nearly independent of distance ($m = 0$) and that repulsion varies inversely with the square of the distance ($n = 2$) as in Coulomb's repulsion of electro-magnetism. Hence, (7.12) simplifies to $F = -a + b/r^2$, and the equilibrium distance becomes $r_e = (b/a)^{1/2}$. Observations of various fish schools reveal that r_e is 16% to 25% of the mean body length of fish and that repulsion nearly vanishes at distances beyond one body length. Under normal school-ing conditions, individual fish rarely allow themselves to be separated by more than 40% of their body length.

The attraction–repulsion effects in a swarm or school have been re-visited more recently. For example, Warburton and Lazarus (1991) described another family of attraction–repulsion-distance functions and used simu-lations to investigate the implications of these on group cohesion. In a number of recent Eulerian models for swarming (e.g., in locusts and other species), attraction and repulsion were taken as integral terms with various ranges and density dependence (e.g., Edelstein-Keshet et al., 1998). Mogilner and Edelstein-Keshet (1999) showed that to account for the uniform internal density of most swarms and to prevent collapse of the swarm to a tight cluster, repulsion is essential. Moreover, the density dependence of the repulsion must be more strongly nonlinear than that of attraction.

Crustacea such as mysid shrimp are known to school, although the schooling is by no means organized to the degree of fish schools. Clutter (1969) made detailed observations of the schooling of *Metamysidopsis elongata* and *Acanthomysis* under both natural and laboratory conditions, distinguishing between schools in which the composite individuals swim with similar orientation and swarms with a lesser degree of uniformity in orien-tation. (Generally speaking, fish schools belong to the former class and insect swarms to the latter.)

A bold analogy to molecular motion might proceed as follows: Insect swarms such as *Anarete* correspond to an ensemble of gas molecules con-tained in a box, and fish schools correspond to an ensemble of molecules the state of which is somewhere between that of solids and liquids. This, of course, is nothing more than an analogy.

In 1977, Okubo, Sakamoto, Inagaki, and Kuroki considered the follow-ing dynamical model for fish schooling in one dimension. The equation of motion for each fish was taken to be

$$\frac{d^2 x_i}{dt^2} = -k\frac{dx_i}{dt} - \omega^2 x_i + A_i(t),\qquad(7.14)$$

where $x_i(t)$ is the deviation of the position of the ith-fish from its equilibrium (defined to be the position of the fish at time t at which the attractive and repulsive forces balance each other). The terms on the right-hand side rep-resent the frictional force per unit mass (assumed proportional to the veloc-ity), the mutual attractive or repulsive force per unit mass, (approximated by a springlike restoring force), and the "swimming force" per unit mass.

Okubo et al. considered a two-fish model, with the x_2-fish leading, and x_1-fish following, and responding to the acceleration of the leader with a small but finite time lag, ε (Hunter, 1969; Shaw, 1978). The swimming forces for the two fish were expressed as

$$A_1(t) \equiv A(t), \qquad A_2(t) = A_1(t+\varepsilon) = A(t+\varepsilon), \qquad \varepsilon > 0. \quad (7.15)$$

For small ε, $A(t+\varepsilon)$ can be expanded in a Taylor series, $A(t+\varepsilon) = A(t) + \varepsilon \dot{A}(t) + \left(\dfrac{\varepsilon^2}{2}\right)\ddot{A}(t) + \cdots$, where $\dot{A}(t) = dA/dt$, etc. From (7.14), we can obtain the equation of the centroid velocity, $V(t) \equiv (1/2)(\dot{x}_1' + \dot{x}_2')$ and the equation of the relative separation between two fish, $r(t) \equiv x_2' - x_1'$:

$$\frac{d^2 V}{dt^2} + k\frac{dV}{dt} + \omega^2 V = \dot{A}(t) + \frac{\varepsilon}{2}\ddot{A}(t) + \cdots, \quad (7.16)$$

$$\frac{d^2}{dt^2}\left[\frac{r}{\varepsilon}\right] + k\frac{d}{dt}\left[\frac{r}{\varepsilon}\right] + \omega^2 \frac{r}{\varepsilon} = \dot{A}(t) + \frac{\varepsilon}{2}\ddot{A}(t) + \cdots. \quad (7.17)$$

Hence, V and r/ε obey the same dynamical equation to first order in ε. This is consistent with experimental results due to Inagaki et al. (1976).

The frequency spectrum for the fluctuations of V or r/ε can be calculated from (7.17) by assuming that $\dot{A}(t)$ has a white-noise spectrum, i.e., $\dot{A}$ varies extremely rapidly compared with the variations of V or r/ε. We then obtain the spectral density function, $\psi(f)$, of V or r/ε as

$$\psi(f) = \lambda^2 \{(f^2 - \omega^2)^2 + k^2 f^2\}^{-1}, \quad (7.18)$$

where f is the frequency and λ^2 is the (constant) spectral density of the white-noise variable $\dot{A}$. For small values of f, (7.18) tends to $\psi(f) \sim \lambda^2/\omega^2 =$ constant, whereas for large values of f, (7.18) tends to $\psi(f) \sim \lambda^2 f^{-4}$.

Inagaki et al. (1976) observed the two limiting relations in experiments, using *Gnathpogon elongatus elongatus*. Figure 7.10 shows a comparison between the theory and observations. The values of the parameters used in the theory are estimated for the fish as follows:

$$\omega = 1.3 \text{ s}^{-1}, \qquad k = 2.0 \text{ s}^{-1}, \qquad \varepsilon = 0.2 \text{ s}. \quad (7.19)$$

It may be important to point out that the dynamical model for fish schooling is essentially the same as that applied to insect swarming, (7.6). The only difference appears in the forcing function $A(t)$; in fish schooling, the force between fish is closely related; in insect swarming, each insect is subject primarily to a random force.

Sakamoto et al. (1975) carried out power and bispectral analyses of fluctuation in the separation distance between two fish and thus measured the deterministic (nonrandom) force associated with visual recognition that controls separation distance. The use of the power spectrum facilitates the analysis of the periodic components of fluctuations due to deterministic attraction and repulsion. It is found that the maximum period from one

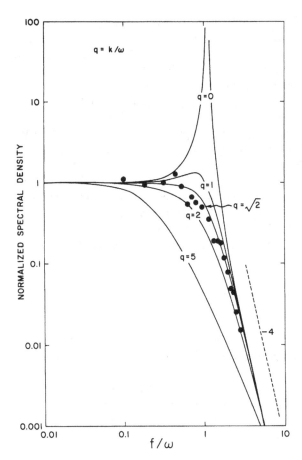

FIGURE 7.10. Frequency spectrum for the fluctuations of the centroid velocity or relative group size for fish schooling. The curves are based on the model, and points are data.

approach to the next is about 100 seconds. Analyses of bispectra show that this periodic component interferes widely with other components of shorter periods.

Kanehiro et al. (1985) performed experiments on the schooling behavior of rose bitterling, *Rhodeus ocellatus ocellatus*, using various fish-school sizes. The frequency distribution of fish swimming speed was fitted to a gamma distribution. Williams (1966) suggested that there is probably little difference between schools containing 19 and 20 individuals but qualitative differences between schools of 2 and 3. Partridge (1980) observed that the internal organization and structure of two-fish schools are quantitatively different from those for larger schools. Time-series analyses with European minnows

(*Phoxinus phoxinus*) show that correlations between instantaneous velocities of individuals increase with school size and as interfish distances decrease. Leader–follower relationships are common in two-fish schools, but they are not seen for schools with more fish. Generalizations about larger schools cannot be made from observations on single pairs of fish.

Aoki (1980) used movies to study behaviors such as internal organization and communication process. In 1986 he demonstrated by simulation that individual differences in schooling behavior play an important role in position preference within a school or leading tendency. Pitcher and Wyche (1983) describe predator-avoidance behaviors of sand-eel schools. Sand-eels exhibit ten characteristic schooling maneuvers: ball, cruise, avoid, herd vacuole, hourglass, split, flash expansion, join, mill. These schooling maneuvers can be generated by combinations of four behavioral mechanisms of individual fish: 1) schooling tendency; 2) startle response; 3) nearest-neighbor distance maintainance; and 4) minimum approach distance allowed for intruders. (For additional work with the juvenile whiting *Merlangius merlangus*, see Hall et al., 1986.)

Aoki (1982, 1986) simulated the movement of a fish school. Two components of movement of an individual, i.e., swimming speed and direction, are assumed to be stochastic variables characterized by certain probability distributions. For simplicity, interactions between individuals are restricted to the directional component (heading). [See also Warburton and Lazarus (1991).]

Recent modeling work for fish schools has been published by Niwa (1994, 1996, 1998), who considers N fish swimming in three dimensions, and includes both deterministic and stochastic effects. In his models, the individual behavior of a fish is modeled by the equation

$$\frac{d\mathbf{v}}{dt} = f(\mathbf{v}) + \eta(t), \qquad (7.20)$$

where $\mathbf{v}$ is the swimming velocity, $f(\mathbf{v})$ a locomotory force, and $\eta(t)$ a fluctuating force assumed to follow a Gaussian white-noise distribution. Niwa assumes that the fish acceleration is proportional to the gradient of a specific energy cost of transport, $E = \alpha(\beta v + 1/v)$, and shows that a reasonable representation of the locomotory force is $f(v) = \kappa(1 - \beta v^2)v$. He then considers the case of N fish swimming in a school, with superimposed forces due to locomotion and to interactions with neighbors. While the grouping force tends to establish a preferred spacing between the neighbors, the arrayal force results in polarized formations.

The nonlinear stochastic equations that result cannot be solved, but various limiting cases are discussed by Niwa. He shows that the main features of the transition between order and disorder can be predicted by the synergetics analysis (Haken, 1983) and that the group behavior changes dramatically as certain parameters are varied. The random motion of the individuals and their nonlinear interactions play a fundamental role in the schooling behavior, and the following results are discussed:

(1) When noise dominates over the tendency to polarize, the center of mass of the group hardly moves, but the group undergoes an amoebiclike motion. When noise is dominated by polarization, the school begins to perform an overall rectilinear motion, with individuals forming a polar school (Niwa, 1994).

(2) If the variance in the individual velocities is beyond some critical level, a school cannot form without some external forces (e.g., effect of a predator). An analogy is drawn with the atomic dipoles in a laser.

(3) Fluctuation of the centroid velocity is needed to lead to school formation, i.e., noise can actually be "beneficial" in school formation: The onset of schooling t_0 is a minimum when the fish "tune" the randomness of the motion to some particular level. For a large school, noise is suppressed, and time of onset lengthens.

7.5 Simulation Model for Animal Grouping

Sakai (1973) and Suzuki and Sakai (1973) constructed a simulation model for animal grouping. Their dynamical model is based on Newton's equation of motion (7.5) under the assumption that the frictional force is proportional to the velocity and the random force is a function of time only, so that for the ith individual of a group ($i = 1, 2, \ldots, N$) it is written

$$m\frac{d^2x_i}{dt^2} + mk\frac{dx_i}{dt} = K_i + A_i(t). \tag{7.21}$$

The nonrandom forces K_i are considered to consist of the following three characteristic forces:

(1) *Forward thrust, K_f.* An animal has a tendency to continue moving forward with its velocity. Thus, the forward thrust is assumed to be expressed as

$$(K_f)_i = a\frac{\dfrac{dx_i}{dt}}{\left|\dfrac{dx_i}{dt}\right|}, \tag{7.22}$$

where a is a positive constant called the *coefficient of thrust*. When an individual is moving with a constant speed under the forward thrust only, its velocity v_0 is given by a/km. We may call v_0 the *proper velocity*.

(2) *Mutual interaction, K_m.* Like Parr-Breder's model, an individual is subject to attraction and repulsion from other individuals in a group. These mutual interaction forces are assumed to depend on the distance between two individuals and are expressed as

$$(K_m)_i = \frac{1}{N-1}\sum_{\substack{j=1 \\ j\neq i}}^{N} Q(R_{ij})(x_j - x_i)/R_{ij}, \qquad (7.23)$$

where $R_{ij} = |x_j - x_i|$ is the distance between the ith and jth animals, and Q is a function of R_{ij} only; Sakai and Suzuki assume that $Q = -c_0(R_0 - R_{ij})$ $(c_0 > 0)$ when $0 < R_{ij} < R_0$, $Q = c > 0$ when $R_0 < R_{ij} < R_1$, and $Q = 0$ when $R_1 < R_{ij}$. Thus, an individual is subject to a linearly varying repulsive force from another animal when two come closer than R_0 and to a constant attractive force when the distance between two animals is larger than R_0 and less than R_1; at distances larger than R_1, no interaction forces belong to this category.

(3) *Arrayal force, K_a.* Neighboring animals in a group have a tendency to settle into an array. In other words, two neighboring animals tend to equalize their velocities. It is assumed that this action takes place only when an individual finds the neighbors within a sphere of influence, and the mean arrayal force on the ith individual is expressed as

$$(K_a)_i = \frac{1}{M_i}\sum_{j\in V_i} h\left(\frac{dx_j}{dt} - \frac{dx_i}{dt}\right), \qquad (7.24)$$

where h is the coefficient of arrayal force, $V_i = (4/3)\pi l_i^3$ represents a sphere of influence around the ith individual (l_i is the radius of the sphere), and M_i is the number of animals within the sphere.

In summary, the equation of motion for the ith individual in a group can be written as

$$m\frac{d^2 x_i}{dt^2} + mk\frac{dx_i}{dt} = a\frac{\frac{dx_i}{dt}}{\left|\frac{dx_i}{dt}\right|} + \frac{1}{N-1}\sum_{\substack{j=1 \\ j\neq i}}^{N} Q(R_{ij})\frac{x_j - x_i}{|x_j - x_i|}$$
$$+ \frac{1}{M_i}\sum_{x_j\in V_i} h\left(\frac{dx_j}{dt} - \frac{dx_i}{dt}\right) + A_i(t). \qquad (7.25)$$

Because of the difficulty in solving (7.25) analytically, Sakai (1973) and Suzuki and Sakai (1973) conducted computer simulation to examine the time variations of grouping pattern. The parameters considered as variables are $a, b \equiv A_i$; c, h; and $l \equiv l_i$; while m, k, and N are fixed constants in the simulation. As a result, the following basic group movements are obtained [Fig. 7.11(a)–(c)]:

(i) *Amoebic movement.* When the random force dominates the forward thrust, the center of mass of animals hardly moves, though the shape of the group fluctuates around a circular pattern [Fig. 7.11(a)].

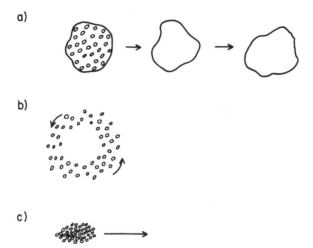

FIGURE 7.11. Basic group movements. (a) Amoebic movement, (b) doughnut pattern, (c) rectilinear movement.

(ii) *Doughnut pattern.* When the forward thrust dominates the random force, a group of animals rotates around an empty center, forming the shape of a doughnut [Fig. 7.11(b)].

(iii) *Rectilinear movement.* When the arrayal force is added to the other forces, animals as a whole tend to perform a rectilinear movement, thus forming a tight (cohesive) group [Fig. 7.11(c)].

The combined effect of these forces on the grouping pattern is shown in Fig. 7.12. We can see that the arrayal force may act to deform a rotating doughnut-shaped group into a tighter group with a rectilinear movement. Details of the temporal change of the group pattern are illustrated in Fig. 7.13.

Sakai and Suzuki also dealt with the movements of a group under the influence of a solid boundary and with the interaction of two moving groups each consisting of an equal number of animals but having different characteristic parameters. Their simulation model may be used to interpret some characteristics displayed by fish schools such as milling (Parr, 1927; Breder, 1951), podding (Breder, 1959), and "waves of agitation" (Radakov, 1973; Sannomiya and Matuda, 1981).

In the models just mentioned, as well as in many others in the literature, "forces" representing behavioral interactions with multiple neighbors are simply summed. This represents an assumption that social interactions with any given neighbor are strictly a function of relative position and are not affected by the presence or absence of other neighbors. While the real behavior of individuals in schools is certain to be more complicated, there is as yet very little experimental information on how fish balance responses

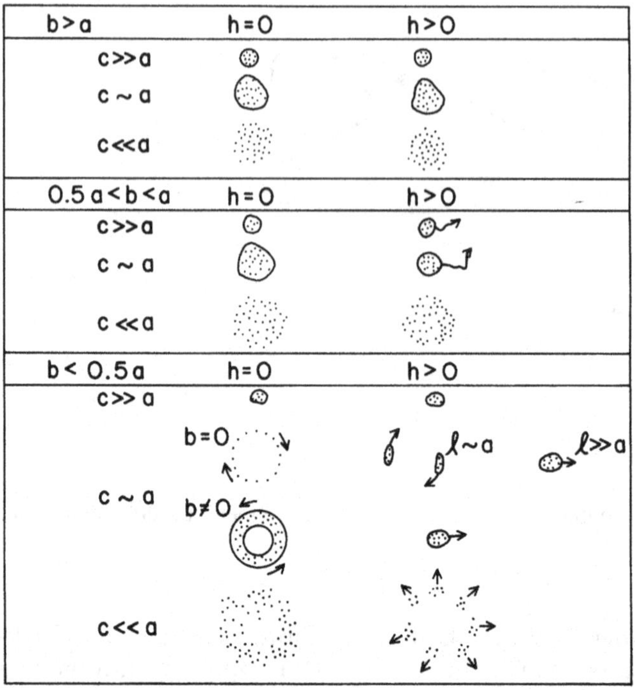

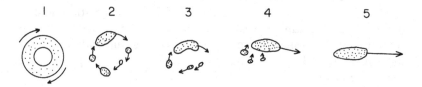

b > a	h = 0	h > 0
c >> a		
c ~ a		
c << a		
0.5 a < b < a	h = 0	h > 0
c >> a		
c ~ a		
c << a		
b < 0.5 a	h = 0	h > 0
c >> a		
c ~ a	b = 0 / b ≠ 0	ℓ ~ a / ℓ >> a
c << a		

FIGURE 7.12. Typical patterns of animal grouping for various values of the parameters characteristic of acting forces.

1   2   3   4   5

FIGURE 7.13. Pattern change from a doughnut-shaped group to a group with rectilinear movement.

to multiple neighbors. However, some preliminary investigations have been carried out using simulation models. Aoki (1982, 1986) made simulation experiments on the movement of a fish school. Two components of movement of an individual, i.e., swimming speed and direction, are assumed to be stochastic variables characterized by certain probability distributions. For simplicity, interactions between individuals were restricted to the directional component (heading). [See also Warburton and Lazarus (1991).] Of particular interest in Aoki's model is the way in which responses to multiple neighbors are explicitly represented: At any one instant, a fish responds to only a single neighbor, chosen at random from nearby fish with probabilities that depend on relative position. Thus, over many behavioral decisions, the effect is akin to a weighted average of responses to neighbors. However, Huth and Wissel (1990, 1992) used Aoki's scheme to explicitly compare the school characteristics that result from this individual-based "decision" algorithm to those resulting from an explicit "averaging" algorithm in which a fish responds to the heading of all nearby neighbors simultaneously. They concluded that the averaging scheme results in greater robustness and more realistic degrees of polarity and that fish were therefore more likely to use such group-level algorithms. Issues surrounding the modulation of responses to multiple neighbors remain a fruitful area for theoretical and experimental research.

Vabo and Nottestad (1997) used a cellular automaton model to examine group-level movements and shape changes resulting from individual-based schooling responses to neighbors and avoidance of predators. They found that many featues of real schools (fusion, fission, vacuole, fountain, etc.) are qualitatively reproduced by such a model. Krakauer (1995) modeled the cognitive effect of schooling on the predator, where a neural net representing the predator's visual processing suffered "confusion effect" from numerous similar prey. Romey (1996) simulated schools composed of fish with variable behavioral parameters. He found that relatively slight behavioral changes in a small minority of school members may nonetheless have a disproportionate effect on group-level movements.

Parrish (1992) discussed the mechanisms by which predators with various attack strategies (marginal versus invasive, solitary versus group predators, etc.) and various responses to school characteristics (such as confusion effect) differentially "shape" the schooling behavior of their prey. Landa (1998) used a "club-theoretic" bioeconomic approach to examine related evolutionary mechanisms, the outcome of which is apparently social behavior on the part of self-interested fish. Nonacs et al. (1994) constructed a behavioral model of the Northern anchovy (*Engraulis mordax*) that included detailed quantitative descriptions of metabolic costs, encounter rates with patches of their zooplankton prey, and size-dependent mortality from predation. They used a Lefkovitch matrix model for population dynamics, and dynamic programming (Mangel and Clarke, 1988) as an optimization tool to specify the preferred swimming speed inside and outside prey patches. They concluded that anchovies do not grow at their maximum potential rate, but instead trade reduced growth for reduced predation risk. These results were extended by Nonacs et al. (1998), who found that measured gut contents of juvenile anchovies (Loukashkin, 1970) were more consistent with modeled survivorship-maximizing than growth-maximizing behaviors.

7.6   The Spilt and Amalgamation of Herds

When several herds occupy a limited space, interaction between the groups may occur so that two of them meet and amalgamate or so that a large herd splits into two smaller ones. The joining and splitting of animal groups is

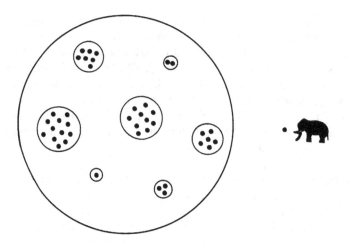

FIGURE 7.14. Mathematical model of the size of elephant herds.

analogous to coagulation–fragmentation processes in physical and chemical sciences, which have been treated quite extensively elsewhere.

An example of a biological treatment of this problem is Holgate's (1967) stochastic model of the size of elephant herds. Suppose that a given area contains a total of N elephants (Fig. 7.14). Let $P_k(t)$ denote the probability that at time t these are grouped into k herds. Clearly, $1 \le k \le N$. The chance that two herds meet and amalgamate will increase with the number of herds, k. There will be no chance of amalgamation when all the animals are grouped into one herd. Hence, if there are k herds at time t, the probability of an amalgamation occurring in a small time interval $(t, t + \Delta t)$ is $\mu(k - 1)\Delta t$, where μ is a coefficient of amalgamation.[7] We next consider the probability of splitting. The chance of a herd of size x_i splitting into two will increase with x_i and may be assumed to be proportional to the excess of the size over unity, $x_i - 1$. Therefore, the probability that one of the herds in existence at time t will split into two herds during the time interval between t and $t + \Delta t$ is $\lambda \sum_{i=1}^{k}(x_i - 1)\Delta t = \lambda(N - k)\Delta t$, where λ is a coefficient of splitting. We furthermore assume that the probability that more than two herds amalgamate or that a herd splits into more than two herds can be ignored.

Thus, the rate of change of $P_k(t)$ during $(t, t + \Delta t)$ is equal to the sum of the probabilities that $(k - 1)$ herds at time t split into k herds during the time interval and that $(k + 1)$ herds at t amalgamate into k herds during Δt, minus the probability that k herds at t split into $k + 1$ herds or amalgamate

---

[7] Boyd (1979) assumes the probability of amalgamation of two herds to be proportional to the number of chance meetings of herds. The probability of amalgamation is then given by $\mu k(k - 1)$. Boyd's stationary distribution has an appreciably smaller variance than that of Holgate.

into $k - 1$ herds during Δt. Accordingly, we obtain

$$\frac{dP_k(t)}{dt} = \lambda\{N - (k - 1)\}P_{k-1} + \mu k P_{k+1}$$

$$- \{\mu(k - 1) + \lambda(N - k)\}P_k \quad (k \neq 1, N), \tag{7.26}$$

$$\frac{dP_1(t)}{dt} = \mu P_2 - \lambda(N - 1)P_1, \tag{7.27}$$

$$\frac{dP_N(t)}{dt} = \lambda P_{N-1} - \mu(N - 1)P_N. \tag{7.28}$$

Although a time-dependent solution of (7.26) has been found (Bunday, 1970; Fielding, 1970), we are interested here in the stationary probability distribution that may be obtained after the passage of considerable time. If we set the left-hand side of (7.26) equal to zero, the stationary probability P_k^* can be obtained; after some reduction, it is found that

$$P_k^* = \frac{(N - 1)!}{(N - k)!(k - 1)!} \left(\frac{a}{1 + a}\right)^{k-1} \left(\frac{1}{1 + a}\right)^{N-k}, \tag{7.29}$$

where $a \equiv \lambda/\mu$. Equation (7.29) is the binomial distribution.

When there are k herds, the average herd size is N/k, and hence the long-term averaged herd size, i.e., the statistical expectation of N/k, is

$$q = \sum_{k=1}^{N} \frac{N}{k} P_k^* = \frac{(1 + a)^N - 1}{a(1 + a)^{N-1}}. \tag{7.30}$$

Clearly,

for $a \to 0$, $q \to N$: the entire population forms one herd;

for $a \to \infty$, $q \to 1$: the entire population divides into single individual herds.

This simple model will require significant refinement or modification if ever applied to real herds. Morgan (1976) has developed several stochastic models that are more general than that of Holgate. Some useful data of the frequency distributions or herd sizes are given by Laws et al. (1975) with regard to elephants, by Sinclair (1977) with regard to the African buffalo in the Serengetti, and by van Lawick-Goodall (1968) with regard to chimpanzees in the Gombe Stream Reserve. For theory of fish-school–size distributions, see Anderson (1981). For data, one can consult Seghers (1981), who found an exponential distribution of spottail shiner groups.

A recent mathematical treatment of group-size distribution related to the group-size–dependent rates of splitting and merging of groups is given in Gueron and Levin (1995) and Gueron (1998). The treatment incorporates stochastic model formulation as an ergodic Markov-chain process together with some Monte Carlo simulations. Determination of the group-size distribution given a set of rates for the coagulation and the fragmentation, and application to animal group sizes, is included.

7.7   The Ecological or Evolutionary Significance of Grouping

Let us pose the following very difficult question: "Why do some animals group?"

Galton (1981) observed the behavior of half-wild herds of cattle in South Africa and noted that if an ox was forced to separate from its herd, the animal strove with all its might to return to the herd. Furthermore, lions apparently preferred to attack isolated or marginal beasts. Galton considered gregarious behavior as a form of cover-seeking in which each animal tries to reduce its chance of being caught by a predator by positioning itself within the herd.

Later, Williams (1964) independently applied this concept to fish schooling. Thus, fish tend to form a "tighter" (denser) school when nearby shelter, such as seaweed and rocks, is absent. Also, according to Breder (1959) and Springer (1957), any sudden stimulus causes schooling fish to cluster more tightly, and in the presence of predators they are observed to pack into a ball so tight that they can hardly swim. The views of Galton and Williams might also be applied to bird flocking and insect swarming.

In his paper about the "selfish herd," Hamilton (1971) presented a model of frogs on the edge of a circular pond inhabited by a water snake that emerges to snatch the nearest frog occasionally (Fig. 7.15). Each frog tries to find the "safest position" along the fringe of the pond. Hamilton assumed

FIGURE 7.15. Model of a colony of frogs and a water snake.

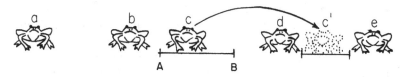

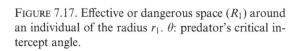

FIGURE 7.16. Frog jumping as a mean of narrowing the "domain of danger."

that a half-length of the gap between the neighbors on either side is the frog's "domain of danger" (Fig. 7.16) and that the frog attempting to improve its situation will jump into a narrower gap to reduce the size of this domain. In Fig. 7.16 the domain of danger for frog c is AB and the frog jumps from c to c' to narrow the domain. Then frog b realizes that its domain has been increased, so it too will jump. This avalanche of jumping results in one or more aggregations of frogs. Hamilton concluded that "selfish avoidance of a predator can lead to aggregation." However, this does not guarantee that aggregation is advantageous for the prey, and indeed, these aggregates may be more easily discovered by a predator than single individuals.

Cannings and Cruz Orive (1975) examined Hamilton's concept and commented that Hamilton's hypothesis is rather convincing in one- and two-dimensional cases but less so in three dimensions. In certain models, an individual may diminish its chance of being caught by a predator by leaving the group. Vine (1971) discussed a similar model of selfish aggregation for the case where the approaching predator is detected while still outside the aggregation. It indicates, as in Hamilton's case, that the most probable dispersion pattern is a tight circular flock of all the prey animals in the vicinity.

From a purely geometrical standpoint, grouped prey tend to be detected more easily than solitary prey. Let a spherical region of radius R_1 around an individual animal denote the effective (or dangerous) space within which the animal is detected by a predator. R_1 is considered to be proportional to the size of the individual, say, its radius r_1 (Fig. 7.17). This is the case where the predator's sight range is limited solely by the critical intercept angle.

For N solitary individuals, the total effective space is given by $(4/3)\pi R_1^3 N$. For one group of N individuals, we may regard the group as a single large

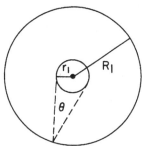

FIGURE 7.17. Effective or dangerous space (R_1) around an individual of the radius r_1. θ: predator's critical intercept angle.

"individual." The radius of the effective space for the group is then given by $N^{1/3}R_1(r_1 + \rho)/r_1$, where ρ represents the mean distance between individual members in the group. Hence, the effective space for the group amounts to $(4/3)\pi R_1^3 N\{(r_1 + \rho)/r_1\}^3$, which is $(1 + \rho/r_1)^3$ times the total effective space for N individuals in solitude. If N_e ($\leq N$) denotes the number of individuals a predator can consume at one meal, the ratio of the probabilities of being eaten when grouped and when solitary is given by $N_e/N(1 + \rho/r_1)^3$. For tight groups consisting of numerous individuals, $\rho \sim r_1$, while $N_e \ll N$, so that grouping is advantageous insofar as reducing predation is concerned.

This is the so-called dilution effect (Foster and Treherne, 1981), i.e., for any one predator attack, the larger the group of prey animals, the smaller is the chance that any particular individual will be the victim. Foster and Treherne (1981) presented field observations of predation on a marine insect in which it was possible, for the first time, to quantify the magnitude of the dilution effect and distinguish[8] it from other benefits such as improved avoidance behavior of group living. By the same token, the detection probability becomes considerably higher for insect swarms, a characteristic of which is the condition $\rho \gg r_1$. This condition may be interpreted to be beneficial for a mating swarm, as it is conducive to the attraction of females to swarming males. In this case the most favorable position for a male is probably not at the center of the swarm, but at the periphery.

Brock and Riffenburgh (1963) developed a similar model for predator detection appropriate for fish schooling. Due to backscatter and light absorption in the sea, the predator's sight detection is limited by the contrast difference between the prey and the background rather than by the intercept angle at the eye. Brock and Riffenburgh thus assumed that regardless of prey size, predators have finite detection ranges. The ratio of probabilities of being eaten when grouped and when solitary is then given by $(1 + bN^{1/3})^3(N_e/N)$, where b denotes a nondimensional factor proportional to the ratio of the average spacing, c, within the school of size N, and the finite detection range R.

For fixed N, as c approaches zero, the ratio of the probabilities approaches N_e/N, which implies that as the prey in a school draw close together, a predator must be able to consume an increasing portion of the entire school in order to nullify the schooling advantage.

Clark (1974) incorporated Brock–Riffenburgh's model into a mortality rate equation of prey and derived the stock-recruitment relationship for certain schooling species.

Pulliam (1973) provided a simple calculation that demonstrates an advantage accrued to flocking birds from a purely defensive standpoint, in that it enhances predator detection. Suppose that each bird in a flock cocks its head

---

[8] The insects could not see their fish predators under the water surface; therefore, group size could not in this respect improve predator detection or enhance avoidance behavior.

TABLE 7.1. The probability of detecting a predator versus flock
size for three values of $\lambda\tau$; λ is the mean rate of head cocking
per minute, and τ is time necessary for a predator to make its
final uncovered approach on the flock

Flock size	Probability		
	$\lambda\tau = \frac{1}{2}$	$\lambda\tau = 1$	$\lambda\tau = 2$
1	0.394	0.632	0.865
2	0.632	0.365	0.982
3	0.777	0.950	0.998
4	0.565	0.952	−1.000
5	0.918	0.993	−1.000
6	0.950	0993	−1.000

instantaneously at a mean rate of λ times per minute and that the individuals
cock independently of each other. For simplicity, we assume that the length
of the intervals between cocks is random, i.e., the cocking takes place according to a Poisson probability law. The probability that the time between
cocks for any one bird is larger than some time t is then given by $e^{-\lambda t}$. Under
the assumption of the independence of cocking between individual birds, the
probability that the time passing between one cock and another in the total
flock (N birds in all) is larger than t is given by $(e^{-\lambda t})^N = e^{-N\lambda t}$. Therefore,
the probability that at least one individual cocks its head in a time interval of
t minutes is $1 - e^{-N\lambda t}$.

If τ is the time necessary for a predator to make his final open assault on
the flock, the probability of detecting the predator is given by $1 - e^{-N\lambda\tau}$. As
birds concentrate more on feeding, the value of λ becomes smaller (see
Cresswell, 1994), so that the flock requires a greater size to detect a predator
(see Table 7.1). At any rate, there is a finite amount of risk of being attacked
by a predator or predators, so that some of the peripheral animals will be
sacrificed. Thus, one would expect more dominant individuals to be located
toward the center of a group. Elgar and Catterall (1981) tested Pulliam's
model using house sparrows (*Passer domesticus*), and their results support its
prediction of an inverse relationship between the time spent in predator surveillance (head cock rate) and the flock size, N. However, their data did not
fit the model exactly, and they attributed this to an unrealistic assumption
of headjerks and of negligible duration. In Pulliam's model: $P = 1 - e^{-\lambda N t}$.
Therefore, $\lambda = -\ln(1 - P)/Nt$.

Treherne and Foster (1980, 1981) made field observations on the ocean
skater, *Halobates robustus* (Hemiptera: Gerridae) and showed a clear effect
of group size on the distance at which an approaching model predator
induced avoidance behavior. The data can be fitted well to an exponential
model for the number of individuals in a group and the distance of response
as predicted by Pulliam (1973).

Elgar (1989) critically reviewed predator vigilance and group size in mammals and birds. Siegfried and Underhill (1975) conducted a field experiment to investigate the initial detection of a predator and the escape reactions by wild laughing doves, *Streptopelia senegalensis*, while feeding as members of flocks of different size. They showed that, generally speaking, there was a linear relationship between flock size and response speed; the larger the flock, the quicker the reaction by the birds. However, there is a complication that birds in very small flocks (3 or fewer birds) and very large flocks (over 15) did not behave "normally" in the experiment. A plausible mechanism by which metabolic requirements of movement might be reduced is a possible biomechanical advantage to coordinated movement within organized groups such as schools and flocks. As another effect of aggregation, a closely packed school of fish may provide an example of collective mimicry, the size and rate of movement of the school perhaps serving to frighten or discourage predators in some circumstances (see Springer, 1957). Wynne-Edwards (1962) suggests that aggregation serves to make individual members aware of the current level of population density and that this awareness produces feedback on reproductive behavior, which serves to control population size.

Thus far, we have viewed the animal grouping primarily from the standpoint of avoidance of predation on the part of the prey. In reality, there is an unremitting struggle between predator and prey, the former seeking to detect and the latter to conceal itself, and this contest generates pressure that may favor group formation. Treisman (1975a and b), examining the contribution of predation to the evolution of gregariousness, suggests that the needs of both predator and prey to conceal themselves and to detect the other may have been a major factor directing the development of social behavior. Some kinetic physical aspects of predator–prey chases, such as the capture of prey or predator avoidance, will be discussed in Chap. 8.

The growth of many biological systems may be limited by the diffusion of nutrients to the system or the diffusion of toxic metabolites away from the system. The diffusion-limited aggregation model (Witten and Sander, 1982) may provide new insight into the (fractal) pattern formation of certain biological systems.

It has recently been shown by Grünbaum (1998) that schooling behavior can, under certain conditions, improve the performance of organisms that seek food by moving up gradients. The alignment with others eliminates some of the stochastic effects that would result from individual sampling errors, but grouping that is too tight may hinder the group's ability to respond to changing gradients or conditions. Analysis of the optimal level of schooling for a given scale of spatial and temporal fluctuation is given. This may explain schooling in Antarctic krill, anchovies, and herring.

The interaction between schooling behavior (that has evolved in the presence of natural predators) and harvesting strategies by novel predators (human beings) is all too often catastrophic for fish populations. The dependence of catch-per-unit-effort and catchability populations of target species may be modeled with various functional responses, some of which are more susceptible to "catchability-led stock collapse" (CALSC) (Pitcher, 1995). For example, the Schaefer production model assumes logistic population growth,

$$\frac{\dot{P}}{dt} = rP\left[1 - \left(\frac{P}{P_\infty}\right)\right] - qEP,$$ (7.31)

where P is population biomass, P_∞ is the carrying capacity, r is the rate of increase, and E is the fishing effort. q is the catchability coefficient, which is a constant in this model. The sustainable yield is

$$Y_e = qEP_\infty[1 - (qE/r)],$$ (7.32)

from which it can be seen that the catch-per-unit-effort (Y_e/E) decreases as the population is depleted. Pitcher argues that, with modern fishing equipment, catch-per-unit-effort is nearly constant through a wide range of population densities. Replacing this functional form into (7.31) gives

$$Y_e = [qEP_\infty^2/P]\left[1 - \left(\frac{qEP_\infty}{rP}\right)\right],$$ (7.33)

implying that catchability is inversely proportional to population (i.e., very large at low population levels). This dramatic increase in the mortality risk (catchability) of an individual fish may lead to sudden and catastrophic population collapse.

7.8 Linking Individuals, Groups, and Populations: The Biological Context of Mathematical Models of Grouping

Among the most interesting and challenging aspects of social grouping behaviors are that multiple organizational levels—individuals, groups, populations, ecosystems—are interrelated and that a coherent theoretical understanding of anyone of these scales must involve an understanding of the others. To draw meaningful conclusions about the evolution of a behaviors, we must have a concrete idea of what types of groups may arise from individuals exhibiting that behavior, how a population may be distributed among groups, and how a population with that particular distribution of groups might interact with other components of a complete ecological system. Because establishing these linkages is such a difficult task, it is worth discussing the importance of each level and identifying what in fact is the biological motivation behind the mathematical modeling. *Individuals* are clearly necessary to consider because behavioral mechanisms operate at the level of the individual and are modified by individual-level factors such as size, physiological state, learning, etc. Furthermore, natural selection acts on individuals' genes, and a behavior will not evolve unless individuals with particular genes are more successful in propagating those genes relative to competing individuals' genes. Finally, many of our observational techniques are based on individuals, and to utilize data from these observations, it is helpful to have models whose parameters relate to individuals.

At the same time, however, the costs and benefits of social behaviors necessarily depend on the characteristics of groups that arise from individuals expressing those behaviors (perhaps in combination with individuals expressing dissimilar behaviors). That is, an individual must alter its foraging, predator avoidance, or migratory behavior to remain part of a group, and whether it gains or loses by doing so depends on how effectively the group performs an analogous set of tasks on behalf of its members. These group characteristics are "emergent" properties of assemblages of individuals, which are, in general, not specified directly by individual behaviors. It seems likely that many desirable combinations of group characteristics are not attainable through individual-level behaviors. For example, some group characteristics may be impossible to realize because sensory limitations make the necessary behaviors physically impossible, or because those behaviors, while favorable at a group level, may be selected against at the individual level (e.g., altruism). Thus, a complete theoretical understanding of social aggregation behavior will explicitly relate constraints on individuals to trade-offs in group characteristics.

A broader biological motivation for studying grouping behavior is that aggregation dramatically modifies the demographic dynamics of a population. Social behaviors that on short time and space scales lead to the formation and maintenance of groups, and at intermediate scales lead to size and state distributions of groups, lead at larger time and space scales to differences in spatial distributions of populations and in rates of encounter and interaction with populations of predators, prey, competitors, and pathogens, and with the physical environment. At the largest time and space scales, aggregation has profound consequences for ecosystem dynamics and for evolution of behavioral, morphological, and life history traits.

Thus, the theoretical goals of biological studies of grouping behavior may be said to include quantitatively associating individual behaviors through group characteristics to the ecological implications of those behaviors. Mathematical modeling approaches are, without question, an essential set of tools to link these disparate organizational levels. Indeed, a principle direction of mathematical innovation in the last two decades has been developing methods to explicitly and quantitatively translate from individuals to groups and groups to populations, as well as to relate empirical observations to models at all these organizational levels. Our theme in the following brief comments is to illustrate these developments. Because individual-based simulations were discussed earlier in this section, we focus now on the models that describe group characteristics and the distributions of groups in populations.

7.9   Continuum Approximations for Density Distributions Within Social Groups

One approach toward understanding how different individual behaviors affect group characteristics is to simulate large numbers of interacting individuals. This approach requires a minimum of complicated mathematical analysis and has the further advantage that the same spatial statistics of groups that apply to real groups can be directly compared to simulated groups (size, speed, degree of directional alignment, robustness, etc.). Flierl et al. (1999) used simulations to characterize the size distribution of groups in a population of schooling or swarming individuals and investigated the effects of environmental flows in modulating fission and fusion of groups. Gueron et al. (1996) explored the consequences of herding behaviors for group shape and coherence. However, the individual-based approach also has limitations. Among the most important of these limitations is that the computational effort required to simulate individual behavior in large social populations may be excessive. Furthermore, individual-based simulations tend to have many parameters, so that an exhaustive series of lengthy simulations would be required to understand how different behavioral parameters affect groups and populations.

An alternative approach is mathematically analyzing individual-level behavior algorithms, in order to derive dynamical equations that explicitly describe groups. These group-level models often involve complicated mathematics and may address a limited subset of possible behaviors. However, group-level models have the advantage that (since the group is the "unit") the computational effort required to simulate large populations may be much smaller than the corresponding individual-based models. The transition from individuals to groups is usually accomplished by means of a parallel transition from stochastic ordinary differential or difference equations for individual position and velocity, to continuum, deterministic, spatially explicit models couched in terms of partial differential equations (PDEs) or their nonlocal analogues, partial integro-differential equations. These PDEs use as their currency the *density distribution* of individuals, $P(t, x)$, which is generally treated as a continuous quantity that can, in principle, take on all real nonnegative values and is defined for all x and t within defined intervals. Such equations take the standard form

$$\frac{\partial P}{\partial t} = -\nabla \cdot J, \tag{7.34}$$

where J is the population flux. The object of the mathematical analysis is to derive J for a particular set of grouping behaviors.

The spatial and temporal scales of continuum grouping models are appropriate to explicitly model group formation, translation, and fragmentation or fusion with nearby groups but are larger than the scales characterizing individual decisions and movements. As a consequence of overlooking individual-level "details," the group-level models typically have fewer parameters than the individual-based models, with the parameters describing groups being combinations of parameters describing individuals. For example, in the advection–diffusion-reaction framework that is frequently used to approximate populations of organisms, the many parameters that describe individuals' social interactions are represented at the group level by just two parameters—the diffusion and advection coefficients. Thus, when it is possible to complete, the analysis leading to these coefficients gives a great deal of insight into which elements, or combinations of elements, are the key determinants of group dynamics. At the same time, the fact that details of individual position, velocity, and prior history are lost in the group-level models raises an interesting problem, because grouping behaviors depend strongly on these details. For example, though we know that schooling behaviors are strongly affected by relative positions of individuals within a school, the population density distributions alone do not provide enough information to specify how school members are geometrically distributed. For these models to work, the internal geometry must be strictly a function of density. Obtaining a "rule of thumb" for the relationship between density and internal geometry is often the most fundamental assumption behind group-level models.

7.9.1   Energy-Potential Analogy

One approach to modeling the dynamics of groups is to view group formation, fission, and fusion essentially as *edge phenomena*. That is, a group is characterized to a large extent by the properties of its edges—whether there is a sharp transition between "inside" and "outside" the group, whether the edge advances or retreats, whether that edge encounters the edge of another group, and so on. In this view, a group is like the air inside of a balloon— determining the geometry of the balloon also determines how various forces of attraction, repulsion, and surface effects affect the state of the air inside. The benefit of an edge-based approach is that it reduces the dimensionality of the problem, e.g., from a two-dimensional herd to a one-dimensional herd front (Gueron and Liron, 1989; Gueron and Levin, 1994).

This analogy with physical forces was used by Cohen and Murray (1981) and Murray (1989) (see also Lara Ochoa, 1984) to derive an *energy-potential theory* of social aggregations. This theory is essentially an algorithm for generating expressions for population flux, starting from assumptions about how a biological analogue of "internal energy" density, $e(P)$, varies with population density. Assuming the existence of this internal energy function allows us to define a potential,

$$\mu(P) = e'(P). \tag{7.35}$$

The population flux is then proportional to the gradient of the potential,

$$J = -D\nabla\mu(P). \tag{7.36}$$

This equation provides a mechanism for systematically investigating the population-level consequences of assumptions on $e(P)$. The simplest case, Fickian diffusion, results from the internal energy $e(P) = P^2/2$. However, Fickian diffusion is not a useful model for aggregation: It is an essentially dispersive process as long as D is positive, and (7.34) is mathematically ill posed if D is negative. More complicated forms of the potential, μ, include forms in which μ depends not only on density but also on spatial derivatives of P. For example, if $\mu(P, \nabla P) = -k\nabla^2 P + e'(P)$ (an expression that follows in part from plausible assumptions on rotational and reflectional invariance; see Murray, 1990, for details), the population-level equation for group formation has the form

$$\frac{\partial P}{\partial t} = -kD\nabla^4 P + D\nabla \cdot (e''(P)\nabla P).$$

Thus, assumptions on e and μ give expressions for J containing higher spatial derivatives of P, potentially in nonlinear combinations. This generalization of standard diffusion can give rise to autonomously generated, steady-state, nonsingular aggregations in which higher derivatives overcome the "negative diffusion" problems associated with sign changes in D in Fickian diffusion.

7.9.2   *Integral Equations for Group Dynamics*

A universal trait of individual-based models of schooling and other social groups is the possibility of interactions between individuals that are separated in space, i.e., *nonlocal interactions*. Group-level models must somehow account for nonlocal interactions, either explicitly or at least at some intermediate stage of analysis. For example, in the energy-potential theory just discussed, nonlocal interactions are manifested as nonlinear functions of spatial derivatives. An explicit way to represent nonlinear interactions at the group level is to specify population fluxes as functions of population density over a *neighborhood*, i.e., in a partial integro-differential equation (PIDE). A relatively large body of theory has been developed concerning PIDEs of the form (in one dimension)

$$\frac{\partial P}{\partial t} = \frac{\partial}{\partial x}\left[D\frac{\partial P}{\partial x} - UP\right],$$

$$U(t, x) = \int_X k(x - x')P(t, x')\,dx', \tag{7.37}$$

where U is an "aggregation flux" that represents the average movement of individuals toward or away from population concentrations in their vicinities. U is expressed in terms of a *convolution* of the population distribution in space with an "interaction kernel," $k(x - x')$, representing the effect of population density at x' on the population density at x. Typically, k embodies several behavioral components such as attraction to distant neighbors and repulsion from neighbors that are too close.

Integro-differential equations are analytically challenging. For example, in contrast to PDEs in which steady-state solutions correspond to ordinary differential equations (ODEs) (and are thus relatively tractable), steady-state solutions of integral equations are frequently difficult or impossible to obtain. However, some progress on such models has been made. See, for example, Kawasaki (1978), Mogilner and Edelstein-Keshet (1999), and Flierl et al. (1998). The inclusion of nonlinear terms is key, because there are important limitations to how well linear models can describe the phenomena.

Given the analytical and numerical difficulties of solving PIDE models, it is worth considering to what extent they can be approximated by more tractable PDEs. Grünbaum and Okubo (1994) used linear stability analysis to compare the growth of small perturbations in Kawasaki-type convolution models to PDE approximations to them obtained by Taylor expansions. They found that, because the Taylor expansion results in polynomial approximations to the dispersion relation, the PDE approximation always gives incorrect stability properties at high wavenumbers.

7.9.3   Poisson-Point Assumption

Thus far, the PIDE descriptions of grouping behavior have been largely *heuristic*, in that they are hypothesized de novo at the population level rather than being explicitly derived from individual-level behaviors. An example of a PIDE model in which the connection between individual swarming behavior and group-level dynamics is explicit was developed in one dimension by Grünbaum (1994). In this model, the "details" of individuals' geometrical arrangement arrangements within swarms are defined (in a statistical sense) by assuming that individuals occur as *Poisson points* with a density distribution $P(t, x)$. The Poisson process that generates such distributions of points has a well-defined set of mathematical conditions, which are described in detail elsewhere (e.g., Papoulis, 1984). For our purposes, the important assumption underlying the Poisson-point assumption is that the number of individuals observed within any interval in space at any given observation is independent of the number observed in any nonoverlapping interval. In the case of the grouping behavior of a fixed number of individuals, this assumption is clearly not satisfied exactly: If social individuals avoid overcrowding, observing an individual in one place surely reduces the likelihood of observing an individual a very small distance away. Grünbaum (1994) explores the consequences of assuming that the requirements for a Poisson-point process are *almost* satisfied and that spatial correlations between individuals do not strongly affect group characteristics.

In the individual-based swarming model, each individual is assumed to seek a number of neighbors, specified by a predefined "target density," μ. Individuals are assumed to detect and respond to neighbors only within a maximum interaction distance. Thus, an individual's estimate of local density can depend only on the number of neighbors it observes within the interaction distance in the positive and negative directions, n^+ and n^-. If the observed number of neighbors detected by an individual is lower than the target density, that individual attempts to move in the direction of increasing density, i.e., *to climb the population gradient*. However, again sensory constraints limit the information available to an individual—its only means of estimating the up-gradient direction is by whether n^+ is larger than n^-, or vice versa. When the number of neighbors is small, this is clearly an errorprone estimate. The Poisson-point assumption allows us to estimate exactly how errorprone the estimate is and the overall frequency of decisions to move left or right as a function of the density distribution.

Specifically, an individual's decision to move right or left to reach a target density of neighbors can be written as

$$g(n^-, n^+) = \operatorname{sgn}(n^- - n^+) \operatorname{sgn}\left(\frac{n^- + n^+ + 1}{2}\right), \qquad (7.38)$$

where the sensing distance has been scaled to unity. Here $g = 1$ reflects a decision to move in the positive x direction, and $g = -1$ in the negative x direction. The Poisson-point assumption effectively determines the *joint frequency distribution*, $f(n^-, n^+)$, of observing any combination of n^- and n^+,

$$f(n^-, n^+) = \exp(-(\rho^- + \rho^+)) \frac{(\rho^-)^{n^-}}{n^-!} \frac{(\rho^+)^{n^+}}{n^+!}, \qquad (7.39)$$

where ρ^- and ρ^+ are the expected number of neighbors to the left and right, respectively. ρ^- and ρ^+ are specified by spatial integrals of the population density,

$$\rho^-(t, x) = \int_{x-1}^{x} P(t, x') \, dx', \quad \rho^+(t, x) = \int_{x}^{x+1} P(t, x') \, dx'. \qquad (7.40)$$

The population-level flux resulting from social interactions is then given by

$$J = -D \frac{\partial P}{\partial x} + \gamma \phi(\rho^-, \rho^+) P, \qquad (7.41)$$

where

$$\phi(\rho^-, \rho^+) = \sum_{n^-=0}^{\infty} \sum_{n^+=0}^{\infty} f(n^-, n^+) g(n^-, n^+). \qquad (7.42)$$

Here ϕ represents the "average" decision made by an individual at x in response to the distribution of population density in its vicinity. D is the usual diffusion coefficient, and γ is a characteristic aggregation speed determined by the frequency at which individuals make social decisions and the distance they move as a consequence of each decision. Because ρ^- and ρ^+ are integrals and $\phi(\rho^-, \rho^+)$ is a nonlinear function, we have arrived at a nonlinear partial integro-differential equation representing the individual-based swarming behavior. Grünbaum (1994) compared steady-state and transient population distributions predicted by this PIDE with the original individual-based model and found good agreement for parameter ranges in which there was considerable short-term randomness in relative individual positions. These are precisely the conditions under which the individual-based model most nearly satisfies the Poisson-point assumption. However, for parameters that result in highly ordered swarm geometries, the approximation was less accurate.

7.10   Dynamics of Groups in Social Grouping Populations

Having made a transition in modeling approach from an individual-based, ODE approach to PDEs or PIDEs that describe how groups form and move in space, we now consider grouping from the perspective of the population, and ask, "How are populations subdivided among groups throughout a landscape (or seascape) and what are the consequences for population movement?" Because a social population is typically distributed among numerous groups, even a group-level description is too numerically costly to use to model entire populations. Instead, we can turn to a statistical description of groups in a population that estimates the frequency of groups of various sizes and characteristics, based on the rates at which such groups appear and disappear. Such an accounting might be written as

$$\frac{d}{dt} p(t, n) = \text{fusion to form } n\text{-groups} - \text{fusion of } n\text{-groups}$$

$$- \text{fission of } n\text{-groups} + \text{fission to form } n\text{-groups},$$

where $p(t, n)$ is the frequency of groups of size n in a population at time t. This model allows for two kinds of dynamics: groups can encounter and merge with other groups; and groups can spontaneously undergo fission to form smaller groups. Equation (7.43) describes how each of these processes creates and eliminates n-groups.

Cumulatively, a set of equations such as (7.43) represents a *dynamical system*, and the solutions to (7.43) provide an estimate of the *distribution of group sizes* in a social population. A quantitative version of such a group-size distribution model is

$$\frac{d}{dt}p(t,m) = \frac{1}{2}\sum_{n=1}^{n-1} a(m,n-m)p(t,m)p(t,n) - \sum_{m=1}^{\infty} a(n,m)p(t,n)p(t,m)$$

$$- \frac{1}{2}\sum_{m=1}^{n-1} b(n,m)p(t,n) + \sum_{m=n+1}^{\infty} b(m,n)p(t,m).$$

Equation (7.43) assumes that the fission rate of n-groups is a *linear* function of $p(t,n)$. However, fusion requires interaction between two groups and so the fusion rate in (7.43) is a *quadratic* function of group frequency. Because the total number of individuals in a population is not changed by interactions among groups,

$$\frac{d}{dt}P(t) = \frac{d}{dt}\sum_{n=1}^{\infty} np(t,n) = 0, \tag{7.43}$$

where $P(t)$ is the total number of individuals. Okubo (1986) gives a number of examples of group-size distribution models from the literature. See also recent developments in Gueron and Levin (1995), Gueron (1998), and Niwa (1998).

7.10.1 *Spatially Explicit Group-Size Distribution Models*

We now consider extending the framework of (7.43) to explicitly include the distribution of a spatially variable population into groups, and consider how grouping affects fluxes at the population level. We begin by defining a "spatial dynamical system" in which locally the group-size distribution is described by (7.43) but in which we also allow emigration and immigration from adjacent areas with different group-size distributions. In addition, we consider the case in which movement statistics (e.g., speed, directional persistence, etc.) are functions of group size. We define $p(t,x,v,n)$ to be the density of n-groups at x, t with velocity v. Then [with the notation $p_i \equiv p(t,x,v,i)$ for brevity]

$$\varepsilon\left(\frac{\partial}{\partial t}p(t,m) + v\cdot\nabla_x p\right) = -\lambda(n)p_n + \lambda(n)\int_V p_n(v')T(v,v',n)\,dv'$$

$$+ \frac{1}{2}\sum_{n=1}^{n-1} a(m,n-m)p_m p_n - \sum_{m=1}^{\infty} a(n,m)p_n p_m$$

$$- \frac{1}{2}\sum_{m=1}^{n-1} b(n,m)p_n + \sum_{m=n+1}^{\infty} b(m,n)p_m. \tag{7.44}$$

In (7.44), $\lambda(n)$ is a group-size–dependent turning rate (reflecting a tendency of larger or smaller groups to have greater or lesser persistence in direction or speed). $T(v,v',n)$ is a transition kernel that defines the probability that an n-group initially travelling with velocity v' will change to velocity v. The ε-term represents an assumption of "slow" variation of $p(t, x, v, n)$ in space. This is directly analogous to the assumption of slow variation made in Chap. 5, and we refer to that chapter for a fuller discussion of this assumption.

The small parameter ε is the basis of a perturbation expansion,

$$p(t, x, v, n) = \sum_{i=0}^{\infty} \varepsilon^i \psi_i(t, x, v, n), \qquad (7.45)$$

which can be used to obtain an approximate solution to (7.44). We summarize the results of this analysis; details may be found in Grünbaum (2000). The order $O(1)$ solution is

$$\psi_0(t, x, v, n) = P_0(t, x)\phi_0(n, P_0), \qquad (7.46)$$

where $P_0 \equiv$ total density of individuals and ϕ_0 is the solution to

$$0 = \frac{1}{2} P_0 \sum_{n=1}^{n-1} a(m, n - m)\phi_0(m, P_0)\phi_0(n - m, P_0)$$

$$- P_0 \sum_{m=1}^{\infty} a(n, m)\phi_0(n, P_0)\phi_0(m, P_0)$$

$$- \frac{1}{2} \sum_{m=1}^{n-1} b(n, m)\phi_0(n, P_0) + \sum_{m=n+1}^{\infty} b(m, n)\phi_0(m, P_0) \qquad (7.47)$$

normalized so that

$$\sum_{n=1}^{\infty} n\phi_0(n, P_0) = 1. \qquad (7.48)$$

The $O(\varepsilon)$ solution is

$$\psi_1 = g(v, n, P_0) v \cdot \nabla P_0, \qquad (7.49)$$

where we define $g(v, n, P_0)$ as the solution of

$$\phi_0(n, P_0) + \partial_{P_0}\phi_0(n, P_0)$$

$$= \frac{1}{2} \sum_{m=1}^{n-1} P_0[\phi_0(m, P_0)g(v, n - m) + g(v, m)\phi_0(n - m, P_0)]$$

$$- \lambda(n)g(v, n) + \lambda(n) \int_V g(v', n) T(v, v', n)\, dv'$$

$$- \frac{1}{2} \sum_{m=1}^{n-1} b(n, m) + \sum_{m=n+1}^{\infty} b(m, n)g(v, m)$$

$$- \sum_{m=1}^{\infty} a(n, m)P_0[\phi_0(n, P_0)g(v, m) + \phi_0(m, P_0)g(v, n)]. \qquad (7.50)$$

Then we can approximate the total population flux as

$$\partial_t P = -\nabla_x J = \varepsilon \nabla_x \left[\sum_{n=1}^{\infty} n \int_V v^2 g(v, n, P_0) \, dv \cdot \nabla_x P_0 \right] + O(\varepsilon^2). \qquad (7.51)$$

Putting this equation into a standard advection–diffusion form, we may write an approximate expression for the population flux as

$$\partial_t P = \nabla_x (D(P) \nabla_x P), \qquad (7.52)$$

where the *density-dependent diffusion coefficient* is given by

$$D(P) = \sum_{n=1}^{\infty} n \int_V v^2 g(v, n, P) \, dv. \qquad (7.53)$$

The coefficient $D(P)$ reflects the effects of total population on the group-size distribution as well as the different movement characteristics of differently sized groups. In a conceptual sense at least, Eq. (7.52) closes a circle. Individual-based models of social behaviors lead to models of groups, and models of groups lead to distributions of populations across space and among groups with different numbers of individuals. While each of these steps needs further elaboration and study, it is clearly possible and productive to use models to establish a direct linkage between the behaviors of individuals and their consequences for ecological dynamics.

8
Animal Movements in Home Range

Akira Okubo and Louis Gross

We define the home range of an animal or population to be the area or volume over which it normally travels in pursuit of its routine activities. The home range includes the nest site, shelter, locations for resting, food-gathering, mating, territory, etc. Terrestrial mammals and many species of fish maintain well-defined home ranges. Home ranges of different individuals may, and usually do, overlap. The entire home range, or some part of it, may become territory, i.e., a defended area (Fig. 8.1).

Considering the importance of the home range in the life of animals, one might expect that animal movements in the home range have been studied in detail. In fact, little is known to us. Study of animal movement was initially limited primarily to the delineation of the home range itself (Brown, 1962; Sanderson, 1966; Jewell, 1966; Jennrich and Turner, 1969; Koeppl et al., 1977; Cooper, 1978).

Since home range can depend upon individual status, including life stage, accurate comparisons between individuals require more explicit definitions. One approach is to limit this to a particular time period or life stage (e.g. breeding season home range, lifetime home range). The term "activity range" is applied when considering movements within a certain time period. An animal's "core area" is the portion of the home range most frequently used by an individual, corresponding to an area that an individual will be within at a certain probability or higher. The method used to calculate these is the "utilization distribution" (UD) (Jennrich and Turner, 1969) from which one estimates the smallest area which accounts for some percentage (say 95%) of the individual's space utilization. Due to statistical problems in estimating home range size, it has been argued that for some questions the focus should be on estimation of the UD (Anderson, 1982).

FIGURE 8.1. Model home range. A broken-line circle inside a home range indicates a territory. Small solid circles are nest sites, and open circles are shelters. Note that the home range overlaps (from Burt, 1943).

8.1 The Size of the Home Range and Its Relation to Animal Weight and Energy Requirements

Home ranges differ widely among animal species. In general, it is conceivable that the spatial density of animals might vary according to the energy (food) requirements of individuals or the population. In other words, there may exist a unique relationship between the size of home range and the rate of energy intake required per animal.

Armstrong (1965) and Schoener (1968) investigated the relation between the size of home range of birds and their body weight. Similar studies were made by Turner et al. (1969) with lizards, by McNab (1963) with mammals, and by Milton and May (1976) with primates. All these results can be summarized by the following formula:

$$R = aW^b, \qquad (8.1)$$

where R is the size of the home range, W is body weight, and a and b are constants. The value of b depends on the animal species and ranges from 0.63 to 1.23.

On the other hand, a familiar relation exists between the rate of basal metabolism, M, and body weight, W:

$$M = cW^{0.75}, \qquad (8.2)$$

where c is a constant (Kleiber, 1932, 1961; Schmidt-Nielsen, 1972).

Harestad and Bunnell (1979) find for large mammals that the exponent differs significantly from 0.75, with 95% confidence limits of 0.8 to 1.24 for herbivores, 0.57 to 1.26 for omnivores and 1.04 to 1.68 for carnivores.

For mammals in particular, an approximate mean value of b in Eq. (8.1) is 0.75, so that from (8.1) and (8.2) home range size is seen to be approximately proportional to the basal metabolism rate, or

$$R = kM, \qquad (8.3)$$

where k is a constant. However, still missing is the link between home-range size and the actual energy requirement, E, in the natural environment. Turner et al. (1969) hypothesized that E should be proportional to M. If so, R is then linearly proportional to E.

Certain animal species such as bees, ants, bats, seals, sea lions, and flocking birds aggregate at fixed cores (roosts) from which they disperse each day into the surroundings to secure resources. The pattern of this dispersal should be regulated by the balance between two mutually opposing effects. The first is the advantagous effect of more distant dispersal in reducing pressure on resource exploitation as nearby resources are depleted. The second effect is the disadvantage of increased time and energy expenditure more distant ranging requires.

Hamilton et al. (1967) and Hamilton and Gilbert (1969) used this reasoning to analyze the strategies of starling foraging dispersal. Kiester and Slatkin (1974) modeled a particular strategy of movement and foraging, which was motivated by observations on iguanid lizards. They derived a diffusion like differential equation for the number density of animals in space and time. Some details of their analysis are given in Sect. 8.5.

8.2 Mathematical Models for Animal Dispersal in Home Ranges

Relatively few mathematical models dealing with animal dispersal in home ranges have been developed. We hereby consider a random walk model with bias due to Holgate (1971).

The movements of an animal in its home range can be considered to be a stochastic process in space, i.e., a random walk. However, the walk is not purely random; it must be regarded as a biased walk. That is, the probabilities of taking a step toward the center of activity of the home range are greater than those of moving away from the center. Such a random walk can be termed "centrally biased." The probabilities may be assigned in a variety of ways, each appropriate to the problem of concern. For simplicity, we analyze the walk in a one-dimensional space, i.e., an x-axis, by taking the activity center at the origin, $x = 0$. Two cases are considered.

Case i). Attraction to the center is assumed to depend solely on the animal's memory and familiarity.

It is reasonable to suppose that the probability of a centrally biased step decreases with distance from the center. Thus, the probabilities of the animal moving from x to $x - 1$ and $x + 1$ might be given by, respectively, $(1/2)(1 + \varepsilon/x)$ and $(1/2)(1 - \varepsilon/x)$, where ε is a constant less than 1 (Fig. 8.2).

Let $p(x_0|x_1, \tau)$ be the probability that an animal starting at x_0 arrives at x_1 at time τ. The probability of an animal being located at m at time $\tau + 1$ is the sum of the probability that an animal located at $m + 1$ at time τ moves to m

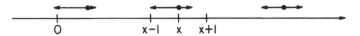

FIGURE 8.2. Centrally biased random walk. The probability of a centrally biased step decreases with distance from the center.

during the next time interval and the probability that an animal located at $m - 1$ at τ moves to m during the next time interval. Thus,

$$p(x_0|m, \tau + 1) = \tfrac{1}{2}\{1 + \varepsilon/(m + 1)\}p(x_0 \,|\, m + 1, \tau)$$
$$+ \tfrac{1}{2}\{1 + \varepsilon/(m - 1)\}p(x_0 \,|\, m - 1, \tau). \qquad (8.4)$$

Equation (8.4) provides a formula for a generalized random walk that corresponds to Eq. (5.9) in Sect. 5.3, where the transition probabilities α and β are functions of position.

Equation (8.4) can be solved subject to the initial condition that the animal started at some known point x_0 at $\tau = 0$ (for details, consult Gillis, 1956). After obtaining $p(x_0|m, \tau)$, we can calculate, for example, the probability that the animal returns to the origin for the first time after $2n$ steps. In particular, the percentage of a given duration that the animal spends at various distances from the center can be evaluated. Although this model incorporates central bias toward the center, the attraction is assumed to diminish hyperbolically with distance, so that the home range may extend over fairly long distances.

Case ii).   The animal is attracted toward its nest, which contains offspring waiting to be fed.

In this case the tendency to move toward rather than away from home becomes greater as the animal moves farther away. Thus, the transition probabilities from x to $x - 1$ and $x + 1$, respectively, might be taken as $(1/2)(1 + x/L)$ and $(1/2)(1 - x/L)$, where $-L \le x \le L$ (Fig. 8.3).

If, as for the previous case i), we let $p(x_0|x_1, \tau)$ be the probability that an animal starting at x_0 arrives at x_1 after time τ, the probability of an animal being located at m after $\tau + 1$ is given by

$$p(x_0|m, \tau + 1) = \tfrac{1}{2}\{1 + (m + 1)/L\}p(x_0 \,|\, m + 1, \tau)$$
$$+ \tfrac{1}{2}\{1 - (m - 1)/L\}p(x_0 \,|\, m - 1, \tau). \qquad (8.5)$$

Kac (1947) obtained a solution to this equation satisfying the initial condition that the animal is located at a known point x_0 at $\tau = 0$. It is found that

FIGURE 8.3. Centrally biased random walk. The probability of a centrally biased step increases with distance from the center.

the mean length of the excursions from the origin is $(\pi L)^{1/2}$ in one dimension and πL in two dimensions.

The distinction between the two models i) and ii) is clear. In the model of case i), the attraction to home diminishes with distance. The diffusive tendency due to random walking overcomes attraction, and the home range extends to infinity in principle. The mean distance of movement from the origin does not converge to a limit; in other words, the animal is more likely to disperse than to return to the original nest. On the other hand, in the model for case ii), the attraction to home increases with distance, and hence the diffusive tendency due to random walking is overshadowed by the attraction of appropriately large distances. The size of the home range thus becomes finite. Holgate suggested that these models could be used to study the probability of an animal being caught in a trap in its home range.

In a random walk, movement is assumed to be discrete. In the limit of increasingly smaller steps, however, the difference equation for probability becomes a generalized diffusion equation describing continuous movement. Thus, in the limit, Eq. (8.5) converges to Eq. (7.11) of the previous chapter, which is a diffusion equation for a population under a central force, the attraction of which is proportional to the distance from the center (Kac, 1947). Equation (8.5) is a difference equation for the probability based on statistical considerations, while Eq. (7.11) is a differential equation, i.e., a Fokker–Planck equation, based on continuum considerations. Both equations describe essentially the same concept. However, the diffusion equation is much more convenient to handle analytically.

Dunn and Gipsen (1977) and Dunn (1978) proposed a multivariate, centrally biased diffusion process as a workable model for the study of home range. The model is characterized in terms of some typical descriptive properties of home range such as activity center, activity radius, and distributions of turning angle and displacement. An extension is made therein to the problem of testing for territorial interaction between two or more individuals in the use of deer, coyote, and bird telemetry data.

8.3   Simulation of Animal Movement in Home Ranges

Siniff and Jessen (1969) attempted to simulate the movement of an animal in its home range on the basis of telemetry data for red foxes (*Vulpes fulva*), snowshoe hare (*Lepus americana*), and raccoons (*Procyon lotor*).

The telemetry data were obtained from the University of Minnesota's Cedar Creek automatic tracking system, which continuously monitors the movement of animals carrying miniature radio transmitters (Cochran et al., 1965). Figure 8.4 shows an example of red fox movement as obtained from telemetry. The following description is restricted to the movement of red foxes, although the method of observation for the other animals is essentially the same.

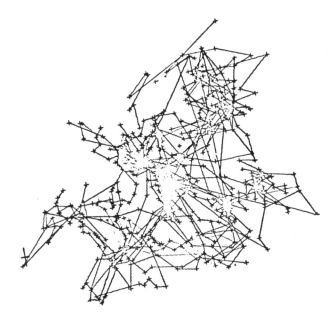

FIGURE 8.4. Example of red fox movement as obtained from telemetry (from Siniff and Jessen, 1969).

Siniff and Jessen analyzed their data in terms of the following three elemental quantities:

(1) The distance of travel, i.e., the distance from an initial location to the point where the next location is recorded, was determined: The mean speed was calculated by dividing the distance traveled by the duration of time between the two locations.
(2) The relative angle was measured in a clockwise direction between 1–2 and 2–3. If an animal reverses its course at location 2, the relative angle is 0° (or 360°). The accuracy of measurement of the relative angle was 0.5°.
(3) The relative durations of rest and movement were also determined.

Figure 8.5 shows some observed distributions of speed (distance traveled in feet per minute) for an adult female, an adult male, and a juvenile red fox. All instances where no movement occurred, i.e., zero distance was traversed, were eliminated from the distribution. For both of the adults and the juvenile, the distribution curve rises rather sharply from the origin and decreases slowly to the right. The generally slower movement of the juvenile during the period June 4 to July 16, 1965, indicates that in this period it was still under parental care.

The observed speed distribution may be approximately represented by the gamma distribution

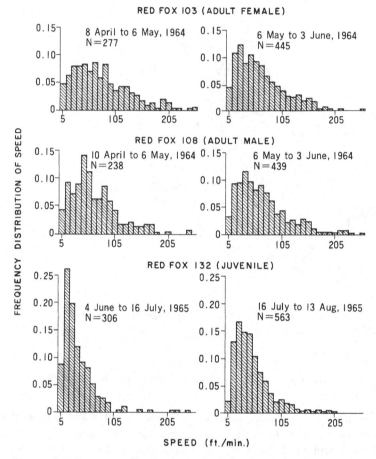

FIGURE 8.5. Frequency distributions of speed for an adult female, an adult male, and a juvenile red fox (from Siniff and Jessen, 1969).

$$f(v) = \frac{1}{\beta^{\alpha+1}\Gamma(\alpha+1)} v^{\alpha}e^{-r/\beta}, \qquad (8.6)$$

where v is speed, α and β are constants satisfying the constraints $\alpha > -1$ and $\beta > 0$, and $\Gamma(v)$ is the gamma function. (It may be of interest to compare this distribution with the speed distribution of swarming insects shown in Fig. 7.5. In fact, the distributions of Fig. 8.5 may not be far from the Maxwell distribution.)

The observed relative angle distributions for red foxes are shown in Fig. 8.6. The data for the adult male and female red fox are similar, with a generally unimodal appearance and a peak at about 180°; the data for the juvenile red fox indicate a nearly uniform distribution. The distribution of the relative angle θ is approximated by the circular normal distribution:

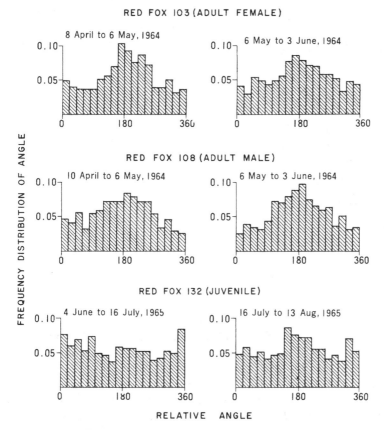

FIGURE 8.6. Frequency distributions of relative angle of movement for an adult female, an adult male, and a juvenile red fox (from Siniff and Jessen, 1969).

$$f(\theta) = \tfrac{1}{2}\pi I_0(c)\,\exp\{c\,\cos(\theta - m)\}, \tag{8.7}$$

where $I_0(c)$ is the modified Bessel function of order zero; m is the angle of maximum probability, i.e., the modal direction; and c is the parameter of concentration of the distribution. As c increases, the spread of the distribution decreases, and the distribution becomes uniform as c approaches zero.

Telemetry data confirm that the red fox, a nocturnal species, moves during the night and remains relatively stationary during the day. For this reason, only data taken during the night were considered in calculating the statistics of activity. The most probable duration of rest is about 60 minutes; rest rarely extends for more than 4 hours. Typical durations of movement differ between the juvenile and the adults. For the juvenile, the duration is generally less than 3 hours, but for the adults it is longer, sometimes exceeding 10 hours.

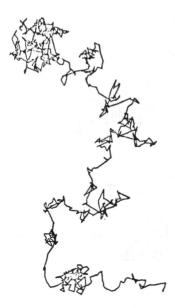

FIGURE 8.7. Example of animal movement pattern obtained by simulation (from Siniff and Jessen, 1969).

For computer simulation, Siniff and Jessen first tried a simple model using the distribution laws (8.6) and (8.7), and the distributions of duration of rest and of movement as obtained from the observed values of the three fundamental quantities. In this model, the movement at each point was assumed to be independent of every other point. For example, the relative angle chosen for movement to the next location was not influenced by the distance chosen or by the location on the home range. Figure 8.7 shows an example of the type of movement pattern one can obtain using this simulation procedure. The overall pattern can be changed by varying the form of the distributions of relative angle and travelled distance.

However, the task of judging the statistical accuracy of the simulated movement patterns, in comparison with telemetrically observed movement patterns, remains. A criterion for goodness of fit is required for this test, the choice of which poses a difficult problem. The issue in question is that of pattern recognition.

Siniff and Jessen applied the method of square sampling unit often used in studies of plant ecology (Greig-Smith, 1964) to measure the differences in animal movement patterns. Thus, the home range was partitioned into square units, and the number of individuals in the unit, n, was assumed to be approximated by the negative binomial distribution. Values of the dispersion parameter k were used to compare the movement pattern. The negative binomial distribution is given by

$$f(n) = \frac{(k+n-1)!}{(k-1)!n!} \left(\frac{m^n}{m+k} \right) \left(1 + \frac{m^{-k}}{k} \right), \quad n = 0, 1, 2, \ldots, \tag{8.8}$$

where m is the arithmetic mean and $k > 0$. Low values of k indicate pronounced clumping, and high values indicate considerable randomness in the distribution.

The values of k for the movement patterns of red foxes derived from telemetry data range from 0.1 to 0.7. This implies that the movement of the red fox in its home range is not entirely random and that certain parts of the home range are used selectively.[1] On the other hand, the values of k obtained from the simulated data present a pattern that is rather more random than the actual data. Such a discrepancy may be due primarily to two reasons: 1) The simulation model does not explicitly account for the boundary effect. An animal typically recognizes the boundary of its home range, beyond which it seldom wanders. 2) An animal moving within a home range tends to reuse preferred areas, including places of rest and foraging sites, while no preferred areas were assigned in the simulation model.

Siniff and Jessen then present an improved simulation model in which home range is defined and movement progresses according to weighted probabilities. The values of k range from 0.12 to 3.6.

The following complexities must be incorporated into simulation models:

1) It appears that for some species the home-range boundary varies on a day-to-day basis. More than one center of activity may exist, and its location may also vary from day to day.
2) The movements of individuals within a community may be influenced by the presence of other individuals. The effect of population pressure should be considered in some fashion.
3) The relationship between animal movement and habitat feature such as vegetation types needs to be clarified.
4) The simulation model should be developed as a tool aiding the capture of animals for census purposes.

As computer capacity increases, the range of simulation may be extended to build a model ever closer to reality. As Watt (1966) claims, the study of biological systems by means of simulation may be the first step toward gaining an understanding of the specific role of the various elements pertinent to the system. However, it must not be assumed that simulation can provide answers to all questions one might wish to pose.

Radio telemetry allows for the tracking of individually marked animals and has proved to be an extremely powerful tool for basic studies of animal movement. Coupled with simultaneous measurements of environmental factors and physiological data, telemetry possesses an inestimable potential.

---

[1] Inoue (1972) suggests that the degree of inhomogeneity of spatial utilization can be clarified by application of the concepts of I_δ due to Morisita (1959), and mean crowding of Pielou (1969), to both time and space distributions, and by evaluation of the adjustability of these parameters.

However, even telemetry has certain significant limitations. First of all, the effect of a transmitter on the animal to which it is attached must be examined thoroughly. Furthermore, although instantaneous positions of an animal can be monitored, telemetry alone provides only indirect clues as to what the animal is actually doing. Also, many individuals must be tracked simultaneously in order to study animal interaction. (In 1980, the Cedar Creek system of the University of Minnesota permitted the tracking of up to 52 animals simultaneously and continuously.)

For very small animals such as insects the use of telemetry may be difficult and tedious. Gary (1973), in studying the distribution of the flight range of honeybees, attached a tiny piece of numbered metal to the back of each bee. Trapping was accomplished with small magnets placed at the hive entrance and also immediately above flowers that bees might visit. After being trapped, a bee was released. The movement of honeybees was recorded in this fashion.

8.4   Animal Dispersal and Settling in New Home Ranges

The process of animal dispersal from an original habitat for the purpose of establishing a new home range plays a crucial role in the survival of species. During the dispersal period, individuals may die from such factors as lack of food, deterioration of environmental conditions, and attack by predators.

Even if an individual reaches a suitable habitat, it may be already occupied by other animals capable of repelling intruders. A few models related to this process are discussed below.

i) The Random Walker Who Collects Food (Energy) as It Wanders.   A one-dimensional random walk problem is considered in which an animal moves at random as it collects food located at each grid point (Fig. 8.8). This provides a model for nomadism.

We suppose that initially an animal is at the origin $x = 0$. Let ℓ, τ, and $m(x)$ be, respectively, the step length, the interval of time between successive steps, and the amount of food located at x. At each point, the animal consumes an amount $f(m)$ dependent on $m(x)$.

Let $p(x, Q, t)\, dx\, dQ$ be the probability that after time t the animal is at x and has gathered amount of food Q. We shall derive an equation for p.

Since, at time $t - \tau$, the animal is located at either $x - \ell$ or $x + \ell$ and has acquired the amount $Q - f(m)$ of food, we obtain

$$p(x, Q, t) = \tfrac{1}{2}p\{x - \ell, Q - f(m(x)), t - \tau\} + \tfrac{1}{2}p\{x + \ell, Q - f(m(x)), t - \tau\}.$$

$$(8.9)$$

In (8.9) it has been assumed that the random walk is isotropic and that the probability that the animal remains at a given grid point for successive time intervals is zero. Expanding the right-hand side of (8.9) in a Taylor series and

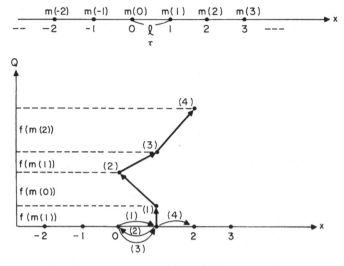

FIGURE 8.8. Random walk model in which an animal moves at random as it collects food (m) located at each grid point. ℓ: step length; τ: step time interval; $f(m)$: rate of food consumption; Q: accumulated amount of food.

rearranging, we obtain

$$\frac{\partial p}{\partial t} = -\frac{f}{\tau}\frac{\partial p}{\partial Q} + \frac{1}{2}\frac{\ell^2}{\tau}\frac{\partial^2 p}{\partial x^2} + \cdots, \tag{8.10}$$

where the higher-order terms of the expansion with respect to ℓ, f, and τ have been dropped.

We now take the limit of (8.10) as f, ℓ, and τ approach zero under the following constraints:

$$\lim_{f,\tau \to 0} f/\tau = F(m(x)), \qquad \lim_{\ell,\tau \to 0} \ell^2/2\tau = D. \tag{8.11}$$

The higher-order terms then converge to zero, and (8.11) becomes

$$\frac{\partial p}{\partial t} = -F(m(x))\frac{\partial p}{\partial Q} + D\frac{\partial^2 p}{\partial x^2} = -\frac{\partial(Fp)}{\partial Q} + D\frac{\partial^2 p}{\partial x^2}. \tag{8.12}$$

Equation (8.12) is formally equivalent to the equation of shear diffusion mentioned in Chap. 2. Thus, $F(m(x)) = F(x)$ may be regarded as a "velocity" component in the direction of Q. Furthermore, this velocity possesses a "shear" that depends on the position x. In other words, individuals diffusing in the x direction with diffusivity D are carried by the respective flow F in the direction of Q. As a result, they are dispersed in the x, Q plane with the probability density p (Fig. 8.8).

Equation (8.12) can be solved for p if $f(m(x))$ is prescribed. Thus, the probability of survival after time t can be calculated given appropriate limits on x and Q.

If (8.12) is integrated over the entire range of Q, the equation for population density $S(x, t) \equiv \int_{-\infty}^{\infty} p \, dQ$ is obtained, and it is found that

$$\frac{\partial S}{\partial t} = D \frac{\partial^2 S}{\partial x^2},$$

as one would expect.

ii) Computer Simulation. Kitching (1971) carried out a simulation of animal dispersal by constructing a grid system in the $x - y$ plane within which a number of habitats are located. "Animal points" originate from the coordinate center. Movement patterns and mortality rates are specified, and dispersal success is evaluated. The simulation run terminates when each animal either reaches one of the habitats available for settlement or dies on the way.

The details of the computer program can be found in the original paper (Kitching, 1971). Herein we present some of the results (Fig. 8.9 A–D). In the figures, INCR represents the distance traveled per unit time, i.e., path length. Doubling INCR increases the chance of success in dispersal (C $\rightarrow$ D). The variance of the angle direction of animal movement is designated by s^2. Larger values of s^2 imply movement that is less directional, and animals show a greater tendency to circle around the origin and a lesser tendency for dispersal (C $\rightarrow$ A). M denotes the mortality percentage.

The simulation allows one to obtain the animal density distribution around the dispersing center. Kitching attempted to compare the simulated distribution with existing empirical laws such as that of Kettle (1951), Eq. (6.9), and that of Wolfenbarger (1959), Eq. (6.6). No definitive conclusion could be reached, however.

French (1971) presented a computer simulation for rodent dispersal. The direction of animal dispersal was set to be random, and an animal moved along a straight line until it intercepted an unoccupied home range, in which case it settled and attempted to repel other animals. The results indicate that the frequency distribution of animal dispersal distances is not necessarily represented by a smooth curve reminiscent of a log-normal distribution; the simulated curve is characterized by a secondary peak at some distance from the origin. However, at this stage it seems uncertain whether or not the simulated distribution is fundamentally similar to the bimodal frequency distribution suggested by some field workers.

8.5 Strategies of Movement for Resource Utilization

Kiester and Slatkin (1974), motivated by observations on iguanid lizards, modeled a specific strategy of movement and foraging known as the "conspecific cueing strategy." In the conspecific cueing strategy, an animal observes the movements, density, and activity of conspecific individuals in addition to seeking food resources and uses these to determine its individual time budgets; thus, the strategy constitutes a type of time minimization strategy.

Out of the total length of a time period (T_0) under consideration (for example, $T_0 =$ one day), an animal must spend some time (T_e) in activities most essential to it; such activities include escape from an avoidance of predators. It must also spend time (T_c) in conspecific interaction such as aggressive encounter. In addition, an animal must spend time (T_f) required to obtain a certain amount of the food available at a given time and space. The rest of the time (E) is considered to be free time available to an individual. Therefore,

$$E = T_0 - T_e - T_c - T_f. \tag{8.13}$$

To complete the model, it is assumed that individuals move through the environment in such a way as to increase free time, i.e., individuals move to areas where E is larger.

We shall derive a differential equation for the population density $S(x, t)$ moving in one-dimensional space. The population flux, J_x, at a point x and time t is given by

$$J_x = uS,$$

where u is an advective velocity, and random animal movement is neglected.

Since individuals move in the direction of increasing E, i.e., "upgradient," u may be expressed as $u = k\partial E/\partial x, k$ being a positive constant. Hence, the population flux is written as

$$J_x = k(\partial E/\partial x)S.$$

Using this expression of the flux, we obtain the differential equation for S:

$$\partial S/\partial t = -\partial J_x/\partial x = -\partial\{k(\partial E/\partial x)S\}/\partial x. \tag{8.14}$$

To render the model specific, we choose particular functional forms for all the components of the time budget. First, T_0 and T_e are assumed to be constant or independent of x. Second, T_c is assumed to be linearly dependent on the population density $S : T_c = cS$, where c is a constant. Third, T_f is assumed to be a function of x and t. Under these assumptions (8.14) can be reduced to

$$\frac{\partial S}{\partial t} = \frac{\partial}{\partial x}\left\{k\frac{\partial T_f}{\partial x}S\right\} + \frac{\partial}{\partial x}\left\{kcS\frac{\partial S}{\partial x}\right\}. \tag{8.15}$$

Equation (8.15) is an advection–diffusion equation with a variable advection $-k\,\partial T_f/\partial x$ and with a density-dependent diffusivity kcS. For a habitat limited by $-L \leq x \leq L$, the conditions at the boundaries become

$$S\,\partial E/\partial x = 0$$

or

$$S\{k\partial T_f/\partial x + kc\partial S/\partial x\} = 0 \quad \text{at } x = -L \text{ and } L. \tag{8.16}$$

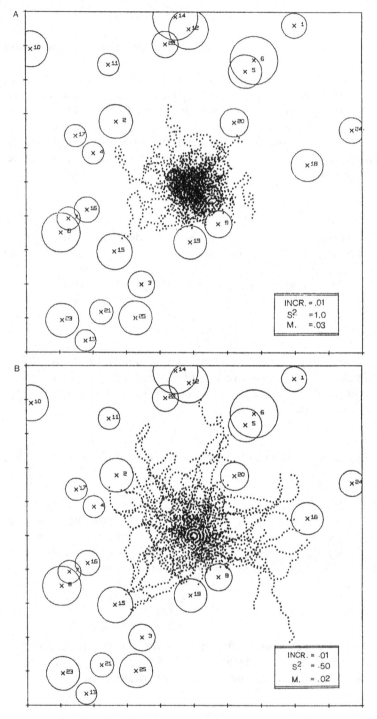

FIGURE 8.9. Simulation of animal dispersal among units of discrete habitats under various conditions of path length, angle of turn, and mortality (A–D) (from Kitching, 1971).

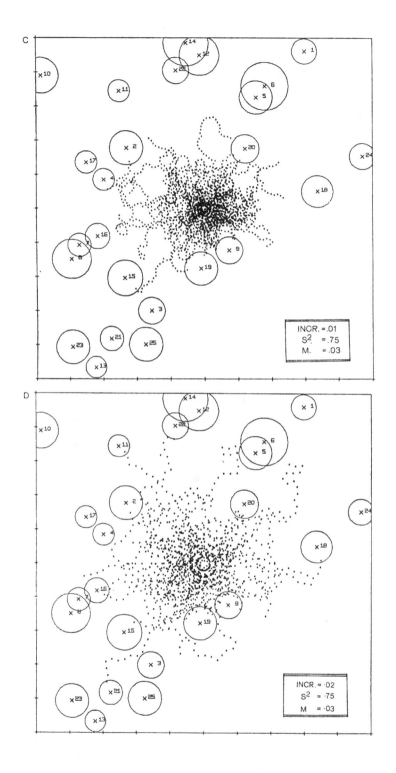

We shall consider an equilibrium distribution of animals for a given distribution of food resources, so that $\partial S/\partial t = 0$ and T_f is a given function of x only. Equating the right-hand side of (8.15) to zero, integration with respect to x and use of the boundary condition (8.16) yield

$$S = A - c^{-1}T_f,$$

where A is a constant of integration that can be evaluated from the constraint condition that

$$\int_{-L}^{L} S\,dx = N,$$

where N is the total number of individuals in the habitat. We finally obtain

$$S(x) = \left\{ N - c^{-1} \int_{-L}^{L} T_f(x)\,dx \right\}/2L - c^{-1}T_f(x). \qquad (8.17)$$

In order to examine a particular example, the following form of T_f is chosen:

$$T_f(x) = a - be^{-\lambda|x|},$$

where a, b, and λ are constants. The solution is then given by

$$S(x) = N/2L + c^{-1}b\{e^{-\lambda|x|} - (1 - e^{-\lambda L})/\lambda L\}. \qquad (8.18)$$

As is seen from (8.18), Kiester and Slatkin's model predicts that the equilibrium population is distributed in space as a mirror image of the resources (Fig. 8.10). (In order to have a biologically acceptable solution, so that $S \geq 0$ in $-L \leq x \leq L$, it is required that

$$N \geq 2Lc^{-1}b\{(1 - e^{-\lambda L})/\lambda L - e^{-\lambda L}\}.$$

Kiester and Slatkin discuss the case where this requirement is not met.)

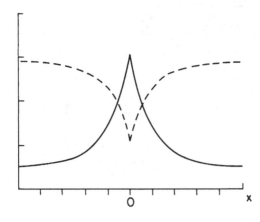

FIGURE 8.10. Spatial distributions of population and resources as a result of the conspecific cueing strategy of movement. (——— animal populations, – – – – resources)

Kiester and Slatkin also studied the development in time of animal dispersal from an initial arbitrary distribution and noted that movement is fastest in regions of high density; in the case of aggregation, clumping is more rapid and the approach to equilibrium is faster in regions of high density, while for dispersal, the reverse is true and dispersal away from regions of high density is more rapid. This is implicit in the nature of the model. Equation (8.15) is a diffusion equation in which the diffusivity is proportional to the population density. Therefore, the rate of diffusion or movement is higher in areas of high density.

Many predators chase their prey and capture them in full pursuit. To name a few examples, bats hunt moths, cheetahs chase gazelles, and dogs chase hares. Thus, movement patterns and chase–escape strategies of predator and prey are another aspect of interest and importance. The problem can be classified primarily into two categories depending on the relative speed between predator and prey. If the speed of the predator exceeds that of the prey, the prey must be equipped with superior maneuverability in order to avoid capture. On the other hand, if the prey's speed exceeds that of the predator, the predator must rely on stalking and attacking by surprise or use its power of initial acceleration in order to catch the prey before it gains its full speed of escape.

As an example of the first category, we shall study a turning gambit involving predator and prey. Howland (1974) discussed optimal strategies for predator avoidance in terms of a simple turning gambit initiated by the prey animal. Suppose (Fig. 8.11) that at some point in a straight-line chase, the prey swerves, turning with its minimum radius of turn, R_2. Simultaneously, the predator also turns with its minimum radius of turn, R_1. The question is, "Can the prey escape?" [The question of pursuit of such a kind appears to have originated in antiquity with Leonardo da Vinci. However,

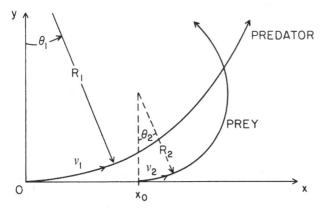

FIGURE 8.11. Courses of predator (R_1, θ_1, v_1) and prey (R_2, θ_2, v_2) during avoidance maneuver (from Howland, 1974).

until recently the problem of pursuit has been treated almost entirely as a mathematical game (Davis, 1960).]

Howland (1974) assumed that if, when the x-coordinates of the predator and prey are equal, the predator is above the prey (see Fig. 8.11), then the predator could outflank the prey, and by turning with a slightly greater radius, capture it. On the other hand, if the prey is above the predator at the same x position, then the prey has escaped at least in this gambit.

Let (x_1, y_1) and (x_2, y_2) be, respectively, the positions of predator and prey at time t, measured from the initiation of the turning gambit. Simple geometric considerations yield

$$x_1 = R_1 \sin \theta_1,$$
$$y_1 = R_1(1 - \cos \theta_1),$$

where θ_1 is the angle swept out by the predator during time t and is equal to $v_1 t / R_1$, where v_1 denotes the speed of the predator (Fig. 8.11). Hence,

$$x_1 = R_1 \sin(v_1 t / R_1),$$
$$y_1 = R_1 \{1 - \cos(v_1 t / R_1)\}. \tag{8.19}$$

Similarly, for the prey,

$$x_2 = x_0 + R_2 \sin(v_2 t / R_2),$$
$$y_2 = R_2 \{1 - \cos(v_2 t / R_2)\}. \tag{8.20}$$

Our criterion is that when $x_1 = x_2$,

$$y_1 \geq y_2 \quad \text{implies capture,}$$
$$y_1 < y_2 \quad \text{implies escape.} \tag{8.21}$$

Using nondimensional quantities, $R \equiv R_2 / R_1 < 1$, $v \equiv v_2 / v_1 < 1$, $\tau \equiv v_1 t / R_1$, and $x_0' \equiv x_0 / R_1$, we can express (8.21) as follows. When $\sin \tau = x_0' + R \sin(v\tau / R)$,

$$1 - \cos \tau \geq R\{1 - \cos(v\tau / R)\} \quad \text{implies capture,}$$
$$1 - \cos \tau < R\{1 - \cos(v\tau / R)\} \quad \text{implies escape.} \tag{8.22}$$

The normalized closure distance d is defined to be the normalized distance between predator and prey:

$$d = \{(x_1' - x_2')^2 + (y_1' - y_2')^2\}^{1/2},$$

where $x_i' = x_i / R_1$ and $y_i' = y_i / R_1$ $(i = 1, 2)$.

It is found that, in order for the prey to escape for all values of x_0, its normalized velocity must be related to its normalized radius of turn by the inequality

$$v^2 > R$$

FIGURE 8.12. Minimum relative closure distance in terms of relative velocity (v) and turning radius (R) (from Howland, 1974).

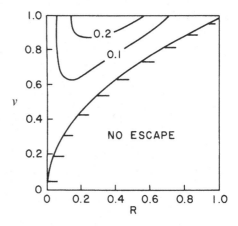

or

$$v^2/R \equiv \alpha > 1, \qquad (8.23)$$

where α is the relative centripetal acceleration of the turn. Figure 8.12 shows the result of capture–escape in terms of v and R. The regions above the no-escape area designate the minimum relative closure distance for an optimal choice of normalized starting distance x_0'. It can be proved (S. Levin in Howland, 1974) that the upper border of the no-escape zone in Fig. 8.12 is given by the equation $v^2 = R$. For small values of τ, the Taylor expansions of the sine and cosine functions in (8.22) yield this relation.

The results of Howland's calculations imply, for example, that the simple advantage of predator speed does not guarantee its ability to catch slower prey and that a prey capable of turning with a sufficiently small radius can escape a faster predator. Howland (1974) applied his theory to specific predator–prey combinations such as falcons versus pigeons, cheetahs versus gazelles, bats versus moths, and pile versus sunfish.

As an example of the second category, Elliott et al. (1977) analyzed in detail the attack of the African lion on its faster prey and estimated the probability of success under various circumstances.

We make use of Newton's equation of motion (5.3). An animal cannot continue to accelerate indefinitely, but exhibits an acceleration that decreases with time until it vanishes and a constant maximum velocity is achieved. Thus, various workers (Furusawa et al., 1928, among others) have proposed the following simple relationship between the acceleration, i.e., the force per unit mass, and the velocity:

$$F/m = k(v_m - v), \qquad (8.24)$$

where k is a positive constant and v_m is the maximum attainable velocity. Combination of (8.24) with (5.3) yields

$$\frac{dv}{dt} = k(v_m - v). \qquad (8.25)$$

TABLE 8.1

Species	k (s^{-1})	v_m (ft/s)
African lion (female)	0.68	45.7
Wildebeest	0.39	46.9
Zebra	0.31	52.5
Thomson's gazalle	0.17	86.8

Equation (8.25) can be integrated with respect to t subject to the condition that an animal starts with some initial velocity v_0. The result is

$$v(t) = v_0 e^{-kt} + v_m(1 - e^{-kt}). \qquad (8.26)$$

Integrating (8.26) with respect to t, we obtain the displacement $x(t)$:

$$x(t) = x_0 + v_m t - (v_m - v_0)(1 - e^{-kt})/k, \qquad (8.27)$$

where x_0 denotes the initial position of the animal.

According to Elliott et al. (1977), the values of the constants k and v_m for the African lion and values of its prey animals are as given in Table 8.1. Note that k^{-1} is a measure of time for the animal to achieve its maximum velocity.

It can be seen that for each of these three prey, the attack strategy of the lion requires that it utilizes its faster acceleration to overtake prey before the prey can achieve maximum velocity and hence escape. On the basis of this dynamic model, the probability of the attack success of lions can be estimated under various conditions.

8.6   Data on Animal Movements

The advent of remote sensing methods for tracking movements of animals has opened opportunities for detailed analysis and comparisons to model assumptions. A variety of texts are available that detail methods for obtaining and analyzing radiotelemetry information (White and Garrott, 1990; Priede and Swift, 1992; Geers et al., 1997). Automated telemetry systems offer the potential to monitor large numbers of individuals with limited observer effort. As with all telemetry methods, differences arise due to variance in observation rates that must be statistically corrected to account for spatial correlations (Johnson et al., 1998). Telemetry data have been used to carry out correlations with biotelemetry data on individual variables such as heart rate and body temperature. These involve attempts to estimate level of "activity," interpreted as a function of the rate of movement. Scaling problems arise in such an estimation since it requires interpretation of discretely sampled values of a continuous variable (location). Methods to account for the temporal dependence of radio-tracking data using semivariograms may be needed to carry out appropriate statistical analyses (Salvatori et al., 1999).

Various texts summarize movement ecology (Swingland and Greenwood, 1983; Turchin, 1998), but there is no across-species compendium of telemetry information or database for such data. Much of the analysis of the available data has focused on estimating survivorship and determining causes of mortality. In these cases, the objective is to look for nonmovement or for habitat use rather than tracking the full dynamics of individual location. This has occurred in part due to the large effort required to regularly follow individuals using a handheld antenna, or through the use of aircraft. Radiotelemetry information has been used as well for translocation studies, to analyze the effectiveness of moving organisms from dense areas to areas of lower density, or for the success of releases of captive-bred individuals. The objectives are to reestablish populations in historic ranges and reduce inbreeding in isolated populations. Sex and age of individuals influence the effectiveness of such translocations. Translocated cougars (*Felis concolor*) that were dominant territorial males, or reproducing females before translocation, had the greatest tendency to return to former home territories (Ruth et al., 1998).

A variety of case studies indicate the dependence of movement on seasonal factors that are species-dependent. Free-ranging large herbivores such as moose (*Alces alces*) have strong seasonal components to their movements driven by seasonal changes in forage availability, quality, and quantity. This leads to very different summer and winter home ranges (Hundertmark, 1997), and detailed energetic models that take account of seasonal dynamics allow comparisons between alternative foraging strategies (Moen et al., 1997). Seasonal changes in movement also may occur due to effects on other environmental factors such as water depth in pulsed systems. Florida panthers shift movements due to changes in seasonal water depths, particularly in wet years (Maehr, 1997b). Major storms such as hurricanes might be expected to cause shifts in movements, but data on white-tailed deer do not indicate any significant impact pre- versus post-hurricane (Labisky et al., 1999).

8.6.1   Case Study: Florida Panther

An example of some of the difficulties associated with realistically modeling movement for a single species may be examined from a case study. Maehr (1997a) summarizes data on the Florida panther (*Felis concolor coryi*) obtained from radio-collar tracking available since 1981. Females typically have home ranges less than half the size of that of males, though this is very dependent on age and social status. There are consistent home-range overlaps between females, whereas the overlap of males' home ranges is minimal. Home-range size is smaller for both genders where large prey, primarily deer, are more abundant. Landscape features play important roles in delineating boundaries of home ranges. Females avoid paved roads and these often form the boundary of a female's home range, but males much more readily cross such roads. No panthers generally cross large rivers, so these can demarcate home ranges.

The typical movement of a Florida panther is to leave a daytime rest site around sunset, stalking or waiting for prey during the night. If a prey is captured, a panther may stay at the kill site for several days or may travel to a new daytime rest site around sunrise. They generally rest in dense cover during the day. Males travel widely throughout their home range to maintain social dominance, covering it completely within a month. Obtaining food is of secondary importance. Females with kittens have very restricted movements within their home range, which is typical of many large felids (Maehr et al., 1989). Only when solitary do females cover the entire home range. The average straight-line distances between consecutive daily readings for males is more than twice that of females (3.4 miles versus 1.4 miles), but panthers often move more than twice these amounts daily when movements during the night are included. There are distinct seasonal components to movements as well, in part due to constraints of water depth which limit movements, particularly for mothers with kittens, under high water conditions.

Female Florida panthers almost always establish home ranges within or greatly overlapping that of their mother. Subadult males disperse out of the natal home range but suffer very high mortality. Of 13 subadult male panthers monitored between 1986 and 1994, 7 were killed by resident adult males during dispersal (Maehr, 1997b). Permanent home-range shifts typically occur only at the death of the resident. When an adult male dies, another takes its place, utilizing essentially the same home range. These behavioral limitations on establishment of new home ranges can make it very difficult for new areas to be occupied, unless females are driven to disperse from the natal range by population pressures.

A model of Florida panthers was constructed for application to Everglades restoration (DeAngelis et al., 2000; Mellott et al., 1999). Since the objective was to compare detailed spatio-temporal alternative plans for hydrology across the region, an individual-based approach was taken. The model includes separate sets of rules for movement to account for (i) search for prey, which are primarily white-tailed deer, (ii) the male's search for mates, (iii) dispersal of juveniles from the natal area, (iv) dispersal of adults when food availability is low, and (v) avoidance of high water depths. Not specifically included were rules for marking of home range and associated avoidance of dominant males by dispersing males. The model was linked to an individual-based deer model (Abbott et al., 1997) modified to allow for the feedback effect of harvesting by panthers. The movement rules were all based on random local neighborhood searches (for prey and mates), constrained by water depths and habitat features, with movement into an expanding range of cells around the current location if the local search was unsuccessful. Comparisons of the model home ranges with data indicated that model movements were considerably more constrained than those observed, pointing out the need to account for territory maintenance rules in the model.

8.7   Home-Range Estimation

Assessment and comparisons of various methods of estimation of home range are difficult in part because it is not possible to determine true sizes. All are estimated from discrete measurements of individuals, leading to sampling and scaling problems. If movement trajectories of individuals were modeled as a stochastic process taking values in 2-space, then a sample from this would correspond to locations obtained from observations. If the process is stationary, with no explicit time dependence due, for example, to diurnal foraging, then there is a spatial autocorrelation function that can be estimated from data. Probabilistic estimators that depend on independence of points cannot be applied to these data directly, and decreasing the sampling frequency to provide independence can decrease the ability to ascertain variables such as within-time interval total movement.

Schoener (1981) derived a ratio that can be used to detect autocorrelation. A particular concern has been that autocorrelation can cause underestimates of home-range size (Swihart and Slade, 1985). The degree of autocorrelation can vary between categories of animals, as, for example, root vole (*Microtus oeconomus*) males have higher autocorrelations than females when sampling intervals were an hour or less (Hansteen et al., 1997), which may arise from the larger home-range sizes of males. Comparisons of various methods for estimation of home-range size indicate significant differences in bias between them (Worton, 1995; Hansteen et al., 1997; Seaman et al., 1999) with home-range estimators differing by as much as 300% (Ostro et al., 1999). A consensus has developed that using individuals as the source of replication rather than locations is appropriate for estimation of habitat use (Otis and White, 1999). Sample size is critical, however, since analysis of many locations from few individuals, which is possible with Global Positioning Systems, may reduce the capacity to make inferences about the population.

8.8   Allometric Relations Between Body Size and Home Range

A large literature has developed around fits of (8.1) to various estimates of home-range size and associated estimates of density (Peters, 1983; Brown and Maurer, 1986; Nee et al., 1991; Blackburn and Gaston, 1997). Indeed, these issues may be considered the starting point for the field of macro-ecology (Brown, 1995), dealing as it does with the division of food and space at large spatial extents. The estimates of the parameter b in (8.1) vary widely, with inherent statistical problems due to lack of complete sampling. Home-range estimates may have been made for only some species within a taxonomic group, and these missing species are undoubtedly not a random subset of the group. Thus, comparisons between taxonomic groups, e.g., statements such as "the rate of home range increase with body size is more pronounced in carnivores than in herbivores," (Peters, 1983) must be interpreted with extreme care. Given the problems with home-range estimation, it is no wonder that there has been continuing controversy regarding allometric relations between body size, home-range size, and the implications of this. Statistical analyses that account for the variance in home-range estimation seem not to have been carried out, since most across-taxon studies use the mean reported species home-range estimates. It is not at all clear how meaningful such comparisons can be when home-range estimates vary so widely as a function of season, age, and individual social status. Cross-taxa comparisons must therefore deal with the fact that the within-taxa variance due to these factors may well not be similar across taxa.

8.9   Individual-based Models of Movement

Individual-based models incorporating movement have become standard tools for the analysis of populations (DeAngelis and Gross, 1992; Grimm, 1999) and are important components of spatially explicit population models, which can include landscape details from geographic information systems (Dunning et al., 1995). The complexities listed in Sect. 8.3 have all been incorporated in models, and simulations have taken the lead of Kitching (1971) described in Sect. 8.4 to compare a variety of measures of movement to available data. Grimm (1999) and Turchin (1998) provide a review of the literature. A summary of applications of this approach to several species in a particular applied context, Everglades restoration, is given in DeAngelis et al. (2000).

   Simulation models have allowed for the inclusion of habitat features so that information obtainable from GIS can be incorporated. Rule sets are considerably more complicated than those described in Sects. 8.3 and 8.4, but generally take the form of a random walk with movement probabilities dependent on local habitat characteristics, the state of the individual, and possibly the states of other individuals (including predators as well as con-specifics). Movement model assumptions may be categorized into local departure and destination rules, which can be derived from models for optimal foraging, taking account of risks and benefits (Railsback et al., 1999). An updated version of the model described in Sect. 8.3 for red fox (Carter and Finn, 1999) compares a random movement model similar to that of Siniff and Jensen (1969) to that derived using expert-system rules. Though much superior to the random movement model in both qualitative and quantitative comparisons to field data, the expert-system model also displays some behaviors not observed in the field data.

8.9.1   Corridors and Movement

Examples of simulation approaches include efforts to account for habitat fragmentation, with an objective being to determine how potentially different corridors of appropriate habitat may enhance dispersal (Brooker et al., 1999). There is considerable disagreement about the utility of corridors in conservation biology due in part to lack of evidence that individuals disperse within corridors more readily than outside them (Hobbs, 1992; Simberloff et al., 1992). Evidence is accumulating though that some species do utilize corridors (Machtans et al., 1996; Sutcliffe and Thomas, 1996); they are not essential for dispersal (Haas, 1995; Dodd and Cade, 1998), nor are they necessarily utilized in similar ways for taxonomically similar species (Hill, 1995). The probability of successful dispersal, as a function of the fraction of the landscape that is appropriate habitat, is dependent quite strongly on patch geometry (Tiebout and Anderson, 1997).

Models can play an important role in determining potential benefits for maintaining corridors within highly fragmented landscapes. In the simulations of Brooker et al. (1999), an individual moves randomly through the available appropriate habitat from a designated origin to a designated target, with a fixed time limit. Individuals are followed by their location on a rectangular grid. There is an assumed gap tolerance, the maximum distance without being in an appropriate habitat that a particular species might cross during dispersal. This relies on the assumption that the species in question prefers certain habitats during dispersal, but will, to some extent, cross others if necessary. Movement then follows a set of rules by which an initial direction is chosen randomly, the individual moves to a cell in that direction with the most appropriate habitat, and if no cell with appropriate habitat is in that direction, another direction is chosen at random. If all directions are exhausted with no appropriate habitat found, then successively larger distant cells from the current location are scanned until either an appropriate cell is found or the gap tolerance is reached. In the latter case, the individual retraces its previous step.

Brooker et al. (1999) parameterize their model using data from banded individuals of the Blue-breasted Fairy-wren (*Malurus pulcherrimus*) and the White-browed Babbler (*Pomatostomus superciliosus*), which occupy remnant patches of native vegetation in western Australia. The model is applied to quantify corridor use by estimating gap tolerance and using this to estimate dispersal mortality. The model is critical for such estimation since these factors interact and cannot be estimated directly from the data, which consist of sitings of individual birds within habitat patches. Estimation of dispersal mortality is considered critical by some due to supposedly high sensitivity of this mortality on the per-movement-step mortality parameter in spatially explicit population models (Ruckelshaus et al., 1997). Mooij and DeAngelis (1999) derive an analytical bound on the prediction error for dispersal mortality due to uncertainty in per-movement-step mortality. Their results indicate that individual-based models are not nearly as sensitive to estimation of dispersal parameters as Ruckelshaus et al. (1997) claim. Thus, unreasonably accurate parameter estimation is not needed in order for individual-based models to be useful in conservation ecology.

8.9.2   Food Density and Home-Range Size

Negative relationships between home-range size and food density have been observed in several mammal species in addition to Florida panther (Kenward, 1985; Tufto et al., 1996; Powell, et al, 1997). Determining whether such differences are due to direct responses to food availability and energy requirements as described in Sect. 8.1, avoidance of conspecifics, or defense of territories is not easy. Determination of the processes driving spacing of individuals is complicated considerably in the case of overlapping home ranges. For several large ungulates, declines in food abundance lead to shrinkage of home range (Harestad and Bunnell, 1979; Georgii, 1980; Ilse and Hellgren, 1995).

In an attempt to determine which patterns of home-range size and overlap arise from individual forager movements, South (1999) carried out a simulation parameterized for the red squirrel (*Sciurus vulgaris*). This evaluated changes in spacing projected when varying food density for a fixed number of individuals moving among hexagonal grid cells in a fixed area. The movement rules assume individuals maintain a memory of food availability as well as a tendency not to move far from a nest site, which is returned to daily. Locations of modeled individuals were sampled in the same manner as would be done in the field and modeled home ranges were computed from these. The general pattern that arose was of small, similarly sized home ranges at high food densities and larger, different-sized overlapping ranges at low food density. Both mean home-range size and mean range overlap were highly negatively correlated with food density, but from the limited comparisons to field data one cannot conclude that the mechanisms included in the model are sufficient to explain spacing in the field. Such models generate predictions from relatively simple behavioral rules and provide a mechanism, by including factors such as individual interactions for mating, to compare the relative effects of different processes on spatial patterning of individuals.

8.10   Diffusion Models

Many extensions of the diffusion techniques discussed in Sects. 8.2 to 8.4 have been made, with reviews in Turchin (1998). Shigesada and Kawasaki (1997) provide a comprehensive review of applications to invasion biology.

Discussed here are some recent efforts using random walk and diffusion models to take account of boundary effects, particularly to analyze reserve design, animal search methods with heterogeneous resource distributions, and attempts to factor in multiple scale approaches for animal foraging movements.

8.10.1   Boundary Effects

Behaviors at habitat boundaries are potentially useful measures to assess possible individual use of corridors between habitats (Wiens et al., 1985). In a mixture of an experimental and simulation approach, Haddad (1999) investigated several butterfly species, following movements and estimating turning angles, speeds, and relationship to habitat. The model was an extension of a correlated random walk (see Sect. 5.3) to include (i) sampling of step distances and turning angles from habitat-dependent empirical distributions derived from the data, (ii) moving distance and turning angle were sampled from a joint distribution rather than independently, and (iii) turning angles near the boundary were sampled from empirical distributions of angle relative to distance from edge, corner, or corridor. The simulations used three patches arranged linearly, two of which were connected by a corridor, with distance between patches, corridor width, and patch size allowed to vary. Results showed that successful dispersal to new patches was increased by addition of a corridor only when the model was parameterized for the two species that had turning behaviors biased near the boundary. This result was consistent across all spatial patch arrangements used and indicates that different species show different responses to corridors, which can be estimated in part by behavioral analysis near boundaries.

A diffusion approach to the problem of biased movement near boundaries within patches has been derived by Cantrell and Cosner (1999). While not dealing with corridor-use issues, it allows for different individual movement behavior near boundaries as Haddad (1999) does. This relies on skew Brownian motion, which is the continuous analogue to a random walk with the modification that at a boundary a parameter s determines the likelihood of crossing the boundary versus being reflected. Their objective is to investigate, for one-dimensional spatial problems, the effect of the boundary behavior on total population size when there are two regions, one of which is more favorable to population growth than the other. The unfavorable region is viewed as a buffer zone, designed to isolate the population. This is an extension of earlier work on the effect of shape and spatial arrangement of favorable and unfavorable regions on population growth (Cantrell and Cosner, 1991). The model is a diffusion equation with a matching condition required at the interface boundary between the two regions. Assuming different linear population growth rates in the two regions, they derive an overall average population growth rate and determine how this depends on the parameter s and the difference in growth rates in the two regions. A main conclusion is that the benefits of a buffer zone are not estimable just from the quality of habitat in it, but also by its effect on dispersal. The model can be useful then in comparing the relative benefits of increasing the refuge (favorable area) versus increasing the buffer zone.

8.10.2   Movement in Heterogeneous Environments

How foragers move in a heterogeneous environment in order to maximally gain resources is a key topic in foraging theory. Much of this theory does not take into account the tactics that a forager must use, based on its knowledge of the environment and response to local cues, to accumulate resources. One approach to analyze a variety of tactical behaviors for movement and foraging in environments with spatially varying resources is to extend the models of taxis (see Sect. 5.4) and build a continuous approximation to a biased random walk (Grünbaum, 1999). The model assumes that turning rate depends on a local cue, and the objective is to determine how local, individual-level behavioral responses to environment translate to population-level responses. The formulation involves the derivation of an advection–diffusion equation with a given turning rate defined as a function of time, location, current orientation, and time since last turn. The model is applied to foragers on a spatial distribution of discrete resource patches. It is found that taxis is possible if turning rate is a function of the integrated resource density encountered along the movement trajectory, while taxis is not possible if the turning-time distribution is a function of this integrated density.

8.10.3   Multiple Scales and Foraging

The use of random walk and diffusion models to describe spatial patterning of organisms contrasts with the body of behavioral literature that indicates, at least for higher organisms, the capacity for individual organisms to be aware of environmental properties beyond the range of their perception. Individuals can potentially make foraging decisions at a wide variety of spatial scales (Bailey at al., 1996) based on different motivations (e.g., for food uptake, mate search, territorial marking).

Farnsworth and Beecham (1999) develop a model that takes account of multiscale processes by extending a diffusion approach to include biased movements due to different factors operating at different scales. The bias is a function of the difference in some measure (quality of shelter, mate availability, food density, etc.) between the current location and the average around this location. By varying the region over which this average is taken, these biases operate at different scales, and several are combined in one model. A simulation approach is taken to investigate how movement patterns are affected by whether long-range or short-range information dominates in affecting movement. Results show that the spatial scale of individual decision bias can have large impacts on both intake rate and resource spatial distribution.

An alternative approach to hierarchical foraging (Fauchald, 1999) assumes a nested sequence of patches with high-quality patches at small spatial extents nested within low-quality patches at larger spatial extents. A stochastic simulation based on a biased random walk allows foragers to change their search radius so as to maximize the short-term encounter rate of prey, based on their history of intake. This requires an individual updating rule that discounts information too distant in the past. The tracking efficiency of the forager (e.g., its success in following changes in prey availability) is maximized at intermediate prey aggregations. The significance of this will depend on efforts to investigate how foragers respond to different scales of patchiness and utilize past experience in determining current behavior.

9
Patchy Distribution and Diffusion

Akira Okubo and James G. Mitchell

Biotic populations are usually distributed heterogeneously in their habitats, and the distribution itself is often patchy. In ecology, we conventionally classify the structure of dispersion into three categories: Frequency distributions may be random, uniform (regular), or clumped (contagious) (Fig. 9.1).

This classification is based on the frequency distribution of individual organisms in the statistical sense, e.g., based on quadrat counts. If organisms disperse according to a purely random process, their frequency distribution should be represented by the binomial distribution (Ito, 1963).

As the mean number of individuals per sampling unit becomes small, the binomial distribution tends to a Poisson distribution, for which the mean and variance are equal. Therefore, we can obtain an empirical classification of the dispersal structure by comparing the sample mean $\bar{n}$ with variance $\overline{s^2}$ of observed frequency distributions of the number of organisms per sampling unit. We call the distribution *random* if $\overline{s^2}/\bar{n} = 1$, *uniform* if $\overline{s^2}/\bar{n} < 1$, and *clumped* if $\overline{s^2}/\bar{n} > 1$, provided that these obtained relations are statistically significant (see Pielou, 1977, p. 125). Various attempts have been made to devise a measure of aggregation (Morisita, 1959; Pielou, 1969, 1975; Iwao, 1968; Iwao and Kuno, 1971; Hurlbert, 1990). Based on field data comprising 3840 samples from 102 species (range from protozoa to human populations and plants), Taylor et al., (1978) examined the relations between $\overline{s^2}$ and $\bar{n}$. Only two data sets were judged random at all population densities, and most of the data were judged significantly more clumped than random. From the fact that an observed frequency distribution can often be fitted to a theoretical contagious distribution (Bliss, 1971), we usually conclude that the population's *spatial* pattern is patchy, i.e., the organism is distributed in a patchy manner in space. However, it is still not possible to draw any conclusions regarding the mechanism that gives rise to the distribution; in any case, the organism distribution obtained by this method provides us with nothing but sample statistics, or at most a static picture of the spatial pattern of organisms. Demand has driven the development of a variety of methods for spatial

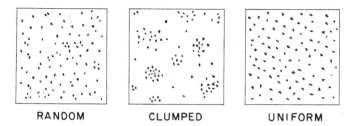

FIGURE 9.1. Distribution patterns of biotic populations: random, clumped, and uniform.

pattern analysis (Davis, 1993; Koehl et al., 1993; García-Moliner et al., 1993; Maurer, 1994; Strutton et al., 1997), and increased computer speed has led to useful computer packages (e.g., S + Spatial Stats) as well as models and predictions of dynamics (Hassell et al., 1991; Moloney, 1993). But what really determines the *spatial distribution* of organisms are the ecological processes represented by such factors as intra- and interspecific relations, dispersal, and migration, to mention a few. In this chapter, we examine the role of biodiffusion in determining the patchy distribution of organisms.

9.1 Role of Diffusion in Plankton Patchiness

Patchy distributions of plankton in the sea and lakes have been well documented (Cassie, 1963; Tonolli and Tonolli, 1960; Steele, 1976a, 1977, 1978; Haury et al., 1978; Okubo, 1984; Powell and Okubo, 1994). The mechanism that maintains this patchiness is still, however, a subject of considerable controversy (Abraham, 1998). A few of the various possibilities that have been proposed are (i) mechanical retention in wind-generated convective cells, or frontal regions, (ii) behavioral reaction to, or association with, temperature, salinity, and nutrient distributions, (iii) exclusion of certain zooplankton by phytoplankton, (iv) food-chain association in predator–prey relationships, and (v) aggregative behavior (swarming, schooling) for breeding and feeding.

The patchiness itself most likely arises from a variety of mechanisms and processes under various conditions. However, many cases share a single process that acts as an "antipatchiness" agent. This mechanism is diffusion due to turbulence in surrounding media or random movements of organisms. The process of diffusion tends to counteract the formation of organism aggregates and to give rise to a uniform distribution.

The entire story is not so simple, however. When, for instance, non-linearity in predator–prey interactions is combined with diffusion, a new

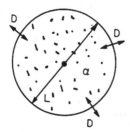

FIGURE 9.2. A water mass within which phytoplankton grow and diffusion takes place: α: growth rate; D: diffusivity; L: size of water mass.

instability may arise which leads to a self-generated "patchiness", or enhances existing aggregations (see Sect. 10.6). Keeping this instability in mind, for the present we focus our attention on the balance between the aggregative process of organism growth or reproduction and the anti-aggregative process of diffusion.

Consider a water mass within which occurs a phytoplankton bloom (Fig. 9.2). We may assume that this water mass is surrounded by ambient water in which plankton cannot survive. Since a part of the population is continuously lost to the surrounding due to diffusion, the plankton patch would cease to exist unless reproduction within it counterbalances this loss.

Now the loss of organisms due to diffusion takes place through the boundary of the patch; hence, its rate is proportional to surface area. On the other hand, reproduction takes place locally within the patch, and hence its rate is proportional to volume. Since the ratio of surface area to volume for a sphere is inversely proportional to the radius of the sphere, a larger sphere carries less surface area relative to its volume than a smaller sphere. As the volume of water mass decreases, diffusion plays an increasingly important role, and eventually a limit is reached beyond which reproduction can no longer compensate for loss due to diffusion.

Thus, there exists a minimum critical size for the water mass, or simply *critical size*, below which no increase of plankton population is possible. We now estimate this critical size. Let D and α be the diffusivity and the rate of growth, respectively. Suppose that the critical size L_c, can be determined by two parameters, D and α; $L_c = f(D, \alpha)$. Dimensionally, $D = [L^2 T^{-1}]$, $\alpha = [T^{-1}]$, $L_c = [L]$. Dimensional analysis leads to

$$L_c = c(D/\alpha)^{1/2}, \qquad (9.1)$$

where c is a nondimensional constant of the order unity.

Dimensional analysis is a technique by which an unknown dependent variable, i.e., the quantity of concern, is expressed in terms of known basic parameters that control the phenomenon of concern, simply by making use of their dimensions. In some cases, the method works almost magically, quickly producing an answer, which is, up to an arbitrary constant usually of order unity, identical to that which could be obtained by using sophisticated mathematics. Of course, the success of dimensional analysis depends on

FIGURE 9.3. One-dimensional model of patchiness. The region of bloom is limited by $(0, L)$.

knowledge and experience in choosing the fundamental parameters as well as on intuition. We should not always expect the analysis to immediately lead to a solution. (9.1) is one of the examples that "works well." One can see its convenient simplicity in comparison with (9.4), which depends on higher mathematics. The reader who wishes to learn more about dimensional analysis should consult a reference text such as Bridgman (1963).

We next present a more "serious" method for estimating L_c. For simplicity, consider a one-dimensional (x-axis) region of bloom limited by $(0, L)$ (Fig. 9.3). Our model consists of exponential growth and Fickian diffusion with constant diffusivity.[1] The diffusion equation of the organism concentration, S, is then given by

$$\frac{\partial S}{\partial t} = \frac{\partial^2 S}{\partial x^2} + \alpha S. \tag{9.2}$$

We seek the solution of (9.2) that vanishes at $x = 0$ and $x = L$ and corresponds to a given initial concentration, $S(x, 0) = f(x)$. Kierstead and Slobodkin (1953) and Skellam (1951) obtained the solution

$$S(x, t) = \sum_{n=1}^{\infty} A_n \sin(n\pi x/L) \exp\{(\alpha - Dn^2\pi^2/L^2)t\}, \tag{9.3}$$

where $A_n = (2/L) \int_0^L f(x) \sin(n\pi x/L)\, dx$, are constants dependent on the initial condition.

We shall focus on the first term ($n = 1$) of the right-hand side of (9.3). The argument of the exponential function is time-dependent, being given by $(\alpha - D\pi^2/L^2)t$. If $\alpha < D\pi^2/L^2$, this exponential function approaches zero as time progresses. The second- and higher-order terms converge to zero even faster. The population will then be unable to maintain itself against diffusion, and the patch will disappear. On the other hand, if $\alpha > D\pi^2/L^2$, at least the first term will increase indefinitely with time; we may say that plankton bloom occurs. Therefore, the critical size, L_c, is determined from the condition $\alpha = D\pi^2/L^2$, i.e.,

$$L_c = \pi(D/\alpha)^{1/2}. \tag{9.4}$$

Comparing (9.1) with (9.4), we identify c with π. Thus, for this case c proves to be of the order unity; remember that the exact value of c could not be

---

[1] This model for the critical size was given by *Kierstead* and *Slobodkin* (1953) and *Skellam* (1951) independently. Hereafter, we abbreviate it as the KISS model.

determined from dimensional analysis. Kierstead and Slobodkin (1953) and Skellam (1951) also calculated the critical size for the two-dimensional case, for which $c = 4.81$.

The critical size given by (9.4) will serve as a reference scale for patchiness when we consider more complex models later on (particularly in Sect. 10.2). Hereafter it refers to the KISS size and is designated as L_0.

Slobodkin (1953a) attempted to apply the KISS model to red tides. He considered it likely that red-tide outbreaks were initiated by the occurrence of discrete masses of water with low salinity originating from coastal drainage, which slowly mixed with coastal waters.

Although the KISS model helps to clarify the role of diffusion in patchiness, the lack of inclusion of other important factors renders the prediction of plankton bloom size quite crude. In particular, when a patch of plankton is placed in a converging current such as will be discussed later, a flow pattern that tends to act against diffusion is also present. It then becomes reasonable to expect a plankton bloom the size of which is smaller than the KISS size. Okubo (1972; 1978a) dealt with a mathematical model for simple growth and diffusion under an attractive force toward the center of a patch. He found the critical size L_c to be $L_c = L_0 f(v^2/4\alpha D)$, where v is convergence velocity. For $v = 0$, $f = 1$, as it should be. As v increases, the value of the function f decreases slowly toward zero.

The above models do not consider the details of the surrounding of the patch, i.e., the region of space outside the bloom. This region is merely considered unsuitable for organism survival. It may be a region of strong predation pressure or a "region of sudden death" with strong currents, so that once plankton have wandered into it, they can no longer return to the original patch. The model that takes into account boundary conditions that are more general is presented in Sect. 10.2.2.

The above models are not necessarily appropriate when fronts are present or define the edge of a patch. As fronts are regions of convergence, suspended matter, detritus, phytoplankton, and zooplankton may be advected to them. This is not only limited to the passive accumulation of material at the front. Bacterial mineralization of organic detritus will stimulate phytoplankton growth, which stimulates zooplankton activity, thereby increasing the recycling rate.

Govoni and Grimes (1992) observed in neuston collections in and about the Mississippi plume front that larval fishes aggregated within the convergence zone of the plume. The convergent velocity was estimated as 0–80 cm/s, associated with lateral shear. The distribution of larvae was modeled by the balance of advection and diffusion. Olson and Backus (1985) used an advection–diffusion model to show that an increase in fish concentration of 100 times at the front of a warm-core ring could be explained by convergence.

As mentioned in Sect. 2.6.2, a mathematical model appropriate for oceanic diffusion cannot be of the Fickian type with constant diffusivity; a more appropriate model that accounts for the scale dependence of diffusion, such as that of Joseph and Sendner (1958), must be employed (Okubo, 1978b). Let us now discuss the critical-size problem by combining a growth mechanism and a diffusion mechanism based on the Joseph–Sendner model.

We again utilize the dimensional arguments first. Since Joseph and Sendner's (1958) model is uniquely characterized by the diffusion velocity, P, one can consider the critical size L_c to be dependent on P and α;

$$L_c = f(P, \alpha).$$

The dimensions of P, α, and L_c are, respectively, $[LT^{-1}]$, $[T^{-1}]$, and $[L]$. Hence,

$$L_c = c_1 P/\alpha, \tag{9.5}$$

where c_1 is a nondimensional constant of the order unity.

A more rigorous derivation employs the diffusion equation

$$\frac{\partial S}{\partial t} = \frac{P}{r}\frac{\partial}{\partial r}\left(r^2\frac{\partial S}{\partial r}\right) + \alpha S, \tag{9.6}$$

which must be solved for S under the same boundary conditions as before. After some manipulations (Okubo, 1978b), which are not presented here, it is found that

$$L_c = 7.34 P/\alpha. \tag{9.7}$$

Again, the constant c_1 is of the order "unity." One might query as to the expression for L_c if the diffusion model of Ozmidov (1958) is employed. The reader may obtain a quick result via dimensional analysis.

By and large, estimated values for the critical size from these models range from 1 to 50 km (Okubo, 1978b). These theoretical models support the general observations that plankton patches in the open sea appear to occur at scales of the order of 10 to 100 km (Steele, 1974a, 1976a). It is interesting to note that no matter what the model may be, the essential feature is preserved in the formulation of the critical size; it is determined by the balance of diffusion rate and reaction rate (or net growth rate) (McMurtie, 1978; Okubo, 1978b).

Patchiness smaller than the KISS scale does exist, indicating an improved model is needed (Owen, 1989; Duarte and Vaquè, 1992; Krembs et al., 1998). The mechanisms of sub-KISS-scale patches are discussed in later sections of this chapter. However, within the context of critical length, there are four important considerations for these patches in that they

i) are short-lived compared to those of the KISS model, so that the assumption that time goes to infinity is unrealistic,

ii) exist in an environment where the background concentration is nonzero, leading to variation from patch–patch fusion and a nonabsorbing boundary condition,

iii) are at scales where aggregation and vertical migration behavior may be important,

iv) imply reduced importance of turbulent eddy diffusivity and the existence of significant fluid dynamic intermittency.

As of yet there is no model that unifies the four considerations into a sub-KISS model estimating critical length. Dimensionally analogous processes can be used to balance two parameters. For example, choosing bacterial growth rate and the Kolmogorov scale, where $L = \pi(D/\alpha)^{1/2}$, $\alpha \sim 10^{-4}/s$ if the doubling time is about 3 hours, $D = 10^{-4}$ cm^2/s gives $L \sim 1$ cm. Similarly, for bacterial chemotaxis, where $\alpha \sim 1$ turn/s and D is 10^{-5} cm^2/s (molecular diffusion of attractant), (Berg, 1983) gives $L \sim 10^{-2}$ cm. Experimental support is available for both estimates (Duarte and Vaquè, 1992; Blackburn et al., 1998), but the broad formalism for a KISS-style model with broad predictive powers has yet to be developed. More of the critical-size problem is presented in Sects. 10.2.2 and 10.4.

9.2   Spectra of Turbulence and Patchiness

As described in Chap. 2, the motion in the surface layer of the sea is generally turbulent. Superposed on the mean flow, i.e., currents, are eddylike motions, i.e., turbulence. These eddies consist of a spectrum of various scales and frequencies. The worldwide prevailing wind system and tide-generating force feed their energy to the large-scale oceanic eddies. These eddies tend to transfer their kinetic energy into eddies of smaller scales as a result of mutual interaction. The smaller eddies feed their energy to yet smaller eddies. This sequence of events gives rise to a cascade of eddies of ever-decreasing size. Eventually, extremely small eddies (of the order of 1 mm) utilize fluid viscosity to dissipate the energy of turbulent motion into heat, i.e., molecular kinetic energy (Fig. 9.4). The wide spectrum of eddies thus produced may reach a steady state if the rate of energy supply balances the rate of dissipation. [For reviews of oceanic turbulence, see Bowden (1964, 1970), Faller (1971), Powell and Okubo, (1994)].

The concept of expressing turbulence in terms of a conglomeration of eddies of various sizes (wavelengths) can be formalized in terms of the

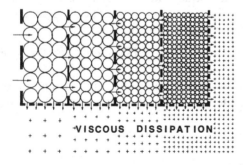

FIGURE 9.4. A conglomeration of eddies of various sizes in turbulence and the flow of kinetic energy from large eddies to small eddies. The energy is eventually dissipated by extremely small eddies into heat (viscous dissipation).

Fourier spectrum. Consider a box of locally statistically uniform turbulent flow within the patch. Flow velocity at any point can be decomposed into a mean component and a fluctuating component, the latter possessing a mean value of zero by definition. Although this fluctuating velocity varies from point to point and in time, the square of the magnitude of velocity possesses a nonzero average, i.e., the variance of the velocity, $\overline{v'^2}$. The average kinetic energy K of the turbulence in the box of volume V is thus nonzero given by

$$K = \frac{1}{2}\rho V \overline{v'^2} \quad (\rho: \text{fluid density})$$

even though turbulent velocities themselves average to zero. We conveniently consider the average turbulent energy per unit mass, i.e.,

$$K_1 = K/\rho V = \frac{1}{2}\overline{v'^2}.$$

Now we should like to explore the distribution of turbulent energy among the various eddy sizes, or wavelengths λ. It proves more convenient to use wavelength and vice versa. We now define $E(k)\,dk$ to be the contribution to $1/2\overline{v'^2}$ of the wavenumbers between k and $k + dk$;

$$\int_0^\infty E(k)\,dk = \frac{1}{2}\overline{v'^2}.$$

We call $E(k)$ the spectral density of turbulent energy or energy spectrum function (Batchelor, 1953).

Now a broad range of eddies exists between the large eddies, which directly receive the external supply of energy, and the small eddies, which dissipate energy into heat. It might be expected there exists a range of intermediate eddy sizes, i.e., the well-known "inertial subrange," the eddies of which act only to transfer energy from those just larger than themselves to those just smaller than themselves. Hence, the spectral density, $E(k)$, of the inertial subrange eddies depends only on the energy transfer rate (also termed the energy dissipation rate) ε and the wavenumber k; $E(k) = f(\varepsilon, k)$.

As E, ε, and k have dimensions of $[L^3 T^2]$, $[L^2 T^3]$, and $[L^{-1}]$, respectively, dimensional arguments lead to

$$E(k) = A\varepsilon^{2/3}k^{-5/3}, \tag{9.8}$$

where A is an absolute constant. This expression is essentially due to Kolmogorov (1941) and constitutes one of the most important predictions in the theory of turbulence [see Batchelor (1953) or Ogura (1955) for details]. Figure 9.5 shows a typical spectrum of turbulence.

Now one can speculate that the statistical time-space structure of phytoplankton, being essentially passive, should be closely related to that of the turbulent motion. However, the connection need not be direct. For example, the structure of some of the environmental factors controlling plankton

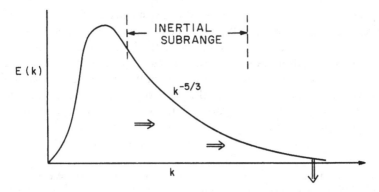

FIGURE 9.5. Energy spectrum of turbulence: E vs. k. The inertial subrange is characterized by $E \sim k^{-5/3}$. White arrows indicate the direction of energy flow.

physiology such as temperature, salinity, and nutrients can also be related to
that of the turbulence. In addition, biological activity can play a controlling
role in the distribution of plankton. The entire problem is quite complex.

This difficulty has perhaps tended to discourage biological oceanographers
from attempting to relate the spectrum of plankton or plankton patchiness
to the structure of turbulence, although the concept has attracted interest for
some time. However, Platt (1972) made a novel discovery; he found that the
dispersion of phytoplankton for scales of 10 to 1000 m is determined purely
by the structure of oceanic turbulence. Platt (1972) measured chlorophyll
(assumed to be an index of phytoplankton abundance) by means of continuous fluorometry using a Turner Model III fluorometer and processed the
resulting data by power spectral analysis [see Platt and Denman (1975a) for
the application of spectral analysis to ecological problems]. Measurements
were made in the Gulf of St. Lawrence, Canada, in July 1970 by pumping
water samples from a depth of 8 m. The time record of the measured chlorophyll content was analyzed to obtain the variance of the chlorophyll fluctuations, i.e., the mean squared fluctuations, in terms of frequencies. The
resultant spectra follow a $-5/3$ power relation; the spectral density of chlorophyll concentration was found to vary as $f^{-5/3}$ ($f =$ frequency), the same
as the law deduced on theoretical grounds by Kolmogorov (1941) for the
spectral density of turbulent energy (Fig. 9.6).

The mean current was observed to be 20 cm/s. The relevant range in time
scales of the analyzed spectra covers approximately 1 to 10 min. If the time
scale is multiplied by the mean flow, the corresponding length scales are seen
to range from approximately 10 m to 1 km. This conversion from time scales
to length scales is based on the usual assumption that the sequence of
changes in time of chlorophyll observed at a fixed point are simply due to the
passage of an unchanging statistical pattern of fluctuations over the point
with a speed of, in this case, 20 cm/s. This assumption is known as Taylor's

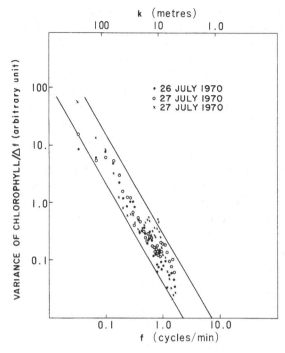

FIGURE 9.6. Chlorophyll-variance spectrum. It follows a $-5/3$ power relation (from Platt, 1972).

"frozen pattern" hypothesis (Tennekes and Lumley, 1972). Thus, in terms of wavenumber, k, the spectrum of variance of chlorophyll fluctuations $E_c(k)$ also obeys the $-5/3$ power law in k:

$$E_c(k) = B\chi\varepsilon^{-1/3}k^{-5/3}, \qquad (9.9)$$

where χ is the rate of dissipation of the chlorophyll variance and B is a dimensionless constant.

Let us speculate on the implications of Platt's spectrum. For the range of scale sizes considered, the spectrum of chlorophyll fluctuations follows essentially the same law as the spectrum of turbulent energy, which suggests that the distribution of the variance of phytoplankton abundance for the range is controlled primarily by hydromechanical processes rather than by biological activities. In other words, the concentration of plankton in many eddies of large size (at least 1000 m in spatial scale) is maintained even as the eddies themselves divide into eddies of smaller and smaller size through the cascade process; i.e., the plankton concentration in these eddies is subjected to no essential change within their lifetime. Thus, the observed statistical ensemble of plankton concentration densities may be attributed to the associated spectrum of eddy energy and is determined by the same spectral law as that of energy density of turbulence. The turbulent kinetic energy is eventually dissipated into heat, i.e., kinetic energy of water molecules; the chlorophyll fluctuations seem to be dissipated by a yet unknown process that occurs at very small scales.

It stands to reason that hydrodynamical mechanisms should dominate Platt's spectrum, as the characteristic time scale of the eddies ranges from 1 to 100 min. This is much shorter than the characteristic time scale of biological activities such as phytoplankton growth and grazing, which is of the order of at least one day.

Physical environmental factors such as wind stresses may influence even large-scale patchiness. Thus, Therriault and Platt (1981) found that during periods of low turbulent mixing (wind speeds ≤ 5 m/s) the phytoplankton patchiness was induced by local differences in production efficiency. On the other hand, for periods of high winds (≥ 5 m/s) the physical turbulent transport processes were responsible for damping out the spatial variations in the phytoplankton distribution. We include the influence of the wind as

$$\frac{\partial S}{\partial t} = r(x)S + D\partial^2 S/\partial x^2,$$

where $r(x)$ is the biological scale variation ($= r_0 \sin kx$) and D is wind speed. If we let $\ln S = q$, then

$$\frac{\partial q}{\partial t} = r(x) + D\left(\frac{\partial^2 q}{\partial x^2} + \left(\frac{\partial q}{\partial x}\right)^2\right).$$

For $q \ll 1$, we ignore the last term and,

$$\frac{\partial q}{\partial t} = r(x) + D\frac{\partial^2 q}{\partial x^2}.$$

In his experiments, Platt did not measure the current velocity fluctuations. Nakata and Ishikawa (1975) and Powell et al. (1975) simultaneously measured water velocity and chlorophyll concentration. Spectral analyses indicate that in the region of wavenumbers exceeding 10^{-4} cm^{-1}, or wavelengths smaller than 100 m, both velocity and chlorophyll spectra show similar forms that follow approximately the $-5/3$ law. However, as scale increases above approximately 100 m, chlorophyll behaves successively less like a simple passive quantity distributed by turbulence.

Denman and Platt (1976) and Denman et al. (1977) have obtained a theoretical expression for the chlorophyll fluctuation spectrum in a range of wavenumbers that covers not only small scales but also large scales where biological activity becomes important. The theory predicts that above a certain characteristic wavenumber, the chlorophyll spectrum follows the velocity spectrum; and that below this wavenumber, the chlorophyll spectrum is flatter than the spectrum of a conservative passive property of the flow field. The characteristic wavenumber corresponds to the wavenumber at which the effect of convective mixing due to turbulence balances the effects of reproduction of phytoplankton. Some observational data (e.g., Lekan and Wilson, 1978) appear to support this theoretical prediction, while others (e.g., Horwood, 1978) do not necessarily support the theory. Support may be scale dependent: Denman and Abbott (1994) applied two-dimensional fast Fourier transforms to mesoscale data from the coastal zone color scanner and advanced very high resolution radiometer. Spectra indicated that phytoplankton behaved as passive scalars with biological processes, such as growth and death, having a minimal influence on spectra. Above all, the steady-state assumption for the theoretical spectrum may not be realistic. Wilson et al. (1979) have discussed time-dependent spectra for chlorophyll variance. This may be particularly important at small scales, where behavior could in theory cause significant deviation from the theoretical spectrum. The extent to which deviations occur at these smaller scales is unclear.

Owen (1989) reported microscale and fine scale patchiness of small plankton. Patchiness occurred below, as well as at, scales of 20 centimeters (microscale) and a few meters (fine scale). Owen (1989) found no significant correlation of patchiness between degree of motility or size across plankton ranging from diatoms to anchovy larvae. At a smaller microscale, Duarte and Vaqué (1992) found evidence for the existence and scale dependence of bacterioplankton patchiness at approximately 1 to 3 cm. It appears that high density ($>10^7$ cells/ml) bacterial patches are embedded within a matrix of low bacterial density (10^5 cells/ml), implying that the processes regulating bacterial abundance may operate at the centimeter scale. This is supported by direct measurements of bacterial spatial distributions (Mitchell and Fuhrman, 1989; Blackburn et al., 1998). Further support for this work comes from the realization that bacteria may exist in a dynamic and heterogeneous organic polymer matrix rather than a classical fluid with widely interspersed discrete particles (Azam, 1998).

For zooplankton, the polymer matrix and low concentrations at small scales has made spectra for this group difficult to generate and interpret. Early work used a Longhurst–Hardy plankton recorder to obtain information on the small-scale patchiness of oceanic zooplankton (Wiebe, 1970; Fasham et al., 1974), but as a method for estimating zooplankton concentration it was not comparable in resolution and transect length to that for estimating phytoplankton abundance. An electronic particle-counting sampling system was developed that simultaneously measured chlorophyll fluorescence and counts particles (presumably individual zooplankton). The device measured various other parameters such as temperature, salinity, and nutrient concentrations, in a sample stream, as well (Pugh, 1978; Mackas and Boyd, 1979).

Mackas and Boyd (1979) have applied spectral analysis to their data taken along 10- to 100-km transects of the North Sea. Some important and consistent results are found such as i) the near-surface zooplankton patchiness is extremely intense, ii) the zooplankton variance spectra are flatter than the corresponding chlorophyll (phytoplankton) spectra, which indicates that the relative intensity of the zooplankton patchiness is greater than that of the phytoplankton at all scales observed and that the difference becomes more noticeable at smaller scales; it is suggested that zooplankton should have special mechanism for formation of small-scale aggregations against the environmental turbulent diffusion, and iii) the concentrations of phyto-plankton and zooplankton consistently show negative spatial correlations. Abraham (1998) provides an alternative, nondiffusive approach to this work. Horwood (1980) from observations in the North Sea, resolved that i) par-ticles (zooplankton) between 30 to 80 µm had prominent patches about 25 km in diameter, ii) fluorescence (phytoplankton) profiles frequently diverged from those of the particle-size groups, iii) the spectra of the particle-size groups could not be distinguished from those of white noise, indicating a Poisson-type distribution. Five of the six points above suggest the impor-tance of behavior for zooplankton patchiness. Richter (1985) found strong evidence for clustering among the entire zooplankton community. For the medusae *Mastigias* sp. migration and swarming is more focused, with indi-vidual medusae migrating up to a kilometer per day back and forth across marine, lagoon-associated lakes (Hamner and Hauri, 1981).

In spite of the usefulness of spectral analysis, Fasham (1978) comments that considerable caution should be taken in the interpretation of spectral slopes. For a given model this slope may change with time (Wilson et al., 1979) in the case of a spring bloom of phytoplankton, and also, in the case of zooplankton counts, the slope may be critically affected by the sampling methods.

9.3   Plankton Distributions in Convection Cells

Langmuir (1938) observed in the sea and lakes the existence of wind-induced cellular motions of water accompanied by upwelling and downwelling regions. A long streak at the water surface running nearly parallel to the wind direction represents a zone of convergence with a sinking motion of water; this convergence acts to collect objects such as chips of wood, sea-weed, and oil. Between these streaks exist regions of divergence charac-terized by upwelling, completing a cellular pattern of circulation (Fig. 9.7). In fact, the circulation pattern is asymmetrical in the transverse vertical plane with the axes of circulation located closer to the surface and the zones of convergence.

Weller et al. (1985) and Weller and Price (1988) have made the first direct observations of three-dimensional flow within the surface mixed layer of the ocean to provide evidence of Langmuir circulation (Fig. 9.8). Note that

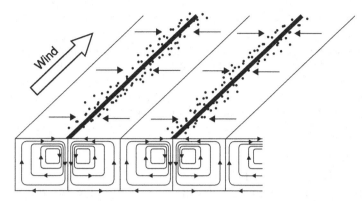

FIGURE 9.7. Langmuir circulation, showing asymmetric cell structure. The extent of asymmetry was interpreted from Weller and Price (1988).

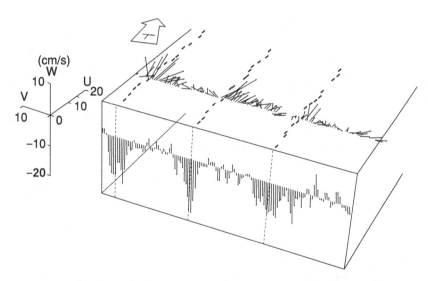

FIGURE 9.8. A three-dimensional representation of Langmuir circulation at the sea surface based on time series by a current profiling system (from Weller et al., 1985). Parallel arrows across the front of the diagram are vertical water velocity vectors (**W**-axis). Nonparallel arrows on the top of the diagram are horizontal water velocity vectors. Evenly spaced black rectangles indicate approximate conditions of computer punch cards floating on the surface. Punch card alignment is east–west (**U**-axis). Wind direction is indicated by the large arrow marked T. **V** is the north–south axis. Horizontal and vertical velocities were plotted after a 3-point running mean was applied to remove wave motion.

average vertical velocities across the horizontal line are not zero, as might be expected. The reason is that the data were taken from a drifting platform with instruments, and thus the instrument spent more time in the narrow regions of convergence and downwelling than in the broad regions of weak divergence and up-welling, which resulted in the nonzero averages of vertical velocity. The magnitude of the observed vertical flow is large compared with the terminal velocities of phytoplankton (Smayda, 1970) and eggs (Sundby, 1983). Hence, Langmuir cells should play an important role in the distribution of these organisms in the surface mixed layer.

Evans and Taylor (1980) discuss the movement and retention of dinoflagellates swimming upward and the implications for red tides. For a species having a daily rhythm of up-and-down movement, the time of onset of Langmuir circulation is important. This extends up to large zooplankton such as medusae, which, upon upward migration, aggregate in convergence zones for feeding and possibly reproduction (Hamner and Schneider, 1986). Barstow (1983) gives a comprehensive review of the ecology of Langmuir circulation.

It is likely that wind plays a primary role in the generation of Langmuir circulations, as streaks appear whenever the wind speed exceeds about 3 m/s. As to their precise mechanism, however, there are a number of debatable theories (Faller, 1971; Faller and Caponi, 1978). Cellular circulations can result not only from wind but also from thermal instability.

The lateral width of Langmuir circulation is of the order of 10 m in lakes. In the sea, widths of the order of 100 m are observed (Assaf et al., 1971). The ratio of width to depth of Langmuir cells is roughly 2:1. Observations show that the speed of downwelling is rather strong and is approximately proportional to wind speed; values of 4 cm/s and 8 cm/s under wind speeds of 5 m/s and 10 m/s, respectively, have been measured (Faller, 1971; Scott et al., 1969).

This intense convective motion enhances mixing of water in the surface layer, and environmental quantities of a passive nature tend to be homogenized. On the other hand, Langmuir circulation tends to enhance the patchiness of particles such as plankton whose density is different from that of water. Those particles that are lighter than water gather at convergences, leaving the region between neighboring convergences with a paucity of particles. Those that are heavier than water would settle at a constant rate in the absence of the convection cells, but, in their presence, are subject to a sorting process, as demonstrated by Stommel (1949).

Let us examine Stommel's simple model. Consider a cross section of square convection cells of size L (Fig. 9.9). A system of coordinates is selected such that the x-axis is horizontal and the z-axis is vertical, and the circulation pattern is assumed to be as in Fig. 9.9. This circulation is described by the stream function

$$\psi = -\psi_0 \sin(\pi x/L) \sin(\pi z/L).$$

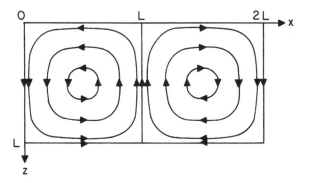

FIGURE 9.9. Stommel's model of convection cells.

The horizontal and vertical velocities of the *fluid*, u, w, are $\delta\psi/\delta z$ and $-\delta\psi/\delta x$, respectively (see Imai, 1970; Batchelor, 1967):

$$u = \delta\psi/\delta z = -\pi/L\psi_0 \sin(\pi x/L) \cos(\pi x/L) \equiv -\Phi_0 \sin(\pi x/L) \cos(\pi z/L),$$

$$w = \delta\psi/\delta x = \pi/L\psi_0 \cos(\pi x/L) \sin(\pi z/L) \equiv \Phi_0 \cos(\pi x/L) \sin(\pi z/L).$$

The horizontal velocity of a particle in the fluid, dx/dt, is set equal to the horizontal velocity of fluid, u, while the vertical velocity of the particle, dz/dt, is the sum of the vertical velocity of fluid, w, and the settling velocity (Sect. 3.31), v_s. Therefore,

$$dx/dt = u = -\Phi_0 \sin(\pi x/L) \cos(\pi z/L), \tag{9.10}$$

$$dz/dt = w + v_s = \Phi_0 \cos(\pi x/L) \sin(\pi z/L) + v_s. \tag{9.11}$$

Dividing (9.11) by (9.10), we obtain

$$dz/dx = -\{\cos(\pi x/L) \sin(\pi z/L) + R\}/\sin(\pi x/L) \cos(\pi z/L), \tag{9.12}$$

where $R \equiv v_s/\Phi_0$.

Equation (9.12) is a differential equation describing the trajectory of a particle in the x–z plane. The motion of particles placed in the flow field, such as described in Fig. 9.9, can be determined by solving (9.12), which contains the single parameter R. This parameter is the ratio of settling velocity to the maximum speed of upward water motion in a Langmuir cell. A detailed discussion of Langmuir circulation dynamics is given by Leibovich (1983).

Figure 9.10 shows the trajectories of particles for various values of R. For $R \geq 1$ (larger settling velocity), all the particles eventually settle out, though the cellular circulation tends to shift particles from their straight course. For $R < 1$ (smaller settling velocity), some particles cannot fall out of the convection cells. Dotted lines shown in Fig. 9.10 describe the *region of retention* within which particles circulate endlessly. If particles of different settling

R > l

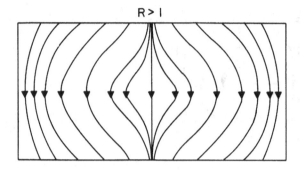

R < l

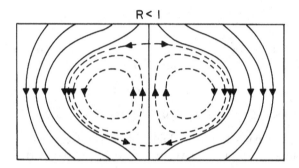

FIGURE 9.10. Trajectories of settling particles involved in convection cells. For $R \geq 1$ (larger settling velocity), all the particles eventually settle out. For $R < 1$ (smaller settling velocity), some particles are caught in the region of retention indicated by broken lines.

velocities are present, the convection cells provide a sorting mechanism by which heavier particles become more concentrated near the central part of the upward currents than lighter ones.

In reality, there is always a degree of turbulence superposed on this regular pattern of convection cells. Thus, a region of retention must gradually lose the particles within it through turbulent exchange with the adjoining region, in which particles are not retained. Nevertheless, a certain degree of inhomogeneity in the distribution of particles can be expected to be maintained.

In addition to lakes and oceans, Stommel's model may be applied to thermal convection cells in the atmosphere, with the implication that the distribution of "aeroplankton" such as insects is to some extent controlled by this process (Schaefer, 1976). Being purely physical, Stommel's model needs further improvement before it can be applied quantitatively to living organisms.

Stavn (1971) discussed and the movement and distribution of planktonic

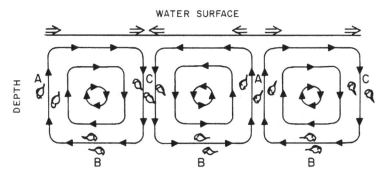

FIGURE 9.11. Distribution of *Daphnia* in the Langmuir circulation (from Stavn, 1971).

organisms that exhibit the dorsal light reaction, such as *Daphnia*, in natural Langmuir spiral circulations. In the daytime, the animal utilizes the dorsal light reaction to orient itself toward, and swim into, the current. Therefore, when current velocities are low in the Langmuir cells, one can expect upward-swimming *Daphnia* to be clumped in the downwellings (locations C in Fig. 9.11), whereas downward-swimming *Daphnia* should be clumped in upwellings when current velocities are high (locations B in Fig. 9.11).

On the other hand, at night when the dorsal light reaction does not operate, the force of gravity and the tendency for upward vertical migration become the principal factors affecting the distribution of *Daphnia*. Thus, the upward-moving plankton tend to aggregate in downwelling regions near the water surface. Stavn (1971) studied the distribution of *Daphnia* placed in a plastic chamber where spiral currents were introduced and artificial illumination was provided. His observations revealed aggregation tendencies that are consistent with his model.

George and Edwards (1973) conducted observations in a reservoir and found that when winds were moderately strong, with speeds above 400 cm/s, *Daphnia* were concentrated near the water surface midway between the foamlines (convergences) associated with Langmuir circulations. They argued that these aggregations of *Daphnia* might possibly be produced by a behavioral response similar to that of Stavn (1971).

The aggregative behavior of *Daphnia* is certainly relevant to the niche diversification in planktonic crustacea. The existence of different species with different adaptations to light intensity can ensure coexistence by means of vertical separation. Coexistence may also be possible for species that do not tend to separate vertically through differing light adaptation, but separate horizontally by being aggregated in different portions of a Langmuir circulation by means of variations in swimming behavior and dorsal light reaction. This can perhaps allow for small refuges for each of two species even though both are feeding on the same food resources.

Ragotzkie and Bryson (1953) discussed the distribution of *Daphnia pulex* in Lake Mendota in relation to lake currents. A study of zooplankton distribution revealed a close relation to the currents. In particular, the effect of divergence and convergence of currents and the phototactic response of the animals were found to be closely synchronized.

In general, the sinking rates of nutrient-depleted cells of phytoplankton are higher than those of nutrient-replete cells (Smayda, 1970; Titman and Kilham, 1976). They also sink earlier in the day and so are more likely to reach the nitracline while light is still available than cells that leave the surface layers later in the day (Heaney and Eppley, 1981). When such nutrient-dependent changes in sinking rate take place within a Langmuir circulation pattern, loss rates of phytoplankton from the upper mixed layer should be much smaller than loss rates without Langmuir circulations, as argued by Titman and Kilham (1976). Thus, as nutrient-depleted cells suffer an increase in sinking rate, they enter trajectories that bring them closer to the bottom of the circulation system, where these cells have an increased probability of encountering water of high nutrient content. Their sinking rate will then decrease, and they will be entrained into trajectories, which take them closer to the surface. A theoretical net potential growth curve, combining both loss from sinking and growth from sinking-dependent nutrient uptake, demonstrates that the growth rate of nutrient-depleted cells may be optimized at high sinking rates.

According to Owen (1966), positive phototactic zooplankton, *Oikopleura longicauda*, tend to concentrate along lines of convergent surface-water flow. In the case studied by him, a thermohaline driving force was inferred to produce the circulation system. Since the plankton would swim upward with an estimated speed of 0.2 cm/s in still ambient water, the downwelling speed of the convergent water could be estimated to be of the same magnitude. In the absence of wind, Owen attributed the probable cause of the observed row formation to convection cells driven by thermohaline forces. The row spacing of 1.5 m was an order of magnitude smaller than the spacing associated with Langmuir circulations.

Kenney (1993) described observation in several lakes of regularly spaced bands of cyanobacteria with horizontal spacing between the bands at 5 to 10 cm being much smaller than any previously reported. In each instance, the small-scale organized structures in the algae were associated with thin shear layers (1 to 2 cm thick) at the lake surface under near-calm conditions. In this thin surface layer, virtually all of the algae were confined to the bands, and the spaces between bands were devoid of algae. The amount of shear was estimated to be about 0.5/s. The spacing of these bands is an order of magnitude smaller than that of surface streaks associated with Langmuir circulation. The generation mechanism of these bands is still unknown.

At the larger scale of Langmuir circulation, Alldredge (1982) reported spawning *O. longicauda* swimming upward at a mean velocity of 4.7 cm/s. Upon approaching the surface they were swept to a convergence by the horizontal current of a Langmuir cell according to the mechanism proposed by Stavn (1971). Additional evidence for the importance of Langmuir circulation in the life history of marine animals is found in Larson (1992). The thimble jelly, *Linuche unguiculata*, forms conspicuous patches in the Caribbean Sea and the Indo-Pacific Ocean. The shape of patches and the inter-medusa distances varied with wind velocity. At low wind speed (<4 m/s) patches were elliptical (or circular) and the individual medusae were separated by distances of 0.5 m, whereas at higher speeds, windrows were evident and animals were close together. At these higher speeds, medusae occurred mostly in the upper 10 to 25 cm, with some occurring down to 5 m. The medusae swam actively and clockwise (viewed from above), in circular patches of 25 to 100 m in diameter. Larson (1992) suggests that these patches result primarily from physical convergence and biological aggregation. Patches may stay together for 4 months, and maintenance behavior appears to be the primary form of social swarming. For *O. longicauda* and *L. unguiculata*, Langmuir circulation may be valuable for increasing reproductive success by reducing inter-mate distances. Langmuir circulation represents a mechanical force that brings order to the distribution of floating objects (sargassum, for instance) and repeatedly draws pelagic animals and plankton together (Fedoryako, 1982).

Internal waves and tidal mixing can also create zooplankton patches. Zeldis and Jillett (1982) found internal waves caused juvenile galatheid crabs (*Munida gregaria*) to aggregate in a manner similar to *L. unguiculata*. Cobb et al. (1983) made field behavioral observations and took plankton tows that showed the American lobster larvae (*Homarus americanus*) are concentrated in downwelling regions of surface convergence, apparently caused by tidal current mixing. Small fish and zooplankton from surface waters were at higher densities in surface slicks than in the rippled water adjacent to them (Kingsford and Choat, 1986). Surface slicks caused by internal waves tend to move in the direction of shore. This is one of the physical mechanisms that may promote the onshore movement of small fish. Careful analysis is required in these studies as water circulation and surface slicks associated with internal waves may closely mimic Langmuir circulation. Shanks (1986) found concentrations of larvae of the intertidal crab, *Pachygrapsus crassipes*, collected in surface slicks above tidally forced internal waves. These slicks may then be transported shoreward producing episodic recruitment. Internal-wave slicks have also been observed in *Randallia ornata* and *Cancer* sp. zoea, as well as the barnacle *Chthamalus* sp. cyprids (Shanks, 1986a and b, 1995).

9.4   Bioconvection

Dense populations of free-swimming organisms such as ciliates and flagellates often exhibit curious patternlike variations in spatial concentration. For example, in dense cultures of *Tetrahymena pyriformis*, the organisms form polygonal cellular patterns that are characterized by the following: upward movements in the central part of the polygons; vigorous downward movements at the edge of the polygons; and between these regions, horizontal movements from the center to the edge on the surface, and from the edge to center at the bottom (Loefer and Mefferd, 1952; Winet and Jahn, 1972; Levandowsky et al., 1975). These patterns resemble Bénard cells, well-known features of thermal convection. Platt (1961) has called this phenomenon "bioconvection."

Certain conditions are required for the occurrence of bioconvection. The patterns are formed only when the average density of organisms is higher than a critical value (experiments suggest a concentration of 150,000 individuals per ml) and when the depth of the culture is at least several millimeters. On the other hand, the patterns are not directly affected by temperature, pH, osmotic pressure, the dissolved oxygen content, etc. Typical speeds of fluid motion in bioconvection are higher than the swimming speed of the organisms; in fact, the values are high enough for an individual organism to be forced to move with the fluid. R. J. Donnelly (in Platt, 1961) suggested a dynamical instability as a possible explanation. When heavy organisms tend to swim to the top of the liquid, an unstable density inversion is produced in a way similar to surface cooling, and beyond a certain concentration of organisms a dynamical instability can occur, followed by the formation of patterns analogous to Bénard cells.

By pure analogy to thermal convection, the instability criterion could be determined by a Rayleigh number, R_a, i.e.,

$$R_a = \Delta \rho g h^3 / \rho K v \qquad (9.13)$$

Chandrasekhar (1961), where $\Delta \rho$ is the excess of density of the surface water containing plankton, ρ_p, over the density of the medium itself, ρ, g is the acceleration of gravity, h is the depth of the container, K is the diffusivity of the organisms, and v is the kinematic viscosity of the medium. For the case of a free surface at the top, the critical value of R_a should be about 1100.

Plesset et al. (1976) and Plesset and Winet (1974) utilized this line of thought to develop a physical model for bioconvection. As the microorganisms swarm toward the surface, they form a well-defined layer of small thickness adjacent to a lower, sparsely populated layer. Since the density of the upper layer exceeds that of the underlying layer, the physical situation is always gravitationally unstable and is of the sort familiar in fluid mechanics as the Rayleigh-Taylor instability (Chandrasekhar, 1961).

An initial disturbance can be assumed to vary in time as e^{nt}, where t is time. Here n can be determined as a function of the wavelength of the disturbance. There exists a characteristic wavelength at which n has positive maximum, corresponding to the fastest growing pattern. Plesset and Winet

(1974) compared the internodal distance, a quantity that can be converted to wavelength, observed in bioconvection in *T. pyriformis* cultures with the values predicted by the theory. The agreement was good.

The models of Plesset et al. (1976) and Plesset and Winet (1974) include direct consideration of neither the effect of negative geotaxis of the organisms, nor even their motility, except implicitly in the condition initiating instability. Levandowsky et al. (1975) and Childress et al. (1975) developed an improved model that incorporates negative geotaxis. It is shown that the equilibrium state associated with a simple layered suspension of organisms is unstable to infinitesimal perturbations when either the layer depth or the mean concentration of the organisms exceeds a critical value.

Although these observations of bioconvection are limited to undisturbed laboratory cultures, a similar phenomenon may occur in the natural world. In particular, when an extremely dense population of planktonic organisms exists in a pond, or a calm lake, or even in a layer near the sea surface under calm conditions, the occurrence of bioconvection seems likely. In this context, more detailed observations on spatial patterns in red tides, and in particular cellular patterns, are necessary.

Crisp (1962) noted that when a dense culture of the first stage nauplius larvae of *Elminius modestus* was placed in an aquarium tank and illuminated evenly from above, it formed a single, large swarm. The movement of organisms within the swarm resembled a convection current.

Classical bioconvection is a fundamentally dispersive process in which high surface concentrations of organisms are mixed with subsurface, low-concentration water. As a result, the maximum organism concentration is at or near the upper surface of water movement. Kessler (1985) recognized that an additional mechanism was required to explain his observation of maximum algal cell concentrations occurring in the vertical "streamers" or columns of convective cells. This mechanism was the balance of shear and gravitational torques on cells. Kessler (1985) named this balance "gyrotaxis" and used *Chlamydomonas* and *Dunaliella* to confirm experimentally that the interaction of these two torques would produce clustering of motile cells in streamers. Formally, the total torque on the cell is

$$\mathbf{T} = 8\pi\mu r^3[(\nabla \times \mathbf{v})/2 - \omega] + m\mathbf{L} \times \mathbf{g}$$

Where μ is dynamic viscosity, $(\nabla \times \mathbf{v})$ is the vorticity, ω is the angular velocity, m is mass, $\mathbf{L}$ is the lever vector, and $\mathbf{g}$ is the gravitational acceleration vector. Pictorially, this is represented in Fig. 9.12.

The oblique movement of cells along AA in Fig. 9.12 is stable for a given shear and gravitational torque in the sense that whatever direction it begins pointing it will quickly rotate around to the orientation shown in Fig. 9.12. This behavior is radially symmetrical around the downward column of a convection cell, and, in fact, around any downward velocity gradient. The result is that cells cluster in the region of maximum downward velocity.

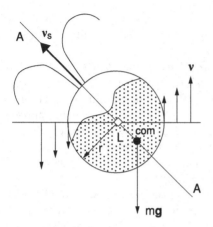

FIGURE 9.12. The vectors acting on a spherical algal cell. The stippling represents a cup-shaped chloroplast that displaces the center of mass (• com) rearwards relative to the direction of movement from the geometric center of the cell (←). This creates a lever arm (L) upon which gravity acts (mg) to produce a gravitational torque. Opposing the gravitational torque is the shear torque produced by the velocity gradient across the cell. The gradient is indicated by 3 arrows on each side of the cell. The shear velocity is indicated by v and the swimming velocity by v_s. When the torques balance, the cell moves along AA. After Kessler (1985).

Kessler (1985) used this to create a cell concentrator and separator. Pedley and Kessler (1987) extended the work from spherical to elongated cells. Mitchell et al. (1990) noted that shear from Kolmogorov eddies is of the correct order of magnitude to balance the estimated gravitational torques for some phytoplankton and therefore gyrotaxis may be relevant in the ocean. Modeling by Timm and Okubo (1994) indicated that the self-focusing streamers associated with gyrotaxis require cell concentrations of $>10^3$ cells/ml, limiting them to blooms of motile phytoplankton. They point out, however, that gyrotaxis is independent of cell concentration and so will occur wherever motile phytoplankton and turbulence co-occur. Timm and Okubo (1995) expand on this later point by modeling the contribution of gyrotaxis to clustering of flagellates around nonmotile, sinking particles. They found that for a particle with a diameter of 200 µm, and sinking 1100 µm/s, gyrotaxis would attract motile cells from up to 400 µm away. The Reynolds number (Sect. 9.5) for their variables is about 0.5. Since uniform shear is needed for sustained gyrotactic migration and begins to break down as a Reynolds number of 1 is passed, Timm and Okubo (1995) may have helped define the upper limit of practical gyrotactic migration.

Turbulence, physical convection, and biological convection combine to transport plankton across the entire mixed layer. The controlling biological factor for organisms ranging from phytoplankton to fish eggs is buoyancy. Sundby (1983) calculated a vertical eddy diffusivity of 100 to 400 cm^2/s for fish eggs using a one-dimensional model. Lande and Wood (1989) found, through modeling, that rapid sinking through the mixed layer, but reducing sinking near the pycnocline, ensured phytoplankton were resuspended back up into the mixed layer. The mechanism of resuspension is through eddy scouring at the pycnocline–mixed layer interface, but depends on careful buoyancy control. When physical conditions are such that this method is reliable, it is an effective method of migrating between high-light–low-nutrient shallow water and high-nutrient–low-light deeper water, even though the model assumes buoyancy is controlled but always negative.

Villareal et al. (1993) found that some diatoms, particularly *Rhizosolenia*, can be positively buoyant. These diatoms form colonies, referred to as mats, in the open ocean, where the pycnocline and elevated nutrients are >80 m deep. The utility of combining mat formation with buoyancy control is evident from Stokes' formula

$$U = \frac{2r^2 g(\rho_c - \rho_w)}{9\eta},$$

where U is positive or negative vertical velocity, r is the equivalent spherical radius of the cell or mat, g is the gravitational acceleration, ρ_c and ρ_w are the buoyant densities of the cell and the surrounding water, and η is the dynamic viscosity. The squared term makes mat size a key parameter in achieving millimeter to centimeter per second speeds. These magnitudes are necessary to make a round trip that may greatly exceed 100 m/day in the open ocean and may be a major source of nutrients to the mixed layer (Villareal et al., 1993, Villareal et al., 1999). Increasing sinking or rising speed is also useful for creating shear that can increase nutrient uptake from the surrounding water (Confer and Logan, 1991; Jackson, 1989).

9.5   Microscale Patchiness

From the Kolmogorov scale downward, patchiness is increasingly controlled by behavior and molecular diffusion rather than eddy diffusion and growth. The use of buoyancy in the previous section is one example of this. These changes represent the transition from inertially to viscously dominated environments. Formally, this balance is:

$$\frac{F_i}{F_v} = \frac{1/2\pi\rho r^2 u^2}{6\pi\eta r v} = \frac{Re}{12},$$

where F_i and F_v are inertial and viscous forces, ρ is density, r is radius, η is dynamic viscosity, u is velocity, and Re is the Reynolds number (Roberts, 1981). The constant of 12 is usually ignored and the dimensional argument is made that

$$\mathrm{Re} = \frac{lu}{v},$$

where l is the characteristic length, u is the characteristic speed, and v is the kinematic viscosity (also written as η/ρ). As Re decreases below 1, the viscosity (represented by v) increasingly dominates over inertia (represented by lu) and flow becomes laminar. As v is 0.01 cm^2/s, the transition to laminar flow occurs just below the Kolmogorov scale when the inertial variables have an average magnitude of 0.1. Equating the two inertial variables implies movement of one body length per second. To an order of magnitude this approximation works because few marine organisms have sustained speeds approaching 10 body lengths per second (Okubo, 1987).

As the dominant force changes with smaller scale, so too does the dominant diffusivity. The ratio of mixing (moving parcels of water) to diffusion (molecular motion of molecules through water) is

$$\mathrm{Sh} = \frac{lu}{D_m},$$

where D_m is molecular diffusivity and Sh is the Sherwood number (Purcell, 1977; Mann and Lazier, 1991). At parameters that give Re = 1, Sh ~ 10^{-3}. Indeed, for a molecular diffusivity of 10^{-5} cm^2/s, Sh does not equal 1 until speed and length average about 3×10^{-3} (cm/s and cm). The difference between the balances of Sh and Re divide the microscale into three functional regimes. The environment from ~1 mm to the Kolmogorov scale is turbulent (weakly) and dominated by inertial transport. From 30 μm to 1 mm the environment is laminar but still dominated by inertial transport. Below 30 μm the environment is laminar and dominated by molecular transport. Patches are observed on all of these scales (Azam, 1998; Blackburn et al., 1998; Bowen et al., 1993).

At these small scales, motility and buoyancy control can be used to overcome limitations of molecular diffusion. Classically, this is achieved by chemotaxis rather than increasing uptake by collision. Berg and Purcell (1977) provide an extensive and elegant explanation of this process using bacterial examples. A shorter, dimensional argument based on the Sh can be made. Set transport by stirring $\propto u_s$, the stirring velocity and $\propto l_t$, the transport length. The stirring time, t_s, is then

$$t_s \sim l_t/u_s.$$

Comparing this to the length-specific diffusion time,

$$t_D \sim l_t^2/D_m,$$

and assuming stirring is significant if $t_s < t_D$ give

$$u_s > D_m/l_t.$$

For small molecules, such as sugar and amino acid monomers, D is about 10^{-5} cm^2/s. At 1 μm u_s is 0.1 cm/s, a speed within the range of speeds for Kolmogorov eddies, but well beyond what bacteria and most flagellates can achieve by swimming. Thus, for small molecules swimming is not an efficient strategy for increasing nutrient uptake.

Confer and Logan (1991) and subsequently Vetter et al. (1998) have appreciated that in natural systems there is a continuous-size range of molecules with nutrient value and thus a wide range of values for D. As a consequence of decreased diffusivity, u_s drops to the point where shear from even weak flow increases uptake. In fact, u_s is so low that taxis toward and swimming in regions of significant concentrations of larger molecules becomes an effective strategy for uptake.

Patches, however, are also known to form based on small molecules (Blackburn et al., 1998; Srinivasan et al., 1998). The explanation here is that small differences in diffusivity are used to migrate toward high concentrations of nutrients (Berg and Purcell, 1977). Motility creates a fine balance between molecular and bacterial diffusivities, presumably permitting the bacteria to control the direction and magnitude of overbalance. The ability of bacteria to control their diffusivity by swimming, D_s, is given by (Chap. 5 and Berg, 1983)

$$D_s = \frac{u^2 \tau}{3(1 - \phi)},$$

where τ is the mean time between turns and ϕ is the mean of the cosine of the angle, in radians, between two runs. By altering u in particular, but also ϕ, bacterial diffusivity can vary from 10^{-6} to 10^{-4} cm^2/s (Berg, 1983; Mitchell et al., 1991, 1995; Blackburn et al., 1998). For contrast, nonmotile marine bacteria diffuse at 10^{-9} cm^2/s by

$$D = \frac{kT}{6\pi\eta r},$$

where k is Boltzmann's constant and T is absolute temperature (Berg, 1983). The low diffusivity of immotile bacteria compared with molecules makes diffusion-limited uptake of nutrients a possible growth limitation for many bacteria. In diffusionally constrained environments ranging from marine aggregates and sediments to gut contents and infections, production of extracellular enzymes may help raise local nutrient concentrations (Vetter et al., 1998). In such cases, motility and swarming may be wasteful expenditures of energy because the environment, e.g., the walls of a single sediment pore, minimize advective loss of nutrients, defining a patch or cluster size.

Over the last two to three decades work on patchiness has expanded from studies of a few kilometers upward to encompass whole oceans (Bennett and Denman, 1989) and downward to encompass micrometer-sized swarms (Blackburn et al., 1998). Satellite technology has taken us to the upper limit of patch size in the ocean, by showing the whole globe. Electron and fluorescent microscopy, flow cytometry, and chemistry have begun to reveal the lower macromolecular limit of patchiness by permitting the mapping of virioplankton and the organic polymer matrix distributions (Proctor and Fuhrman, 1990; Suttle, 1994; Azam, 1998; Chin et al., 1998).

The existence of abundant virioplankton impacts plankton patch dynamics through a number of mechanisms. The simplest is that as phytoplankton concentrations reach a critical level viral epidemics may spread through the population causing the growth rate to slow or reverse (Suttle et al., 1990, 1992). This will alter the critical length, L_c, but the extent will depend on the level of turbulence, sinking behavior of lysing cells, and whether bursts are sudden or occur as continuous leakage. The nonhost surface area available to free virus will determine efficacy of disease propagation (Gonzalez and Suttle, 1993). Particularly, a model is needed that incorporates the observation that about 1% of the bacterial community is infected in coastal waters where contact rates are high, and 18% to 34% of the bacterial community is infected in oceanic waters where contact rates are low (Wilhelm et al., 1998). Murray and Jackson (1992) have begun modeling ocean viral dynamics. Future efforts will need to consider UV inactivation of viruses above 10 m and the photoreactivation from 10 to 50 m (Murray and Jackson, 1993; Weinbauer et al., 1997). This means accurate KISS length estimates for viruses require incorporation of the effect of light fluctuations on phytoplankton growth rate and viral dynamics. To understand host dynamics it may be necessary to merge standard epidemic models (Antonovics et al., 1995; Lloyd and May, 1996) with the KISS model.

9.6   Diffusion and the Entropy of Patchiness

We shall now briefly examine diffusion from the viewpoint of statistical mechanics. Consider a system composed of a number of particles, each subject to a random force and moving in an infinitely large space. Initially $(t = 0)$ let all the particles be concentrated in a small region (Fig. 9.13). As time elapses $(t = t_1)$, the particles diffuse to occupy a larger volume. Later $(t = t_2)$ the particles diffuse even further. The system, once having evolved from state A to state B, will never return by itself from B to A. (The reader should refer to Sect. 2.1 for the probabilistic interpretation of this statement.)

An external force must operate on the system in order to return it from B to A; in this case, the particles must be compressed into a denser concentration. Otherwise, state B will evolve to state C by itself. The random force acting on each particle vanishes when averaged. Nevertheless, a group of particles appears to be pulled apart, as if some deterministic force were in operation. It may be concluded that as a result of diffusion the system tends to change from A, "an ordered state," to B and C, "disordered states."

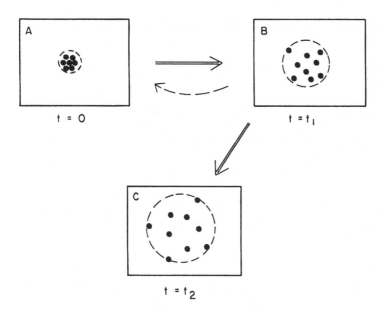

FIGURE 9.13. Diffusion and order-and-disorder $(t = 0 < t_1 < t_2)$.

We now consider the problem of quantification of the order and disorder of the individual states. We take a one-dimensional space and conveniently divide it into as many segments as the total number of particles, M. Let $p_i = m_i/M$, $i = 1, 2, \ldots, M$, to be the relative frequencies of particles in the segments; here m_i is the number of particles found in the ith segment (Fig. 9.14). Clearly, $\sum_{i=1}^{M} p_i = 1$. Figure 9.14 shows one state of the system.

Based on p_i, we define a quantity H as a *measure of disorder*:

$$H = -a \sum_{i=1}^{M} p_i \log p_i, \tag{9.14}$$

where a is a positive constant. The form of H is in fact that of entropy of the set of probabilities, p_i, as defined in statistical mechanics, where p_i is the probability that some physical isolated system is in some state I (Morowitz, 1972; Gallucci, 1973).

One can see that H has the following properties that substantiate it as a

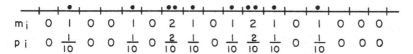

FIGURE 9.14. Probability of states and order-and-disorder.

FIGURE 9.15. Two distinctly different spatial distributions that have the same value of H (measure of disorder).

reasonable measure of disorder (Shannon and Weaver, 1949): (i) If all the particles are gathered in one line segment ($i = 1$, say), corresponding to a state of maximum order, $p_i = 1, p_i = 0$ ($i \neq 1$). Hence, $H = -a \log 1 = 0$, that is, the disorder is zero. (ii) If each segment contains exactly one particle, corresponding to the state of maximum disorder, $p_i = 1/M$ ($i = 1, 2, \ldots, M$). Thus, for a given M, H has a maximum value; $H = -a1/M(\log 1/M)M = a \log M$. This is the most disordered state. As $M \to \infty, H \to \infty$. (iii) Any change toward equalization of the probabilities p_i increases H. As is evident in Fig. 9.13, H increases as A progresses to B and then C.

Equation (9.14) may also be used as the information measure of "species diversity" (Pielou, 1975) according to which species correspond to segments and individuals to particles. In this respect, we can say that H is a measure of information in the spatial distribution of particles, or a measure of patchiness in the distribution. Note that although H measures the patchiness of particles dispersing from a point by a diffusion process, it cannot completely specify the distribution in a general sense. For instance, Fig. 9.15 shows two distinctly different spatial distributions that have the same value of H. This is in fact exactly what we observed in the beginning of this chapter; patchiness estimated from quadrat counts does not necessarily determine uniquely the spatial distribution (Ito, 1963).

Replacing the summation in (9.14) by an integral, we obtain the measure of disorder of a continuous distribution with the density distribution function $p(x)$:

$$H = -a_1 \int_{-\infty}^{\infty} p(x) \log P(x) \, dx. \qquad (9.15)$$

Note that a_1 is a constant the dimensions of which differ from those of a.

Suppose a group of particles undergoes a simple random walk from $x = 0$. The solution of the associated diffusion equation was given by (2.3). Using the solution for p and substituting it into (9.15), we obtain

$$H(t) = a_1 \log(4\pi e D t)^{1/2}. \qquad (9.16)$$

Since $2Dt = \sigma^2$ (variance), (9.16) can be rewritten as

$$H(t) = a_1 \log\{2\pi e\sigma^2(t)\}^{1/2}. \qquad (9.17)$$

Thus, H is seen to be uniquely related to the variance of particle displacements, but not directly connected to the spatial distribution itself. As diffusion progresses, σ^2, and accordingly H, increases. Among all the spatial distributions having the same value of the variance, $\sigma^2(t)$, the Gaussian distribution gives the highest value of $H(t)$ (Shannon and Weaver, 1949; Hida, 1975).

10
Population Dynamics in Temporal and Spatial Domains

Akira Okubo, Alan Hastings, and Thomas Powell

Until now we have been concerned primarily with diffusional aspects of animal movements. When such problems as the dispersal of muskrats or the critical size of plankton patches were discussed, only the simplest aspects of population dynamics, i.e., exponential growth and simple diffusion, were taken into account.

The fundamental importance of the spatial distribution of organisms, emphasized by Skellam (1951) in his classic work, has been recognized to a noted degree in the literature of theoretical biology. In fact, a large number of population models that include organism dispersal have appeared during the last thirty years. In this chapter some aspects of these models are presented. An exhaustive survey of this rapidly developing field is nearly impossible in the context of this book. The reader may consult Levin (1976a, b; 1978a, b) and McMurtrie (1978) for older reviews on the modeling of spatial patterns, stability, and community structure in both homogeneous and heterogeneous environments.

As the relations describing the growth, decay, and interaction of biological populations are very closely analogous to the laws of chemical kinetics in general, the mathematical methods and analyses underlying the reaction–diffusion systems in chemistry may often be applicable to an ecosystem with diffusing populations. In this context books by Lin and Segel (1974), Aris (1975), Glansdorff and Prigogine (1971), Nicolis and Prigogine (1977), and Denn (1975) are very useful. A very extensive review from a physical point of view is in Cross and Hohenberg (1993). More recent general books include Temam (1997), Murray (1989), Brown and Lacey (1990), Logan (1994), and Grindrod (1996). The general methods fall into several categories: comparison theorems and the maximum value theorem and sometimes more specialized approaches to look at asymptotic solutions. In some special cases, one can solve models exactly (e.g., West, 1976), which can help provide insight into more general cases.

10.1 Mathematical Representation of Intra- and Interspecific Interactions in Temporal and Spatial Domains: Advection–Diffusion-Reaction Equations

A simple and formal expression of advection–diffusion-reaction equations for multispecies ecosystems may be obtained by adding intra- and interspecific population interaction terms, or "reaction" terms, to the advection–diffusion equation utilized heretofore to treat animal dispersal. In three-dimensional space, we thus have,

$$\frac{\partial S_i}{\partial t} = -\frac{\partial}{\partial x}(u_i S_i) - \frac{\partial}{\partial y}(v_i S_i) - \frac{\partial}{\partial z}(w_i S_i)$$

$$+ \frac{\partial}{\partial x}\left(K_{x,i}\frac{\partial S_i}{\partial x}\right) + \frac{\partial}{\partial y}\left(K_{y,i}\frac{\partial S_i}{\partial y}\right) + \frac{\partial}{\partial z}\left(K_{z,i}\frac{\partial S_i}{\partial z}\right)$$

$$+ F_i(S_j, t, x, y, z), \qquad i, j = 1, 2, \ldots n, \qquad (10.1)$$

where S_i is the number density or concentration of the ith species, which may be biotic and abiotic, (u_i, v_i, w_i) are the advection velocities of the species in the x, y, and z directions, respectively, $(K_{x,i}, K_{y,i}, K_{z,i})$ are the diffusivities of the ith species in the x, y, and z directions, respectively, and F_i $(i = 1, \ldots, n)$ are the intra- and interspecific population interaction terms of the ith species [the reader should compare (10.1) with (2.16)].

Reaction–diffusion equations of the form (10.1) only allow local movement. In many biological situations, local movement is also coupled with long-range influences, such as the combination of clonal growth and a dispersing phase like seeds. One example of a model for this situation was developed by Furter and Grinfeld (1989), who examined diffusion-reaction models of single-species dynamics that incorporate nonlocal effects. Thus, they replace (10.1) with a model incorporating a reaction term dependent on characteristics of the population as a whole, leading to the system

$$\partial S/\partial t = D\Delta S + F(S, S^*) \quad \text{in } \Omega, \quad t > 0,$$

$$\partial S/\partial n = 0 \quad \text{on } \partial\Omega,$$

$$S(\mathbf{x}, 0) = S_0(\mathbf{x}) \quad \text{in } \Omega,$$

$$S^* = \int_\Omega G(\mathbf{x}, \mathbf{y}) S(\mathbf{y}, t)\, d\mathbf{y}.$$

More recent variations on this theme include the integro-difference equation approaches discussed below.

In general, the advection velocity and diffusivity are functions of tile concentrations of all the species, and possibly the spatial gradients of the concentrations, as well as of time and space. More rational formulations of the fundamental equation for population dynamics have been given by various investigators, including Dubois (1975a, b), Gurtin and MacCamy (1977), Platt et al. (1977), and Rosen (1977).

Kerner (1959), Gurtin (1974), Jorné (1977), and Rosen (1977) consider a more general representation of the diffusion terms. The diffusive flux of the ith species in the x direction, for example, is expressed as

$$J_{x,i} = -(K_x)_{ii}\partial S_i/\partial x - \sum_{j \neq i}^{n}(K_x)_{ij}\partial S_j/\partial x. \qquad (10.2)$$

The first term represents the "self-diffusion" with which we normally deal. It is required that $(K_x)_{ii} \geq 0$. The other terms represent "cross-diffusion" and express the population fluxes of the ith species due to the presence of other species, subject to the conditions $j \neq i$. The coefficients $(K_x)_{ij}$ are the cross-diffusivities in the x direction; the value of the coefficient may be positive, negative, or zero.

The condition $(K_x)_{ij} > 0$ implies that the flux in the x direction is directed toward decreasing population density of the jth species, while $(K_x)_{ij} < 0$ implies that the flux in the x direction is directed toward increasing population density of the jth species. If, as we might surmise, a prey (i) tends to diffuse in the direction of lower concentrations of its predator (j), and the predator tends to diffuse in the direction of higher concentrations of prey, $K_{ij} > 0$ and $K_{ji} < 0$. Of course, K_{ij} or K_{ji} may vanish for a nonmotile or nonresponsive prey or predator.

In (10.1), F_i may be called the reaction term, nonconservative term, or biological term, according to the user's discipline. In this book we shall call it the reaction term although, in essence, it expresses the local rate of growth or death of the ith species due to intra- and interspecific interactions.

Any deterministic formulation of biological interaction can be realized in the reaction term by regarding S_j as the *mean* population density of the jth species, i.e., S_j should be interpreted as S_j (see Chap. 2). In fact, (10.1) was derived on the basis of averaging the original equations for instantaneous population densities in a way similar to the derivation of (2.16). As a result, the reaction term generally contains not only the interaction between mean population densities but also the averaging of interaction between fluctuating population densities. Thus, for a predator–prey interaction of the Lotka–Volterra type with population densities S_1 and S_2, the averaging procedure applied to the second-order reaction term gives rise to the following two terms, $\overline{S_1 S_2} = \overline{(\bar{S}_1 + S_1')(\bar{S}_2 + S_2')} = \bar{S}_1\bar{S}_2 + \overline{S_1' S_2'}$. Under certain circumstances the correlation function, $\overline{S_1' S_2'}$, may be ignored in comparison with the product of the averaged concentrations, but this is not always true. Perhaps at the present stage of ecological modeling it may be wise to disregard these correlation terms simply because we cannot handle them easily. For more details, consult Dubois (1975a, b) and Hill (1976).

Since real ecosystems must date back to some previous time and have a finite spatial extent, it is necessary to specify the initial and boundary conditions that the system satisfies in order to solve the advection–diffusion-reaction equations. In some cases, however, these conditions are considered to be irrelevant to the problem. Thus, if the equation admits of a unique stable steady-state solution to which the system will tend for all initial conditions, we may ignore initial conditions in obtaining it (May, 1977). Kramer (1982) provides conditions where there is only a single stable steady state for the spatial problem. Also, when we are interested in the dispersal pattern of a population in an early period during which individuals remain away from boundaries, we may ignore the presence of the boundary and treat the domain as if it extends to infinity.

Ordinarily the population densities throughout the region of study are specified at the instant which we take to be the origin of time, i.e.,

$$\lim_{t \to 0} S_i(\mathbf{x}, t) = S_i(\mathbf{x}, 0) = f_i(\mathbf{x}), \quad \text{given at } t = 0. \tag{10.3}$$

On the other hand, spatial boundary conditions can be specified in various manners, such as those described below [see, e.g., Fife (1979)].

1) Prescribed population densities at the boundary; these densities may be constant or functions of time.

$$S_i(\mathbf{x}, t) = G_i(t) \quad \text{given at } \mathbf{x} = \mathbf{x}_b \text{ (boundary).} \tag{10.4}$$

A population reservoir at the boundary may be represented by the condition (10.4). When a habitat is surrounded by a completely hostile environment, the population density can be considered to be zero at the boundary, i.e., an *absorbing* boundary. In this particular case,

$$G_i = 0 \quad \text{at} \quad \mathbf{x} = \mathbf{x}_b.$$

2) Prescribed flux across the boundary; the flux may be constant or a function of time.

$$\mathbf{J}_i = \mathbf{H}_i(t) \quad \text{given at } \mathbf{x} = \mathbf{x}_b. \tag{10.5}$$

Immigration or emigration across the boundary may be represented by the condition. When a habitat boundary is completely closed to a population, e.g., a fenced population, its flux can be considered to be zero across the boundary, i.e., a *reflecting* boundary. In this particular case,

$$\mathbf{J}_i = 0 \quad \text{at} \quad \mathbf{x} = \mathbf{x}_b. \tag{10.6}$$

If the flux is represented by a diffusion process, (10.5) can be written as

$$-K_i \partial S_i / \partial \mathbf{x} = \mathbf{H}_i(t) \quad \text{at} \quad \mathbf{x} = \mathbf{x}_b. \tag{10.7}$$

3) Radiation boundary conditions (Crank, 1975, p. 9).

If the flux across the boundary of a domain of interest is proportional to the difference of the population density in the domain extrapolated to the boundary and the population density of the surroundings of the domain extrapolated to the boundary, we may write

$$\mathbf{J}_i = \mathbf{h}_i \{ S_i(\mathbf{x}_b, t) - S_{io}(t) \}. \tag{10.8}$$

Here $S_{io}(t)$ is the population density of the surroundings, and $\mathbf{h}_i$ is an exchange constant. As $\mathbf{h}_i \to 0$, this condition tends to a form of 2), and as $\mathbf{h}_i \to \infty$, it tends to a form of 1).

4) Continuity conditions at a boundary separating two regions of interest.

The population density and its flux must be continuous at the boundary of separation between regions 1 and 2 (Skellam, 1951). Thus, the conditions at a common boundary are expressed mathematically as follows:

$$(S_i)_1 = (S_i)_2,$$
$$(\mathbf{J}_i)_1 = (\mathbf{J}_i)_2. \tag{10.9}$$

More complex coupling at boundaries (Olmstead, 1980) has also been considered, where rather than just matching at the boundary as in (10.9), there is a function describing exchange across the boundary.

Often, for problems with spatial boundary conditions, or in infinite regions, the interesting biological questions are the asymptotic behavior of the reaction–diffusion equations, as reviewed in Fife (1978) and Cross and Hohenberg (1993). Comparison methods (e.g., Hallam, 1979) can be used to obtain bounds to solutions of the reaction–diffusion system in terms of the pure reaction system.

Boundary conditions may play a crucial role in the solution of diffusion–reaction equations. Thus, Arcuri and Murray (1986) investigate the effects of a variety of boundary (and initial) conditions on Turing instability. Sprigler and Zanette (1992) propose a Fokker–Planck-type model to describe the kinetics of certain chemical reactions in dealing with reaction and diffusion. The competing effect between transport and reaction processes is analyzed. The lowest-order density in the asymptotic expansion obeys a reaction–diffusion equation. For other interaction scalings, the prevalence of chemical processes implies that the lowest-order density is determined by the equations of chemical equilibrium. In contrast, when transport prevails, the reaction term affects only higher-order densities.

The primary emphasis for initial-value problems has been on the dynamics of and convergence to travelling wave solutions. This is a very difficult area, and one of the few cases where this has been done for nonlinear problems is the work of Grundy and Peletier (1987). They studied the short-time behavior of the nonlinear problem

$$\partial S / \partial t = \partial^2 S / \partial \mathbf{x}^2 - S^p,$$
$$S(0, t) = 1, \quad t > 0,$$
$$S(\mathbf{x}, 0) = 0, \quad \mathbf{x} > 0,$$

using a matched asymptotic expansion to look at the dynamics of the interface that arises. Needham (1992) used a similar method of asymptotic expansions to look at the dynamics of convergence to travelling fronts with general reaction terms, allowing determination of the long-term propoagation speed. Start with the system

$$\partial S/\partial t = \partial^2 S/\partial \mathbf{x}^2 + F(S),$$

$$S(\mathbf{x}, 0) = \varepsilon G(\mathbf{x}), \quad \mathbf{x} \geq 0,$$

$$\partial S/\partial \mathbf{x}(0, t) = 0, \quad t \geq 0,$$

$$S(\mathbf{x}, t) \to 0 \quad \text{as} \quad \mathbf{x} \to \infty, \ t > 0,$$

$$F(0) = 0, \quad F'(0) > 0,$$

and examine the problem when $\varepsilon \ll 1$, assuming

$$F(S) = \varepsilon f'(0) + \cdots.$$

The state $S = 0$ is locally unstable. The long-time asymptotic behavior of the system with $\varepsilon \ll 1$ is reduced to studying the long-time development of a Fourier integral representing the solution of a pure diffusion problem. This method provides a useful tool for rapidly assessing the possible wave-front formation in a given reaction–diffusion model and provides an analytical expression for the leading-order wave-front propagation speed.

Another method for determining approximate solutions was presented by Rosen (1984) based on the connection between diffusion equations and probabilistic processes. The method works for the initial-value problem for the general system

$$\partial S/\partial t = D\Delta S + R(S) \tag{10.9a}$$

where the initial population is positive, so at $t = 0$,

$$S(\mathbf{x}, t) = S_0(\mathbf{x}) > 0 \tag{10.9b}$$

and the reaction term satisfies

$$d^2 R/dS^2 \leq 0 \quad \text{for all } S,$$

which is satisfied by, for example,

$$R = aS - bS^{1+m}$$

or

$$R = -kS^2.$$

The solution to (10.9a) subject to (10.9b) can be approximated analytically by considering the associated Brownian-motion (stochastic-convection) equation

$$\partial \Psi/\partial \tau = -v(\tau)\text{grad}(\Psi) + P(\Psi) \tag{10.9c}$$

where $C(\mathbf{x}, t)$ is an auxiliary stochastic variable and $\mathbf{u}(t)$ is a Gaussian random velocity with mean 0 and a covariance given by the δ-function,

$$\langle \mathbf{u} \rangle = 0,$$

$$\langle \mathbf{u}_i(t')\mathbf{u}_i(t'') \rangle = 2D\delta_{ij}\delta(t' - t''),$$

and the variable

$$C(\mathbf{x}, t) = S_0(\mathbf{x}).$$

The exact solution for C can be found from the integral relationships

$$\int_{S_0(\mathbf{x}-\mathbf{y})}^{C} R(z)^{-1} \, dz = t, \qquad (10.9d)$$

where

$$\mathbf{y} = \mathbf{y}(t) = \int_0^t \mathbf{u}(t') \, dt',$$

as can be shown by differentiating (10.9d). Rosen (1984) then proves, using a maximum principle, that a lower bound for S is given by

$$S_1 = \langle C \rangle \leq S.$$

In an exactly analogous fashion, an upper bound for $S(S_u \geq S)$ can be found from

$$\int_{S_0^*}^{S_u} R(z)^{-1} \, dz = t \quad \text{with } S_0^* = \langle S_0 \rangle.$$

Finally, the average of the upper and lower bounds provides a good approximation to the solution of the original problem (10.9a).

10.2   Single-Species Population Dynamics with Dispersal

We shall focus our discussion of the single-species dispersive system on travelling frontal waves of a growing population, the critical minimum habitat size for the survival of a dispersing population, and invasion and settlement of new species. A simple diffusion process is assumed for the dispersal of the population. The effect of density-dependent dispersal on population dynamics will be discussed in Sect. 10.4.

10.2.1   Some Simple Population Models and Associated Travelling Frontal Waves

We first consider a simple model of populations of exponential growth (Malthusian populations) with diffusion:

$$\frac{\partial S}{\partial t} = D \frac{\partial^2 S}{\partial x^2} + \alpha S, \qquad \alpha > 0. \qquad (10.10)$$

As in the case of muskrat dispersal (see Sect. 6.4), the solution of (10.10) for a population initially concentrated at the origin, $x = 0$, and diffusing in an unbounded space is given by

$$S = \frac{M}{2(\pi Dt)^{1/2}} \exp\left\{\alpha t - \frac{x^2}{4Dt}\right\}, \tag{10.11}$$

where M denotes the total number of individuals in a one-dimensional space at $t = 0$.

In order to see how a front of isoconcentration, $S = S_f$, travels with time, we solve (10.11) with respect to x/t for a fixed value of S_f (Kendall, 1948). This results in

$$x/t = \pm[4\alpha D - 2Dt^{-1}\ln t - 4Dt^{-1}\ln\{2(\pi D)^{1/2}S_f/M\}]^{1/2}. \tag{10.12}$$

Asymptotically, (10.12) gives

$$x/t = \pm 2(\alpha D)^{1/2}.$$

In other words, the population wave front from the initial disturbance propagates outward with a velocity ultimately equal to $2(\alpha D)^{1/2}$ (Fig. 10.1). This result is consistent with that obtained in our previous analysis of muskrat dispersal (Sect. 6.4).

The general solution of (10.10) for an arbitrary initial distribution, $S_0(x)$, can be found by superposition of the fundamental solution (10.11) (see Carslaw and Jaeger, 1959)

$$S(t, x) = \int_{-\infty}^{\infty} \frac{S_0(x')}{2(\pi Dt)^{1/2}} \exp\left\{\alpha t - \frac{(x - x')^2}{4Dt}\right\} dx'. \tag{10.13}$$

If, for example, $S_0(x)$ is Gaussian, with the form

$$S_0(x) = \frac{M}{(2\pi)^{1/2}\sigma_0} \exp\{-x^2/2\sigma_0^2\},$$

where σ_0^2 is the variance, then (10.13) yields

$$S(t, x) = \frac{M}{(2\pi)^{1/2}\sigma} \exp\{\alpha t - x^2/2\sigma^2\},$$

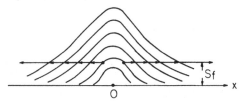

FIGURE 10.1. Population frontal waves of dispersing Malthusian populations.

in which $\sigma^2 = \sigma_0^2 + 2Dt$. An argument similar to that used for the point source solution shows that the ultimate speed of propagation again becomes $2(\alpha D)^{1/2}$.

One might conclude that the same would be true for any initial distribution. However, it is true only for initial distributions the amplitudes of which fall zero with sufficient rapidity as $|x|$ tends to infinity. Thus, Kendall (1948) showed that an initial Gaussian distribution of population density of the form $\exp(-\alpha x^2)$ attains the same velocity of propagation asymptotically, whereas for an initial distribution of an exponential type of the form $\exp(-b|x|)$, $b > 0$, the velocity of propagation may attain any value greater than $2(\alpha D)^{1/2}$ by dint of a suitable choice of b that is smaller than $(\alpha/D)^{1/2}$; Mollison (1977b) shows in a simpler way how the wave velocity depends on the value of b. More recently, Needham (1992) examined the long-term asymptotic wave-front propagation speed with general initial distribution of small amplitude.

We now proceed to discuss the problem of logistic growth of a population by the following model equation:

$$\frac{\partial S}{\partial t} = D\frac{\partial^2 S}{\partial x^2} + \alpha S - \beta S^2 \qquad (10.14)$$

where α and β are positive constants. This model was originally proposed by Fisher (1937) as a means of analyzing the wave of advance of advantageous genes, and it has received continuous attention in the search for progressive wave solutions (Kolmogorov et al., 1937; Barakat, 1959; Landahl, 1957; Skellam, 1973; Hadeler and Rothe, 1975; Kametaka, 1976; Fife and Peletier, 1977). Kuno (1991) criticizes the expression in the logistic equation that has

$$dS/dt = rS(1 - S/K)$$

and suggests using

$$dS/dt = (r - hS)S$$

instead, which is equivalent to taking

$$dS/dt = \alpha S - \beta S^2$$

in (10.14).

Thus far, no exact analytical solution exists for (10.14) in space and time, although some approximate solutions have been attempted (Montrol, 1967; Rotenberg, 1982). In particular, the addition of a nonlinear term to diffusion makes an analytical solution possible even with temporal and spatial varations in the carrying capacity (Rotenberg, 1982). Thus, the model becomes

$$\partial S/\partial t = D\{\partial^2 S/\partial x^2 - 2/S(\partial S/\partial x)^2\} + rS(1 - S/K), \qquad (10.14a)$$

with carrying capacity $K = K(x, t)$. Equation (10.14a) is made linear by means of the substitution $S = 1/G$, to give

$$\partial G/\partial t = D\partial^2 G/\partial x^2 - rG + r/K. \tag{10.14b}$$

Equation (10.14b) is solved by standard Fourier method (see Carslow and Jaeger, 1959) to give

$$G(x,t) = (4\pi Dt)^{-1/2} e^{-rc} \int_{-\infty}^{\infty} G_0(y)\exp\{-(x-y)^2/4Dt\}\, dy$$

$$+ r \int_{0}^{t} dt' \int_{-\infty}^{\infty} K(y,t')^{-1}\exp\{-r(t - t'(x-y)^2/4D(t-t')\}/$$

$$\{4\pi D(t-t')\}^{1/2}\, dy\, dt'$$

in an infinite space, where $G_0(x) = 1/S_0(x)$ is the reciprocal of the initial population. Bergman (1983) developed a mathematical model for cell growth, which takes into account cell–cell interactions and thus leads to nonexponential inhibited growth of the number of cells. The resultant difference equation leads to, as a continuous limit, a nonlinear diffusion equation for cell density function, which looks like (10.14a):

$$\partial S/\partial t = rS(1 - S/2) + D\{\partial^2 S/\partial x^2 - (\partial S/\partial x)^2\},$$

which can be linearized by a similar substitution.

We look for a solution of (10.14) that represents a wave of *stationary* form propagating in the direction of positive x with velocity c. Thus, we assume

$$S(x,t) = S(x - ct) = S(\xi), \tag{10.15}$$

where $\xi \equiv x - ct\ (c > 0)$.

Substituting (10.15) into (10.14) and observing that $\partial/\partial t = -cd/d\xi$ and $\partial/\partial x = d/d\xi$, we obtain the following ordinary differential equation for $S(\xi)$:

$$Dd^2 S/d\xi^2 + cdS/d\xi + \alpha S - \beta S^2 = 0. \tag{10.16}$$

By letting

$$g = -dS/d\xi,$$

(10.16) can be cast in the form

$$Dg\, dg/dS - cg + \alpha S - \beta S^2 = 0.$$

At the point of inflection, $dg/dS = 0$ and $cg = \alpha S - \beta S^2$, and in advance of this point dg/dS is positive. If g/S tends to a limit k as S tends to zero, i.e., at the very end of the advancing front, then k must satisfy the equation (Kendall, 1948)

$$Dk^2 - ck + \alpha = 0.$$

Since k may tend to neither zero nor infinity, solutions exist *only* for

$$c^2 \geq 4\alpha D$$

or

$$c^2 \geq 2(\alpha D)^{1/2},$$

which guarantees that k has real roots. Fisher (1937) suggested that ultimately only the minimum velocity of advance, $c_m = 2(\alpha D)^{1/2}$, is possible. It is interesting to observe that this minimum velocity of propagation of a logistic population is equal to the ultimate speed of propagation of a Malthusian population; the carrying capacity of the resources, $\alpha/\beta \equiv S_e$, for the logistic population has no contribution to the wave speed. Aronson (1979) [see also Aronson and Weinberger (1975)] proves that for an initial condition that

$$S(x, 0) = 0 \qquad \text{for all } x \leq x_0$$
$$= S_0 \geq 0 \quad \text{for } x < x_0,$$

with S_0 being not zero identically, (10.14) has an asymptotical solution of travelling wave type for which

$$\lim S(x - ct, t) = \begin{cases} 0 & \text{if } c > c_m = 2(\alpha D)^{1/2} \\ \alpha/\beta & \text{if } 0 \leq c \leq c_m. \end{cases}$$

Roughly speaking, if we start from a point of the x-axis and run toward $+x \to \infty$ with speed c, then we will outrun the solution of (10.14) if $c > c_m$, but it will outrun us if $c \leq c_m$. Note (Aronson, 1979) that the fact that c_m is the asymptotic speed of propagation of disturbances from rest does not imply that disturbances from rest propagate with a finite speed. In fact, for Fisher's equation, as for the diffusion equation, all disturbances are propagated with infinite speed.

Exact solutions for (10.16) are known only for special cases, and these are typically unstable solutions. Newman (1980) discussed an exact solution for 10.16) (see also Ablowitz and Zeppetella, 1979; Kametka, 1977). Kalioppan (1984) obtained an exact solution for more general cases where

$$\partial S/\partial t = D\partial^2 S/\partial x^2 + r\{S - (S/K)^m\}, \quad m > 1.$$

Canosa (1973) obtained an approximate solution of (10.16) that is strictly valid in the limit of infinitely large propagation speed, but that is quite accurate for smaller propagation speeds.

Since, as noted above, in some sense the reaction–diffusion formulation allows an infinite speed of propagation, descriptions of movement other than diffusion have been studied. Holmes (1993) compared a diffusion model with a telegraph model (see Eq. 5.15) of invasion. She used simple logistic growth for the population dynamics to calculate rates and patterns of invasion for the two models. As in the reaction–diffusion model, travelling wave solutions also exist in the reaction-telegraph model at the following velocities:

$$c_{tel}^2 \geq \{8ra/(r+2a)^2\}v^2 \quad \text{for } 0 < (r/2a)^{1/2} \leq 1$$

$$\geq v^2 \qquad\qquad \text{for } (r/2a)^{1/2} \geq 1. \qquad (10.16a)$$

To compare the speed of propagation to the diffusion model, take the minimum speed in (10.16a),

$$c_{m,tel}^2 = 8ra/(r+2a)^2v^2 \quad \text{for } 0 < (r/2a)^{1/2} \leq 1$$

$$= v^2 \qquad\qquad \text{for } (r/2a)^{1/2} \geq 1.$$

Then, with $1/a = T$, $r = \alpha$, which corresponds to a diffusion equation,

$$c_{m,tel}^2 = 8\alpha T/(\alpha T + 2)^2 v^2$$

$$= 4\alpha/(1+\alpha T/2)^2 \quad \text{for } 0 \leq (\alpha T/2)^{1/2} \leq 1$$

$$= 4\alpha D/(1+\alpha T/2)^2 \quad \text{for } 0 \leq (\alpha T/2)^{1/2} \leq 1, \quad \text{where } D = v^2 T/2$$

$$= v^2 \qquad\qquad \text{for } (\alpha T/2)^{1/2} \geq 1, \quad \text{or } \alpha T/2 \geq 1,$$

$$\qquad\qquad\qquad\qquad \text{or } \alpha \geq T/2, \quad \text{or } T_{gr} \leq T, \quad \text{where } T_{gr} = 2/\alpha.$$

Thus, for $0 \leq (\alpha T/2)^{1/2} \leq 1$,

$$C_{tel}/C_{diff} = 1/\{1+(\alpha T/2)\}.$$

Consequently, for large intrinsic rates of increase, the rate of spread is much less for the telegraph equation than for the corresponding reaction–diffusion equation.

In general, the diffusivity and growth rate that appeared in (10.10) may vary with time, which will also affect speed of propogation. Matsuda and Akamine (1994) consider $D = D(t)$, $\alpha t = -\beta(t)$, i.e., a population subject to time-varying dispersal rate and mortality. For a population initially concentrated at the origin, $x = y = 0$, and diffusing in an unbounded two-dimensional space (x, y) is given by

$$S = \frac{M_0}{2\{\pi W(t)\}^{1/8}} \cdot \exp\left\{\frac{-(x^2+y^2)}{4W(t)} - \int_0^t \beta(t')\,dt'\right\}, \qquad (10.16b)$$

where $W(t) = \int_0^t D(t')\,dt'$, and M_0 is the total number of individuals released at $t = 0$. Based on this model, Matusda and Akamine estimated the mortality and dispersal rates of an artificially cultured abalone (*Haliotis discus hannai*) population.

In their classic paper, Kolmogorov et al. (1937) presented a more rigorous analytical approach to the travelling wave problem on the basis of a class of nonlinear reaction–diffusion equations

$$\frac{\partial S}{\partial t} = D\frac{\partial^2 S}{\partial x^2} + F(s), \tag{10.17}$$

where $F(S)$ belongs to a certain appropriate class of functions and includes the logistic growth function as a special case. Kolmogorov et al. showed that $2\{dF/dS(0)D\}^{1/2}$ is the (critical) minimum speed of propagation for non-trivial solutions of (10.17). In general, the behavior of the initial distribution, $S(x, 0)$, as $|x|$ goes to infinity is crucial in the temporal behavior of the travelling wave. For more recent developments of travelling wave solutions, the reader may consult papers by Aronson and Weinberger (1975), Fife and McLeod (1975, 1977), Sects. 5.3 and 5.4 of Murray's book (Murray, 1977), and a book by Hoppensteadt (1975).

A variety of other special cases of waves in reaction–diffusion equations illustrate the complexity of the dynamics that are possible. Kawahara and Tanaka (1983) consider the bistable system

$$\partial S/\partial t = D\partial^2 S/\partial x^2 \pm S(a - S)(1 - S), \qquad -1 \le a < 1,$$

and focus on the behavior of two travelling fronts using a formal perturbation method, An exact solution that describes the coalescence of two travelling fronts of the same sense into a front connecting two constant states $(S = 1, a)$ is found. Turyn (1986) examined the behavior of a forced Fisher's equation

$$\partial S/\partial t = D\partial^2 S/\partial x^2 + S(1 - S) + mg(x - ct),$$

with small m. Using the same method of looking for travelling waves by converting to an ordinary differential equation system we have employed previously, the author looks at two cases that have different phase plane behavior. In the case of a node-saddle connection in the phase plane, "pushed" waves are found, while in the case of a saddle-saddle connection, arguments are made that chaotic solutions are obtained.

More recently, formulations for travelling waves in terms of integro-difference equations (Kot et al., 1996) that are discrete in time but continuous in space have shown that very different speeds of propagation can be found if the description of movement is not diffusive.

10.2.2   The Critical Size Problem

In Sect. 9.1 we briefly discussed the critical patch size of plankton, i.e., the minimum scale of water mass within which a dispersing Malthusian population of plankton can survive. We now extend the critical size problem to more general situations.

Let us first consider a population undergoing logistic growth and diffusion, and seek a minimum size of habitat for the survival of that population when it is surrounded by a totally hostile environment. Thus, our task is to obtain the solution of (10.14) that vanishes at $x = 0$ and $x = L$ and corresponds to a given initial concentration, and then to find the critical size; this analytical procedure is the same as that applied to the Malthusian population of Sect. 9.1. However, only approximate solutions have been attempted for the logistic case (Landahl, 1957, 1959; Barakat, 1959; Montroll, 1968; Wilhelm, 1972, among others).

For the purpose of determining the critical size, it is sufficient to consider a steady-state solution (Skellam, 1951; Levandowsky and White, 1977; Ludwig et al., 1979). Thus, the population distribution in a habitat, $-L/2 \leq x < L/2$, satisfies

$$D\frac{d^2 S}{dx^2} + \alpha S - \beta S^2 = 0 \tag{10.18}$$

subject to the boundary conditions that $S = 0$ at $x = -L/2$ and $L/2$. Since $S(x)$ is symmetric with respect to the origin $(x = 0)$; we replace the boundary conditions by

$$dS/dx = 0 \quad \text{and} \quad S = S_m \quad \text{at } x = 0$$

and

$$S = 0 \quad \text{at} \quad x = L/2, \tag{10.19}$$

where S_m is the population density at the origin, i.e., central density.

Equation (10.18) can be written as

$$\frac{D}{2}\frac{d}{ds}\left(\frac{dS}{dx}\right)^2 + \alpha S - \beta S^2 = 0. \tag{10.20}$$

Integrating (10.20) first with respect to S and then with respect to x and using (10.19), we obtain (Skellam, 1951)

$$L/L_0 = 4\{\pi(1 + 2\gamma)\}^{-1/2} F[\arcsin\{(u_1 - u_2)/(1 - u_2)\}^{1/2},$$
$$\{u_1(1 - u_2)/(u_1 - u_2)\}^{1/2}], \tag{10.21}$$

where $L_0 \equiv \pi(D/\alpha)^{1/2}$, the KISS size (see Sect. 9.1). Here

$$\gamma = \beta S_m/\alpha = S_m/S_e, \qquad u_1 = [-1 + \{(3 + 6\gamma)/(3 - 2\gamma)\}^{1/2}],$$
$$u_2 = [-1 - \{(3 + 6\gamma)/(3 - 2\gamma)\}^{1/2}],$$

$$F\{\arcsin \gamma, k\} \equiv \int_0^\gamma \{(1 - z^2)(1 - k^2 z^2)\}^{-1/2} dz \quad \text{(an elliptic integral)}.$$

The relationship between L/L_0 and γ is shown in Fig. 10.2. From (10.21) it can be shown that as $\gamma \to 0$, i.e., $S_m \to 0$ or $\beta \to 0$, $L \to L_0$ and as $\gamma \to 1$, i.e., $S_m \to S_e$, $L \to \infty$. In other words, the KISS size, L_0, is also the minimum size required for survival of a logistic population. To accommodate

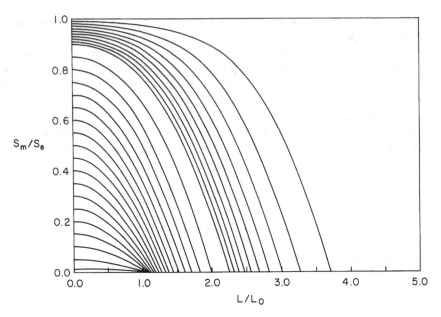

FIGURE 10.2. Relationship between the minimum habitat size and central population density for a population undergoing logistic growth and diffusion. The habitat size L is normalized by the KISS size (L_0), and the central population density S_m is normalized by the carrying capacity (S_e).

more individuals participating in logistic growth, the habitat size must be larger than L_0, and to maintain the carrying capacity of the resources, the habitat size must be infinitely large.

Skellam (1951) has proved that the steady-state distribution of a diffusing logistic population is stable. This justifies the use of the steady-state solution in the discussion of the critical size.

McMurtrie (1978) has obtained non-steady-state numerical solutions of (10.14) for two habitat sizes; one is smaller than the critical size for exponential growth and the other is slightly larger than the critical size. For a smaller habitat, the population steadily declines, as expected. For a larger habitat, i.e., a favorable habitat, the population tends to a stable equilibrium distribution the central value S_m of which is well below the apparent carrying capacity, S_e.

We next consider the critical patch size of "asocial populations" (Philip, 1957). The growth function for such populations at small densities is negative due either to a paucity of reproductive opportunity or other causes, and the growth function at large densities is again negative due to overcrowding. The reduction of the per-capita net growth rate for both sparse and overcrowding populations, and the existence of an optimal population density at

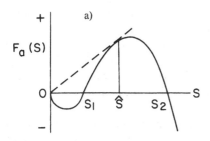 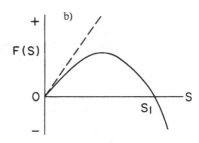

FIGURE 10.3. **a** Growth function of asocial populations. **b** Growth functions of Malthusian (– – –) and logistic (——) populations.

which the per-capita growth rate is maximum, is known as the *Allee effect* (Allee, 1938).

The diffusion–reaction equation of an asocial population is given by

$$\frac{\partial S}{\partial t} = D\frac{\partial^2 S}{\partial x^2} + F_\alpha(S), \tag{10.22}$$

where the growth function $F_\alpha(S)$ has the form

$$F_\alpha(S) < 0 \quad \text{for } 0 < S < S_1,$$
$$F_\alpha(S) > 0 \quad \text{for } S_1 < S < S_2, \tag{10.23}$$
$$F_\alpha(S) < 0 \quad \text{for } S_2 < S,$$

with $F_\alpha(0) = F_\alpha(S_1) = F_\alpha(S_2) = 0$. This is illustrated in Fig. 10.3(a). The growth functions of Malthusian and logistic populations are also shown in Fig. 10.3(b).

Bradford and Philip (1970a, b) examined the stability of steady-state solutions of (10.22). They demonstrated that in order for asocial populations to survive in a habitat with absorbing boundaries it is required not only that the habitat size be larger than $c(D/\alpha_1)^{1/2}$, where $\alpha_1 = dF_\alpha/dS(S)$ and c is a numerical constant of order unity, but also that the central density must exceed certain critical values. This contrasts with the criterion for the survival of logistic populations; steady distributions of logistic population are possible for all central densities less than the carrying capacity, no matter how small, provided the habitat size is larger than L_0. The difference is a consequence of the fact that asocial populations are characterized by negative growth at small population densities. In fact, Yoshizawa (1970) analyzed numerically the transient behavior of the solution of (10.22), and showed that for sufficiently small initial population densities, the solution tends to zero as time progresses.

Manoranjan and van den Driessche (1986) model the sterile insect release method for pest control by the reaction–diffusion equation

$$\partial S/\partial t = D\partial^2 S/\partial x^2 + \{aS/(S+N) - b\}S - 2cS(S+N),$$

where $2N$ is the constant number density of sterile insects (sex ratio is assumed to be one to one), S is the fertile female density, a is the birthrate, b is death rate, and $S/(S + N)$ is the reduction in birthrate due to sterilization. The last term on the right-hand side indicates a density-dependent death rate. Populations with the above growth function are termed *asocial*. They discussed the critical strip size for control, which is essentially a version of the critical size problem.

We now return to plankton patches and discuss some more complicated models. Wroblewski et al. (1975) and Platt and Denman (1975b) have included the effect of herbivore grazing in the diffusion–reaction equation for phytoplankton

$$\frac{\partial S}{\partial t} = D\frac{\partial^2 S}{\partial x^2} + \alpha S - R_m[1 - \exp\{-\lambda(S - S^*)\}], \quad \text{for } S \geq S^*, \quad (10.24)$$

where the last term on the right-hand side of (10.24) represents an Ivlev-type grazing function (Parsons et al., 1967; Parsons and Takahashi, 1973). R_m is the maximum grazing rate of the herbivores, A is the Ivlev constant, and S^* is the threshold of food availability.

On the basis of this model, Wroblewski et al. and Platt and Denman find the resulting critical size to be

$$L = \pi\left(\frac{D}{\alpha - R_m\lambda}\right)^{1/2} \quad \text{for } \alpha > R_m\lambda, S^* = 0. \quad (10.25)$$

If $\alpha \leq R_m\lambda$, all patches must vanish as time progresses. Thus, the effect of herbivore grazing is to reduce the net growth rate of the population, so that the population requires a critical patch size larger than L_0.

Levandowsky and White (1977) present a model that incorporates logistic growth as well as the grazing effect of the Ivlev type. The steady-state distribution associated with this model exhibits an interesting feature. Levandowsky and White predict the existence of a critical patch size beyond which there is sudden dynamic growth to a stable, steady-state level of high population; if the patch of suitable water were to shrink, we might also expect a relatively sudden disappearance of the population. This situation might be interpreted in terms of catastrophe theory (Thom, 1975; Zeeman, 1976).

Ludwig et al. (1979) have constructed a diffusion–reaction model to analyze the outbreak patch size of the spruce budworm, a defoliating insect. It deals with a dispersing population that obeys the logistic growth law and is subject to predation with a functional response of Holling's type (Holling, 1966); quite independently, Teramoto (1978) has used the same model to discuss harvesting strategies for the conservation of animal resources. The one-dimensional diffusion–reaction equation for budworm population is thus

$$\frac{\partial S}{\partial t} = D\frac{\partial^2 S}{\partial x^2} + \alpha S\left(1 - \frac{S}{K}\right) - \beta\frac{S^2}{H^2 + S^2}, \quad (10.26)$$

where K is the carrying capacity, β is the maximum predation rate, and H^2 is the functional response constant.

This model exhibits a feature somewhat similar to that of Levandowsky and White. If for any value of K/H the value of $\alpha H/\beta$ ($\equiv R$) is sufficiently small, then no steady state associated with outbreak is possible; however, a low endemic state may be present if the patch size is larger than L_0. If R is relatively large, there is a critical length $L > L_0$ beyond which either a low endemic state or an outbreak steady state occurs, depending on the initial budworm density. If R is sufficiently large and the habitat size is larger than L_0, the only stable stationary state becomes that associated with outbreak. Using this model, Clark et al. (1978) interpreted the behavior of the spruce budworm–coniferous forest ecosystem of eastern Canada.

Truscott and Brindley (1994) present a model for phytoplankton out-breaks, so-called red tides or spring blooms. A feature of phytoplankton populations leading to the outbreak is the occurrence of rapid population explosions and often equally rapid declines, separated by periods of low, quiescent population levels. The phytoplankton population dynamics of the same form as (10.26) coupled with (grazer) zooplankton population dynamics can yield results that resemble the red tide phenomenon. Models incorporating spatial diffusions are in progress by the same authors.

A review of the reaction–diffusion models for critical patch size is in Okubo (1984). Also, as in the case of propagation, the same biological problem can also be studied using integro-difference formulations (e.g., van Kirk and Lewis, 1997). Here similar results are obtained for both formulations.

Thus far, we have assumed that the habitat of a given species is surrounded by a completely hostile environment. In reality, the transition from favorable to unfavorable conditions is gradual with organisms more or less free to move both into and out of the favorable region.

Evans (1978a) and Ludwig et al. (1979) calculated critical sizes assuming that organisms can survive but not grow outside the suitable area, so that the diffusion–reaction equation outside is expressed as

$$\frac{\partial S}{\partial t} = D\frac{\partial^2 S}{\partial x^2} - \beta S \quad \text{for } |x| > L/2, \tag{10.27}$$

where $\beta \geq 0$ denotes the death rate of organisms. The continuity conditions (10.9) at the boundaries, located at $|x| = L/2$, are applied to the population density and its flux. The combination of the "inside" and "outside" equations leads to the following criterion for critical size L_c:

$$L_c = L_0\{(2/\pi)\tan^{-1}(\beta/\alpha)\},$$

where $\tan^{-1}(\beta/\alpha)$ takes the value of the principal branch. For infinite β/α, the critical scale is the KISS scale. For finite β/α, L_c is less than the KISS scale, decreasing to zero as β/α approaches zero; this limiting situation corresponds to a Malthusian population surrounded by reflecting barriers, e.g., fenced boundaries.

Evans (1978a) applied this criterion to phytoplankton patches; a minimum reasonable value for β/α might be 0.05 when population decay is simply due to respiration in the absence of nutrients. For this value, L_c is 14% of the KISS scale for a linear patch and 42% for a circular patch; thus, the critical length of the KISS model does not change greatly when the boundary conditions are relaxed in this way. Of course, in the limiting case of a population completely surrounded by reflecting barriers, the very concept of critical size is lost.

Gurney and Nisbet (1975) consider a model in which the growth rate α varies with distance from a habitat center in a parabolic fashion (for detail, see Sect. 10.4). Their result implies that the essential feature of critical size is preserved in this formulation; it is again determined by the balance of diffusion and growth.

The concept of minimum critical scales for population survival may also be applied to the design of national parks and wildlife refuges (McMurtrie, 1978) as well as to the pest-control problems. Using the concept to obtain design criteria for natural reserves, Diamond and May (1976) suggest that i) one large reserve is usually better than several smaller reserves whose areas add up to the same total area as the single larger reserve, ii) if other circumstances allow, any given reserve should be as nearly as circular in shape in order to maximize the area-to-perimeter ratio, iii) the effective area of a park may be halved if it is bisected by narrow dispersal barriers, and iv) if one must settle for several smaller parks, one way to raise the equilibrium number of species in any one such park is to raise the immigration rate into it. These ideas have been used to consider the shape of parks (Game, 1980), though the validity of some of these rules (i–iii) has been recently questioned (Higgs and Usher, 1980; Gilpin, 1980). Game examined Diamond's rule that reserves should be as round as possible and concludes that in certain circumstances the optimal shape may be other than circular. Data on the role of park size in extinctions has been reviewed by Newmark (1987).

Thus far, we have focused on whether a population can survive in a given habitat. The next step would be to use a nonlinear (bifurcation) approach to determine what the population levels would be if the population could survive. A first step toward this problem was analyzed by Rubenfeld (1979), beginning with

$$\partial S/\partial t = D\partial^2 S/\partial x^2 + b(t)f(S), \quad 0 < x < L, \quad 0 < t, \qquad (1)$$

where $b(t)$ is a time-dependent bifurcation parameter and the boundary conditions are

$$S(0,t) = S(L,t) = 0.$$

Essentially, one can show that the behavior is determined by the time average of $b(t)$ with survival possible if the average is positive, and the only mathematically interesting result arises in the biologically uninteresting case where the time average is zero. The analysis also provides an approximation to the population levels.

10.2.3   Population Balance with Diffusion and Direct Immigration-Emigration: Invasion and Settlement of New Species

In Sect. 9.1 the problem of critical habitat size for population survival was discussed by means of a simple diffusion–reaction model. In some cases, the balance between the rates of net growth and of random dispersal may determine the invasive potential of a new species and success or failure of species colonization. In other cases, another factor may appear; a species may migrate to and from the new habitat not only by advective and diffusive processes through the habitat boundary, but also by direct immigration or emigration. Strictly speaking, such direct immigration and emigration could be incorporated into the advective and diffusive processes. However, when we consider a one- or two-dimensional space, but model three-dimensional advection and diffusion, we are often forced to represent the higher-dimensional process by direct immigration or emigration. Thus, for animals such as ants, which normally dwell on the ground, i.e., a two-dimensional space, a flow of fellow animals materializing from the sky (e.g., winged queens) would of necessity be treated as immigration rather than advection or diffusion.

Here we deal with a simple model of diffusion–reaction with the inclusion of immigration and emigration:

$$\frac{\partial S}{\partial t} = D \frac{\partial^2 S}{\partial x^2} + \alpha(S)S + I(S), \qquad (10.28)$$

where α is the net growth rate, generally dependent on S and the density of other species; and I denotes the rate of immigration ($I > 0$) or of emigration ($I < 0$), which generally depends on S. Equation (10.28) must be solved subject to boundary conditions; we distinguish the following two cases.

Case 1.   The habitat is confined to the interval $0 \le x \le L$, and the surroundings are totally hostile, i.e., $S = 0$ at $x = 0$ and $x = L$.

When α and $I > 0$ are constants, (10.28) is easily solved for appropriate initial conditions. We should like to compare the result of this model with that of Sect. 9.1. Without the immigration effect, the population of invaders can increase indefinitely when α is positive and the habitat dimension is larger than the KISS size: $L_0 = \alpha(D/\alpha)^{1/2}$. With a constant rate of immigration, on the other hand, the species can survive even if $L < L_0$; its population ultimately settles to a finite value, irrespective of the initial condition, given by

$$S(x, \infty) = I \left\{ \cos \frac{\pi}{2} \left(\frac{L}{L_0} - \frac{2x}{L_0} \right) \middle/ \alpha \cos \left(\frac{\pi}{2} \frac{L}{L_0} \right) \right\} - I/\alpha. \qquad (10.29)$$

Equation (10.29) represents a population density that is positive everywhere except at $x = 0$ and L, where the density is zero. The result shows that direct immigration, no matter how small, plays a crucial role in the maintenance of equilibrium populations in small habitats and acts to preserve them against the risk of extinction. This conclusion is unchanged even when both local immigration (for $S < I_0/E$) and emigration (for $S > I_0/E$) are present in such a manner that $I = I_0 - ES$ and $\alpha > E$.

Case 2. The habitat is confined, i.e., $0 \le x \le L$, and no diffusion to the surroundings takes place, i.e., $D\partial S/\partial x = 0$ at $x = 0$ and $x = L$.

Islands provide examples of this case. When α is constant and $I = I_0 > 0$, or $I = I_0 - ES$, (10.28) can easily be solved analytically. We shall, however, choose to discuss the total number of individuals in the habitat, $N = \int_0^L S\,dx$. We consider the case where $I = I_0 - ES$. Integrating (10.28) with respect to x from $x = 0$ to $x = L$ and making use of the boundary conditions, we obtain

$$\frac{dN}{dt} = (\alpha - E)N + I_0 L. \tag{10.30}$$

Let $S(x, 0) = f(x)$ be the initial population density. Then

$$N(0) = \int_0^L f(x)\,dx \equiv F(\text{constant}) \ge 0. \tag{10.31}$$

Equation (10.30) is easily solved subject to (10.31); the result is

$$N(t) = -I_0 L/(\alpha - E) + \{F + I_0 L/(\alpha - E)\}\exp\{(\alpha - E)t\}. \tag{10.32}$$

When $\alpha > E$, growth dominates the effect of emigration, and the population increases indefinitely. A net emigration, $I_0 = 0$, cannot check the population explosion unless the carrying capacity effect acts to limit population growth.

Only when $E > \alpha$ does (10.32) asymptotically reaches a limiting equilibrium population $\hat{N}$, given by

$$\hat{N} = I_0 L/(E - \alpha). \tag{10.33}$$

Even if $\alpha = -\mu < 0$, i.e., mortality exceeds reproduction, the effect of immigration tends to balance the internal loss of population, and the equilibrium value

$$\hat{N} = I_0 L/(E + \mu) \tag{10.34}$$

is attained (see Fig. 10.4).

The basic concept behind (10.34) is the same as that used by MacArthur and Wilson (1967) in island biogeography to discuss equilibrium species number; therein N is interpreted to be the number of species. MacArthur and Wilson and MacArthur (1972) derived a relation between island area and the

FIGURE 10.4. Equilibrium population number $\hat{N}$ determined by the balance of the effect of immigration $I(N)$ and mortality μN.

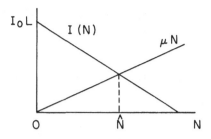

number of species present, assuming that μ depends on the size of the island and I_0 and E depend on the distance between the source of immigration and the island.

Finally, we examine a study due to Kuno (1968), in which he stochastically analyzed the processes of invasion into and colonization (settlement) of a new habitat by means of repeated reproduction and dispersion of successive generations. As an ecological application, Kuno devised a simulation model for the spatial structure of populations of rice leafhoppers in paddy fields; he based the analysis on the following assumptions regarding four assumed fundamental processes.

Process A. The initial spatial distribution of individuals of the invading population obeys the Poisson distribution, i.e., the probability of finding n individuals settled on each rice stalk hill is given by

$$P_1(n) = m^n e^{-m}/n!, \tag{10.35}$$

where m denotes the mean value of individuals on each rice stalk hill.

Process B. The adult immigrants, after settlement or after emergence, lay eggs at a constant rate, during which adult mortality occurs at an equal probability in time and for each individual. Thus, the number of eggs per adult, q, is distributed geometrically with the probability given by

$$p_2(q) = f^q/(1+f)^{q+1},$$

where f is the mean number of eggs laid by one adult.

Process C. The mortality of the progeny population in the period from hatching to emergence occurs at an equal probability for each individual. Thus, the probability that b individuals of the next generation are produced from q individual eggs obeys the binomial distribution

$$p_3(q,b) = \{q!/b!(q-b)!\}\lambda^b(1-\lambda)^{q-b}, \tag{10.37}$$

where λ is the mean survival rate of individuals.

Process D. Individual insects disperse according to a random walk. Thus, the probability of finding adult individuals of the second-generation population in a circular region with radius r from the location of settlement of a first-generation adult is given by

$$p_4(r) = 1 - \exp(-r^2/\sigma_r^2), \qquad (10.38)$$

where σ_r^2 is the variance of the displacement of the adult progeny.

It can be seen from (10.36) and (10.37) that the probability $p_3(b)$ of producing b individual adults of the second generation from one individual adult is given by the geometric distribution

$$p_3(b) = \sum_q p_2(q)p_3(q,b) = \beta^b/(\beta+1)^{b+1}, \qquad (10.39)$$

where $\beta \equiv f^\lambda$, a parameter that may be regarded as the mean reproduction rate per generation.

In terms of our model (10.28), process A corresponds to the stochastic representation of the term $I(S)$, processes B and C correspond to the stochastic representation of $\alpha(S)S$, and process D corresponds to the term $D\partial^2 S/\partial x^2$.

Kuno's simulation model provides the following results.

1) The spatial structure of the descendants of the invaders is not random, but is approximated by a negative binomial distribution characteristic of contagious patterns (Pielou, 1969), even though the initial population of invaders is assumed to be Poisson-distributed.
2) The degree of aggregation in the descendent populations varies widely depending on the values assigned to four parameters used in the model: m, β, σ_r^2, and N (= the number of generations).
3) For each generation, the degree of aggregation[1] depends inversely on m but very weakly on, β; also it depends almost inversely on σ_r^2 when the size of quadrat is small, but very weakly on σ_r^2 when the size of quadrat is large.

Kuno also attributes the emergence of the contagious distribution of the population of progeny from the randomly distributed pioneer parents to the combined effects of random dispersal and reproduction. A similar feature was observed in the compound distribution of eggs laid by dispersing parents (see Sect. 6.2).

Iwasa and Teramoto (1977) succeeded in treating Kuno's model analytically. Their mathematical model enables expressing the index of aggregation as a function of characteristic parameters involved in the model. The dependence of C_A on the parameters is then discussed in comparison with the results obtained by Kuno's simulation method.

---

[1] Kuno defines the index of aggregation, C_A, to be $C_A = (V - m)/m^2$, where V and m are respectively the variance and mean of the frequency distribution (see Chap. 9).

10.3 Two- and Multispecies Population Dynamics with Dispersal

For proper modeling of an ecosystem, it is often necessary to deal with interacting populations of several species. However, as the mathematical treatment is simplest when only two species are considered in the problem of interacting population dynamics with dispersal, we shall restrict our attention primarily to this case. Most of the results can presumably be extended to systems involving more species at the expense of added mathematical detail and complexity.

Perhaps the zeroth order "interacting populations with dispersal" problem is the critical size problem (see Sect. 9.1). Namely, when populations are interacting, is there a minimum patch size (the critical size) below which no increase of any species is possible? We model the population, S_i, in one spatial dimension, using Fickian diffusion plus an interaction term that is proportional to S_i. [n.b., see the first two terms on the r.h.s. of Eq. (10.28).]

$$\frac{\partial S_i}{\partial t} = D_i \frac{\partial^2 S_i}{\partial x^2} + G_i(S_j)S_i, \quad i, j = 1, 2, \ldots, m, \tag{A}$$

$$t = 0; \quad S_i = C_i(x), \tag{B}$$

$$x = 0, L; \quad S_i = 0. \tag{C}$$

Pursuing the analysis that follows Eq. (9.2), the initial conditions and boundary conditions at $x = 0, L$ are given by Eqs. (B) and (C).

We can consider this set of m equations as a local stability problem, with the initial densities of the interacting populations being infinitesimally small. In this way it will be possible to reduce the m coupled equations in Eq. (A) to a set of uncoupled critical size problems for individual species, regardless of the interaction between the populations.

Let

$$C_i(x) = \varepsilon_1 c_i(x), \quad |\varepsilon_1| \ll 1. \tag{D}$$

To first order in ε_1, $c_i(x)$ satisfies

$$0 = D_i \frac{\partial^2 c_i}{\partial x^2} + G_i(0)c_i, \tag{E}$$

where $[G_i(0) = G_i(S_j)|_{S_j=0}]$. Now let

$$S_i(x) = \varepsilon_1 C_i(x) + \varepsilon_2 s_i(x, t), \quad |\varepsilon_2| \ll 1. \tag{F}$$

Substitution of Eq. (F) into Eq. (A), with the use of Eq. (E), leads to first order in ε_2,

$$\frac{\partial s_i}{\partial t} = D_i \frac{\partial^2 s_i}{\partial x^2} + G_i(0)s_i, \qquad (G)$$

$$t = 0; \quad s_i = h_i(x),$$

$$x = 0, L; \quad s_i = 0,$$

where $h_i(x)$ is the initial condition for the time-dependent part of population s_i.

Note that Eq. (G) depends only on s_i. More importantly, it is also identical to Eq. (9.2), with $G_i(0)$ analogous to α, the exponential rate of growth. Thus, we have reduced the critical size problem of the interacting, multispecies assemblage to a critical size problem for each individual species. Accordingly, the critical size for the entire group of species depends on the minimum value of $[D_i/G_i(0)]^{1/2}$; and a lower bound for the critical size is

$$(L_c)_{\min} = \pi[D_{\min}/G_{\max}(0)]^{1/2}. \qquad (H)$$

10.3.1   Lotka–Volterra Predator–Prey Model

Although the classical Lotka–Volterra model provides an oversimplified representation of interacting populations, calculations based on the model have provided much insight and have proved to be of practical value. It would thus seem natural to extend the model by including diffusional effects, allowing for investigation of the dynamics of interacting populations with dispersal. Nevertheless, this must be undertaken with the utmost precaution due to the fact that the Lotka–Volterra system is structurally unstable (May, 1973), allowing the possibility of rather peculiar behavior (Levin, 1978a; Murray, 1977; Nicolis and Prigogine, 1977).

Let $S_1(x, t)$ and $S_2(x, t)$ be the population densities of prey and predator, and D_1 and D_2 be the diffusivities of the two populations, respectively. The Lotka–Volterra predator–prey model with diffusion can then be written as

$$\frac{\partial S_1}{\partial t} = D_1 \frac{\partial^2 S_1}{\partial x^2} + a_1 S_1 - b_1 S_1 S_2, \qquad (10.40)$$

$$\frac{\partial S_2}{\partial t} = D_2 \frac{\partial^2 S_2}{\partial x^2} - a_2 S_2 + b_2 S_1 S_2, \qquad (10.41)$$

where a_1 and a_2 are, respectively, the linear rates of birth and death for the individual species; and b_1 and b_2 are, respectively, the linear decay and growth factors due to binary interactions. Equations (10.40) and (10.41) are subject to appropriate initial and boundary conditions.

We shall study the solutions of (10.40) and (10.41) for several special cases. Clearly, the case $S_1 = S_2 = 0$, and the case $S_1 = a_2/b_2$, $S_2 = a_1/b_1$ constitute two steady, homogeneous solution sets, i.e., equilibrium solutions.

It is known that without diffusion, the state $S_1 = S_2 = 0$ is unstable, i.e., if it is perturbed, the predator and prey populations tend to deviate from the zero state. However, with diffusion in bounded region $(0, L)$ which is surrounded by a totally hostile environment, (10.40) reduces to (9.2) for an infinitesimal perturbation S_1 above zero values of predator–prey populations, so that stability depends on the size of the region. According to Sect. 9.1, the prey population never grows if $L < \pi(D_1/a_1)^{1/2}$, while the predator population never grows above its zero value from an infinitesimal perturbation.

It is also known that the state $S_1 = a_2/b_2$ and $S_2 = a_1/b_1$ is neutrally stable without diffusion (the predator–prey cycle). To examine the stability of this state in the presence of diffusion, let $S_1 = a_2/b_2 + S_1'(x, t)$ and $S_2 = a_1/b_1 + S_2'(x, t)$. Substituting these values into (10.40) and (10.41) and ignoring the second- and higher-order terms of S_1' and S_2', we obtain

$$\frac{\partial S_1'}{\partial t} = D_1 \frac{\partial^2 S_1'}{\partial x^2} - \frac{a_2 b_1}{b_2} S_2', \qquad (10.42)$$

$$\frac{\partial S_2'}{\partial t} = D_2 \frac{\partial S_2'}{\partial x^2} + \frac{a_1 b_2}{b_1} S_1'. \qquad (10.43)$$

Appropriate boundary conditions are either

$$S_1' = S_1' = 0 \quad \text{at} \quad x = 0 \quad \text{and} \quad x = L \qquad (10.44)$$

or

$$D_1 \partial S_1'/\partial x = D_2 \partial S_2'/\partial x = 0 \quad \text{at} \quad x = 0 \quad \text{and} \quad x = L. \qquad (10.45)$$

The former boundary conditions imply that the surroundings always accommodate the equilibrium population densities of predator and prey, and the latter boundary conditions imply that there are no fluxes of the populations through the boundary. The effect of boundary conditions different from these two types on the population dynamics has been discussed by Hadeler et al. (1974) and Gopalsamy (1977a).

Because of the nature of the boundary conditions, finite sine and cosine transformations prove useful (Carslaw and Jaeger, 1959). Thus for (10.44) we consider the finite sine transform

$$(S_i')_{si} \equiv \Gamma_i(t, m) \equiv \int_0^L S_i'(t, x) \sin(m\pi x/L)\, dx, \quad i = 1, 2; \ m = 1, 2, \ldots.$$

Then the sine transform of $\partial^2 S_i'/\partial x^2$ becomes, after integrating by parts,

$$(\partial^2 S_i'/\partial x^2)_{si} = (m\pi/L)\{S_i'(t, 0) - (-1)^m S_i'(0, L)\} - (m^2\pi^2/L^2)\Gamma_i$$
$$= -(m^2\pi^2/L^2)\Gamma_i,$$

where the boundary conditions (10.44) are invoked. Therefore, taking the sine transform of (10.42) and (10.43), we find

$$\frac{\partial \Gamma_1}{\partial t} = -\frac{m^2 \pi^2}{L^2} D_1 \Gamma_1 - \frac{a_2 b_1}{b_2} \Gamma_2, \tag{10.46}$$

$$\frac{\partial \Gamma_2}{\partial t} = -\frac{m^2 \pi^2}{L^2} D_2 \Gamma_2 - \frac{a_1 b_2}{b_1} \Gamma_1. \tag{10.47}$$

In order to perform a stability analysis, we assume

$$\Gamma_i = A_i e^{\lambda t}, \quad i = 1, 2. \tag{10.48}$$

Substituting (10.48) into (10.46) and (10.47) and seeking solutions for which A_1 and A_2 do not vanish simultaneously, we are led to the equation for the eigenvalues, λ,

$$(\lambda + m^2 \pi^2 D_1 / L^2)(\lambda + m^2 \pi^2 D_2 / L^2) + a_1 a_2 = 0$$

or

$$\lambda = \tfrac{1}{2}[-m^2 \pi^2 (D_1 + D_2)/L^2 \pm \{m^4 \pi^4 (D_1 - D_2)^2 / L^4 - 4 a_1 a_2\}^{1/2}]. \tag{10.49}$$

As both roots of λ have negative real parts, infinitesimal perturbations tend to decay as $t \to \infty$; thus, the populations must return to the constant equilibrium state $(a_2/b_2, a_1/b_1)$.

For the boundary conditions (10.45), on the other hand, we employ the finite cosine transform

$$(S_i')_{co} \equiv \Lambda_i(t, m) \equiv \int_0^L S_i'(t, x) \cos(m\pi x/L)\, dx, \quad i = 1, 2; \ m = 0, 1, 2 \dots.$$

The cosine transform of $\partial^2 S_1'/\partial x^2$ reduces to

$$(\partial^2 S_i'/\partial x^2)_{co} = -(-1)(\partial S_i'/\partial x)_{x=L} - (\partial S_i'/\partial x)_{x=0} - (m^2 \pi^2/L^2)\Lambda_i$$
$$= -(m^2 \pi^2/L^2)\Lambda_i,$$

where the boundary conditions (10.45) are invoked. Therefore, we obtain the same equations for Λ_i as (10.46) and (10.47) for Γ_i with the exception that m may now take the value zero as well as positive integers. Since the largest real part of λ occurs when $m = 0$, i.e., spatially uniform perturbations, and the values of λ are $\mp i(a_1 a_2)^{1/2}[i = (-1)^{1/2}]$, the equilibrium state corresponds to neutral stability. Spatially inhomogeneous components of the perturbed populations tend to "die out" asymptotically to leave uniform populations that oscillate around the reference state with a period of $2\pi(a_1 a_2)^{-1/2}$.

Although the above stability argument relies on a linear analysis, investigators, including, Murray (1975), Chow and Tam (1976), Jorné and Carmi (1977), Rosen (1975), Rothe (1976), and Mimura and Nishida (1978) have proved in general that for $t \to \infty$, there exist no temporally periodic spatially nonuniform solutions for the Lotka–Volterra diffusion equation in a finite domain, subject to either of the boundary conditions that the population densities are fixed at their equilibrium values, or that no fluxes of population exist. Thus, the population densities either converge to their equilibrium values everywhere or admit only spatially uniform temporal oscillations; no patchiness arises.

Conway et al. (1978) and Othmer (1977) have extended this asymptotic decay of spatial inhomogeneities to more general situations. They show that every solution of weakly coupled multispecies systems undergoing advection, diffusion, and nonlinear reactions with arbitrary initial conditions, and subject to reflecting boundary conditions, decays exponentially with time to a spatially homogeneous function of time if the minimum value of diffusivities for multispecies diffusion exceeds a certain critical value. That is, when the smoothing effect of diffusion is sufficient to overcome the source and sink effect of the reaction term and the effective advection, no spatial patterns can emerge asymptotically. Shukla and Das (1982) extended this result to more general food webs. They showed that if the food web was stable in the absence of dispersal, it remained stable with dispersal. Sullivan (1988) explored the case of Fickian diffusion with logisticlike responses (i.e, $\propto \pm c_i S_i \pm d_i S_i^2$, where c_i and d_i are positive constants) plus Monod interactions between species [see Hilborn (1979), who attributed this model to Bazykin (1974)]. He found that temporally averaged spatial means of predator and prey abundances were robust to changes in diffusion rates; but that temporally averaged variances changed dramatically. Finally, using general mathematical techniques, Beretta et al. (1988) identified conditions when a two-species Lotka–Volterra system diffusing between two patches "ends up" in a bounded region of phase space (i.e., the phase space trajectories are confined to a bounded region).

In contrast to these results, Kishimoto (1982) and Kishimoto et al. (1983) have found stable, nonuniform oscillating states for diffusive Lotka–Volterra systems with three species. Detailed mathematical aspects that determine when one can expect to encounter the Kishimoto and Kishimoto et al. results, as well as similar findings from others (e.g., Shukla and Verma, 1981), are addressed by Kishimoto and Weinberger (1985) and Voorhees (1982).

However, if the diffusivities are so small that spatial inhomogeneities in population densities may not decay in time intervals of our concern, the existence of slowly varying wave solutions of the diffusive Lotka–Volterra model is possible, and these might be adequate to explain certain observed spatial patterns [Wyatt, 1973 (in McMurtrie, 1978)]. Likewise when the dimensions of the habitat are large compared with a characteristic distance of dispersal, one is, for all practical purposes, dealing with infinite space, and the system may admit wavelike solutions. Indeed, such wave solutions can appear as reaction fronts in chemical systems of the type $A + B \to C$, when these are modeled using first-order reaction kinetics and Fickian diffusion (Araujo et al., 1992; Havlin et al., 1992). It is to this topic in an ecological context that we now turn.

Edelstein-Keshet (1986, 1988) proposed a theory for plant–herbivore interactions where vegetative features (called *plant quality*) are affected by herbivores and, in turn, influence the effectiveness of herbivores. Spatial patterns arise when the herbivores are mobile and can attack the plants on the basis of "quality" and the density of attacking herbivores. Vail (1990, 1993) constructed a reaction–diffusion approximation to this plant-quality–herbivore system. He found that in an experimental mimic of this system, large-scale differences were more effective in producing patterns than small-scale perturbations. Vail argued that "diffusion" (i.e., herbivore mobility) "smoothed out" differences at the smaller scales. Walsh (1996), on the other hand, using an area-restricted search algorithm [see Kareiva and Odell (1987)] for herbivore movement, found that small-scale patterns were amplified in the plant-quality–herbivore systems. And Lewis (1994) found a wide diversity of behaviors, including travelling waves. These plant-quality–herbivore investigations are examples of detailed mechanistic formulations that can develop complex spatio-temporal patterns. Many can be approximated by reaction–diffusion systems, thus providing a useful tool to determine how such patterns arise.

10.3.2 Travelling Waves

In an unbounded domain, $-\infty < x < \infty$, it is reasonable to inquire as to the existence of travelling waves in the diffusive Lotka–Volterra system (Chow and Tam, 1976). For simplicity, we first consider the case where the prey is sedentary ($D_1 = 0$), and integrate (10.40) with respect to t to yield

$$S_1(t,x) = f_1(x) \exp\left\{ a_1 t - b_1 \int_0^t S_2(t',x)\,dt' \right\}, \qquad (10.50)$$

where $f_1(x)$ is the initial prey distribution in space. Substitution of (10.50) into (10.41) yields

$$\frac{\partial S_2}{\partial t} = D_2 \frac{\partial^2 S_2}{\partial x^2} - a_2 S_2 \left[1 - a_2^{-1} b_2 f_1(x) \exp\left\{ a_1 t - b_1 \int_0^t S_2(t',x)\,dt' \right\} \right]. \qquad (10.51)$$

To obtain asymptotic travelling waves, we take the initial prey distribution $f_1(x)$ to be

$$A(x) = Ae^{-k|x|}, \quad A,k > 0. \qquad (10.52)$$

We look for stationary wave-form solutions of (10.50) and (10.51) propagating in the positive direction of x with speed C, i.e.,

$$S_1(t, x) = S_1(\xi),$$
$$S_2(t, x) = S_2(\xi),$$

(10.53)

where $\xi = x - ct$.

Substituting (10.53) into (10.50) and (10.51), and looking for asymptotic situations as $t \to \infty$ and $x \to \infty$, Chow and Tam (1976) obtain

$$S_1(\xi) = A \exp\left\{ -k\xi + b_1 c^{-1} \int_\xi^\infty S_2(\xi') \, d\xi' \right\},$$

(10.54)

$$D_2 \ddot{S}_2 + c\dot{S}_2 - a_2 S_2 \left[1 - A a_2^{-1} b_2 \exp\left\{ -k\xi + b_1 c^{-1} \int_\xi^\infty S_2(\xi') \, d\xi' \right\} \right] = 0,$$

(10.55)

where $\dot{S}_2 = dS_2/d\xi$ and $\ddot{S}_2 = d^2 S_2/d\xi^2$. Also, the speed of wave propagation must be

$$c = a_1/k.$$

(10.56)

For large values of ξ, (10.54) and (10.55) yield

$$S_1(\xi) \sim A \exp(-k\xi),$$

(10.57)

$$S_2(\xi) \sim \frac{1}{\Gamma(v+1)} \left(\frac{Ab_2}{k^2 v^2} \right)^{v/2} \exp\{-(a_2/D_2)^{1/2}\xi\},$$

(10.58)

where $v = 2k^{-1}(D_2/a_2)^{1/2}$ and Γ is the gamma function. The solutions (10.57) and (10.58) show that the wave fronts are sharp and descending with the exponential rates in space.

Chow and Tam carried out a numerical integration of (10.54) and (10.55). The result is shown in Fig. 10.5, which reveals that the prey population wave propagates ahead of the predator population wave.

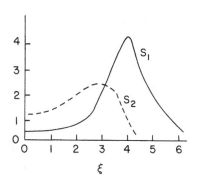

FIGURE 10.5. Travelling waves of predator (S_2) and prey (S_1) populations, when the prey is sedentary and the predator undergoes diffusion (from Chow and Tam, 1976).

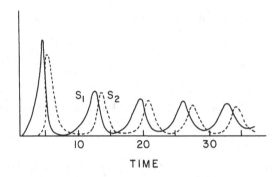

FIGURE 10.6. Oscillations of the predator (S_2) and prey (S_1) populations when the prey undergoes small diffusion and the predator undergoes diffusion (from Chow and Tam, 1976).

Similar patterns of the population wave are obtained in the case of non-zero but small values of D_1, with the only exception that the trailing section of the wave is oscillatory in space. Thus, the population waves observed at a fixed location as time progresses look like those shown in Fig. 10.6. We observe the successive passage of the frontal waves of population density, the prey front being followed by the predator front, after which the local population densities settle down to periodic oscillations as in the case of a bounded domain.

On the basis of (10.40) and (10.41), Dubois (1975a) numerically calculated the non-steady-state distributions of predator and prey densities within favorable habitats with absorbing boundary conditions. As is shown in his Fig. 1, an initially concentrated patch of population spreads across the habitat in the form of travelling waves. Although the solutions of (10.40) and (10.41) eventually settle down to a spatially uniform steady state, the waves persist for a considerable length of time. Dunbar (1983, 1986) demonstrated the existence of travelling-wave solutions for more complicated species interactions. He first modified the Lotka–Volterra prey equation to allow logistic growth and then extended his analysis to a case with Monod functional response for the prey (and numerical response for the predator). In this second case, the prey did not diffuse [i.e., $D_1 = 0$ as in Chow and Tam (1976)]. Murray (1976) analyzed a "diffusion plus (birth – death)" system that was designed to mimic predator–prey communities. When there is a delay in the (birth – death) term, travelling waves result. Gopalsamy (1980) and Gopalsamy and Aggrawala (1981) also found travelling waves in predator–prey and logistic systems. Bonilla and Linan (1984) explored a diffusive Volterra situation with time delay for the effect of predation. Their model formulation is close (but not identical) to that of Eq. (10.51). That is, S_2 in the integral in (10.51) appears as $G(t')S(t - t', x)$, where $G(t')$ is a delay kernal. The behavior of the Bonilla–Linan system is diverse—with relaxation oscillations, travelling waves, etc., and exhibits the main features of excitable media.

The geographic spread of epidemics is another example of travelling waves in interacting populations with dispersal. Noble (1974) has developed a mathematical model of the spread of plagues in human populations.

For simplicity, we assume only two interacting populations, taken to be infectives and susceptives, although we might equally well consider hosts and parasites. The local rate of change of the number of infectives consists of three terms: the rate of transition from the susceptive population; the death rate; and the net influx of infective population by means of diffusion. Similarly, the local rate of change of the number of susceptives consists of two terms: the rate of transition into the infective population; and the net inflow of the susceptive population by diffusion. The transition rate is assumed to be proportional to the rate of binary encounter between infective and susceptive populations.

We thus formulate

$$\frac{\partial S_1}{\partial t} = kS_1 S_2 - \mu S_1 + D\frac{\partial^2 S_1}{\partial x^2},$$ (10.59)

$$\frac{\partial S_2}{\partial t} = -kS_1 S_2 + D\frac{\partial^2 S_2}{\partial x^2},$$ (10.60)

where S_1 and S_2 are infective and susceptive population densities, respectively; k is the transmissibility coefficient; μ is the death rate; and D is the diffusivity of the population, which is assumed to be the same for both groups. It can be seen that Noble's model for plagues is a special case of the diffusive Lotka–Volterra model.

We assume the initial condition to be that of a uniform distribution of susceptive population in space,

$$S_2(0, x) = N.$$

The infective population is not specified beyond the assumption that the population is sufficient to trigger the plague.

We now assume wavelike solutions of the form $S_1(t, x) = S_1(\xi)$ and $S_2(t, x) = S_2(\xi)$, where $\xi \equiv x - ct$, and rewrite (10.59) and (10.60) as follows:

$$\ddot{S}_1 + (c/D)\dot{S}_1 - (\mu/D)S_1 + (k/D)S_1 S_2 = 0,$$ (10.61)

$$\ddot{S}_2 + (c/D)\dot{S}_2 - (k/D)S_1 S_2 = 0,$$ (10.62)

where $\dot{S}_1 = dS_1/d\xi'$, etc.

Noble (1974) obtained a criterion for the existence of wavelike solutions satisfying the boundary conditions $S_1(-\infty) = S_1(\infty) = 0$, and the conditions $0 \le S_2(-\infty) < S_2(\infty) = N$ and S_1, S_2, $c > 0$. Adding (10.61) and (10.62) together, integrating the resulting equation once with respect to ξ from $-\infty$ to ξ, and using the boundary conditions, we find

$$\frac{d}{d\xi}(S_1 + S_2) = \frac{\mu}{D}e^{-c\xi/D}\int_{-\infty}^{\xi} e^{c\xi'/D}S_1(\xi')\,d\xi' > 0$$

so that

$$S_1(\xi) + S_2(\xi) < S_1(\infty) + S_2(\infty) = N.$$ (10.63)

If we integrate (10.61) in ξ from $-\infty$ to ∞ and utilize the boundary conditions, noting that they also imply that $S_1(-\infty) = S_1(\infty) = 0$, we obtain

$$\int_{-\infty}^{\infty} S_1 \, d\xi = \frac{k}{\mu} \int_{-\infty}^{\infty} S_1 S_2 \, d\xi. \tag{10.64}$$

Since $0 < S_2(\xi) < N$ for all ξ from (10.63),

$$\int_{-\infty}^{\infty} S_1 \, d\xi > \int_{-\infty}^{\infty} S_1(S_2/N) \, d\xi = \frac{1}{N} \int_{-\infty}^{\infty} S_1 S_2 \, d\xi. \tag{10.65}$$

Comparing (10.64) with (10.65), a criterion for plague wave propagation is seen to be

$$\gamma \equiv kN/\mu > 1 \tag{10.66}$$

This criterion has an obvious interpretation. There is a minimum population density, $N_c \equiv \mu/k$, necessary for the outbreak of an epidemic. For a given initial population multiplied by the transmission coefficient, there is a maximum mortality beyond which the disease cannot spread.

Noble solved numerically the original partial differential equations (10.59) and (10.60). A typical wavelike solution is shown in Fig. 10.7, where the pulselike character of the plague wave is clear. The speed of the wave is of the order of $(kDN)^{1/2}$. Noble applied his model to the Black Death of 1347

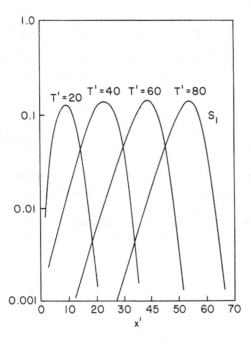

FIGURE 10.7. The plague wave of infectives according to Noble's model. The infective population, time, and space are nondimensionalized (from Noble, 1974).

FIGURE 10.8. Propagation of the
Black Death in Europe, 1347 to
1350 A.D.

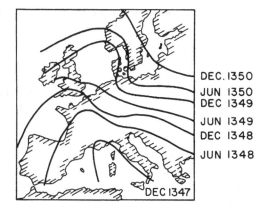

DEC. 1350
JUN 1350
DEC 1349

JUN 1349
DEC 1348

JUN 1348

DEC 1347

A.D. (Fig. 10.8). At the time of the Black Death, the population density N
of Europe was about one person per 50 miles^2. From the spread of news and
gossip, D is estimated to be 10^4 miles^2/yr. The values of k and μ are assumed
to be $kN \sim 20$/yr and $\mu \sim 15$/yr. These estimates give $c \sim 300$ miles/yr. This
velocity is in good agreement with that estimated from Fig. 10.8. In addition, these values are consistent with the criterion for the outbreak of plague,
Eq. (10.66), since $\gamma = 20/15 > 1$.

Webb (1981) considered mathematical details of the epidemic describied
by Eqs. (10.59) and (10.60). He showed that on a finite interval there is
a unique solution to the equations, and that $S_1(t = +\infty) \rightarrow 0$, while
$S_2(t = +\infty) \rightarrow$ (positive constant). Hence, the infection must die out as
$t \rightarrow \infty$ because some susceptible individuals never contract the disease. This
asymptotic behavior is analogous to that found in spatially uniform models
for epidemics (Waltman, 1974). See also de Mottoni et al. (1979) for an
extension of the epidemic modeled by (10.59) and (10.60) to some nonlocal
effects in the "transition" term ($\propto S_1 S_2$) and unequal diffusivities.

During the last few hundred years, continental Europe has been swept
repeatedly by rabies epidemics similar to the plague epidemics of the Middle
Ages and later. Murray (1993) presents several models of varying complexity, as well as much background information on these epidemics, which
are believed to be generated by rabid fox vectors. For example, the more
detailed model (Murray et al., 1986) is a reaction–diffusion generalization of
Eqs. (10.59) and (10.60) for three species S, I, and R; $S =$ susceptible foxes;
$I =$ infected, but not infectious, foxes; and $R =$ infectious, rabid foxes. Since
the rabid foxes spread the disease organisms, only the equation for the rate
of change of R contains a diffusion term $D\partial^2 R/\partial x^2$. See Murray (1993) for
details on this model, and other S–I–R epidemic models, as well as a
simpler, two-species rabies model (Kallen et al., 1985).

Disease phenomena in insect systems present excellent examples of ecological travelling waves. Dwyer (1992, 1994) analyzed diffusive an "S–I–R-like"

model constructed with the help of data from a virus (nuclear polyhedrosis virus, NPV) that attacks a Douglas-fir tussock moth. Equations are written for the rate of change of 1) S, susceptible host moth density, 2) I, infectious host density, and 3) P, pathogen (virus) density. Dwyer found threshold behavior (for travelling waves) in his model and was able to estimate the rate of spread of the NPV infection. Agreement with field observations was encouraging, especially when density dependence was introduced into the host population dynamics.

Manoranjan and van den Driessche (1986) and Lewis and van den Driessche (1993) modeled the release of sterile females released into a population of fertile females (= density of fertile males when an equal sex ratio is assumed). Lewis and van den Driessche found that when the sterile insect density exceeds a critical threshold, a travelling-wave solution can reverse and give rise to local extinction.

Kopell and Howard (1973) approached the problem of travelling waves quite generally, using the following set of diffusion–reaction equations for n-species:

$$\frac{\partial S_i}{\partial t} = D_{ij} \nabla^2 S_j + F_i(S_j), \quad i, j = 1, 2, \ldots, N, \tag{10.67}$$

where $S_i(\mathbf{x}, t)$ is the population density of the ith species, D_{ij} is the matrix of diffusivities (see Eq. (10.2)], and F_i denotes the nonlinear reaction kinetics for the ith species. Kopell and Howard then proved that the system possesses periodic travelling wave-train solutions of small amplitudes under the following conditions: The cross-diffusivities $D_{ij}(i \neq j)$ must be small; the self-diffusivities D_{ii} must be sufficiently close to each other; and (10.67) with $D_{ij} \equiv 0$, i.e., the reduced system in which there are no spatial variations in S_i, must either have a limit cycle solution (May, 1973) or be unstable to growing oscillations about a steady state given by $F_i(S_j^*) = 0$. Howard and Kopell (1977) discussed slowly varying wave-train solutions of diffusion–reaction equations of this class.

We conclude this section with references to several general articles and books that the reader is urged to consult for further discussion of topics in the broad area of biological waves. Odell (1980) gives a useful tutorial on mathematical subtleties that underlie this subject. Fife (1981) discusses models of phenomena like fronts and target patterns in chemical systems that have analogies to ecological situations. Renshaw (1982) analyzes a stepping-stone model for the case of Lotka–Volterra dynamics within discrete patches that also leads to a travelling-wave solution. Monk and Othmer (1990) consider travelling waves in the much-studied slime mould system. Murray (1993) presents many carefully explained studies of travelling waves found in a number of biological examples. Grindrod (1996) discusses the detailed (and advanced) mathematics behind nonlinear travelling waves in particular and reaction–diffusion equations in general.

10.3.3   The Effect of Cross-Diffusion

Kerner (1959) and Jorné (1977) examined the effect of cross-diffusion on the diffusive Lotka–Volterra system. Their treatment is based on the following equations:

$$\frac{\partial S_1}{\partial t} = D_1 \frac{\partial S_1}{\partial x^2} + D_{12} \frac{\partial^2 S_2}{\partial x^2} + a_1 S_1 - b_1 S_1 S_2, \tag{10.68}$$

$$\frac{\partial S_2}{\partial t} = D_2 \frac{\partial^2 S_2}{\partial x^2} + D_{21} \frac{\partial^2 S_1}{\partial x^2} - a_2 S_2 + b_2 S_1 S_2, \tag{10.69}$$

where the cross-diffusivities D_{12} and D_{21} can be either positive or negative, but the (self-) diffusivities D_1 and D_2 are nonnegative. Otherwise, the notation is the same as that of (10.40) and (10.41).

We again consider small perturbations S_1' and S_2' from the steady, uniform population densities a_2/b_2 and a_1/b_1 and linearize to obtain the following equations for the perturbations:

$$\frac{\partial S_1'}{\partial t} = D_1 \frac{\partial^2 S_1'}{\partial x^2} + D_{12} \frac{\partial^2 S_2'}{\partial x^2} - \frac{a_2 b_1}{b_2} S_2', \tag{10.70}$$

$$\frac{\partial S_2'}{\partial t} = D_2 \frac{\partial^2 S_2'}{\partial x^2} + D_{21} \frac{\partial^2 S_1'}{\partial x^2} + \frac{a_1 b_2}{b_1} S_1'. \tag{10.71}$$

The linear stability problem can be investigated in a way similar to that of Sect. 10.3.1. Thus the solution for the eigenvalues is given by

$$\lambda = \tfrac{1}{2}[-m^2\pi^2(D_1 + D_2)/L^2 \pm \{m^4\pi^4(D_1 - D_2)^2/L^4$$
$$- 4(a_2 b_1/b_2 + m^2\pi^2 D_{12}/L^2)(a_1 b_2/b_1 - m^2\pi^2 D_{21}/L^2)\}^{1/2}], \tag{10.72}$$

where $m = 1, 2, \ldots$ for absorbing boundary conditions and $m = 0, 1, \ldots$ for reflecting boundary conditions at $x = 0$ and L.

In the absence of cross-diffusion ($D_{12} = D_{21} = 0$), the solution for the eigenvalues (10.72) reduces to (10.49). In the absence of self-diffusion ($D_1 = D_2 = 0$), if $D_{12} > 0$ and $D_{21} < 0$, i.e., the sole diffusive mechanisms are chase and escape of predator and prey, the eigenvalues become pure imaginary for all modes of wavenumber $m\pi/L$. In this case the wave frequency $\omega = \lambda i$ depends on the wavenumber so that the animal density fluctuations propagate as a dispersive wave:

$$|\omega| = (a_1 a_2)^{1/2}(1 + m^2\pi^2 b_2 D_{12}/a_2 b_1 L^2)^{1/2}(1 + m^2\pi^2 b_1 D_{21}/a_1 b_2 L^2)^{1/2}.$$

As discussed in Sect. 10.3.1, self-diffusion tends to damp out all spatial variations in the Lotka–Volterra system. On the other hand, cross-diffusion may give rise to instability in the system, although this situation seems quite rare from an ecological point of view (Jorné, 1977). More detail is presented in Sect. 10.6.

Gurtin (1974) developed some mathematical models for population dynamics with the inclusion of cross-diffusion as well as self-diffusion, and he showed that the effect of cross-diffusion may give rise to the segregation of two species.

10.3.4   The Effect of Advective Flow on Predator–Prey Systems

Heterogeneity in environments influences not only parameters such as the birthrate, which controls intra- and interspecies relations, but also the parameters relevant to dispersal. An example of the effect of heterogeneity on the latter is embodied in the refuge behavior of prey as they seek to avoid predators. The effect of environmental variation can be modeled in the following way. The animals are allowed to disperse throughout the environment, but are allowed sufficient perception of environmental heterogeneity so that they tend to migrate toward more desirable regions (Comins and Blatt, 1974; Shigesada and Teramoto, 1978; Kawasaki, 1978).

Suppose the animals tend to drift toward the favorable region centered at $x = 0$. This biased dispersal can be modeled as an advective process directed uniformly toward the center of attraction, e.g., Eq. (6.18). If we combine the Lotka–Volterra diffusion model with biased dispersal, the following differential equations are obtained:

$$\frac{\partial S_1}{\partial t} = D_1 \frac{\partial^2 S_1}{\partial x^2} + u_1 \frac{\partial}{\partial x}\{\mathrm{sgn}(x)S_1\} + a_1 S_1 - b_1 S_1 S_2 \qquad (10.73)$$

$$\frac{\partial S_2}{\partial t} = D_2 \frac{\partial^2 S_2}{\partial x^2} + u_2 \frac{\partial}{\partial x}\{\mathrm{sgn}(x)S_2\} - a_2 S_2 + b_2 S_1 S_2 \qquad (10.74)$$

where u_1 and u_2 are the magnitudes of the advective flow for prey and predator, respectively, and $\mathrm{sgn}(x)$ is defined as $+1$ if $x > 0$ and -1 if $x < 0$. [Comins and Blatt (1974) formulated the advection term in a slightly different manner by taking $\mathrm{sgn}(x)$ outside the derivative operator in Eqs. (10.73) and (10.74). However, no difference results when only a semiinfinite domain, $x \geq 0$, is considered.]

The relative importance of advective flow versus diffusion is determined by the parameters θ_1 and θ_2, defined by

$$\theta_1 \equiv u_1/(a_1 D_1)^{1/2}$$

and

$$\theta_2 \equiv u_2/(a_2 D_2)^{1/2}.$$

Comins and Blatt (1974) performed a numerical analysis of (10.73) and (10.74) with appropriate initial conditions. They found that, in contrast to the Lotka–Volterra case without biased dispersal, the animal density distribution may converge to a stable spatial distribution with a decay time that

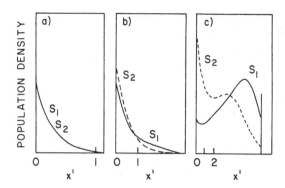

FIGURE 10.9. Stable spatial distributions of prey (S_1) and predator (S_2) for various values of the parameter θ. (a) $\theta = 4$; (b) $\theta = 2$; (c) $\theta = 1$ (from Comins and Blatt, 1974).

may be as small as a single population cycle. The result is shown in Fig. 10.9 for three cases of $\theta_1 = \theta_2 \equiv \theta$; the conditions $a_1 = a_2$ and $D_1 = D_2$ are also assumed therein.

For large values of θ, i.e., the case for which the centrally directed flow is predominant, both predator and prey populations concentrate toward the central region. As the value of θ decreases, i.e., diffusion becomes more important, the prey are partially excluded from the center and concentrate in regions of large x. The region away from the center thus tends to act as a refuge for the prey even though it is accessible to the predators as well. In other words, when both species have a weak preference for a particular spatial region, the prey tend to occupy less favorable regions than the predators.

Another case studied by Comins and Blatt (1974) is that for which only the prey has a biased dispersal; this case is perhaps more realistic than the previous case. The equilibrium distributions of predator and prey for $a_1 = a_2$, $D_1 = D_2$, $\theta_1 = 0.8$, and $\theta_2 = 0$ are shown in Fig. 10.10. In this case there is no remote region of refuge for the prey since the predators have no preference in dispersal. As the prey population concentrates toward the

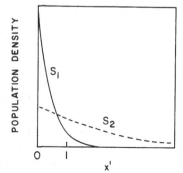

FIGURE 10.10. Equilibrium distributions of predator (S_2) and prey (S_1) when predator diffusion is unbiased (from Comins and Blatt, 1974).

attractive center, the predator population acts likewise to partake of the abundance of prey. McMurtrie (1978) suggests that motion induced by attraction to or repulsion from particular points in space as a rule tends to confer stability on predator–prey interactions.

If the ratio of the advective effects to diffusion is large, the population dynamics equations are to a first approximation represented only by advection and reaction terms. Some interesting work on an advective predator–prey system has been done by Hashimoto (1974) and Yoshikawa and Yamaguti (1974). Also, Yamaguti (1976) has discussed in some detail Volterra's competition system with migratory effects, i.e., advection.

We shall consider the following advective predator–prey equations:

$$\frac{\partial S_1}{\partial t} + u_1 \frac{\partial S_1}{\partial x} = -aS_1 S_2, \tag{10.75}$$

$$\frac{\partial S_2}{\partial t} + u_2 \frac{\partial S_2}{\partial x} = bS_1 S_2, \tag{10.76}$$

where S_1 and S_2 are, respectively, the prey and predator densities; u_1 and u_2 are, respectively, the advective velocities of prey and predator; and a and b are positive constants $(u_2 > u_1 > 0)$.

Hashimoto (1974) obtained analytically the general solution of (10.75) and (10.76):

$$S_1 = b^{-1}(u_1 - u_2)F'(\xi_1)/\{F(\xi_1) + G(\xi_2)\}, \tag{10.77}$$

$$S_2 = a^{-1}(u_1 - u_2)G'(\xi_2)/\{F(\xi_1) + G(\xi_2)\}, \tag{10.78}$$

with arbitrary functions F and G. Here $\xi_1 = x - u_1 t$, $\xi_2 = x - u_2 t$, $F' = dF/d\xi_1$, and $G' = dG/d\xi_2$.

In particular, for $F(x) = e^{-\lambda x}/c$ and $G(x) = e^{-\mu x}$ at $t = 0$, the solutions (10.77) and (10.78) become

$$S_1 = b^{-1}\lambda(u_1 - u_2)/[1 + c \exp\{-k(x - ut)\}],$$

$$S_2 = a^{-1}\mu(u_1 - u_2)/[1 + c^{-1} \exp\{k(x - ut)\}],$$

where $k > 0$ and $u > 0$, and these solutions represent the invasion of the predator from the region of negative x into the region of prey extending toward positive x. The whole pattern of predator and prey distributions moves with a uniform velocity u in the positive direction of x; the phase speed u is different from the advective velocities of predator and prey.

10.4   The Effect of Density-Dependent Dispersal on Population Dynamics

In Sects. 5.6 and 6.2 we presented models of density-dependent dispersal and inspected some features of the spatial distribution of individual animals. Herein we investigate the combined effect of density-dependent dispersal and species interactions on the distribution of biological populations.

As discussed in Chaps. 5 and 6, density-dependent dispersal can be modeled macroscopically in the advection–diffusion process by letting the advection velocity and diffusivity be functions of the population density. Thus, Gurney and Nisbet (1975, 1976) derived from microscopic considerations the following density-dependent population flux:

$$\mathbf{J} = -D\nabla S - cS\nabla S,$$

or in the one-dimensional case,

$$J_x = -D\partial S/\partial x - cS\partial S/\partial x, \tag{10.79}$$

where D and c are constants.

The density-dependent part of the dispersal arises either because individuals of a population can directly obtain information about the spatial distribution of their comrades or because the direct response of individuals to local population density can cause their average behavior to reflect gradients in population density.

Gurtin and MacCamy (1977) proposed a similar model in which the population flux is expressed as

$$\mathbf{J} = -kS^m\nabla S,$$

or in the one-dimensional case,

$$J_x = -kS^m\partial S/\partial x,$$

where $k > 0$ and $m \geq 1$.

Gurtin and MacCamy (1977) combine this flux with the population reaction term, $F(S)$, and consider diffusion–reaction problems in one dimension. Their basic equation,

$$\frac{\partial S}{\partial t} = K\frac{\partial^2 S^{m+1}}{\partial x^2} + F(S), \qquad K = k(m+1) > 0,$$

is subject to the initial condition

$$S = S_0(x) \quad \text{at} \quad t = 0.$$

It is proved therein that if the population is initially confined to a finite interval, then the entire space can never be populated in a finite time; since diffusivity D is expressed as kS^m, wherever $S = 0$, there will be no flux of population. This contrasts with the case of density-independent random dispersal, and even in this respect alone, the model of density-dependent dispersal provides a more realistic pattern of biodiffusion than that of a simple random walk.

For a birth-dominant population, i.e., $F(S) \geq 0$, the population fronts tend to reach infinity as time becomes large, so that the entire spatial domain is ultimately populated. On the other hand, for a death-dominant population, i.e., $F(S) < \mu S$ for some $\mu < 0$, the population fronts have finite limits as time goes to infinity so that only a finite domain of the entire space is ultimately populated.

To see how the population spreads into the empty spatial domain, we consider an example wherein $F(S) = \alpha S$ and $S_0(x) = N\delta(x)$, where α is constant, N is the total number of individuals at an initial instant, and $\delta(x)$ is the Dirac function. The diffusion–reaction equation becomes

$$\frac{\partial S}{\partial t} = K \frac{\partial^2 S^{m+1}}{\partial x^2} + \alpha S \qquad (10.80)$$

subject to

$$S = N\delta(x) \quad \text{at} \quad t = 0.$$

Let

$$S(t, x) = \rho(t, x)e^{\alpha t}. \qquad (10.81)$$

Substituting (10.81) into (10.80) and letting $\tau = \{e^{\alpha(m-1)t} - 1\}/\{\alpha(m - 1)\}$, we obtain

$$\frac{\partial \rho}{\partial t} = K \frac{\partial^2 \rho^{m+1}}{\partial x^2} = K(m + 1)\frac{\partial}{\partial x}\left(\rho^m \frac{\partial \rho}{\partial x}\right) \qquad (10.82)$$

subject to the condition

$$\rho = N\delta(x) \quad \text{at} \quad \tau = 0.$$

Equation (10.82) has the same form as (6.21), the solution of which is given by (6.22). We obtain the following solution for S:

$$S(t, x) = e^{\alpha t}\left(\frac{t_0}{K\tau}\right)^{1/m+2}\left\{1 - \frac{x^2}{x_1^2}\right\}^{1/m} \quad \text{if } |x| \leq x_1, \quad S(t, x) = 0 \quad \text{if } |x| > x_1$$

where

$$t_0 = \frac{N^2\Gamma^2(3/2 + m^{-1})}{2(m + 1)\pi\Gamma^2(1 + m^{-1})}$$

$$x_1 = \left(\frac{t_0}{K\tau}\right)^{1/m+2}\frac{N\Gamma(3/2 + m^{-1})}{\pi^{1/2}\Gamma(1 + m^{-1})}$$

$$\tau = \{e^{\alpha(m-1)t} - 1\}/\alpha(m - 1)$$

and Γ is the gamma function.

Figure 10.11 shows the distribution of population density as a function of position x for various values of time t. The population-wave fronts are also shown in broken lines. For $\alpha > 0$, the population front tends to infinity as time goes to infinity, and the central concentration $S(0, t)$ is infinite as $t = 0$, decreases to a minimum, and then increases to infinity monotonically with time. For $\alpha < 0$, the population front approaches a position that is a finite distance from the center as time goes to infinity, and the central concentration decreases monotonically with time.

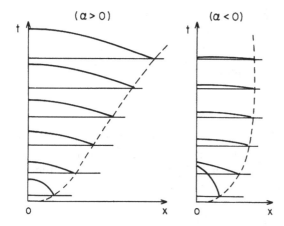

FIGURE 10.11. Distribution of population density as a function of position x for various values of time t. α: growth rate of population (from Gurtin and MacCamy, 1977).

Gurney and Nisbet (1975), using models for a density-dependent dispersal as well as for a random walk, examined the mechanism whereby a population in a spatially variable environment might regulate its numbers through dispersal. Their basic equation for the one-dimensional case may be written as

$$\frac{\partial S}{\partial t} = -\frac{\partial}{\partial x} J_x + r(x)S, \tag{10.83}$$

where J_x is given by (10.79) and r is the local net rate of increase, which depends only on position.

We shall restrict our attention to a largely hostile environment $(r(x) < 0)$ containing a single central region of viable habitat $(r(x) > 0)$; the growth function is chosen to have the form

$$r(x) = r_0\{1 - (x/x_0)^2\}, \tag{10.84}$$

where r_0 and x_0 are positive constants; r_0 denotes the maximum growth rate at the center of the viable habitat that extends from $x = -x_0$ to $x = +x_0$. The favorable region might be conceived to be an area of refuge from some predator or an area of abundant food.

Three distinct cases of dispersal are examined: purely random, random plus density-dependent dispersal, and purely density-dependent dispersal.

i) Purely Random, i.e., Density-Independent Dispersal. For this case, $J_x = -D\partial S/\partial x$, where D is a constant. Then Eq. (10.83) together with (10.84) becomes

$$\frac{\partial S}{\partial t} = D\frac{\partial^2 S}{\partial x^2} + r_0\left\{1 - \left(\frac{x}{x_0}\right)^2\right\}S, \tag{10.85}$$

which is subject to the boundary conditions that $S \to 0$ at $x = \pm\infty$. We assume

$$S = e^{\lambda t}\phi(x).$$

Substituting this into (10.85), we obtain

$$\frac{d^2\phi}{dx^2} + \{(r_0 - \lambda)/D - (r_0 D^{-1}x_0^{-2})x^2\}\phi = 0.$$

This is Schrödinger's wave equation for a harmonic oscillator (Kemble, 1937). The appropriate eigenvalues and normalized eigenfunctions are available. Thus, the solution to (10.85) is given by a discrete set of eigenfunctions $\{\phi_i(x)\}$ with a corresponding set of eigenvalues $\{\lambda_i\}$

$$S(t, x) = \sum_{i=0}^{\infty} c_i e^{\lambda_i t}\phi_i(x). \tag{10.86}$$

Here the constants c_i are determined to be the coefficients of the eigenfunction expansion of the initial distribution $S_0(x)$. The eigenvalues and eigenfunctions are given by

$$\lambda_i = r_0 - (2i + 1)(Dr_0)^{1/2}x_0^{-1},$$

$$\phi_i(x) = H_i(r_0^{1/4}D^{-1/4}x_0^{1/2}x)\exp\{-r_0^{1/2}D^{-1/2}x_0^{-1}x^2/2\},$$

where $i = 0, 1, \ldots$ and $H_i(\xi)$ are the Hermitian polynomials.

After a sufficiently long time elapse, the solution (10.86) is approximated by the first term in the series, which has the largest eigenvalue, λ_0. Thus,

$$S \sim c_0 e^{\lambda_0 t}\phi_0(x) = c_0 e^{\lambda_0 t}\exp\{-r_0^{1/2}D^{-1/2}x_0^{-1}x^2/2\},$$

and hence the population density is highest at the center of the habitat, decreasing rapidly away from the center. The population ultimately grows or decays exponentially depending on whether λ_0 is positive or negative, i.e., whether r_0 is larger or smaller than D/x_0^2. A critical size of the habitat is therefore given by $(2x_0)_{cr} = 2(D/r_0)^{1/2}$, which should be compared with L_0 in Sect. 9.1.

Gurney and Nisbet's work represents an improvement over the KISS model, as the assumption of a gradual transition from favorable to unfavorable conditions is more realistic than that of a suitable habitat surrounded by a totally hostile environment.

ii) Random Plus Density-Dependent Dispersal. For this case, $J_x = -D\partial S/\partial x - \beta S\partial S/\partial x$, where D and b are constants. Then Eq. (10.83) reads

$$\frac{\partial S}{\partial t} = D\frac{\partial^2 S}{\partial x^2} + \frac{\beta\partial}{\partial x}\left(S\frac{\partial S}{\partial x}\right) + r_0\left\{1 - \left(\frac{x}{x_0}\right)^2\right\}S. \tag{10.87}$$

Gurney and Nisbet (1975) proved that if $r_0 < D/x_0^2$, the only steady-state solution of (10.87) is the trivial solution $S(x) = 0$ everywhere, i.e., the population proceeds exponentially to extinction; likewise, if $r_0 > D/x_0^2$, there exists a nontrivial, nonnegative, steady-state solution $S(x)$, which is stable with respect to small fluctuations in population density.

iii) Purely Density-Dependent Dispersal. For this case, $J_x = -\beta S \partial S/\partial x$, where β is a constant, and (10.83) reads

$$\frac{\partial S}{\partial t} = \beta \frac{\partial}{\partial x}\left(S \frac{\partial S}{\partial x} \right) + r_0 \left\{ 1 - \left(\frac{x}{x_0} \right)^2 \right\} S.$$

This case can be regarded as a special case of ii), and there always exists a nontrivial, nonnegative, steady-state solution. This steady state is also stable under small perturbations. Namba (1980a) found an explicit solution for the steady state, $u_{ss}(x)$, in the case that the population density was nonzero in a finite interval $(-L_c, L_c)$ and vanished outside this interval. L_c is of the same order of magnitude as x_0 [see Eq. (10.84)]. Namba's $u_{ss}(x)$ is also globally stable, i.e., stable against large amplitude fluctuations. Because of the nature of purely density-dependent dispersal, the population tends to avoid non-populated areas where the environment is unfavorable, thus having less risk of extinction than a population subject to random movement as well as density-dependent dispersal.

Gurney and Nisbet further investigated by numerical analysis the transient behavior of the diffusion–reaction equation for these three cases of dispersal. A typical set of results is reproduced in Fig. 10.12, which shows the time development of the total population number, N, for two cases of r_0: $r_0 < D/x_0^2$ and $r_0 > D/x_0^2$. It can be seen that in those cases where a stable

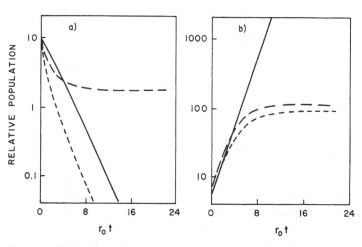

FIGURE 10.12. Time development of total population number in (a) $r_0 < D/x_0^2$, (b) $r_0 > D/x_0^2$. Three types of dispersal are considered: purely random (———), random plus density-dependent (------), and purely density-dependent (– – – –) (from Gurney and Nisbet, 1975).

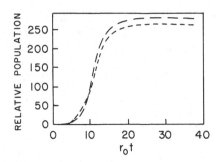

FIGURE 10.13. Population build-up as a function of normalized time (r_0t) for random plus density-dependent dispersal (------) and purely density-dependent dispersal (----) (from Gurney and Nisbet, 1975).

asymptotic state exists, it is reached within ten times of the maximum growth time, r_0^{-1}.

More detail of the build-up of a population is shown in Fig. 10.13 as a function of the relative time when the habitat size is 7.7 times the critical. It is interesting to observe that the population growth curves shown in Fig. 10.13 resemble the logistic growth curve of a population in a uniform environment of limited capacity. Gurney and Nisbet's study demonstrates that the introduction of a suitable density effect into the dispersal of population living in an inhomogeneous environment leads to a quite similar regulatory behavior.

The essential effect of introducing a density-dependent dispersal into the system is to increase the diffusion rate at high population density. If the growth function is such that a population undergoing purely random dispersal would tend to increase its number, then the density-dependent dispersal will have the effect of stabilizing the population by increasing the diffusion rate in densely populated regions. Extending the Gurney and Nisbet (1975) model, Namba (1980b) presented some detailed special cases that illustrate these general remarks as well as clarify the role played by a finite, viable region (as opposed to a hostile environment).

Investigations of multispecies dispersion in which groups disperse to avoid crowding have been pursued by Gurtin and Pipkin (1984) and Bertsch et al. (1984). Gurtin and Pipkin derive general expressions for N biological groups in which each group disperses (locally) toward lower values of the total population. In this formulation, the advection velocity, u_j, for the jth species is proportional to the local gradient of the total population, S. Namely, $u_j = -K_j$ grad S, where $S = \Sigma_j S_j$ and S_j is the population of the jth species. Gurtin and Pipkin explore their expressions for an age-structured population and then calculate a simple two-species case in which the species never disperse through one another, but remain segregated for all time. Bertsch et al. (1984), limiting their studies to two species, explore the case where one population is sedentary and the other is mobile. They find that when the mobile population is sufficiently large relative to the sedentary population, the mobile species eventually populates the entire habitat; i.e., the sedentary and mobile populations do *not* remain segregated. Bertsch et al. also investigated the case where the mobile population was not large (relative to the maximum value of the sedentary population). Under these conditions, the sedentary population acts as a barrier, and the mobile individuals cannot penetrate a sedentary colony.

Travelling waves appear as solutions to single-species reaction–diffusion equations (Sect. 10.2.1) and multispecies communities described by reaction–diffusion equations as well (Sect. 10.3.2). Newman (1980) and Hosono (1986, 1989) have also found travelling waves in single-species populations when the diffusivity is density-dependent. Hosono (1987) extended this analysis to two (competing) species and found travelling wave solutions in this system, too.

10.5   The Effect of Dispersal on Competing Populations

The usual analysis of competition between two species in spatially uniform distribution starts with the ordinary differential equations of Lotka–Volterra type

$$\frac{dS_1}{dt} = (a_1 - b_{11}S_1 - b_{12}S_2)S_1, \tag{10.88}$$

$$\frac{dS_2}{dt} = (a_2 - b_{21}S_1 - b_{22}S_2)S_2, \tag{10.89}$$

where all the parameters are nonnegative constants; a_1 and a_2 are the instrinsic growth rates of the populations of two species (S_1 and S_2), b_{11} and b_{22} are the coefficients of intraspecific competition, and b_{12} and b_{21} are those of interspecific competition.

As is well known, coexistence of these two species becomes possible only when the conditions $b_{11}/a_1 > b_{21}/a_2$ and $b_{22}/a_2 > b_{12}/a_1$ are satisfied; otherwise, only one of the species can survive and the other species is led to extinction (Maynard-Smith, 1971, 1974). We shall investigate the effect of population dispersal on the coexistence of the competing species.

Levin (1974) has considered competition between two species over two patchy habitats where interhabitat migration is allowed. He shows that when there are at least two patches in the habitat, coexistence of competing species that would otherwise exclude each other is possible, provided the rate of migration is small enough. When the migration rate is too high, rapid mixing produces effectively a single-patch situation, and hence coexistence is no longer possible (Fig. 10.14). A number of authors have expanded upon the results of Levin for different detailed boundary conditions (Harada and Fukao, 1978; Pao, 1981; Zhou and Pao, 1982) with different expressions for dispersion (Takeuchi, 1989), different forms for the interaction between species (Cosner and Lazer, 1984), and extending the analysis to more than two species (Lopez-Gomez and Pardo, 1992). As a result of these and other studies, we knew a great deal more about coexistence, extinction, and stability

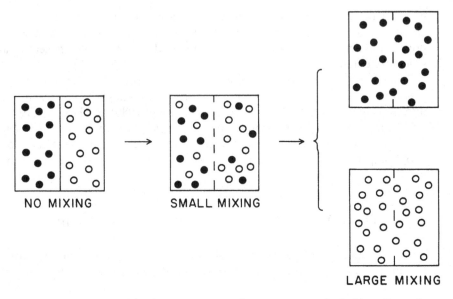

FIGURE 10.14. Competition between two species over two patchy habitats for various interhabitat migration rates ranging from no migration to small migration and to large migration (based on Levin, 1974).

in these model competition systems with dispersing organisms. Many of the results depend on the detailed relationships that the authors chose in their calculations. Nonetheless, Levin's essential insights—that dispersion can allow species that would otherwise be competitively excluded to persist with their competitors, so long as dispersion is neither too high nor too low, seems to be true. And "too high" and/or "too low" depends on the details of the interaction between species.

Levin's concept can be extended to more complex situations of dispersal. Gopalsamy (1977b) constructs a mathematical model of a spatially heterogeneous habitat in the sense that the resources for which two species compete are nonuniformly distributed over the habitat. It is assumed that the individuals of two species are capable of tracking the resources by a mechanism of biased random walk toward the most enriched part of the habitat; the attractive velocities of the individuals of two species are proportional to their distances from the center.

The basic equations for the competing species are then written as

$$\frac{\partial S_1}{\partial t} = D_1 \frac{\partial^2 S_1}{\partial x^2} - c_1 x \frac{\partial S_1}{\partial x} + (a_1 - b_{11}S_1 - b_{12}S_2)S_1, \tag{10.90}$$

$$\frac{\partial S_2}{\partial t} = D_2 \frac{\partial^2 S_2}{\partial x^2} - c_2 x \frac{\partial S_2}{\partial x} + (a_2 - b_{21}S_1 - b_{22}S_2)S_2, \tag{10.91}$$

where D_1 and D_2 are the diffusivities of two species, and c_1 and c_2 are the coefficients of attractive velocity for two species. Equations (10.90) and (10.91) are subject to appropriate initial conditions and boundary conditions at $x = 0$ and $x = L$.

Gopalsamy analyzed the model to obtain the conditions for the existence of nonnegative solutions at all t. We shall summarize only the result. Let $\hat{D}$, $\hat{a}$, and $\hat{c}$ be, respectively, the minimum value of D_1 and D_2, maximum value of a_1 and a_2, and maximum value of c_1 and c_2.

When the habitat is surrounded by a totally hostile environment, i.e., $S_1 = S_2 = 0$ at $x = 0$ and L, the global extinction of the populations is possible if

$$\hat{D} > (\hat{a} + \hat{c}/2)L^2.$$

In other words, if the dispersal rates of the populations are predominant over the species growth rates and their attraction for resources, then the loss of the populations into the hostile environment cannot be compensated and the populations head to their eventual extinction. If the spatial extent of the habitat is too small, a similar phenomenon can happen.

On the other hand, if

$$\hat{D} \leq (\hat{a} + \hat{c}/2)L^2, \tag{10.92}$$

the species can survive; (10.92) is a necessary condition for the coexistence of two competing species in a spatially heterogeneous environment. The minimum habitat size for species survival is given by $L_c = \{\hat{D}/(\hat{a} + \hat{c}/2)\}^{1/2}$, which[2] should be compared with the critical size for a single-species case.

For a habitat, the boundaries of which are closed to the populations, i.e., reflecting boundaries, Gopalsamy concludes that if

$$\hat{D} > (\hat{a} - \hat{c}/2)L^2/4\pi^2 > 0,$$

i.e., if the dispersal rate is high enough or the habitat size is small enough, the spatial distributions of two species approach spatially uniform densities. Under such circumstances, initial nonuniformities in population distributions tend to disappear asymptotically, and the spatially uniform population densities are governed by the corresponding equations of competition, (10.88) and (10.89). Thus, a necessary condition for nonuniform habitats to continue to support nonuniform population densities is given by

$$(\hat{a} - \hat{c}/2)L^2/4\pi^2 \geq \hat{D}.$$

These results are in essence consistent with that of Levin's patch model.

Shigesada et al. (1979) combine the advection–diffusion process described in Sect. 5.6 with competitive interactions of the Lotka–Volterra type to dis-

---

[2] Somehow a factor π is missing.

cuss the coexistence problem of two species. Their basic equations are written as

$$\frac{\partial S_1}{\partial t} = \frac{\partial^2}{\partial x^2}\{(\alpha_1 + \beta_{11}S_1 + \beta_{12}S_2)S_1\} + \gamma_1\frac{\partial}{\partial x}\left(\frac{\partial\phi}{\partial x}S_1\right)$$

$$+ (a_1 - b_{11}S_1 - b_{12}S_2)S_1, \tag{10.93}$$

$$\frac{\partial S_2}{\partial t} = \frac{\partial^2}{\partial x^2}\{(\alpha_2 + \beta_{21}S_1 + \beta_{22}S_2)S_2\} + \gamma_2\frac{\partial}{\partial x}\left(\frac{\partial\phi}{\partial x}S_2\right)$$

$$+ (a_2 - b_{21}S_1 - b_{22}S_2)S_2. \tag{10.94}$$

As shown in Sect. 5.6, the habitat segregation occurs between two species in the absence of competitive effects. By numerical computation of (10.93) and (10.94), Shigesada et al. (1979) demonstrated that the spatial segregation effect caused by the advection–diffusion mechanisms can stabilize the competing populations and lead to coexistence.

Figure 10.15 shows the numerical result for the spatial distributions of two species, which are established after a sufficiently long time elapses starting

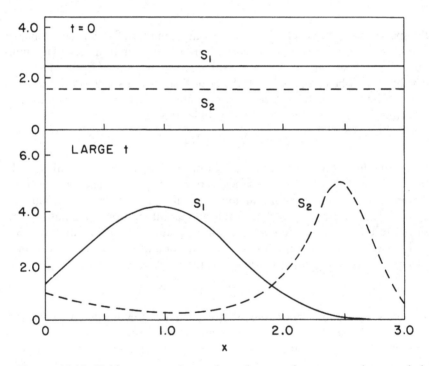

FIGURE 10.15. Habitat segregation and coexistence of two competing populations undergoing advection and density-dependent diffusion (from Shigesada et al., 1979).

from initially uniform distributions, under the conditions that $b_{11}/a_1 < b_{21}/a_2$ and $b_{22}/a_2 < b_{12}/a_1$, i.e., only one of the species can survive and the other is led to extinction in spatially uniform situations. Thus, it becomes clear that if we take into account the environmental heterogeneity and the nonlinear dispersal, the coexistence of two similar and competing species becomes possible at least under some conditions, even though the two species have the same affinity for the environment.

Pacala and Roughgarden (1982) and Shigesada and Roughgarden (1982) further explored the role of an *explicitly* heterogeneous habitat in determining the outcome of competition in two-species communities. Using the different carrying capacities to characterize a (one-dimensional) region, Pacala and Roughgarden determined when invasions will succeed if one dispersing (diffusing) species invades the region inhabited by another. One interval with the region may be "suitable," while the remainder may be "unsuitable," in terms of carrying capacities, for the invading species. The conditions for success, coexistence, exclusion, etc., depend on the details of the carrying capacities, growth rates, and diffusivities of the two interacting species, and on the size of (length) of suitable habitats. Shigesada and Roughgarden focused their attention on the mode of dispersal of the two species, one of which was considered a "generalist," while the other was a "specialist." They found that both directed movement toward favorable habitat and density-dependent random motion led to coexistence. Namba (1989) asked whether or not two species could coexist when they compete for space, and there is a (nonlinear) effect of population pressure upon dispersal. For some special cases, he was able to determine when coexistence was possible.

Since aggregation and clustering of organisms are thought to play significant roles in ecological processes like food acquisition, predator avoidance, etc. (for further discussion see Sect. 7.7.), Grindrod (1988, 1996) addressed models that are commonly invoked to display clustering in single- and multi-species assemblages. Though these models show transient aggregationlike effects, when intraspecific competition dominates interspecific competition, *no* clustering can arise in the solutions. Grindrod noted the value of (future) models that would demonstrate the effects of deterministic dispersal in producing aggregation-type behavior. Britton (1989) presented a model for aggregation of a single animal species with diffusion. Intraspecific competition at a point (e.g., carrying capacity) was chosen to depend on both population density at that point and on the *average* population density near the point. Britton argued that this was a useful formulation since animals move through space foraging for resources, perhaps comparing the suitability of one spot against others nearby. The resulting aggregation may lead to the coexistence of populations, one of which would otherwise be excluded by the other. We note the similarity to earlier reaction–diffusion studies in which nonlocal mechanisms, generally attributed to Othmer (1969), are invoked [see Murray (1993) for discussion].

10.6   Pattern-Developing Instability in Diffusion–Reaction Systems: The Development of Patchiness

In general, a diffusion process in an ecosystem tends to give rise to a uniform density of population in space. As a consequence, it may be expected that diffusion, when it occurs, plays the general role of increasing stability in a system of mixed populations and resources, as discussed in previous sections.

However, there is an important exception, known as "diffusion-induced instability" or "diffusive instability." This exception might not be a rare event especially in aquatic systems. Herein we shall explore the Turing effect (Turing, 1952) in ecosystems.

Turing (1952) proposed a diffusion–reaction theory of morphogenesis on the basis of well-known laws of physical chemistry. He demonstrated theoretically how, in an embryonic tissue with an initially homogeneous distribution of diffusible reacting substances, a regular, stable, patterned distribution of the substance may nevertheless emerge. Mechanistic explanations of these patterns were dominated by the idea of "positional information" associated with the sources and sinks of some biochemical agent (a morphogen) that was thought to influence the molecular differentiation of the cells (but see also Cocho et al., 1987a, b; Meakin, 1986). Many mechanistic explanations and Turing's earlier notions can be rationalized (Livshits et al., 1981); and we now know that morphogenetic patterns may arise spontaneously as bifurcations to the nonlinear parabolic partial differential equations that are thought to describe chemically reacting, diffusing systems (Hunding and Sorenson, 1988). These general "reaction–diffusion" approaches have been applied to a number of patterns in living organisms: *Drosophila* wings (Kauffmann et al., 1978); unicellular algae (Lacalli, 1981); and mammalian coats (Murray, 1981; Bard, 1981). Harrison (1987) presented a useful review of the experimental situation as it relates to the case for and against Turing's initial ideas; and Berding (1987) suggested a "test" measure whereby various theoretical explanations might be compared against data. In addition, Brenner et al. (1981) and Murray (1993) give information on many other applications of these ideas. General theoretical discussions can be found in Segel (1984), Edelstein-Keshet (1988), and Murray (1993); and mathematical details related to reaction–diffusion equations are available in Grindrod (1996).

The concept can be extended to any set of individuals undergoing reaction (interaction) and diffusion (Rosen, 1971). The substances involved may be considered as interacting populations, and the diffusion process may be considered as a dispersal of population. Segel and Jackson (1972) were the first to call attention to the fact that the diffusive instability can appear in an ecological context. They presented an example of a predator–prey interaction where the addition of random dispersal results in an instability of the "normal" uniform steady-state distribution to perturbations of a certain wavelength. The diffusive instability is further studied in an ecological context by Levin (1974, 1977), Levin and Segel (1976), Okubo (1978b), Mimura (1978), and Mimura et al. (1979). Levin and Segel (1985) present a general, unified review of pattern formation incorporating, for example, ideas from both ecological and morphogenetic investigations.

The mathematical basis of this diffusive instability in ecology is presented in detail by Segel and Jackson (1979), Levin (1978a), and Mimura et al. (1979); techniques for discrete-time models are discussed by Kot and Schaffer (1987). Here we briefly review the matter by considering two interacting species, S_1 and S_2, which are subject to one-dimensional diffusion:

$$\frac{\partial S_1}{\partial t} = D_1 \frac{\partial^2 S_1}{\partial x^2} + F_1(S_1, S_2), \qquad (10.95)$$

$$\frac{\partial S_2}{\partial t} = D_2 \frac{\partial^2 S_2}{\partial x^2} + F_2(S_1, S_2), \qquad (10.96)$$

where t is time, x is the one-dimensional coordinate, and D_1 and D_2 are diffusivities. The functions F_i ($i = 1, 2$) denote arbitrary ecological interaction terms.

We assume the existence of a spatially uniform steady state $S_1 = S_1^*$ and $S_2 = S_2^*$, such that

$$F_i(S_1^*, S_2^*) = 0, \qquad (10.97)$$

and furthermore assume that this state is stable with respect to spatially homogeneous perturbations, i.e., stable in the absence of diffusion.

A basically stable ecosystem that is perfectly homogeneous would continue indefinitely to be homogeneous. In practice, however, irregularities and stochastic fluctuations in population size or environmental parameters continuously introduce small local fluctuations to the spatially uniform steady state.

To examine the stability of the uniform steady state to perturbations, we write

$$S_i(x, t) = S_i^* + S_i'(x, t). \qquad (10.98)$$

It is assumed that the perturbation is infinitesimal; in other words, we examine the local stability of the system. Substituting (10.98) into (10.95) and (10.96), using (10.97), and linearizing the equations, we obtain

$$\frac{\partial S_1'}{\partial t} = D_1 \frac{\partial^2 S_1'}{\partial x^2} + a_{11} S_1' + a_{12} S_2', \qquad (10.99)$$

$$\frac{\partial S_2'}{\partial t} = D_2 \frac{\partial^2 S_2'}{\partial x^2} + a_{21} S_1' + a_{22} S_2', \qquad (10.100)$$

where a_{ij} are the values of $\partial F_i / \partial S_j$ evaluated at the uniform steady state, S_i^* [Levins (1968) calls (a_{ij}) the "community matrix"].

For an examination of linear stability, it is sufficient to assume solutions of (10.99) and (10.100) in the form

$$S_1' \sim \exp(\lambda t + ikx),$$
$$S_2' \sim \exp(\lambda t + ikx),$$

where λ and k are the frequency and wavenumber, respectively. The eigenvalue equation then reads

$$\begin{vmatrix} \lambda + D_1 k^2 - a_{11} & -a_{12} \\ -a_{21} & \lambda + D_2 k^2 - a_{22} \end{vmatrix} = 0 \qquad (10.101)$$

or solving for λ

$$\lambda = \tfrac{1}{2}(\hat{a}_{11} + \hat{a}_{22}) \pm \tfrac{1}{2}[(\hat{a}_{11} + \hat{a}_{22})^2 - 4(\hat{a}_{11}\hat{a}_{22} - a_{12}a_{21})]^{1/2},$$

where

$$\hat{a}_{11} \equiv a_{11} - D_1 k^2,$$
$$\hat{a}_{22} \equiv a_{22} - D_2 k^2.$$

The condition $k = 0$ corresponds to the neglect of diffusion and, by definition, perturbations of zero wavenumber are stable when diffusive instability sets in. It is thus required that

$$a_{11} + a_{22} < 0, \qquad (10.102)$$
$$a_{11} + a_{22} - a_{12}a_{21} > 0. \qquad (10.103)$$

If $D_1 = D_2 = D > 0$, i.e., the two species diffuse with the same diffusivity, it can be seen that if a homogeneous steady state is stable, the perturbed nonhomogenous state is also stable. To this end, let $\lambda + Dk^2 = v$; (10.101) then becomes

$$\begin{vmatrix} v - a_{11} & -a_{12} \\ -a_{21} & v - a_{22} \end{vmatrix} = 0.$$

From (10.102) and (10.103), the two roots for v have negative real parts, so the same must be true for λ. Hence, the perturbed state is stable for all values of k. This also means that for diffusive instability to occur, *the diffusivity must not be the same for both species*.

Diffusive instability sets in when at least one of the following conditions is violated subject to the conditions (10.102) and (10.103):

$$\hat{a}_{11} + \hat{a}_{22} < 0, \qquad (10.104)$$
$$\hat{a}_{11}\hat{a}_{22} - a_{12}a_{21} > 0. \qquad (10.105)$$

(Here we exclude neutral instability.) It is seen that the first condition (10.104) is not violated when the requirement (10.102) is met. Hence, only violation of the second condition (10.105) gives rise to diffusive instability. Reversal of the inequality (10.105) yields

$$H(k^2) \equiv D_1 D_2 k^4 - (D_1 a_{22} + D_2 a_{11})k^2 + a_{11}a_{22} - a_{12}a_{21} < 0.$$

The minimum of $H(k^2)$ occurs at $k^2 = k_m^2$, where

$$k_m^2 = (a_{22}D_1 + a_{11}D_2)/2D_1 D_2 > 0. \tag{10.106}$$

Thus, a sufficient condition for instability is that $H(k_m^2)$ be negative. Therefore,

$$(a_{11}a_{22} - a_{12}a_{21}) - (a_{22}D_1 + a_{11}D_2)^2/4D_1 D_2 < 0. \tag{10.107}$$

Combination of (10.102), (10.103), (10.106), and (10.107) leads to the following final criterion for diffusive instability (Segel and Jackson, 1972):

$$a_{11}D_2 + a_{22}D_1 > 2(a_{11}a_{22} - a_{12}a_{21})^{1/2}(D_1 D_2)^{1/2} > 0. \tag{10.108}$$

When inequality (10.108) is barely satisfied, diffusive instability is incipient. The critical conditions for the occurrence of the instability are obtained when the first inequality of (10.108) is an equality. The critical wavenumber k_c of the first perturbations to grow is found by evaluating k_m from (10.106).

We note from the required conditions (10.104), (10.105), and (10.107) that $a_{11}a_{22} < 0$ and $a_{12}a_{21} < 0$. A diffusive instability can be immediately excluded in the absence of the opposition of signs between a_{11} and a_{22} and between a_{12} and a_{21}. That is, diffusive instability may occur only in a system of predator–prey type that exhibits certain density effects on intraspecific coefficients. Also, condition (10.108) requires that the diffusivity for the species with negative a_{ii}, i.e., the "stabilizer," must be larger than that with positive a_{ii}, i.e., the "destabilizer." The same condition also implies that the ratio of the two diffusivities has a critical value beyond which diffusive instability sets in.

The concept of diffusion-driven instability in ecosystems may be verbally understood as follows. Again in two-species ecology, let us consider the case for which $a_{11} > 0$, $a_{12} < 0$, $a_{21} > 0$, $a_{22} < 0$, and a_{ij} satisfy conditions (10.102) and (10.103). In other words, species 1 is the destabilizer and species 2 is the stabilizer: S_2 decays faster than S_1 grows, while an increase in S_1 causes a growth in S_2 (since $a_{21} > 0$), which in turn causes decay of S_1 (since $a_{12} < 0$), and the relative stabilizing tendency $(a_{21}/|a_{22}|)$ is larger than the relative destabilizing tendency $(a_{11}/|a_{12}|)$.

To this basically stable system, we now introduce spatially inhomogeneous perturbations in population densities and also allow for population dispersal. If the two species disperse at the same rate, the fluctuations tend to be smoothed out by diffusion without disturbing the balance between the interacting species. The stability of the system is preserved.

On the other hand, if the stabilizing species (e.g., a predator species) diffuses faster than the destabilizing species (e.g., a prey species), the balance may be upset in the presence of fluctuations (Fig. 10.16). Thus, in a relatively populated region, i.e., a region where population fluctuation is positive, the prey number starts to increase in the presence of a relative lack of predation,

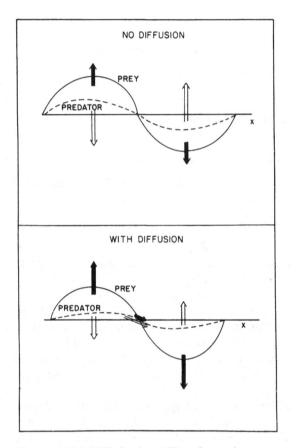

FIGURE 10.16. Diffusive instability of a predator–prey system.

while in a relatively less populated region, i.e., a region of negative fluctuation, the prey number begins to decrease, accompanied by an increase in predation. This situation is followed by a feedback process. That is, the growth of the prey population favors the growth of the predator population in the former region, whereas diminishing prey population leads to a reduction in the number of predators in the latter region.

We note that this situation is precisely that which the phrase "local (short-range) activation—long-range inhibition" describes. This phrase is commonly used to describe the most popular "morphogenesis via reaction–diffusion" mechanism. See, e.g., Gierer and Meinhardt (1972), Levin and Segel (1985), and Murray (1993). Merely substitute "stabilizer" for "inhibitor" and "destabilizer" for "activator." Finally, we associate "long-range" with high diffusivity because effects are felt more quickly at great distances in this case.

Of course, the homogenizing effect of diffusion is always inherent, so that perturbations of relatively short spatial period (wavelength) may be effaced by diffusion at a faster rate than that at which the diffusive instability operates. Hence, these perturbations never grow in time. For perturbations of relatively long spatial period, on the other hand, so much time is required for the predator species to disperse from one part of the perturbation to the other that a net destabilizing effect on the prey is not produced. [The perturbation theorem (Levin, 1974) guarantees that small amounts of migration or diffusion do not destabilize an equilibrium that is stable without migrations.] Thus, only fluctuations in a certain wavenumber range may exhibit the diffusion-driven instability.

We shall now examine some specific predator–prey models and inquire as to whether diffusion instability is possible.

1) Lotka–Volterra Model (S_1: prey, S_2: predator)

$$F_1 = \alpha_1 S_1 - \beta_1 S_1 S_2,$$

$$F_2 = -\alpha_2 S_2 + \beta_2 S_1 S_2,$$

where α_1, α_2, β_1, and β_2 are positive constants. For this model, the elements of the community matrix are found to be $a_{11} = 0$, $a_{12} < 0$, $a_{21} > 0$ and $a_{22} = 0$. Thus, conditions (10.102) and (10.108) are not met, and hence no diffusive instability occurs.

2) Phytoplankton–Herbivore System (Levin and Segel, 1976)

Let S_1 and S_2 be the densities of phytoplankton and herbivore, respectively. The following reaction terms are considered:

$$F_1 = aS_1 + eS_1^2 + b_1 S_1 S_2,$$

$$F_2 = -cS_2^2 + b_2 S_1 S_2,$$

where a, b_1, b_2, c, and e are all positive constants. Two new concepts are incorporated in this predator–prey model; one is an autocatalytic effect on the phytoplankton's growth rate and the other is the density-dependent mortality of herbivore. For this model, the elements of the community matrix are given by

$$a_{11} = eS_1^* > 0,$$

$$a_{12} = b_1 S_1^* < 0,$$

$$a_{21} = b_2 S_2^* > 0,$$

$$a_{22} = -cS_2^* < 0,$$

where S_1^* and S_2^* are equilibrium densities such that

$$S_1^* = ac/(b_1 b_2 - ce) > 0,$$

$$S_2^* = ab_2/(b_1 b_2 - ce) > 0.$$

The system is seen to have a diffusive instability if $b_1 b_2 > ce$, $b_2 > e$, and furthermore

$$D_2/D_1 > \{(b_1/c)^{1/2} - (b_1/c - e/b_2)^{1/2}\}^{-2} \equiv \theta_c^2.$$

Segel and Levin (1976) have used nonlinear stability analysis (see Sect. 10.7) to show that for D_2/D_1 slightly greater than θ_c^2, the uniform state is replaced by a new steady state in which phytoplankton and zooplankton are more concentrated in certain regions, giving patchiness.

3) Predator–Prey Systems with Functional Response (Levin, 1977)

Let S_1 and S_2 be the densities of prey and predator populations, respectively. The following reaction terms are considered:

$$F_1 = aS_1 - f(S_1)S_2,$$
$$F_2 = -g(S_2)S_2 + bf(S_1)S_2,$$

where a and b are positive constants, $f(S_1)$ is the consumption rate of prey by an individual predator (the "functional response"; Holling, 1965, 1966), and $g(S_2)$ represents the density-dependent death rate of the predator. For this model, the elements of the community matrix are given by

$$a_{11} = \{f(S_1^*)/S_1^* - f'(S_1^*)\}S_2^*,$$
$$a_{12} = -f(S_1^*) < 0,$$
$$a_{21} = bf'(S_1^*)S_2^* > 0,$$
$$a_{22} = -g'(S_2^*)S_2^*,$$

where S_1^* and S_2^* are equilibrium densities, $f' = df/dS_1 > 0$, and $g' = dg/dS_2$.

The functional response often possesses the property that $f(S_1^*)/S_1^* > f'(S_1^*)$, as Qaten and Murdock (1975) have noted. If this is the case, diffusive instability requires self-damping in the predator, $g'(S^*) > 0$, and a sufficiently high rate of dispersal of the predator.

4) Nutrient-Phytoplankton System (Okubo, 1974, 1978b)

Let S_1 and S_2 be the nutrient concentration and phytoplankton density, respectively. The following model for the reaction terms is assumed:

$$F_1 = Q - \alpha(S_1)S_2,$$
$$F_2 = -Z + \beta(S_1)S_2,$$

where Q and Z are, respectively, constant rates of nutrient supply and grazing, and α and β are arbitrary functions of S_1 only.

This provides a model of a well-mixed upper layer of the ocean, where nutrient is supplied at a constant rate from the lower layer. Grazing generally depends on both phytoplankton density and herbivore density; for simplicity, we regard the grazing term as a constant sink. This implies that a great number of phytoplankton exist and that a constant population of herbivores graze on them. A more rigorous treatment would include the herbivores in the system (three-species system).

The elements of the community matrix are given by

$$a_{11} = -\alpha'(S_1^*)S_2^* < 0,$$

$$a_{12} = -Q/S_2^* < 0,$$

$$a_{21} = \beta'(S_1^*)S_2^* > 0,$$

$$a_{22} = Z/S_2^* > 0,$$

where S_1^* and S_2^* are equilibrium densities, $\alpha' = d\alpha/dS_1 > 0$, and $\beta' = d\beta/dS_1 > 0$.

It turns out that a diffusive instability is possible if the following conditions are satisfied:

(i) The ratio of the nutrient supply per unit concentration of nutrient and the grazing loss per unit density of phytoplankton must be greater than the ratio of the mean slope of the uptake rate, α, to the gradient of the uptake coefficient at equilibrium (i.e., the functional response).

(ii) The latter ratio must also be greater than the ratio of the mean slope of the growth rate, β, to the gradient of the growth coefficient at equilibrium (the functional response).

(iii) The ratio of diffusivities of nutrient and of phytoplankton must be greater than a certain critical value that itself exceeds unity.

The critical wavenumber k_c of the first perturbations to grow is found to be

$$k_c = \{Z/2D_2S_2^* - \alpha'(S_1^*)S_2^*/2D_1\}^{1/2}.$$

Measurements of the horizontal diffusivity of planktonic organisms are quite scanty (Talbot, 1974). We might intuitively assume that the horizontal diffusivities of soluble nutrient and of phytoplankton should be the same. However, the shear effect on horizontal dispersion (see Sect. 2.6.2) suggests that this is not necessarily true. If phytoplankton concentrate in the vicinity of the water surface, while nutrients are distributed more or less uniformly in the upper mixed layer, the shear effect on the plankton dispersion may be smaller by an order of magnitude than that of the soluble substance (Okubo, 1968). In this connection a model that takes into account the vertical distributions and shear in currents is highly desirable (Steele, 1976a; Evans, 1978b; Kullenberg, 1978; Evans et al., 1977).

5) A Generalized Predator–Prey Model

Mimura et al. (1979), Mimura (1978), and Mimura and Murray (1979) consider the following generalized predator–prey model with dispersal to study

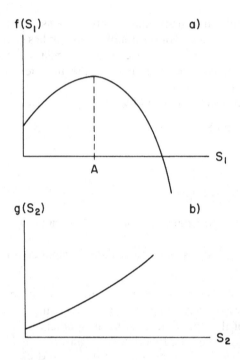

f(S₁)   a)

A

S₁

g(S₂)   b)

S₂

FIGURE 10.17. Mimura's (1978) predator–prey model. (a) Prey's per-capita
growth rate which exhibits the Allee
effect, being maximum at A. (b) Predator's mortality rate.

in detail the development of patchiness:

$$\frac{\partial S_1}{\partial t} = D_1 \frac{\partial^2 S_1}{\partial x^2} + \{f(S_1) - bS_2\}S_1, \qquad (10.109)$$

$$\frac{\partial S_2}{\partial t} = D_2 \frac{\partial^2 S_2}{\partial x^2} + \{-g(S_2) - bS_1\}S_2. \qquad (10.110)$$

Here the prey's per-capita growth rate, $f(S_1)$, exhibits the Allee effect, being maximum at $S_1 = A$ (Fig. 10.17a), and predator's mortality increases with the population density (Fig. 10.17b).

For this model the elements of the community matrix are given by $a_{11} = f'(S_1^*)S_1^*$, $a_{12} = -(bS_1^*) < 0$, $a_{21} = bS_2^* > 0$ and $a_{22} = -g'(S_2^*)S_2^* < 0$. For diffusive instability to occur, it is first of all required that $f'(S_1^*) > 0$. In other words, the equilibrium density of prey must be smaller than A. Further requirements are $f(S_1^*)g'(S_2^*) > f'(S_1^*)g(S_2^*)$, $b^2 > f'(S_1^*)g'(S_2^*)$, and $D_2 > D_1$ in such a way that inequality (10.108) is satisfied. A nonlinear analysis of this model will be presented in Sect. 10.7.

These special examples represent only a small portion of the many conceivable models. Although conditions for the occurrence of diffusive instabilities in ecological problems seem rather stringent, these examples provide impetus for more conclusive laboratory and field work.

The case of three or more species can be treated in a similar way although an increasing degree of complexity in mathematical analysis is certainly implied (Gmitro and Scriven, 1966; Othmer and Scriven, 1971; Shukla et al., 1981; Shukla and Shukla, 1982). The addition of each new species should produce a richer range of possibilities for diffusive instability. For example, although no diffusive instability is possible for a two-species competition model (Levin, 1974), Evans (1980) found that a Turing-type instability can occur for three competing species, and Powell and Richerson (1985) showed that a popular two-species, two-nutrient resource competition model (mathematically equivalent to four species—see, e.g., Tilman, 1982) can have a diffusive instability.

The concept of diffusive instability has been extended to include the effects of advection (Jorné, 1974; Shukla and Verma, 1981) and of cross-diffusion (Jorné, 1975, 1977; Rosen, 1977; Shukla and Verma, 1981; Almirantis and Papageorgiou, 1991). Thus, Jorné (1974) showed that the inclusion of species migration (advection) as an additional transport process increases the possibility of diffusive instability and of oscillatory behavior in the perturbation. The inclusion of negative cross-diffusivities can result in the appearance of a stable wavelike pattern in a homogeneous Lotka–Volterra system.

The linear stability analysis presented in this section is useful for studying the onset of diffusive instability and for gaining insight into the matter as to what kinds of instability occur in a specific diffusion–reaction system. However, only a nonlinear analysis can address the question as to what spatial pattern ultimately emerges from such an instability. An outline of (weakly) nonlinear methods is given in Sect. 10.7 (see, e.g., Wollkind et al., 1994).

10.7 Nonlinear Stability Analysis for Diffusion– Reaction Systems

The treatment described in Sect. 10.6 is essentially based on linear stability theory, i.e., the discussion of the fate of infinitesimally small perturbations about a reference state. Thus, it does not provide us with more practical answers concerning the effect of perturbations of finite amplitude.

Although linear analysis may be considered to be the first step in any stability problems, we often need to investigate further what happens to the unstable perturbations when they have grown so large that linearization can no longer be considered valid. Thus, in studying the nonlinear problem, we may find that the perturbation amplitude, instead of growing without limits, tends to a finite value asymptotically.

It was probably Steele (1974b) who first recognized the importance of nonlinear stability analysis in ecological diffusion–reaction systems, although he did not actually pursue the analysis. Steele's model is the Lotka–Volterra system for predator–prey species with the inclusion of diffusional effects such as (10.40) and (10.41). We now assume that at some time there are random perturbations within the spatial limits $(0, L)$ of x, and express the population densities in terms of Fourier series

$$S_1(t, x) = \sum_{m=0}^{\infty} A_m(t) \cos(2\pi mx/L),$$ (10.111)

$$S_2(t, x) = \sum_{m=0}^{\infty} B_m(t) \cos(2\pi mx/L),$$ (10.112)

where $A_m(t)$ and $B_m(t)$ are, respectively, the Fourier coefficients for S_1 and S_2 and are functions of time only. Note that the cosine expansion implies that the nonflux boundary conditions are satisfied at $x = 0$ and L (see Sect. 10.3.1). By substituting (10.111) and (10.112) into (10.40) and (10.41), the following differential equations for $A_m(t)$ and $B_m(t)$ are obtained:

$$dA_m/dt = (a_1 - 4\pi^2 m^2 D_1/L^2)A_m - b_1 C_m$$ (10.113)

$$dB_m/dt = (-a_2 - 4\pi^2 m^2 D_2/L^2)B_m + b_2 C_m$$ (10.114)

where

$$C_m = \frac{1}{2}\left\{ \sum_{k+\ell=m} A_k B_\ell + \sum_{|k-\ell|=m} A_k B_\ell \right\}.$$

The terms C_m arise from the nonlinear predator–prey interaction, and for a fixed value of m, an infinite number of combinations of A_k and B_L contribute to the term. In other words, the nonlinear interaction consists of the exchange of "energy" between different Fourier modes; thus, even if initial population patterns consists only of a few Fourier modes, the nonlinear process tends to create new modes of smaller and larger wavelengths as time progresses. Obviously, without C_m, (10.113) and (10.114) rule out the creation of new Fourier modes.

From his analysis, Steele conjectured that diffusion effects may not be able to damp out spatial population fluctuations if the nonlinear terms are included. Unfortunately, however, Steele chose an inappropriate model for his discussion. As presented in Sect. 10.3.1, the Lotka–Volterra system with diffusion bounded by the nonflux condition cannot admit temporally periodic, spatially nonuniform solutions. Nevertheless, Steele's concept remains worthy of attention.

Bhargava and Saxena (1977) have recently revisited the Lotka–Volterra diffusion model in an infinite spatial domain. They found periodic solutions that oscillate about the constant equilibrium values and that are stable with respect to the nonlinear perturbations for small values of diffusivities. However, Gopalsamy and Aggarwala (1981) later showed that temporally fluctuating periodic solutions to the type of equations considered by Bhargava and Saxena (1977) cannot exist in one-dimensional space in the limit as the length of the space expands to infinity. In fact, Gopalsamy and Aggarwala proved that the constant (uniform) equilibrium solution is globally asymptotically stable.

As was seen in Sect. 10.6, the Lotka–Volterra model does not permit the occurrence of diffusion instability. Thus, if we are interested in studying the effects of nonlinearity on the possible destabilizing mechanism of diffusion, we must choose a model other than the Lotka–Volterra.

Segel and Levin (1976) developed a nonlinear analysis for the following diffusion–reaction system (Levin and Segel, 1976):

$$\frac{\partial S_1}{\partial t} = D_1 \frac{\partial^2 S_1}{\partial x^2} + a_1 S_1 + c_1 S_1^2 - b_1 S_1 S_2 \tag{10.115}$$

$$\frac{\partial S_2}{\partial t} = D_2 \frac{\partial^2 S_2}{\partial x^2} - c_2 S_2^2 + b_2 S_1 S_2, \tag{10.116}$$

where a_1, b_1, b_2, c_1, and c_2 are positive constants.

This system admits a spatially uniform stable equilibrium, S_1^* and S_2^*, in the absence of diffusion,

$$S_1^* = a_1 c_2 / (b_1 b_2 - c_1 c_2),$$

$$S_1^* = a_1 b_2 / (b_1 b_2 - c_1 c_2),$$

provided $b_1 b_2 > c_1 c_2$ and $b_2 > c_1$ (see Sect. 10.6).

To examine the stability of the equilibrium state in the presence of diffusion, let $S_1 = S_1^* + S_1'$ and $S_2 = S_2^* + S_2'$. Substituting these values into (10.115) and (10.116) and introducing dimensionless variables by setting $t = \tau/a_1$, $x = \eta (D_1/a_1)^{-1/2}$, $D_2/D_1 = \theta^2$, $c_1/b_2 = a^2 - b^2$, $b_1/c_2 = a^2$, $S_1^* = a_1/(b_2 b)$, $S_2^* = a_1/(c_2 b)$, $S_1' = \gamma a_1/b_2$, and $S_2' = \sigma a_1/c_2$, we obtain

$$\frac{\partial \gamma}{\partial \tau} = \frac{\partial^2 \gamma}{\partial \eta^2} + \left(\frac{a^2}{b^2} - 1 \right) \gamma - \frac{a^2}{b^2} \sigma + (a^2 - b^2) \gamma^2 - a^2 \gamma \sigma, \tag{10.117}$$

$$\frac{\partial \sigma}{\partial \tau} = \theta^2 \frac{\partial^2 \sigma}{\partial \eta^2} + \frac{1}{b^2} \gamma - \frac{1}{b^2} \sigma - \sigma^2 + \gamma \sigma. \tag{10.118}$$

The equilibrium is stable with respect to perturbations of the form

$$\gamma, \sigma \sim \exp(\lambda \tau + i k \eta)$$

if

$$\theta^2 k^4 + b^{-1} \{ 1 - (a^2 - b^2)\theta^2 \} k^2 + b^{-1} > 0, \tag{10.119}$$

where

$$0 < a^2 - b^2 < 1.$$

The region expressed by (10.119) is bounded by the curve

$$\theta^2 k^4 + b^{-1} \{ 1 - (a^2 - b^2)\theta \} k^2 + b^{-1} > 0.$$

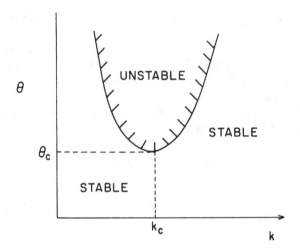

FIGURE 10.18. Stability diagram for diffusive instability. θ: ratio of diffusivities, k: wavenumber, θ_c and k_c: critical values (from Segel and Levin, 1976).

The **minimum** value of θ associated with this curve occurs at

$$\theta = \theta_c = (a - b)^{-1} \quad \text{and} \quad k^2 = k_c^2 = (a/b) - 1 \quad \text{(Fig. 10.18)}.$$

Thus, θ_c^2 is the critical diffusivity ratio; the steady-state solution is stable to small perturbations when $\theta < \theta_c$ but unstable when $\theta > \theta_c$. If θ is increased slowly beyond θ_c, instability is expected to occur for a perturbation having the *critical wavenumber*, k_c. Given an **unstable** situation for which $\theta \approx \theta_c$, we now seek an approximate solution to the nonlinear perturbation equations, (10.117) and (10.118).

Let λ_1 and λ_2 be the two eigenvalues ($\lambda_1 \geq \lambda_2$) for the linearized version of (10.117) and (10.118). When $\theta \approx \theta_c$, and $k = k_c$, we have

$$\lambda_1 \approx 2(a - b)^2 (\theta - \theta_c)/b\{1 - (a - b)^2\},$$
$$\lambda_2 \approx -k_c^2(1 + \theta_c^2) - \{1 - (a^2 - b^2)\}/b^2.$$

We shall limit our discussion to the nonlinear development of the "most dangerous sinusoidal mode" of the critical wavenumber, which is associated with the eigenvalue λ_1. Thus, the initial conditions are assigned such that at $\tau = 0$,

$$\gamma = \varepsilon A \cos k_c \eta,$$
$$\sigma = \varepsilon B \cos k_c \eta,$$

(10.120)

where ε is a small parameter of the order of $(\theta - \theta_c)^{1/2}$, and $A = a\theta_c$, $B \approx 1$.

To find solutions that are of the order of ε *uniformly in time*, Segel and Levin utilized a successive approximation procedure (γ_1, σ_1), (γ_2, σ_2),

$(\gamma_2, \sigma_3), \ldots$ in such a way that

$$\frac{\partial \gamma_j}{\partial \tau} - \frac{\partial^2 \gamma_j}{\partial \eta^2} - \left(\frac{a^2 - b^2}{b^2}\right)\gamma_j + \frac{a^2}{b^2}\sigma_j = (a^2 - b^2)\gamma_{j-1}^2 - a^2\gamma_{j-1}\sigma_{j-1} + 0(\varepsilon^{j+1})$$

(10.121)

$$\frac{\partial \sigma_j}{\partial \tau} - \theta^2 \frac{\partial^2 \sigma_j}{\partial \eta^2} - \frac{1}{b^2}\gamma_j + \frac{1}{b^2}\sigma_j = -\sigma_{j-1}^2 + \gamma_{j-1}\sigma_{j-1} + 0(\varepsilon^{j+1}),$$

(10.122)

with $j = 1, 2, \ldots$.

As is well known, this successive approximation scheme tends to produce terms like $\exp \lambda\tau$, $\exp 2\lambda\tau$, etc., for some positive λ, so that the higher approximations are valid over no more than the same limited period as the first-order approximation. To avoid this difficulty, Segel and Levin (1976) adopt the recently popular strategy of multiple scales (two-timing); this involves seeking solutions to (10.120), (10.121), and (10.122) that are functions of the time variables $\tau_0 = \tau$, $\tau_1 = \varepsilon\tau$, $\tau_2 = \varepsilon^2\tau$, etc., such that the approximations (γ_j, σ_j) are $0(\varepsilon)$ uniformly in time.

Thus, the third-order approximations are found to be

$$\gamma_3 = \varepsilon A\phi \cos k_c\eta + 2/9\{\varepsilon^2\phi^2 a^2 b^3 k_c^2\theta_c^2/(a-b)\} \cos 2k_c\eta + 0(\varepsilon^3),$$

$$\sigma_3 = \varepsilon B\phi \cos k_c\eta + \varepsilon^2\phi^2\{b^3\theta_c/2 + 4a^2 k_c^2\theta_c^2(1 + 4k_c^2)\} \cos 2k_c\eta + 0(\varepsilon^3),$$

where ϕ satisfies the following equation

$$d\phi/d\tau_2 = p^2\{(\theta - \theta_c)/|\theta - \theta_c|\}\phi - q^2\phi^3.$$

Here p^2 and q^2 are of the order of unity.

From the successive approximations, Segel and Levin (1976) draw the following important conclusions as to the amplitude $\varepsilon A\phi$ of the Fourier mode of the critical wavelength $2\pi/k_c$, when θ is slightly larger than the critical value θ_c.

1) ϕ is constant with respect to the fast time scale τ_0, and also with respect to the intermediate time scale τ_1.
2) Linear theory predicts that θ grows exponentially as $\exp(p^2\varepsilon^2\tau)$ with $p^2 = 0(1)$, but the nonlinear theory shows that on the slow time scale τ_2, which is of the order of ε^{-2}, ϕ actually approaches the equilibrium value p/q, which is $0(1)$.

In other words, random dispersal can destabilize a predator–prey interaction in the presence of an autocatalytic effect in the prey population. Thus, if the predator dispersal rate becomes sufficiently high compared with that of prey, the uniform steady state becomes unstable. Perturbations initially grow exponentially with time, but sooner or later nonlinear effects come into play and give rise to a new steady but spatially inhomogenous species distribution where the prey particularly tends to be clumped in a "patchy" fashion. Because of the conditions on diffusivities that the prey must diffuse more slowly than the predator, concentration peaks are sharper for the prey than for the predator. (Such behavior might explain the patchiness of oceanic plankton.) It is also shown (Segel and Levin, 1976) that although the average prey density is unaltered by the nonlinear effect, the average predator density is elevated by it.

Mimura and his colleagues (Mimura, 1978; Mimura et al., 1979; Mimura and Murray, 1979) studied the initial-value problem for the model equations (10.109) and (10.110) from a standpoint of nonlinear stability analysis, i.e., the problem of whether or not the system with reflecting boundary conditions exhibits asymptotically stable patchiness when an arbitrary heterogeneous distribution is given initially.

The main difference between Mimura's model and Segel and Levin's model is that the former incorporates the Allee effect in the prey per-capita growth rate, while the latter incorporates prey autocatalysis. In the neighborhood of the homogeneous steady state that leads to diffusive instability, both systems have a similar property of local instability. On the other hand, if we are interested in large amplitude perturbation solutions, the two models may exhibit a different asymptotic behavior because of the different nonlinearities at large prey population densities.

According to the general proof of Conway et al. (1978), the solution of (10.109) and (10.110) tends to be homogeneous at large times if both D_1 and D_2 are sufficiently large. Consequently, as a sufficient condition for patchiness to occur, we require that D_1 must be sufficiently small (note that $D_2 > D_1$ for the occurrence of diffusive instability). When D_1 is zero or sufficiently small, Mimura et al. (1979) demonstrate the existence of large-amplitude steady-state solutions of (10.109) and (10.110) subject to the above-mentioned boundary conditions. Also, these solutions exhibit remarkable spatial heterogeneity.

As specific numerical examples, we take $f(S_1) = (35 + 16S_1 - S_1^2)/9$, $g(S_2) = 1 + 2S_2/5$, $b = 1$, $L = 5$, $D_2 = 4$, and two values of $D_1 = 1/20$ and $1/2000$ (Fig. 10.19). Both cases, starting with small population densities at $t = 0$, give rise to asymptotically stable patterns of spatial heterogeneity in the population densities. For $D_1 = 1/20$, prey and predator coexist in the spatially inhomogeneous pattern. For $D_1 = 1/2000$, on the other hand, the whole space is composed of two different regions: one in which prey and predator coexist and the other in which prey is completely depleted and only predator populations are present. Mimura et al. (1979) extended the numerical computation to a two-dimensional space. The result shows that the stable population distributions at large times exhibit strikingly complex patterns of spatial heterogeneity.

Fife (1976) described a mechanism by which stationary spatial forms and structures can emerge from a mixture of reacting and diffusing substances when the diffusivity of one of the reactants is much smaller than the other. Fife discussed large-amplitude pattern formation in this system. The development of the structured stationary state from an arbitrary initial distribution occurs

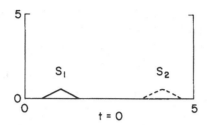

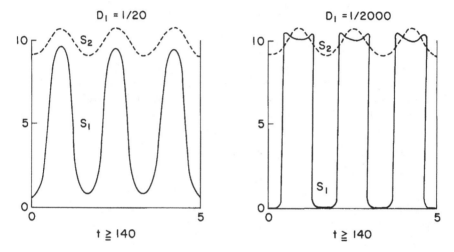

FIGURE 10.19. Development of spatial structures (patchiness) for a predator–prey system. S_1: prey, S_2: predator.

in two stages. The first involves the formation of sharply differentiated sub-regions, bounded by layers within which one of the substance distributions has a large spatial gradient, and the second stage involves the migration of the boundaries of the subregion into a stable final configuration that is relatively independent of the initial distribution of substances.

Bhargava et al. (1984) discussed predator–prey diffusive models using techniques to look at nonlinear stability. They considered the model

$$\partial S_1/\partial t = D_1 \partial^2 S_1/\partial x^2 + S_1\{1 - c_1 S_1 - b_1 S_1/(1 + S_2)\},$$

$$\partial S_2/\partial t = D_2 \partial^2 S_2/\partial x^2 + a_2 S_2(1 - S_2/S_1),$$

or

$$\partial S_1/\partial t = D_1 \partial^2 S_1/\partial x^2 + a_1 S_1(1 - S_1/K) - b_1 S_2/(1 - e_1^S),$$

$$\partial S_2/\partial t = D_2 \partial^2 S_2/\partial x^2 + S_2\{-a_2 + b_2(1 - e_1^S)\}.$$

Using standard methods of asymptotic expansions, they constructed non-constant solutions in terms of the small parameter $\varepsilon = |\mathrm{Re}\, \sigma_{1-}|^{1/2} \ll 1$, where σ_{1-} is the dominant eigenvalue of the linearization of the full problem.

A natural way to obtain approximations to periodic patterns in reaction–diffusion equations is to use a bifurcation approach, which essentially is analogous to the work of Segel and Levin. Rothe (1979) presented a very simple diffusion–reaction model (the only nonlinearity appears in one term)

$$\partial S_1/\partial t = D_1 \Delta S_1 + f(S_1) - S_1,$$

$$\varepsilon \partial S_2/\partial t = D_2 \Delta S_2 + S_1 - S_2.$$

Later Rothe (1981) and Rothe and deMottooi (1979) used the same system with

$$f(S_1) = aS_1 - g(S_1),$$

where $g(S_1)$ is nonlinear and $a = f'(0)$ is a bifurcation parameter, and investigated spatially inhomogeneous equilibrium solutions, i.e., stationary patterns, and showed the existence of infinitely many global solution branches.

Yan (1993) looked at periodic patterns in reaction–diffusion equations in the coupled system,

$$\partial S_1/\partial t = D_1 \partial^2 S_1/\partial x^2 - f_1(S_1, S_2, k),$$

$$\partial S_2/\partial t = D_2 \partial^2 S_2/\partial x^2 - f_2(S_1, S_2, k),$$

in the finite region $0 \leq x \leq 2L$, with zero-flux boundary conditions

$$\partial S_1/\partial t = \partial S_2/\partial t = 0 \quad \text{at} \quad x = 0, 2L.$$

One can use the diffusion coefficients as bifurcation parameters to determine both the existence and stability of the periodic steady state. This method can be applied to a variety of functional forms, including models of insect embryogensis and the Fitzhugh–Nagumo system. The approximation reveals some characteristic features of the solutions and explains the essential pattern of the periodic steady states. For a general discussion on nonlinear behavior of solutions near equilibrium when the equilibrium becomes unstable, see Catalano et al. (1981). When homogeneous state (without diffusion) becomes unstable, different kinds of symmetry-breaking structure may arise in systems with diffusion. In general, the mathematical theory based on linear perturbation theory is a good guide to the nonlinear results observed except that the theory is inadequate at very small diffusivities (Fujii et al., 1982).

10.8   Models for Age-Dependent Dispersal

Nearly all attempts to construct mathematical models of populations have assumed that all animals of a species are physiologically identical regardless of age or size. However, numerous field and laboratory populations have served to demonstrate that the internal heterogeneity of a population profoundly affects its dynamics.

Mathematical models should be more realistic if they incorporate the effect of age dependence. For this purpose we consider instead of the usual population number density $S(t, \mathbf{x})$, the age density function $f(t, a, \mathbf{x})$, i.e., the population density, per unit volume and age, of numbers of age a at time t and position $\mathbf{x}$. Integration of f over age yields the population density at time t and position $\mathbf{x}$.

$$S(t, \mathbf{x}) = \int_0^\infty f(t, a, \mathbf{x})\, da. \tag{10.123}$$

We first derive the fundamental equation for an age-dependent dispersing single-species population. Let $Q_j(t, a, \mathbf{x})$ denote the population flux per unit age in the jth direction across a unit area at $\mathbf{x}$ in a unit time ($j = x, y, z$). By analogy to the equation for $S(t, \mathbf{x})$, e.g., (2.10), we obtain the balance equation for $f(t, a, \mathbf{x})$:

$$\left(\frac{df}{dt}\right)_{\mathbf{x}} = \sum_j \left(-\frac{\partial}{\partial j} Q_j\right) - \mu_a f, \tag{10.124}$$

where the left-hand side of (10.124) denotes the time rate of change of the age density function per unit volume at position $\mathbf{x}$, and μ_a is the death-rate density at age a. Since in a small time interval Δt the age of each individual is increased by Δt, we have

$$\left(\frac{df}{dt}\right)_{\mathbf{x}} = \lim_{\Delta t \to 0} \frac{f(t + \Delta t, a + \Delta t, \mathbf{x}) - f(t, a, \mathbf{x})}{\Delta t}. \tag{10.125}$$

By making use of the expansion,

$$f(t + \Delta t, a + \Delta t, \mathbf{x}) = f(t, a, \mathbf{x}) + \frac{\partial f}{\partial t} \Delta t + \frac{\partial f}{\partial a} \Delta t + 0(\Delta t^2),$$

(10.125) becomes, in the limit of $\Delta t \to 0$,

$$\left(\frac{df}{dt}\right)_{\mathbf{x}} = \frac{\partial f}{\partial t} + \frac{\partial f}{\partial a},$$

so that we obtain from (10.124)

$$\frac{\partial f}{\partial t} + \frac{\partial f}{\partial a} = \sum_j \left(-\frac{\partial}{\partial j} Q_j\right) - \mu_a f. \tag{10.126}$$

Without the Q-terms, or in a homogeneous population, (10.126) is known as Von Foerster's equation (Von Foerster, 1959; Trucco, 1965a, b; Weiss, 1968). The formulation (10.126) describing the dispersion of age-dependent populations has been considered by Rotenberg (1972), Gurtin (1973), and Gopalsamy (1976, 1977c).

A reasonable assumption for the age population flux seems to be an analogy to the ordinary population density flux, i.e.,

$$Q_j = v_j f - \kappa_j \partial f / \partial j, \tag{10.127}$$

where v_j and κ_i are, respectively, the advection velocity and diffusivity for the age density function.[3] This assumption enables the closure of (10.126);

$$\frac{\partial f}{\partial t} + \frac{\partial f}{\partial a} = -\sum_j \frac{\partial}{\partial j}\left(\kappa_j \frac{\partial f}{\partial j}\right) - \mu_a f \tag{10.128}$$

where v_j, κ_i, and μ_a are, in general, functions of t, a, and $\mathbf{x}$ (see, e.g., MacCamy, 1981).

The physical nature of population dynamics demands that the following additional conditions be satisfied by the age density function:

i) The initial age and space distribution must be specified; i.e., when $A(a, \mathbf{x})$ is a given function,

$$f(0, a, \mathbf{x}) = A(a, \mathbf{x}) \quad \text{at} \quad t = 0. \tag{10.129}$$

ii) The age-boundary condition representing the rate of addition of individuals of age zero by birth; at $a = 0$,

$$f(t, 0, \mathbf{x}) = \int_0^\infty \lambda_a(t, a, \mathbf{x}) f(t, a, \mathbf{x})\, da \equiv B(t, \mathbf{x}), \tag{10.130}$$

where B is the birthrate density, and $\lambda_a \geq 0$ is the age and volumetric density of the rate of production of offspring. This condition represents an integral feedback to the population. Also, at $a = \infty$,

$$f(t, \infty, \mathbf{x}) = 0. \tag{10.131}$$

We note that $B(t, \mathbf{x})$ is not arbitrarily assigned but must be related to the age density function itself; the kinematic (bookkeeping) condition that $A(0, \mathbf{x}) = B(0, \mathbf{x})$ must be satisfied.

iii) Appropriate spatial boundary conditions
The integration of (10.128) with respect to age from 0 to ∞ and the use of (10.123), (10.130), and (10.131) lead to the dynamic equation for population density. We thus find

$$\frac{\partial S}{\partial t} = \int_0^\infty \lambda_a f\, da - \sum_j \frac{\partial}{\partial j}\int_0^\infty v_j f\, da + \sum_j \frac{\partial}{\partial j}\int_0^\infty \kappa_j \frac{\partial f}{\partial j}\, da - \int_0^\infty \mu_a f\, da. \tag{10.132}$$

We now define

---

[3] Gurtin (1973) proposed that the diffusive part of the age population flux should be proportional to the gradient of the *total* population density $Q_j = -Kj\partial S/\partial j$ so that the local flow of population always lies in the direction of decreasing *total* density.

$$u_j(t, \mathbf{x}) \equiv \int_0^\infty v_j f \, da \bigg/ \int_0^\infty f \, da = \int_0^\infty v_j f \, da / S$$

$$\lambda(t, \mathbf{x}) \equiv \int_0^\infty \lambda_a f \, da \bigg/ \int_0^\infty f \, da = \int_0^\infty \lambda_a f \, da / S$$

$$D_j(t, x) \equiv \int_0^\infty \kappa_j \frac{\partial f}{\partial j} \, da \bigg/ \int_0^\infty \frac{\partial f}{\partial j} \, da = \int_0^\infty \kappa_j \frac{\partial f}{\partial j} \, da / (\partial S / \partial j)$$

$$\mu(t, x) \equiv \int_0^\infty \mu_a f \, da \bigg/ \int_0^\infty f \, da = \int_0^\infty \mu_a f \, da / S$$

Equation (10.132) can then be written as

$$\frac{\partial S}{\partial t} = -\sum_j \frac{\partial}{\partial j} (u_j S) + \sum \frac{\partial}{\partial j} \left(D_j \frac{\partial S}{\partial j} \right) + (\lambda - \mu) S \tag{10.133}$$

We thus recover the basic equation for a dispersing population including simple birth and death processes, with the understanding that the (macroscopic) parameters, u_j, D_j, λ, and μ represent quantities averaged over the (microscopic) age structure.

Gopalsamy (1976, 1977c) investigated several solutions of Eq. (10.128) for the one-dimensional case $(j = x)$, i.e., a linear habitat. For simplicity, it is assumed that no advective processes occur and that κ, λ, μ, are positive constants. We thus have

$$\left(\frac{\partial}{\partial t} + \frac{\partial}{\partial a} \right) f = \kappa \frac{\partial^2 f}{\partial x^2} - \mu_a f \tag{10.134}$$

subject to the boundary conditions

$$f(0, a, x) = A(a, x) \tag{10.135}$$

$$f(t, 0, x) = \lambda_a \int_0^\infty f(t, a, x) \, da \tag{10.136}$$

$$f(t, a, \pm\infty) = 0 \tag{10.137}$$

The solution of (10.134) subject to (10.135)–(10.137) is obtained via Fourier transforms (Gopalsamy, 1977a), yielding the following solution; for $t > a$,

$$f(t, a, x) = \lambda_a \exp\{(t - a)\lambda_a - \mu_a t\} (4\pi\kappa t)^{-1/2} \int_0^\infty da'$$

$$\times \int_{-\infty}^\infty A(a', x') \exp\{-(x - x')^2 / 4\kappa t\} \, dx' \tag{10.138}$$

and for $t < a$;

$$f(t, a, x) = \exp(-\mu_a t)(4\pi\kappa t)^{-1/2} \int_{-\infty}^{\infty} A(a - t, x')\exp\{-(x - x')^2/4\kappa t\} \, dx'$$

$$(10.139)$$

This simple case provides some insight as regards the structure of the age distribution in space and time. The condition $\lambda_a = \mu_a$ is not a sufficient condition for the age density function to attain a steady state. Also, if the initial age structure is spatially homogeneous within a habitat, the age density function remains homogeneous for all subsequent times in spite of dispersal of the population.

Gopalsamy (1976) also obtained a formal solution for the cumulative age density function,

$$F(t, a, \mathbf{x}) = \int_0^a f(t, a', \mathbf{x}) \, da',$$

in a finite linear habitat. He argued that the asymptotic age distribution in dispersive populations is uniform in the range of the habitat and that this uniform age distribution is asymptotically stable with respect to perturbations.

The dynamics of the age distribution function in a spatially uniform populations have been generalized by Rotenberg (1975) to n interacting species, and the dynamics of two interacting species in a one-dimensional habitat is considered by Gopalsamy (1977c).

Di Blasio (1979) and Di Blasio and Lamberti (1978) deal with a more general form than (10.134), where the death rate is not a constant,

$$\partial f/\partial t + \partial f/\partial a = D\partial^2 f/\partial x^2 - \mu_a(S)f$$

subject to

$$f(O, a, \mathbf{x}) = A(a, \mathbf{x}),$$

$$\partial f/\partial x(t, a, \mathbf{x}) = 0 \quad \text{at boundary},$$

$$f(t, 0, \mathbf{x}) = \int_0^{\infty} \lambda(a, S)f(t, a, \mathbf{x}) \, da.$$

It is proved that if A is nonnegative, then this problem admits a unique nonnegative solution. The solutions are shown to depend continuously on the initial data. Gurtin and MacCamy (1974) were the first to have studied the case where λ and μ depend not only on a but also on the total population. This overcomes the chief disadvantage of the Lotka–Von Foerster model, which does not take into account overcrowding and shortage of food. Gurtin

and MacCamy (1981) further discussed age-dependent, dispersing population models for which there are stable age distributions, i.e., $f(y = t, a, \mathbf{x})$
admits product solutions $f = Q(a)P(t, \mathbf{x})$.

Langlais (1985, 1987, 1988) and Garromi-Langlais (1982) analyzed some
mathematical problems (large-time behavior, stability, etc.) of

$$\partial f/\partial t + \partial f/\partial a = K\Delta f - \mu f + G,$$

$$f(t, 0, \mathbf{x}) = \int_0^\infty \beta\{a, \mathbf{x}, P(\mathbf{x}, t)\} f(t, a, \mathbf{x})\, da,$$

$$P(t, \mathbf{x}) = \int_0^\infty f(t, a, \mathbf{x})\, da,$$

$$f(t, a, \mathbf{x}) = 0 \quad \text{on the boundary.}$$

The problem remains to get beyond mathematical results and achieve biologically meaningful solutions. A first, important step in this direction was
taken by Busenberg and Lannelli (1982a, b, 1983a, b, 1985), who described
a method for treating a broad class of nonlinear (density-dependent) age-
dependent population problems that may involve spatial diffusion (linear or
nonlinear). This method effectively decouples the age-dependent part of the
problem from the population interaction terms, allowing the treatment of a
number of age-structured problems that had not been otherwise accessible.
Smith and Wollkind (1983) performed a stability analysis of an arthropod
predator–prey model incorporating age structure and passive diffusion. The
so-called paradox of enrichment is discussed. The destabilizing effect of age
structure is seen to occur most dramatically when interspecific interactions
are large, while the effect of dispersal is to offset that tendency and restabilize
the system.

Slobodkin (1953b, 1954) noted that age structure alone is not adequate to
explain the population dynamics of some species. The size of individuals
could also be used to distinguish cohorts. In principle, there are many ways to
differentiate individuals in addition to age. Thus, the physiological features
of an individual, such as body size and dietary requirements, may influence
the population processes. Schlesinger et al. (1981) measured in laboratory
cultures the specific growth rate of freshwater algae in relation to cell size
and light intensity. There was an inverse relationship between cell size and
specific growth rate. As light intensity was decreased, the relationship weak-
ened. Some field data also suggest this kind of relation. Size-dependent
growth, as it relates to light intensity and nutrient concentrations, should be
considered as a factor influencing patterns of algal succession. Kirkpatrick
(1984) emphasized that the size of an organism is more important than age
in determining life history characteristics, and developed a model incorporating these features.

Let $\mathbf{g} = (g_1, g_2, \ldots, g_m)$ be a complete set of internal characteristics that influence the population growth rate, such as physiological variables and behavioral parameters. We define the generalized density function $h(t, a, \mathbf{g}, \mathbf{x})$ to be the population density of members of age a and internal variables $\mathbf{g}$ at time t and position $\mathbf{x}$. Clearly, if we integrate $h(t, a, \mathbf{g}, \mathbf{x})$ over the entire range of $\mathbf{g}$, we obtain the age density function, $f(t, a, \mathbf{x})$:

$$\int h(t, a, \mathbf{g}, \mathbf{x}) \, d\mathbf{g}.$$

The dynamic equation for the generalized density function for spatially uniform populations has been derived in several contexts (Oster, 1976, 1978; Streifer, 1974; Sinko and Streifer, 1967). It may be written as

$$\frac{\partial h}{\partial t} + \frac{\partial h}{\partial a} + \frac{\partial}{\partial \mathbf{g}}(\mathbf{G}h) = -\mu_g h, \qquad (10.140)$$

where μ_g is the death rate associated with population density h, and $\mathbf{G}$ is the average time rate of change of the quantities $\mathbf{g}$ for a single individual.

$$d\mathbf{g}/dt = \mathbf{G}(t, a, \mathbf{g}). \qquad (10.141)$$

Generally speaking, the internal variables $\mathbf{g}$, unlike the chronological age a, may be subject to stochastic variation. On the basis of a Markovian hypothesis, Weiss (1968) derived a Fokker–Planck equation for the spatially uniform generalized density function:

$$\frac{\partial h}{\partial t} + \frac{\partial h}{\partial a} + \frac{\partial}{\partial \mathbf{g}}(\mathbf{M}_g h) = \frac{\partial^2}{\partial \mathbf{g}^2}(\mathbf{V}_g h) - \mu_g h,$$

where

$$\mathbf{M}_g = \lim_{\Delta t \to 0} \overline{\mathbf{g}(t + \Delta t) - \mathbf{g}(t)}/\Delta t$$

$$\mathbf{V}_g = \lim_{\Delta t \to 0} \overline{\{\mathbf{g}(t + \Delta t) - \mathbf{g}(t)\}^2}/2\Delta t.$$

Here $\mathbf{M}_g$ is the mean rate of increase of the internal variables, and $\mathbf{V}_g$ is the second moment of the increase of the variables. In the absence of stochastic variations in the internal variables, $\mathbf{V}_g \equiv 0$ and $\mathbf{M}_g$ reduces to $\mathbf{G}$ of Eq. (10.141).

Inclusion of advective and diffusive processes in the fundamental equation for $h(t, a, \mathbf{g}, \mathbf{x})$ is straightforward (Streifer, 1974). The following equation is obtained for a one-dimensional space:

$$\frac{\partial h}{\partial t} + \frac{\partial h}{\partial a} + \frac{\partial}{\partial \mathbf{g}}(\mathbf{M}_g h) = \frac{\partial^2}{\partial \mathbf{g}^2}(\mathbf{V}_g h) - \mu_g h - \frac{\partial}{\partial x}(\omega h) + \frac{\partial}{\partial x}\left(q \frac{\partial h}{\partial x}\right) \qquad (10.142)$$

where ω and q are, respectively, the advective velocity and diffusivity for the generalized density function.

Very little work has been done using Eqs. (10.128) and (10.142); however, the dynamical equation of age-size-specific populations without dispersal has been used by several investigators in an ecological context

Levin and Paine (1974, 1975) applied the age-size population equation to the structure of marine rocky intertidal communities. In the rocky middle intertidal zone, empty patch areas form in the background of mussels (*Mytilus californianus*) through the action of ecological disturbances such as waves and predation. These patches are characterized by age a and size g. Let $h(t, a, g) \, da \, dg$ be the number of patches with age between a and $a + da$ and size between g and $g + dg$ at time t. Then the patch density function h is governed by

$$\frac{\partial h}{\partial t} + \frac{\partial h}{\partial a} + \frac{\partial (Gh)}{\partial g} = -\mu h, \tag{10.143}$$

where G describes the rate of expansion of patch size g and μ is the rate of extinction of patches of age a and size g at time t.

Levin and Paine solved (10.143) in closed form, given the initial distribution of patches $h(0, a, g)$ and the birthrate of patches $h(t, 0, g) \sim b(t, g)$ due to disturbances. Thus, for given functions of G and μ (both obtainable from data fits) the size distribution of patches at $t \geq a$ may be related in a simple way to the new patch distribution at time $t - a$. In effect, for sufficiently large t, the entire age-size distribution can be related to the data concerning $b(t, g)$ at previous times. In later work Paine and Levin (1981) applied this model successfully to an extensive set of experiments in the intertidal.

Under same circumstances, e.g., when it is difficult to determine the age of patches, we may wish only to know the size distribution $n(t, g) = \int_0^\infty h(t, a, g) \, da$. If G and b are functions of g only and $\mu = 0$, it is found that

$$n(t, g) \simeq n(g) = G^{-1} \int_g^\infty b(g') \, dg'$$

for large values of t. Thus, the equilibrium distribution of patch size is related to the intensity of disturbances, i.e., $b(g)$. Models of this kind relate environmental disturbances to pattern, and in turn relate environmental disturbance to species diversity for a wide range of conditions of patch formation.

Silvert and Platt (1978) analyzed the size structure of the pelagic chain in the sea on the basis of the biomass density function $B(g, t)$, which is defined to be the mass of organisms per unit volume of water with size g at time t. The function B obeys the equation

$$\frac{\partial B}{\partial t} + \frac{\partial (GB)}{\partial g} = -\mu B, \tag{10.144}$$

where G and μ are respectively the growth function and the loss-rate coefficient.

The steady-state solution $B_0(g)$ is obtained by setting $\partial B/\partial t = 0$ and integrating (10.144); it is found that

$$B_0(g) = AG^{-1}(g) \exp\left\{ -\int \mu(g)/G(g)\, dg \right\},$$

where A is a constant. If the growth and loss functions are given by the particular forms used by Platt and Denman (1977), i.e., $G \sim g^{1-\alpha}$ and $\mu \sim g^{1-\beta}$, where α and β are constants and $\alpha + \beta = 1$, the solution is found to be

$$B_0(g) \propto g^{\alpha-1}.$$

This result corresponds to Platt and Denman's equation for the biomass spectrum.

Silvert and Platt (1978) also discuss the non-steady-state solutions of (10.144), which possess the feature that a disturbance of the biomass spectrum propagates *intact* through the spectrum. Intuitively one would expect the disturbance to spread with time over a broader spectral band. This suggests the need for more refined models.

The size of organisms may play a crucial role in determining population numbers. DeAngelis and Mattice (1979) consider a partial differential equation for $B(g, t)$, which is somewhat more complex than (10.144), i.e.,

$$\partial B/\partial t + \partial(GB)/\partial g = -\mu(g, t)B + F_0(g - g_0)I(t), \qquad (10.145)$$

where $\mu(g, t)$ is the mortality rate dependent, in general, on the time and the average size of organisms in the cohort, $F_0(g - g_0)$ is the distribution of the size of newly born organisms about some mode, $g = g_0$, and $I(t)$ is the number of the new cohort. (10.145) is subject to the initial condition that at $t = 0$,

$$B(g, t) = B_0(g). \qquad (10.146)$$

If G and μ are independent of t, the solution of (10.145) is obtainable by the method of characteristics to give

$$B(g, t) = I\{t - T(g)\} \exp\{-J(g)\}/G(g), \quad t > 0, \qquad (10.147)$$

where

$$T(g) = \int_{g_0}^{g} G(z)^{-1}\, dz, \qquad (10.148)$$

$$J(g) = \int_{g_0}^{g} \mu(z)/G(z)\, dz. \qquad (10.149)$$

Assuming typical organism growth pattern, DeAngelis and Mattice (1979) show that the size distribution at first broadens and then eventually becomes narrow again as the cohort grows in average size. This conclusion is in agreement with growth data on some marine and freshwater organisms.

Size structure plays an important role in virtually all aquatic systems. The size structure of the phytoplankton plays a crucial role in evaluation of food requirements for suspension-feeding zooplankton (Frost, 1972, 1977; Steele, 1977; Runge and Ohman, 1982). Thompson and Cauley (1979) presented a dynamical model for fish populations with age and size distributions; in Weiss's (1968) equation (10.141a), let $g = L$ (size) for steady-state distribution

$$\partial h/\partial a + \partial/\partial L(M_L h) = V_L \partial^2 h/\partial L^2,$$

where V_L is assumed independent of size and death rate is ignored.

$$dLnL/da = bf(a) = b \exp(-ka), \text{ say,}$$

$$M_L = dL/dt = dL/da \text{ (steady state)} = bL \exp(-ka).$$

Therefore,

$$\partial h/\partial a + \partial/\partial L\{bL \exp(-ka)h\} = V_L \partial^2 h/\partial L^2. \tag{10.150}$$

Another version of a model like (10.150) was used in a stochastic model for growth and size distribution in plant populations developed by Hara (1984).

Oster and Takahashi (1974) have examined the dynamical effect of age structure and environmental periodicity on the behavior of two interacting species. Certain novel phenomena emerge as a result of viewing populations as distributed-parameter systems with periodic forcing. Streifer (1974) presented an extensive discussion of age-size-specific models, which he appropriately termed "realistic models in population ecology." To be even more realistic, advective and diffusive processes must be included in age-size-specific models. Our task remains quite difficult, but the rewards of accruing our efforts promise to be great.

References

Abbott, C. A., Berry, M. W., Comiskey, E. J., Gross, L. J., Luh, H.-K. (1997): Computational models of white-tailed deer in the Florida Everglades. *IEEE Computational Science and Engineering* **4**, 60–72.

Abbott, M. R., Denman, K. L., Powell, T. M., Richerson, P. J., Richards, R. C., Goldman, C. R. (1984): Mixing and the dynamics of the deep chlorophyll maximum in Lake Tahoe. *Limnol. and Oceanogr.* **29**, 862–878.

Ablowitz, M. J., Zeppetella, A. (1979): Explicit solutions of Fisher's equation for a special wave speed. *Bull. Math. Biol.* **41**, no. 6, 835–840.

Abraham, E. R. (1998): The generation of plankton patchiness by turbulent stirring. *Nature* **391**, 577–580.

Ackerman, J. D. (1986): Mechanistic implications for pollination in the marine angiosperm, *Zostera marina* L. *Aquat. Bot.* **24**, 343–353.

Ackerman, J. D. (1989): Biomechanical aspects of submarine pollination in *Zostera marina* L. Doctoral dissertation. Ithaca: Cornell Univ. Press.

Ackerman, J. D. (1993): Pollen germination and pollen tube growth in the marine angiosperm, *Zostera marina* L. *Aquat. Bot.* **46**, 189–202.

Ackerman, J. D. (1995): Convergence of filiform pollen morphologies in seagrasses: Functional mechanisms. *Evol. Ecol.* **9**, 139–153.

Ackerman, J. D. (1997a): Submarine pollination in the marine angiosperm, *Zostera marina*: Part I. The influence of floral morphology on fluid flow. *Amer. J. Bot.* **84**, 1099–1109.

Ackerman, J. D. (1997b): Submarine pollination in the marine angiosperm, *Zostera marina*: Part II. Pollen transport in flow fields and capture by stigmas. *Amer. J. Bot.* **84**, 1110–1119.

Ackerman, J. D. (1998a): The effect of turbulence on the functioning of aquatic organisms. In *Engineering Mechanics: A Force for the 21st Century*, 1784–1787. Murakami, H., Luco, J. E. (eds.). Reston, VA: ASCE.

Ackerman, J. D. (1998b): Is the limited diversity of higher plants in marine systems due to biophysical limitations for reproduction or evolutionary and physiological constraints? *Functional Ecol.* **12**, 979–982.

Ackerman, J. D. (1999): The effect of velocity on the filter feeding of zebra mussels (*Dreissena polymorpha and D. bugensis*): Implications for trophic dynamics. *Can. J. Fish. Aquat. Sci.* **56**, 1551–1561.

Ackerman, J. D. (2000): Abiotic pollen and pollination: Ecological, functional, and evolutionary perspectives, *Plant Syst. Evol.* **222**, 167–185.

374

Ackerman, J. D., Loewen, M. R., Hamblin, P. F. (2000): Biological evidence for benthic–pelagic coupling over a zebra mussel reef in western Lake Erie. In *Fifth International Symposium on Stratified Flows*, 249–254. Lawrence, G. A., Pieters, R., Yonemitsu, N. (eds.). Vancouver: UBC Civil Engineering.

Ackerman, J. D., Loewen, M. R., Hamblin, P. F. (2001): Benthic–pelagic coupling over a zebra mussel bed in the western basin of Lake Erie. *Limnol. Oceanogr.* **46**, 892–904.

Ackerman, J. D., Okubo, A. (1993): Reduced mixing in a marine macrophyte canopy. *Funct. Ecol.* **7**, 305–309.

Adachi, K., Kiriyama, S., Yoshioka, N. (1978): The behavior of a swarm of particles moving in a viscous fluid. *Chem. Engineering Sci.* **33**, 115–121.

Aikman, D., Hewitt, G. (1972): An experimental investigation of the rate and form of dispersal in grasshoppers. *J. Appl. Ecol.* **9**, 807–817.

Alcarez, M., Paffenhofer, G.-A., Strickler, J. R. (1980): Catching the algae: First account of visual observations on filter-feeding calanoids. In *Evolution and Ecology of Zooplankton Communities*. Kerfoot, W. C. [ed.]. Dartmouth, NH: The University Press of New England.

Alexander, R. M. (1968): *Animal Mechanics*. Seattle: Univ. Washington Press.

Allan, J. D. and A. S. Flecker (1989): The mating biology of a mass-swarming mayfly. *Anim. Behav.* **37**, 361–371.

Allanson, B. R., Skinner, D., Imberger, J. (1992): Flow in prawn burrows. *Est. Coast. Shelf Sci.* **35**, 253–266.

Alldredge, A. L. (1982): Aggregation of spawning appendicularians in surface windrows. *Bull. Mar. Sci.* **32**, 250–254.

Alldredge, A. L., Granata, T. C., Gotschalk, C. C., Dickey, T. D. (1990): The physical strength of marine snow and its implications or particle disaggregation in the ocean. *Limnol. Oceanogr.* **35**, 1415–1428.

Alldredge, A. L., Silver, M. W. (1988): Characteristics, dynamics and significance of marine snow. *Prog. Oceanogr.* **20**, 41–82.

Aller, R. C. (1980): Quantifying solute distributions in the bioturbated zone of marine sediments by defining an average microenvironment. *Geochim. Cosmochim. Acta* **44**, 1955–1965.

Aller, R. C. (1988): Benthic fauna and biogeochemical processes in marine sediments:The role of burrow structures. In *Nitrogen Cycling in Coastal Marine Environments*. Blackburn, T. H., Sorensen, J. (eds.). New York: John Wiley and Sons.

Aller, R. C., Aller, J. Y. (1992): Meiofauna and solute transport in marine muds. *Limnol. Oceanogr.* **37**, 1018–1033.

Allee, W. C. (1938): *The Social Life of Animals*. New York: W. W. Norton and Co.

Almirantis, Y., Papageorgiou, S. (1991): Cross-diffusion effects on chemical and biological pattern formation. *J. Theor. Biol.* **151**, 289–311.

Alt, W. (1980): Biased random walk models for chemotaxis and related diffusion approximations. *J. Math. Biol.* **9**, 147–177.

Ambler, J. W., Broadwater, S. A., Buskey, E. J., Peterson, J. O. (1996): Mating behavior in swarms of *Dioithona oculata*. In *Zooplankton: Sensory Ecology and Physiology*, 287–299. P. H. Lenz, D. K. Hartline, J. E. Purcell, D. L. MacMillan (eds.). London: Gordon and Breach Publishers.

Anderson, J. J. (1981): A stochastic model fot the size of fish schools. *Fish. Bull.* **79(2)**, 315–323.

Anderson, D. J. (1982): The home range: A new nonparametric estimation technique. *Ecology* **63**, 103–112.

Anderson, J. J., Robbins, A., Okubo, A. (1978): A reaction–diffusion model for nitrite in the oxygen minium zones of the eastern tropical Pacific. From abstracts of papers submitted for the 41st Annual Meeting, Amer. Soc. Limnol. and Oceanogr., Univ. Victoria, British Columbia, Canada, June 19–22.

Anderson, S. M., Charters, A. C. (1982): A fluid dynamics study of seawater flow through *Gelidium nudifrons*. *Limnol. Oceanogr.* **27**, 399–412.

Andow, D., Kareiva, P., Levin, S., Okubo, A. (1990): Spread of invading organisms. *Landscape Ecol.* **4**, 177–188.

Andrewartha, H. G., Birch, L. C. (1954): *The Distribution and Abundance of Animals.* Chicago, London: Univ. Chicago Press.

Andrews, J. C. (1983): Deformation of the active space in the low Reynolds number feeding current of calanoid copepods. *Can. J. Fish. Aquat. Sci.* **40**, 1293–1302.

Angelakis, A. N., Kadir, T. N., Rolston, D. E. (1993): Analytical solutions for equations describing coupled transport of two solutes and a gaseous product in soil. *Water Resour. Res.* **29**, 945–956.

Antonovics, J., Iwasa, Y., Hassell, M. P. (1995): A generalized model of parasitoid, venereal, and vector-based transmission processes. *Amer. Nat.* **145**, 661–675.

Aoki, I. (1980): An analysis of the shooling behavior of fish: internal oganization and communication process. *Bull. Ocean Res. Inst. Univ. of Tokyo*, **12**, 1–62.

Aoki, I. (1982): A simulation study on the schooling mechanism in fish. *Bull. Jpn. Soc. Sci. Fish.* **48**, 1081–1088.

Aoki, I. (1984): Internal dynamics of fish schools in relation to inter-fish distance. *Bull. Jpn. Soc. Sci. Fish.* **50(5)**, 751–758.

Aoki, I. (1986): *Bull. Jpn. Soc. Sci. Fish.* **52**, 1115–1119.

Araujo, M., Larralde, H., Harlin, S., Stanley, H. E. (1992): Reaction-front dynamics of $A + B \rightarrow C$ with initially separated reactants. *Physica* A **191**, 168–171.

Arcuri, P. (1986): Pattern sensitivity to boundary and initial conditions in reaction–diffusion models. *J. Math. Biol.* **24**, no. 2, 141–165.

Aris, R. (1975): *The Mathematical Theory of Diffusion and Reaction in Permeable Catalysts.* Vols. 1 and 2. Oxford: Oxford Univ. Press.

Arkin, G. F., Perrier, E. R. (1974): Vorticular air flow within an open row crop canopy. *Agri. Meteorol.* **13**, 359–374.

Armstrong, J. T. (1965): Breeding home range in the nighthawk and other birds; its evolutionary and ecological significance. *Ecology* **46**, 619–629.

Armstrong, R. A. (1994): Grazing limitation and nutrient limitation in marine ecosystems: Steady-state solutions of an ecosystem model with multiple foodchains. *Limnol. Oceanogr.* **39**, 597–608.

Armstrong, R. A., Siegel, D. A. (2001): A large catchment area for sediment traps is not incompatible with sharply defined patterns of benthic deposition. *Abstracts of the 2001 Aquatic Sciences Meeting*, Albuquerque, NM (Amer. Soc. Limnol. Oceanogr.), p. 18.

Arnold, L. (1974): *Stochastic Differential Equations: Theory and Applications.* New York: J. Wiley & Sons.

Arocena, J., Ackerman, J. D. (1998): Use of statistical tests to describe the basic distribution patterns of iron oxide nodules in soil thin section. *Amer. Soil Sci. Soc. J.* **62**, 1346–1350.

Aronson, D. G. (1979): Density-dependent interaction diffusion systems. In: *Dynamics and modeling of reactive systems.* pp. 161–176. Publ. Math. Res. Center Univ. Wisconsin: Academic Press, New York-London.

Aronson, D. G. (1985): The role of diffusion in mathematical population biology: Skellam revisited. In *Mathematics in Biology and Medicine*, Lecture Notes in Biomathematics, Vol. 57, 2–18. Capasso, V., Grosso, E., Paveri-Fontana, S. L. (eds.). Berlin: Springer-Verlag.

Aronson, D. G., Weinberger, H. F. (1975): Nonlinear diffusion in population genetics, combustion, and nerve propagation. In *Partial Differential Equations and Related Topics*. 5–49. Lecture Notes in Mathematics 446, Goldstein, J. A. (ed.). Heidelberg, New York, Berlin: Springer.

Aronson, D. G., Weinberger, H. F. (1975): Nonlinear diffusion in population genetics, combustion, and nerve pulse propagation. In: *Partial differential equations and related topics*. pp. 5–49. Program, Tulane Univ., New Orleans, La.

Asahina, S., Turuoka, Y. (1970): Records of the insects visited a weather-ship located at the Ocean Weather Station "Tango" on the Pacific. V. Insects captured during 1968. *Kontyu (Insects)* **38**, 318–330 (Japanese with English summary).

Asper, V. L. (1987): Measuring the flux and sinking speed of marine snow aggregates. *Deep-Sea Res.* **34**, 1–17.

Assaf. G., Gerard, R., Gordon, A. L. (1971): Some mechanism of oceanic mixing revealed in aerial photographs. *J. Geophys. Res.* **76**, 6550–6572.

Atema, J. (1996): Eddy chemotaxis and odor landscapes: Exploration of nature with animal sensors. *Biol. Bull.* **191**, 129–138.

Atkinson, R. J. A., Taylor, A. C. (1991): Burrows and burrowing behavior of fish. *Symp. Zool. Soc. Lond.* **63**, 133–155.

Atlas, D., Harris, F. I., Richter, J. H. (1970): Measurement of point target speeds with incoherent non-tracking radar: Insect speeds in atmospheric waves. *J. Geophys. Res.* **75**, 7588–7595.

Augspurger, C. K., Franson, S. E. (1987): Wind dispersal of artificial fruits varying in mass area and morphology. *Ecology* **68**, 27–42.

Aylor, D. E. (1986): A framework for examining inter-regional aerial transport of fungal spores. *Agric. For. Meteorol.* **38**, 263–288.

Aylor, D. E., Taylor, G. S., Raymor, G. S. (1982): Long-range transport of tobacco blue mold *Peronospora tabacina*. *Agric. Meteorol.* **27**, 217–232.

Azam, F. (1998): Microbial control of oceanic carbon flux: The plot thickens *Sci.* **280**, 694–696.

Babcock, R. C., Bull, G. D., Harrison, P. L., Heywood, A. J., Oliver, J. K., Wallace, C. C., Willis, B. L. (1986): Synchronous spawning of 105 scleractinian coral species on the Great Barrier Reef. *Mar. Biol.* **90**, 379–394.

Babcock, R. C., Mundy, C. N., Whitehead, D. (1994): Sperm dilution models and *in situ* confirmation of long-distance fertilization in the free-spawning asteriod *Acanthaster planci. Biol. Bull.* **186**, 17–26.

Bailey, D. W., Gross, J. E., Laca, E. A., Rittenhouse, L. A., Coughenour, M. B., Swift, D. M., Sims, P. L. (1996): Mechanisms that result in large herbivore grazing distribution patterns. *J. Range. Manage.* **49**, 386–400.

Bailey, K. M., Canino, M. F., Napp, J. M., Spring, S. M., Brown, A. L. (1995): Contrasting years of prey levels, feeding conditions, and mortality of larval walleye pollock *Theragra chalcogramma* in the western Gulf of Alaska. *Mar. Ecol. Prog. Ser.* **119**, 11–23.

Bailey, K. M., Stabeno, P. J., Powers, D. A. (1997): The role of larval retention and transport features in mortality and potential gene flow of walleye pollock. *J. Fish Biol.* **51** (Supp. A), 135–154.

Baines, G. B. K. (1972): Turbulence in a wheat crop. *Agric. Meteorol.* **10**, 93–105.

Baker, P. S., Grewecke, M., Cooter, R. J. (1984): Flight orientation of swarming *Locusta migratoria. Physiol. Entomol.* **9**, 247–252.

Baker, R. R. (1978): *The Evolutionary Ecology of Animal Migration.* New York: Holmes and Meier Publishers.

Balazs, G. H. (1976): Green turtle migrations in the Hawaiian Archipelago. *Biol. Conserv.* **9**, 125–140.

Banks, H., Kareiva, P., Zia, L. (1988): Analyzing field studies of insect dispersal using two-dimensional transport equations. *Environ. Entom.* **17**, 815–820.

Banks, H., Kareiva, P., Lamm, P. (1985): Modeling insect dispersal and estimating parameters when mark-release techniques may cause initial disturbances. *J. Math. Biol.* **22**, 259–277.

Barakat, R. (1959): A note on the transient stage of the random dispersal of logistic population. *Bull. Math. Biophys.* **21**, 141–151.

Barber, M. N., Ninham, B. W. (1970): *Random and Restricted Walks.* New York, London, Paris: Gordon and Breach.

Bard, J. B. L. (1981): A model for generating aspects of zebra and other mammalian coat markings. *J. Theor. Biol.* **93**, 363–385.

Barstow, S. F. (1983): The ecology of Langmuir circulation: A review. *Mar. Environ. Res.* **9**, 211–236.

Basil, J., Atema, J. (1994): Lobster orientation in turbulent odor plumes: Simultaneous measurements of tracking behavior and temporal odor patterns. *Biol. Bull.* **187**, 272–273.

Batchelor, G. K. (1950): The application of the similarity theory of turbulence to atmospheric diffusion. *Quart. J. Roy. Meteorol. Soc.* **76**, 133–146.

Batchelor, G. K. (1953): *The Theory of Homogeneous Turbulence.* Cambridge, London, New York: Cambridge Univ. Press.

Batchelor, G. K. (1967): *An Introduction to Fluid Dynamics.* Cambridge, London, New York: Cambridge Univ. Press.

Batchelor, G. K., van Rensburg, R. W. J. (1986): Structure formation in bidisperse sedimentation. *J. Fluid Mech.* **166**, 379–407.

Bateman, A. J. (1947): Contamination of seed crops. II. Wind pollination, *Heredity* **1**, 235–246.

Bazykin, A. D. (1974): In: *Problems in Mathematical Genetics.* Ratner, V. A. (ed.), USSR: Acad. Sci. Novosibirsk (in Russian).

Becker, D. S. (1978): Evaluation of a hard clam spawner transplant site using a dye tracer technique. *Marine Sciences Research Center Special Report No. 10*, Ref. 77-6, Stony Brook: State Univ. of New York.

Bengtsson, G., Lindqvist, R., Piwoni, M. D. (1993): Sorption of trace organics to colloidal clays, polymers, and bacteria. *Soil Sci. Soc. Amer. J.* **57**, 1261–1270.

Bennett, A. F., Denman, K. L. (1989): Large-scale patchiness due to an annual plankton cycle. *J. Geophys. Res.* **94**, 823–829.

Benzie, J. A. H., Dixon, P. (1994): The effect of sperm concentration, sperm:egg ratio and gamete age on fertilization success in Crown-of-Thorns starfish (*Acanthaster planci*) in the laboratory. *Biol. Bull.* **186**, 139–152.

Benzie, J. A. H., Black, K. P., Moran, P. J., Dixon, P. (1994): Small-scale dispersion of eggs and sperm of the crown-of-thorns starfish (*Acanthaster planci*) in a shallow coral reef habitat. *Bio. Bull.* **186**, 153–167.

Berding, C. (1987): On the heterogeneity of reaction-diffusion generated patterns. *Bull. Math. Biol.* **49**, 233–252.

Beretta, E., Fergola, P., Tenneriello, C. (1988): Ultimate boundedness for non-autonomous diffusive Lotka–Volterra patches. *Math. Biosci.* **92**, 29–53.

Berg, H. C. (1983): *Random Walks in Biology.* Princeton: Princeton Univ. Press.

Berg, H. C., Purcell, E. M. (1977): Physics of chemoreceptors. *Biophys. J.* **20**, 193–219.

Bergman, M. O. (1983): Mathematical model for contact inhibited cell division. *J. Theor. Biol.* **102**, 375–386.

Bertsch, M., Gurtin, M. E., Hilhorst, D., Peletier, L. A. (1984): On interacting populations that disperse to avoid crowding: the effect of a sedentary colony. *J. Math. Biol.* **19**, 1–12.

Betts, E. (1976): Forecasting infestations of tropical migrant pests: The Desert Locust and the African Army worm. In *Insect Flight*, 113–134. Rainey, R. C. (ed.). New York: J. Wiley & Sons.

Bhargava, S. C., Saxena, R. P. (1977): Stable periodic solutions of the reactive-diffusive Volterra systems of equations. *J. Theor. Biol.* **67**, 399–406.

Bhargava, S. C., Karmeshu, Unny, T. E. (1984): Role of diffusion in some growth models. *Ecol. Model.* **24**, 1–8.

Bienfang, P. K. (1981): Sinking rates of heterogeneous, temperate phytoplankton populations. *J. Phytoplankton Res.* **3**, 235–253.

Bienfang, P. K., Syper, J., Laws, E. (1993): Sinking rate and pigment responses to light-limitation of a marine diatom: Implications to dynamics of chlorophyll-maximum layers. *Oceanol. Acta* **6**, 55–62.

Blackburn, N., Azam, F., Hagstrom, Å. (1997): Spatially explicit simulations of a microbial food web. *Limnol. Oceanogr.* **42**, 613–622.

Blackburn, N., Fenchel, T., Mitchell, J. (1998): Microscale nutrient patches in planktonic habitats shown by chemotactic bacteria. *Sci.* **282**, 2254–2256.

Blackburn, T. M., Gaston, K. J. (1997): A critical assessment of the form of the relationship between abundance and body size in animals. *J. Anim. Ecol.* **66**, 233–249.

Blair, N. E., Levin, L. A., Demaster, D. J., Plaia, G. (1996): The short-term fate of fresh algal carbon in continental slope sediments. *Limnol. Oceanogr.* **41**, 1208–1219.

Bliss, C. I. (1971): The aggregation of species within spatial units. In *Statistical Ecology*, Vol. 1, 311–335. Patil, G. P., Pielou, E. C., Waters, W. E. (eds.). University Park, PA: Penn. State Univ. Press.

Boicourt, W. C. (1988): Estuarine larval retention mechanisms on two spatial scales. In *Estuarine Comparisons*, 445–458. Kennedy, V. S. (ed.). New York: Academic Press.

Boll, J., Selker, J. S., Shalit, G., Steenhuis, T. S. (1997): Frequency distribution of water and solute transport properties derived from pan sampler data. *Water Resour. Res.* **33**, 2655–2664.

Bonilla, L., Linan, A. (1984): Relaxation oscillations, pulser, and travelling waves in the diffusive Volterra delay differential equation. *SIAM J. Appl. Math.* **44**, 369–391.

Bossert, W. H., Wilson, E. O. (1963): The analysis of olfactory communication among animals. *J. Theor. Biol.* **5**, 443–469.

Bossert, W. H. (1968): Temporal patterning in olfactory communication. *J. Theor. Biol.* **18**, 157–170.

Bouche, M. B., Al-Addan, F. (1997): Earthworms, water infiltration and soil stability: Some new assessments. *Soil Biol. Biochem.* **29**, 441–452.

Boudreau, B. P. (1986a): Mathematics of tracer mixing in sediments: I. Spatially-dependent, diffusive mixing. *Amer. J. Sci.* **286**, 161–198.

Boudreau, B. P. (1986b): Mathematics of tracer mixing in sediments: II. Nonlocal mixing and biological conveyor-belt phenomena. *Amer. J. Sci.* **286**, 199–238.

Boudreau, B. P. (1997): *Diagenetic Models and Their Implementation. Modelling Transport and Reactions in Aquatic Sediments.* Berlin: Springer-Verlag.

Boudreau, B. P. (1998): Mean mixed depth of sediments: The wherefore and the why. *Limnol. Oceanogr.* **43**, 449–457.

Boudreau, B. P. (2000): The mathematics of early diagenesis: From worms to waves. *Rev. Geophys.* **38**, 389–416.

Boudreau, B. P., Imboden, D. M. (1987): Mathematics of tracer mixing in sediments: III. The theory of nonlocal mixing within sediments. *Amer. J. Sci.* **286**, 693–719.

Boudreau, B. P., Marinelli, R. L. (1994): A modelling study of discontinuous irrigation. *J. Mar. Res.* **52**, 947–968.

Bowden, K. F. (1964): Turbulence. In *Oceanogr. Mar. Biol. Ann. Rev.* **2**, 11–30. Barnes, H. (ed.). London: G. Allen and Unwin Ltd.

Bowden, K. F. (1965): Horizontal mixing in the sea due to a shearing current. *J. Fluid Mech.* **21**, 83–95.

Bowden, K. F. (1970): Turbulence II. In *Oceanogr. Mar. Biol. Ann. Rev.* **8**, 11–32. Barnes, H. (ed.). London: G. Allen and Unwin Ltd.

Bowden, K. F. (1975): Oceanic and estuarine mixing processes. In *Chemical Oceanography*, Vol. 1, 1–41. Riley, J. P., Skirrow, G. (eds.). London, New York: Academic Press.

Bowen, J. D., Stolzenbach, K. D., Chisholm, S. W. (1993): Simulating bacterial clustering around phytoplankton cells in a turbulent ocean. *Limnol. Oceanogr.* **38**, 36–51.

Bowles, P., Burns, R. H., Hudswell, F., Whipple, R. T. P. (1958): Sea disposal of low activity effluent. *Proc. 2nd Intl. Conf. on Peaceful Uses of Atomic Energy*, Geneva, **18**, 296.

Boyd, A. V. (1979): The size of animal herds. *Ecol. Modelling* **6**, 91–96.

Bradford, E., Philip, J. P. (1970a): Stability of steady distributions of asocial populations dispersing in one dimension. *J. Theor. Biol.* **29**, 13–26.

Bradford, E., Philip, J. P. (1970b): Note on asocial populations dispersing in two dimensions. *J. Theor. Biol.* **29**, 27–33.

Bradley, W. H. (1965): Vertical density currents. *Science* **150**, 1423–1428.

Bradley, W. H. (1969): Vertical density currents II. *Limnol. Oceanogr.* **14**, 1–3.

Breder, C. M. (1951): Studies on the structure of the fish school. *Bull. Amer. Mus. Nat. Hist.* **98**, Article 1, 1–28.

Breder, C. M. (1954): Equations ascriptive of fish school and other animal aggregations. *Ecology* **35(3)**, 211–229.

Breder, C. M. (1954): Equations descriptive of fish schools and other animal aggregations. *Ecology* **35**, 361–370.

Breder, C. M. (1959): Studies on social groupings in fishes. *Bull. Amer. Mus. Nat. Hist.* **117**, 397–481.

Breder, C. M. (1976): Fish schools as operational structures. *Fishery Bull.* **74**, 471–502.

Brenner, S., Murray, J. D., Wolpert, L. (1981): Theories of biological pattern formation. A Royal Society discussion meeting. *Phil. Trans. R. Soc. Lond.* B **295**, 425–617.

Bridgman, P. W. (1963): *Dimensional Analysis*. New Haven, London: Yale Univ. Press.

Britton, N. F. (1989): *Reaction–diffusion equations and their applications to biology*. New York: Academic Press.

Broadbent, S. R., Kendall, D. G. (1953): The random walk of *Trichostrongylus retortaeformis*. *Biometrics* **9**, 460–466.

Brock, V. E., Riffenburgh, R. H. (1963): Fish schooling: A possible factor in reducing predation. *J. du Cons. Int. Explor. Mer.* **25**, 307–317.

Broecker, W. S. (1991): The great ocean conveyor. *Oceanogr.* **4**, 79–89.

Broecker, W. S., Peng, T. H. (1982): *Tracers in the Sea*. Palisades, NY: Eldigio Press.

Bronk, D. A., Glibert, P. M., Ward, B. B. (1994): Nitrogen uptake, dissolved organic nitrogen release, and new production. *Sci.* **243**, 1843–1846.

Brooker, L., Brooker, M., Cale, P. (1999): Animal dispersal in fragmented habitat: Measuring habitat connectivity, corridor use, and dispersal mortality. *Conservation Ecology* [online] 3(1): 4. Available from the Internet. URL: http://www.consecol.org/vol3/iss1/art4.

Brown, J. H. (1995): *Macroecology*. Chicago: Univ. Chicago Press.

Brown, J. H., Maurer, B. A. (1986): Body size, ecological dominance and Cope's rule. *Nature* **324**, 248–250.

Brown, K. J., Lacey, A. A. (eds.) (1990): *Reaction-diffusion equations: the proceedings of a symposium year on reaction-diffusion equations/organized* by the Department of Mathematics, Heriot-Watt University, 1987–1988. New York: Oxford University Press.

Brown, L. E. (1962): Home range in small mammal communities. In *Survey of Biological Progress*, Vol. 4, 131–179. Class, B. (ed.). New York: Academic Press.

Brown, R. (1828): A brief account of microscopical observations made in the months of June, July and August, 1827, on the particles contained in he pollen of plants; and on the general existence of active molecules in organic and inorganic bodies. *Phil. Mag.* (new series) **4**, 161–173.

Brownlee, J. (1911): The mathematical thoery of random migration and epidemic distribution. *Proc. Roy. Soc. Edinburgh* **31**, 262–289.

Brzezinski, M. A., Alldredge, A. L., O'Bryan, L. M. (1997): Silica cycling within marine snow. *Limnol. Oceanogr.* **42**, 1706–1713.

Buchmann, S. L., O'Rourke, M. K., Niklas, K. J. (1989): Aerodynamic of *Ephedra trifurca*. III Selective pollen capture by pollination droplets. *Bot. Gaz.* **150**, 122–131.

Bunday, B. D. (1970): The growth of elephant herds. *Math. Gazette* **54**, 38–40.

Burrows, F. M. (1975a): Wind-borne seed and fruit movement. *New Phytol.* **75**, 405–418.

Burrows, F. M. (1975b): Calculation of the primary trajectories of dust seeds, spores, and pollen in unsteady winds. *New Phytol.* **75**, 389–403.

Burrows, F. M. (1987): The aerial motion of seeds, fruits, spores, and pollen. In *Seed Dispersal*, 1–47. Murray, D. (ed.). New York: Academic Press.

Burt, W. H. (1943): Territoriality and home range concepts as applied to mammals. *J. Mammal.* **24**, 346–352.

Busenberg, S., Iannelli, M. (1982a): Nonlinear diffusion problems in age-structured population dynamics. *Mathematical Ecology.* **54**, 524–540.

Busenberg, S., Iannelli, M. (1982b): A method for treating a class of nonlinear diffusion problems (English. Italian summary). *Atti. Accad. Naz. Lincei. Rend. Cl. Sce.
Fis. Mat. Natur.* (8) **72**, no. 3, 121–127.

Busenberg, S., Iannelli, M. (1983a): A class of nonlinear diffusion problems in age-
dependent population dynamics. *Nonlinear Anal.* **7**, no. 5, 501–529.

Busenberg, S., Iannelli, M. (1983b): A generate nonlinear diffusion problem in age-
structured population dynamics. *Nonlinear Anal.* **7**, no. 12, 1411–1429.

Busenberg, S., Iannelli, M. (1985): Separable models in age-dependent population
dynamics. *J. Math. Biol.* **22**, no. 2, 145–173.

Buskey, E. J., Peterson, J. O., Ambler, J. W. (1996): The swarming behavior of the
copepod *Dioithona oculata*: In situ and laboratory studies. *Limnol. Oceanogr.* **41**,
513–521.

Butman, C. A., Fréchette, M., Geyer, W. R., Starczak, V. R. (1994): Flume experiments on food supply to the blue mussel *Mytilus edulis* L. as a function of
boundary-layer flow. *Limnol. Oceanogr.* **39**, 1755–1768.

Cain, M. L. (1989): The analysis of angular data in ecological field studies. *Ecology*
70, 1540–1543.

Cain, M., Damman, H., Muir, A. (1998): Seed dispersal and the Holocene migration
of woodland herbs. *Ecol. Monog.* **68**, 325–347.

Calder, K. L. (1949): Eddy diffusion and evaporation in flow over aerodynamically
smooth and rough surfaces: A treatment based on laboratory laws of turbulent
flow with special reference to conditions in the lower atmosphere. *Quart. J. Mech.
& Appl. Math.* **2**, 153–176.

Cambalik, J. J., Checkley, D. M., Jr., Kamykowski, D. (1998): A new method to
measure the terminal velocity of small particles: A demonstration using the ascending eggs of the Atlantic menhaden (*Brevoortia tyrannus*). *Limnol. Oceanogr.* **43**,
1722–1727.

Campbell, G. S., Norman, J. M. (1998): *An Introduction to Environmental Biophysics.*
2nd ed. New York: Springer-Verlag.

Cannings, C., Cruz Orive, L. M. (1975): On the adjustment of the sex ratio and the
gregarious behavior of animal populations. *J. Theor. Biol.* **55**, 115–136.

Canosa, J. (1973): On a nonlinear diffusion equation describing population growth.
IBM J. Res. Develop. **17**, 307–313.

Cantrell, R. S., Cosner, C. (1991): The effects of spatial heterogeneity in population
dynamics. *J. Math Biol.* **29**, 315–338.

Cantrell, R. S., Cosner, C. (1999): Diffusion models for population dynamics
incorporating individual behavior at boundaries: Applications to refuge design.
Theor. Pop. Biol. **55**, 189–207.

Carlson, D. J. (1987): Viscosity of sea-surface slicks. *Nature* **329**, 823–825.

Carpenter, S. R., Cottingham, K. L., Schindler, D. E. (1992): Biotic feedbacks in lake
phosphorus cycles. *TREE* **7**, 332–336.

Carr, A. (1967): Adaptive aspects of the scheduled travel of *Chelonia*. In *Animal
Orientation and Navigation*, 35–55. Storm, R. M. (ed.). Corvallis: Oregon State
Univ. Press.

Carr, A., Coleman, P. J. (1974): Seafloor spreading theory and the odyssey of the
green turtle. *Nature* **249**, 128–230.

Carsel, R. F., Smith, C. N., Mulkey, L. A., Dean, J. D., Jowise, P. P. (1984):
User's manual for the pesticide root zone model (PRZM): Release 1. USEPA
EPA-600/3-84-109. U.S. Govt. Printing Office, Washington DC.

Carslaw, H. S., Jaeger, J. C. (1959): *Conduction of Heat in Solids*, 2nd ed. Oxford: Oxford Univ. Press.

Carter, H. H., Okubo, A. (1978): A study of turbulent diffusion by dye tracers: A review. In *Estuarine Transport Processes*, 95–111. Kjerfve, B. (ed.). Columbia, SC: Univ. South Carolina Press.

Carter, J., Finn, J. T. (1999): MOAB: A spatially explicit, individual-based experts system for creating animal foraging models. *Ecological Modelling* 119, 29–41.

Cassie, R. M. (1963): Microdistribution of plankton. *Oceanogr. Mar. Biol. Ann. Rev.* 1, 223–252.

Catalano, G. Eilbeck, J. C., Monroy, A., Parisi, E. (1981): A mathematical model for pattern formation in biological systems. *Phys. D* 3, no, 3, 439-456.

Chandrasekhar, S. (1943): Stochastic problems in physics and astronomy. *Rev. Modern Phys.* 15, 1–89.

Chandrasekhar, S. (1961): *Hydrodynamic and Hydromagnetic Stability*. Oxford: Oxford Univ. Press.

Chapman, S. (1928): On the Brownian displacements and thermal diffusion of grains suspended in a non-uniform fluid. *Proc. Roy. Soc. Lond. A* 119, 34–54.

Charbeneau, R. J., Daniel, D. E. (1993): Contaminant transport in unsaturated flow. In *Handbook of Hydrology*. Maidment, D. R. (ed.). New York: McGraw-Hill.

Charlesworth, D. (1993): Why are unisexual flowers associated with wind pollination and unspecialized pollinators? *Amer. Nat.* 141, 481–490.

Chatwin, P. C. (1968): The dispersion of a puff of passive contaminant in the constant stress region. *Quart. J. Roy. Meteorol. Soc.* 94, 350–360.

Chen, L. (1985): Long-distance atmospheric transport of dust from the Chinese desert to the North Pacific. *Acta Oceanol. Sin.* 4, 527–534.

Chesson, P. (1986): Environmental variation and the coexistence of species. In *Community Ecology*, 240–254. Diamond, J., Case, T. J. (eds.). NY: Harper and Row.

Chiang, H. C. (1961): Ecology of insect swarms. I. Experimental studies of the behavior of *Anarete* near *felti* Pritchard in artificially induced swarms (Cecidomyiidae, Diptera). *Animal Behav.* 9, 213–219.

Childress, S., Levandowsky, M., Spiegel, E. A. (1975): Pattern formation in a suspension of swimming microorganisms: Equations and stability theory. *J. Fluid Mech.* 63, 591–613.

Childress, S., Percus, J. K. (1981): Nonlinear aspects of chemotaxis. *Math. Biosci.* 56, 217–237.

Childs, E. C., Collis-George, N. (1950): The permeability of porous materials. *Proc. Roy. Soc. Lond. A* 201, 392–405.

Chin, W. C., Orellana, M. V., Verdugo, P. (1998): Spontaneous assembly of marine dissolved organic matter into polymer gels. *Nature* 391, 568–572.

Chow, P. L., Tam, W. C. (1976): Periodic and travelling wave solutions to Volterra–Lotka equations with diffusion. *Bull. Math. Biol.* 38, 643–658.

Christensen, J. P., Smethie, W. M., DeVol, A. H. (1987): Benthic nutrient regeneration and denitrification on the Washington continental shelf. *Deep-Sea Res.* 34, 1027–1047.

Churchill, A. C., Nieves, G., Brenowitz, A. H. (1985): Floatation and dispersal of eelgrass seeds by gas bubbles. *Estuaries* 8, 352–354.

Cionco, R. M. (1965): A mathematical model for air flow in a vegetative canopy. *J. Appl. Meteorol.* 4, 517–522.

Cionco, R. M. (1972): Intensity of turbulence within canopies with simple and complex roughness elements. *Boundary-Layer Meteorol.* **2**, 453–465.

Clark, C. W. (1974): Possible effects of schooling on the dynamics of exploited fish populations. *J. Cons. Intl. Explor. Mer.* **36**, 7–14.

Clark, D. P. (1962): An analysis of dispersal and movement in *Phaulacridium vittatum* (Sjöst) (Acrididae). *Aust. J. Zool.* **10**, 382–399.

Clark, W. C., Jones, D. D., Holling, C. S. (1978): Patches, movements, and population dynamics in ecological systems: A terrestrial perspective. In *Spatial Pattern in Plankton Communities*, 385–432. J. H. Steele (ed.). New York, Plenum Press.

Clarke, R. H. (1970): Observational studies in the atmospheric boundary layer. *Quart. J. Roy. Meteorol. Soc.* **96**, 91–114.

Close, R. C., Moar, N. T., Tomlinson, A. I., Lowe, A. D. (1978): Aerial dispersal of biological material from Australia to New Zealand. *Intl. J. Biometeorol.* **22**, 1–19.

Clutter, R. I. (1969): The microdistribution and social behavior of some pelagic mysid shrimp. *J. Exp. Mar. Biol. Ecol.* **3**, 125–155.

Cloern, J. E. (1991): Tidal stirring and phytoplankton bloom dynamics in an estuary. *J. Mar. Res.* **49**, 203–221.

Cobb, J. S., Gulbransen, T., Phillips, B. F., Wang, D., Syslo, M. (1983): Behavior and distribution of larval and early juvenile *Homarus americanus*. *Can. J. Fish. Aquat. Sci.* **40**, 2184–2188.

Cocho, G., Perez-Pascual, R., Rius, J. L. (1987a): Discrete systems, cell-cell interactions, and color patterns of animals. I. Conflicting dynamics and color patterns. *J. Theor. Biol.* **125**, 419–435.

Cocho, G., Perez-Pascual, R., Rius, J. L. (1987b): Discrete systems, cell-cell interactions, and color patterns of animals. II. Clonal theory and cellular automata. *J. Theor. Biol.* **125**, 437–447.

Cochran, W. W., Warner, D. W., Tester, J. R., Kuechle, V. B. (1965): Automatic radio-tracking system for monitoring animal movements (Bioelemetry). *Bio Science* **15**, 98–100.

Cohen, D. S., Murray, J. D. (1981): A generalized diffusion model for growth and dispersal in a population. *J. Math. Biol.* **12**, 237–249.

Comins, H. N., Blatt, D. W. E. (1974): Prey–predator models in spatially heterogeneous environments. *J. Theor. Biol.* **48**, 75–83.

Confer D. R., Logan, B. E. (1991): Increased bacterial uptake of macromolecular substrates with fluid shear. *Appl. Environ. Microbiol.* **57**, 3093–3100.

Conway, E., Hoff, D., Smoller, J. (1978): Large time behavior of solutions of systems of nonlinear reaction–diffusion equations. *SIAM J. Appl. Math.* **35**, 1–16.

Conway, G. R. (1977): Mathematical models in applied ecology. *Nature* **269**, 291–297.

Cook, C. D. K. (1987): Dispersion in aquatic and amphibious vascular plants. In *Plant Life in Aquatic Amphibious Habitats*, 179–190. Crawford, R. M. M. (ed.). Oxford: Blackwell.

Cook, C. D. K. (1988): Wind pollination in aquatic angiosperms. *Ann. Missouri Bot. Gard.* **75**, 768–777.

Cook, S. M. F., Linden, D. R. (1996): Effect of food type and placement on earthworm (*Aporrectodea tuberculata*) burrowing and soil turnover. *Biology and Fertility of Soils* **21**, 201–206.

Cooper, W. E. (1978): Home range criteria based on temporal stability of areal occupation. *J. Theor. Biol.* **73**, 687–695.

Corrsin, S. (1974): Limitations of gradient transport models in random walks and in turbulence. *Adv. in Geophys.* Vol. **18A**, 25–60. Frenkiel, F. N., Munn, R. E. (eds.). New York: Academic Press.

Cosner, C., Lazer, A. C. (1984): Stable coexistence states in the Volterra-Lotka competition model with diffusion. *SIAM J. Appl. Math.* **44**, 1112–1132.

Coutts, M. P., Grace, J. (1995): *Wind and Trees*. Cambridge: Cambridge Univ. Press.

Cowles, D. L., Childress, J. J., Gluck, D. L. (1986): New method reveals unexpected relationship between velocity nad drag in the bathypelagic mysid *Gnathophausia ingens*. *Deep-Sea Res. Part A—Ocean. Research papers.* **33**, 865–880.

Cox, P. A. (1988): Hydrophilous pollination. *Ann. Rev. Ecol. Syst.* **19**, 261–280.

Cox, P. A., Sethian, J. A. (1985): Gamete motion, search, and the evolution of anisogamy, oogamy and chemotaxis. *Amer. Naturalist* **125**, 74–101.

Craig, H., Hayward, T. (1987): Oxygen supersaturation in the ocean: Biological versus physical contributions. *Sci.* **29**, 363–374.

Craigie, J. H. (1945): Epidemiology of stem rust in Western Canada. *Scient. Agric.* **25**, 285–401.

Crane, P. R. (1986): Form and function in wind dispersed pollen. In *Pollen and Spores: Form and Function*, 179–202. Blackmore, S., Ferguson, I. K. (eds.). London: Academic Press.

Crank, J. (1975): *The Mathematics of Diffusion*, 2nd ed. Oxford: Oxford Univ. Press.

Crenshaw, H. (1996): A new look at microorganisms: Rotating and translating. *Amer. Zool.* **36**, 608–618.

Cresswell, W. (1994): Flocking as an effective anti-predation strategy in redshanks, *Tringa totanus*. *Animal Behav.* **47**, 433–442.

Crisp, D. J. (1962): Swarming of planktonic organisms. *Nature* **193**, 597–598.

Cross, M. C., Hohenberg, P. C. (1993): Pattern formation outside of equilibrium. *Rev. Mod. Phys.* **65**, 851–1112.

Crowley, J. M. (1976): Clumping instability of a falling horizontal lattice. *Phys. Fluids* **19**, 1296–1300.

Crowley, J. M. (1977): Clumping instability which is not predicted by the nearest neighbor approximation. *Phys. Fluids* **20**, 339.

Csanady, G. T. (1973): *Turbulent Diffusion in the Environment*. Boston: D. Reidel Publ. Co.

Dahmen, H. J., Zeil, J. (1984): Recording and reconstructing three-dimensional trajectories: A versatile method for the field biologist. *Proc. Roy. Soc. Lond. B* **222**, 107–113.

Dallavalle, J. M. (1948): *Micromeritics. The Technology of Fine Particles*. New York: Pitman.

Dam, H. G., Drapeau, D. T. (1995): Coagulation efficiency, organic-matter glues and the dynamics of particles during a phytoplankton bloom in a mesocosm study. *Deep-Sea Res. I* **42**, 111–123.

Dame, R. F. (1996): *Ecology of Marine Bivalves*. Boca Raton, FL: CRC Press.

Danielsen, E. F. (1981): Trajectories of the Mount St. Helens eruption plume. *Sci.* **211**, 819–821.

Darwin, C. (1881): *The Formation of Vegetable Mould Through the Action of Worms with Observations on Their Habits*. New York: New York Univ. Press, 1989 (originally published by J. Murray, London).

386   References

David, C. T. (1986): Mechanisms of directional flight in wind. In *Mechanisms in Insect Olfaction*. Payne, T. L., Birch, M. C., Kennedy, C. E. J. (eds.). Oxford: Clarendon Press.

Davies, K. C., Jarvis, J. U. M. (1986): The burrow system and burrowing of the mole-rats *Bathyergus suillus* and *Cryptomys hottentotus* in the fynbos of the southwestern Cape, South Africa. *J. Zool. London A* **209**, 125–147.

Davis, F. W. (1993): Introduction to spatial statistics. In *Patch Dynamics*, Vol. 96, 16–26. Lecture Notes in Biomathematics. Levin, S. A. (ed.). Berlin: Springer-Verlag.

Davis, H. T. (1960): *Introduction to Nonlinear Differential and Integral Equations*. Washington, DC: U.S. Atomic Energy Comm. (also available at Dover Publications, New York).

Dawes, C. J. (1998): *Marine Botany*, 2nd ed. New York: Wiley.

Dawson, F. H., Robinson, W. N. (1984): Submerged macrophytes and the hydraulic roughness of a lowland chalkstream. *Vehr. Intl. Verein. Limnol.* **22**, 1944–1948.

Deacon, E. L. (1949): Vertical diffusion in the lower layers of the atmosphere. *Quart. J. Roy. Meteorol. Soc.* **75**, 89–103.

DeAngelis, D. L., Gross, L. J. (eds.) (1992): *Individual-Based Models and Approaches in Ecology*. New York, NY: Routledge, Chapman and Hall.

De Angelis, D. L., Mattice, J. S. (1979): Implications of a partial-differential-equation cohort model. *Math. Biosci.* **47**, 271–285.

DeAngelis, D. L., Gross, L. J., Wolff, W. F., Fleming, D. M., Nott, M. P., Comiskey, E. J. (2000): Individual-based models on the landscape: Applications to the Everglades. In *Landscape Ecology: A Top-Down Approach*, 199–211. Sanderson, J., Harris, L. D. (eds.). Boca Raton, FL: Lewis Publishers.

DeGrandpre, M. D., Hammar, T. R., Wallace, D. W. R., Wirick, C. D. (1997): Simultaneous mooring-based measurements of seawater CO_2 and O_2 off Cape Hatteras, North Carolina. *Limnol. Oceanogr.* **42**, 21–28.

Deibel, D. (1990): Still-water sinking velocity of fecal material from the pelagic tunicate *Dolioletta gegenbauri*. *Mar. Ecol. Progr. Ser.* **62**, 55–60.

Deirmendjian, D. (1973): On volcanic and other particulate turbidity anomalies. *Adv. Geophys.* **16**, 267–296.

De Marsily, G. (1986): *Quantitative Hydrogeology*. Orlando: Academic Press.

de Mottoni, P., Orlandi, E., Tesei, A. (1979): Asymptotic behavior for a system describing epidemics with migration and spatial spread of infection. *Nonlinear Anal. Theory Math. Appl.* **3**, 663–675.

Den Hartog, C. (1970): *The Sea-Grasses of the World*. Amsterdam: North-Holland.

Dengler, A. T. (1985): Relationship between physical and biological processes at an upwelling front off Peru 15 degrees south. *Deep-Sea Res.* Part a. **32**, 1301–1316.

Denman, K. L., Platt, T. (1976): The variance spectrum of phytoplankton in a turbulent ocean. *J. Mar. Res.* **34**, 593–601.

Denman, K. L., Okubo, A., Platt, T. (1977): The chlorophyll fluctuation spectrum in the sea. *Limnol. Oceanogr.* **22**, 1033–1038.

Denman, K. L., Abbott, M. R. (1994): Time scales of pattern evolution from cross-spectrum analysis of advanced very high resolution radiometer and coastal zone color scanner imagery. *J. Geophys. Res.* **99**, 7433–7442.

Denn, M. M. (1975): *Stability of Reaction and Transport Processes*. Englewood Cliffs, NJ: Prentice-Hall.

Denny, M. W. (1988): *Biology and the Mechanics of Wave-Swept Environments*. Princeton: Princeton Univ. Press.

Denny, M. W. (1993): *Air and Water*. Princeton: Princeton Univ. Press.

Denny, M. W., Gaylord, B., Helmuth, B., Daniel, T. (1998): The menace of momentum: Dynamic forces on flexible organisms. *Limnol. Oceanogr.* **43**, 955–968.

Denny, M. W., Shibata, M. F. (1989): Consequences of surf-zone turbulence for settlement and external fertilization. *Amer. Nat.* **134**, 859–889.

Deuser, W. G., Muller-Karger, F. E., Hemleben, C. (1988): Temporal variations of particle fluxes in the deep subtropical and tropical North Atlantic: Eulerian versus Lagrangian effects. *J. Geophys. Res.* **93**(C6), 6857–6862.

Deuser, W. G., Muller-Karger, F. E., Evans, R. H., Brown, O. B., Esais, W. E., Feldman, G. C. (1990): Surface-ocean color and deep-ocean carbon flux: How close a connection? *Deep-Sea Res.* **37**, 1331–1343.

Diamond, J. M., May, R. M. (1976): Island biogeography and the design of natural reserves. In *Theoretical Ecology*, 163–186. May, R. M. (ed.). Philadelphia-Toronto: W. B. Saunders Co.

Di Blasio, G. (1979): Nonlinear age-dependent population diffusion. *J. Math. Biol.* **8**, no. 3, 265–284.

Di Blasio, G., Lamberti, L. (1978): An initial-boundary value problem for age-dependent population diffusion. *SIAM J. Appl. Math.* **35**, no. 3, 593–615.

Dickey-Collas, M., Brown, J., Fernand, L., Hill, A. E., Horsburgh, K. J., Garvine, R. W. (1997): Does the western Irish Sea gyre influence the distribution of pelagic juvenile fish? *J. Fish Biol.* **51** (Suppl. A), 206–229.

Dill, L. M., Dunbrack, R. L., Major, P. F. (1981): A new sterophotographic technique for analysing the three-dimensional structure of fish schools. *Environmental Biology of Fishes* **6**, 7–13.

Dobzhansky, T., Wright, S. (1943): Genetics of natural populations. X. Dispersion rates in *Drosophila pseudoobscura. Genetics* **28**, 304–340.

Dobzhansky, T., Wright, S. (1947): Genetics of natural populations. XV. Rate of diffusion of a mutant gene through a population of *Drosophila pseudoobscura. Genetics* **32**, 303–324.

Dobzhansky, T., Powell, J. R. (1974): Rates of dispersal of *Drosophila pseudoobscura* and its relations. *Proc. Roy. Soc. London B* **187**, 281–298.

Doall, M. H., Colin, S. P., Strickler, J. R., Yen, J. (1998): Locating a mate in 3D: The case of *Temora longicornis. Phil. Trans. Roy. Soc. Lond.* **353**, 681–689.

Dodd, C. K., Jr., Cade, B. S. (1998): Movement patterns and the conservation of amphibians breeding in small, temporary wetlands. *Conservation Biol.* **12**, 331–339.

Doney, S. C., Glover, D. M., Najjar, R. G. (1996): A new coupled, one-dimensional biological-physical model for the upper ocean: Applications to the JGOFS Bermuda Atlantic Timeseries Study (BATS) site. *Deep-Sea Res. II. Topical Studies in Oceanography* **43**, 591–624.

Doucet, P. G., Drost, N. J. (1985): Theoretical studies on animal orientation II. Directional displacement in kineses. *J. Theor. Biol.* **117**, 337–361.

Downes, J. A. (1955): Observations on the swarming flight and mating of *Culicoides* (Diptera: Ceratopogonidae). *Trans. Roy. Entomol. Soc. London* **106**, 213–236.

Downes, J. A. (1969): The swarming and mating flight of Diptera. *Ann. Rev. Entomol.* **14**, 271–298.

Draper, J. (1980): The direction of desert locust migration. *J. Anim. Ecol.* **49**, 959–974.

Droppo, L. G., Leppard, G. G., Flannigan, D. T., Liss, S. N. (1997): The freshwater flow: A functional relationship of water and organic and inorganic floc constituents affecting suspended sediment properties. *Water Air Soil Poll.* **99**, 43–54.

Du Toit, J. T., Jarvis, J. U. M., Louw, G. N. (1985): Nutrition and burrowing energetics of the Cape mole-rat *Georychus capensis. Oceologia* **66**, 81–87.

Duarte, C. M., Vaquè, D. (1992): Scale dependence of bacterioplankton patchiness. *Mar. Ecol. Prog. Ser.* **84**, 95–100.

Dubois, D. M. (1975a): A model of patchiness for prey–predator plankton populations. *Ecol. Modelling* **1**, 67–80.

Dubois, D. M. (1975b): The influence of the quality of water on ecological systems. In *Computer Simulation of Water Resources Systems*, 535–542. Vansteenkiste, G. C. (ed.). North-Holland Publ. Co. (New York: American Elsevier Publ. Co.).

Duce, R. A., Liss, P. S., Merrill, J. T., Athans, E. L., Buat-Menard, P., Hicks, B. B., Miller, J. M., Prospero, J. M., Atimoto, R., Church, T. M., Ellis, W., Galloway, J. N., Hansen, L., Jickells, T. D., Knap, A. H., Reinhardt, K. H., Schneider, B., Soudine, A., Tokos, J. J., Tsunogai, S., Wollast, R., Ahmou, M. (1991): The atmospheric input of trace species to the world ocean. *Global. Biogeochem. Cycles* **5**, 193–259.

Dugdale, R. C., Wilkerson, F. P., Minas, H. J. (1995): The role of a silicate pump in driving new production. *Deep-Sea Res.* **42**, 697–719.

Dunbar, S. R. (1983): Travelling wave solutions of diffusive Lotka-Volterra equations. *J. Math. Biol.* **17**, 11–32.

Dunbar, S. R. (1986): Travelling waves in diffusive predator-prey equations: periodic orbits and point-to-periodic heteroclinic orbits. *SIAM J. Appl. Math.* **46**, 1057–1078.

Dunbar, S. R., Othmer, H. G. (1986): On a nonlinear hyperbolic equation describing transmission lines, cell movement, and branching random walks. In: H. G. Othmer, ed. *Nonlinear Oscillations in Biology and Chemistry*. Berlin: Springer-Verlag.

Dunn, J. E., Gipson, P. S. (1977): Analysis of radio telemetry data in studies of home range. *Biometrics* **33**, 85–101.

Dunn, J. E. (1978): Optimal sampling in radio telemetry studies of home range. In *Time Series and Ecological Processes*, 53–70. Shugart, H. H. (ed.). Philadelphia: SIAM Inst. for Mathematics and Society.

Dunning, J. B., Jr., Stewart, D. J., Danielson, B. J., Noon, B. R., Root, T. R., Lamberson, R. H., Stevens, E. E. (1995): Spatially explicit population models: current forms and future uses. *Ecol. Appl.* **5**, 3–11.

Durrett, R., Levin, S. A. (1994a): The importance of being discrete (and spatial). *Theor. Pop. Biol.* **46**, 363–394.

Durrett, R., Levin, S. A. (1994b): Stochastic spatial models: A user's guide to ecological applications. *Philosophical Trans. Roy. Soc. Lond. B* **343**, 329–350.

Dusenbery, D. B. (1998): Fitness landscapes for effects of shape on chemotaxis and other behaviors of bacteria. *J. Bacteriol.* **180**, 5978–5083.

Dusenbery, D. B. (1997): Minimum size limit for useful locomotion by free-swimming microbes. *Proc. Natl. Acad. Sci. USA* **94**, 10949–10954.

Dusenbery, D. B., Snell, T. W. (1995): A critical body size for use of pheromones in mate location. *J. Chemi. Ecol.* **21**, 427–438.

Dworschak, P. C. (1983): The biology of *Upogebia pussila* (Petagna) (Decapoda, Thalassinidae) 1. The burrows. *Mar. Ecol.* **4**, 19–43.

Dwyer, G. (1992): On the spatial spread of insect pathogens: theory and experiment. *Ecology* **73**, 479–494.

Dwyer, G. (1994): Density dependence and spatial structure in the dynamics of insect pathogens. *Amer. Nat.* **143**, 533–562.

Eberhard W. G. (1978): Mating swarms of a South American Acropygia (Hymenoptera: Formicidae) *Entomol. News* **89**, 14–16.

Edelstein-Keshet, L. (1986): Mathematical theory for plant-herbivore systems. *J. Math. Biol.* **24**, 25–58.

Edelstein-Keshet, L. (1988): *Mathematical Models in Biology.* New York: Birkhauser, McGraw-Hill.

Edelstein-Keshet, L., Watmough, J., Grunbaum, D. (1998): Do travelling band solutions describe cohesive swarms? An investigation for migratory locusts. *J. Math. Biol.* **36**, 515–549.

Einstein, A. (1905): Über die von der molekularkinetischen Theorie der Wärme geforderte Bewegung von in ruhenden Flüssigkeiten suspendierten Teilchen. *Ann. d. Physik* (4) **17**, 549–560.

Elgar, M. A. (1989): Predator vigilance and group size in mammals and birds: A critical review of the empirical evidence. *Biol. Rev.* **64**, 13–33.

Elgar, M. A., Catterall, C. P. (1981): Flocking and predator surveillance in house sparrows—test of an hypothesis. *Anim. Behav.* **29**, 868–872.

Elliott, J. P., McTaggart Cowan, I., Holling, C. S. (1977): Prey capture by the African lion. *Can. J. Zool.* **55**, 1811–1828.

Elskens, M., Baeyens, W., Goeyens, L. (1997): Contribution of nitrate to the uptake of nitrogen by phytoplankton in an ocean margin environment. *Hydrobiol.* **353**, 139–152.

Elton, C. S. (1958): *The Ecology of Invasions by Animals and Plants.* London: Methuen & Co. Ltd.

Emerson, S., Jahnke, R., Heggie, D. (1984): Sediment-water exchange in shallow water estuarine sediments. *J. Mar. Res.* **42**, 709–730.

Emerson, S., Quay, P., Stump, C., Wilbur, D., Schudlich, R. (1993): Determining primary production from the mesoscale oxygen field. *ICES Mar. Sci. Symp.* **197**, 196–206.

Emlet, R. B. (1986): Larval production, dispersal, and growth in a fjord: A case study on larvae of the sand dollar *Dendraster excentricus. Mar. Ecol. Prog. Ser.* **31**, 245–254.

Enfield, C. G., Carsel, R. F., Cohen, S. Z., Phan, T., Walters, D. M. (1982): Approximate pollutant transport to groundwater. *Groundwater* **20**, 711–722.

Epel, D. (1991): How successful is the fertilization process of the sea urchin egg? In *Proc. 7th Intl. Echinoderm Conf.*, Atami, 1990, 51–45. Yanagisawa, T., Yasumasu, I., Oguro, C., Suzuki, N., Motokawa, T. (eds.). Rotterdam: Balkema.

Evans, G. T. (1978a): The persistence of a phytoplankton patch diffusing into slightly hostile surroundings. *Deep-Sea Res.* (submitted).

Evans, G. T. (1978b): Biological effects of vertical-horizontal interaction. In *Spatial Pattern in Plankton Communities*, 157–179. Steele, J. H. (ed.). New York: Plenum Press.

Evans, G. T. (1980): Diffusive structure: counter examples to any explanation. *J. Theor. Biol.* **82**, 313–315.

Evans, G. T., Steele, J. H., Kullenberg, G. E. B. (1977): A preliminary model of shear diffusion and plankton populations. Scottish Fisheries Research Report No. 9. Aberdeen: Dept. Agriculture and Fisheries for Scotland.

Evans, G. T., Taylor, F. J. R. (1980): Phytoplankton accumulation in Langmuir cells. *Limnol. Oceanogr.* **25**, 840–845.

Fadlallah, Y. H. (1983): Sexual reproduction, development and larval biology in scleractinian corals. *Coral Reefs* **2**, 129–150.

Faegri, K., van der Pijl, L. (1979): *The Principles of Pollination Ecology*, 3rd ed. Oxford: Pergamon Press.

Fagan, W. (1997): Introducing a boundary-flux approach to quantifying insect diffusion rates. *Ecology* **78**, 579–587.

Falkowski, P. G., Raven, J. A. (1997): *Aquatic Photosynthesis.* Oxford: Blackwell.

Falkowski, P. G., Ziemann, D., Kolber, Z., and Bienfang, P. K. (1991): Role of eddy pumping in enhancing primary production in the ocean. *Nature* **353**, 55–58.

Faller, A. J. (1971): Oceanic turbulence and the Langmuir circulations. *Ann. Rev. Ecol. Systematics* **2**, 201–236.

Faller, A. J., Caponi, E. A. (1978): Laboratory studies of wind-driven Langmuir circulations. *J. Geophys. Res.* **83**, 3617–3633.

Farkas, S. R., Shorey, H. H. (1972): Chemical trail-following by flying insects: A mechanism for orientation to a distant odor source. *Sci.* **178**, 67–68.

Farnsworth, K. D., Beecham, J. A. (1999): How do grazers achieve their distribution? A continuum of models from random diffusion to the ideal free distribution using biased random walks. *Amer. Nat.* **153**, 509–526.

Fasham, M. J. R., Angel, M. V., Roe, H. S. J. (1974): An investigation of the spatial pattern of zooplankton using the Longhurst–Hardy plankton recorder. *J. Exp. Mar. Biol. Ecol.* **16**, 93–112.

Fasham, M. J. R. (1978): The application of some stochastic processes to the study of plankton patchiness. In *Spatial Pattern in Plankton Communities*, 131–156. Steele, J. H. (ed.). New York: Plenum Press.

Fasham, M. J. R., Ducklow, H. W., McKelvie, S. M. (1990): A nitrogen-based model of plankton dynamics in the oceanic mixed layer. *J. Mar. Res.* **48**, 591–639.

Fauchald, P. (1999): Foraging in a hierarchical patch system. *Amer. Nat.* **153**, 603–613.

Feller, W. (1968): *An Introduction to Probability Theory and its Applications, Vol. 1.* New York, NY: Wiley and Sons.

Fenchel, T. (1987): *Ecology of Protozoa: The Biology of Free-Living Phagotrophic Protists.* Madison, WI: Science Tech Publishers.

Fedoryako, B. I. (1982): Langmuir circulations and a possible mechanism of formation of fish associations around a floating object. *Oceanol.* **22**, 228–232.

Feller, W. (1968): *An Introduction to Probability Theory and Its Applications*, Vol. 1. New York: Wiley and Sons.

Fenchel, T. (1996a): Worm burrows and oxic microniches in marine sediments. 1. Spatial and temporal scales. *Mar. Biol.* **127**, 289–295.

Fenchel, T. (1996b): Worm burrows and oxic microniches in marine sediments. 2. Distribution patterns of ciliated protozoa. *Mar. Biol.* **127**, 297–301.

Fielding, G. T. (1970): A further note on elephant herds. *Math. Gazette* **54**, 297–298.

Fife, P. C. (1976): Pattern formation in reacting and diffusing systems. *J. Chem. Phys.* **64**, 554–564.

Fife, P. (1978): Asymptotic states for equations of reaction and diffusion. *Bull. Amer. Math. Soc.* **84**, no. 5, 693–726.

Fife, P. C. (1979): *Mathematical aspects of reacting and diffusing systems. Lecture notes in Biomathematics* v. 28. New York: Springer-Verlag, 1979.

Fife, P. C. (1981): Wave fronts and target patterns. In: *Application of nonlinear analysis in the physical sciences.* Amann, H., N. Bazley, and K. Kirchgasser (eds), 206–228. Boston: Pitman.

Fife, P. C., McLeod, J. B. (1975): The approach of solutions of nonlinear diffusion equations to travelling wave solutions. *Bull. Amer. Math. Soc.* **81**, 1076–1078.

Fife, P. C., McLeod, J. B. (1977): The approach of solutions of nonlinear diffusion equations to travelling wave solutions. *Arch. Rat. Mech. Anal.* **65**, 335–361.

Fife, P. C., Peletier, L. A. (1977): Nonlinear diffusion in population genetics. *Arch. Rat. Mech. Anal.* **64**, 93–109.

Finch, S., Skinner, G. (1975): An improved method of marking cabbage root flies. *Ann. Appl. Biol.* **79**, 243–246.

Finnigan, J. J. (1979a): Turbulence in waving wheat. I Mean statistics and Honami. *Boundary-Layer Meteorol.* **16**, 181–211.

Finnigan, J. J. (1979b): Turbulence in waving wheat. II Structure of turbulent transfer. *Boundary-Layer Meteorol.* **16**, 213–236.

Finnigan, J. (2000): Turbulence in plant canopies. *Ann. Rev. Fluid Mech.* **32**, 519–571.

Finnigan, J. J., Brunet, Y. (1995): Turbulent airflow in forests on flat and hilly terrain. In *Wind and Trees*, 3–40. Coutts, M.P., Grace, J. (eds.). Cambridge: Cambridge Univ. Press.

Finnigan, J. J., Mulhearn, P. J. (1978): Modelling waving crops in a wind tunnel. *Boundary-Layer Meteorol.* **14**, 253–277.

Fischer, H. B. (1968): Dispersion predictions in natural streams. *J. Sanitary Eng. Div., Proc. Amer. Soc. Civil Eng.* **94***(SA 5)*, 927–943.

Fischer, H. B. (1973): Longitudinal dispersion and turbulent mixing in open channel flow. *Ann. Rev. Fluid Mech.* **5**, 59–78.

Fischer, H. B., List, E. J., Koh, R. C. Y., Imberger, J., Brooks, N. H. (1979): *Mixing in Inland and Coastal Waters.* New York: Academic Press.

Fisher, R. A. (1937): The wave of advance of advantageous genes. *Ann. Eugen., London* **7**, 355–369.

Flierl, G., Grunbaum, D., Levin, S. A., Olson, D. (1999): From individual to aggregations: The interplay between behavior and physics. *J Theor Biol.* **196**, 397–454.

Flohn, H., Penndorf, R. (1950): The stratification of the atmosphere. (I): *Bull. Amer. Meteorol. Soc.* **31**, 71–78. (II): *Bull. Amer. Meteorol, Soc.* **31**, 126–130.

Foda, M. A. (1983): Dry-fall of fine dust on sea. *J. Geophys. Res.* **88**C, 6021–6026.

Follett, R. F., Keeney, D. R., Cruse, R. F. (eds.) (1991): Managing nitrogen for groundwater quality and farm profitability. *Soil Sci. Soc. Am.* Madison, WI.

Fong, P., Foin, T. C., Zedler, J. B. (1994): A simulation model of lagoon algae based on nitrogen competition and internal storage. *Ecol. Monogr.* **64**, 225–247.

Fonseca, M. S., Fisher, J. S. (1986): A comparison of canopy friction and sediment movement between four species of seagrass with reference to their ecology and restoration. *Mar. Ecol. Progr. Ser.* **29**, 15–22.

Fonseca, M. S., Fisher, J. S., Zieman, J. C., Thayer, G. W. (1982): Influence of the seagrass, *Zostera marina*, on current flow. *Est. Coast. Shelf Sci.* **15**, 351–364.

Fonseca, M. S., Zieman, J. C., Thayer, G. W., Fisher, J. S. (1983): The role of current velocity in structuring eelgrass (*Zostera marina*) meadows. *Est. Coast. Shelf Sci.* **17**, 367–380.

Forgacs, O. L., Mason, S. G. (1958): The flexibility of wood-pulp fibers. *TAPPI* **41**, 695–704.

Foster, W. A., Treherne, J. E. (1981): Evidence for the dilution effect in the selfish herd from fish predation on a marine insect. *Nature* **293**, 466–467.

Foster, W. A., Treherne, J. E. (1982): Reproductive behavior of the ocean skater, Halobates robustus (Hemiptera, Gerridae) in the Galapagos Islands. *Oecologia* **55**, 202–207.

Fowler, S. W., Knauer, G. A. (1986): Role of large particles in the transport of elements and organic compounds through the oceanic water column. *Progr. Oceanogr.* **16**, 147–194.

Fraenkel, G. S., Gunn, D. L. (1961): *The Orientation of Animals*. New York: Dover Publ. Inc.

France, R. L., Holmquist, J. G. (1997): $\delta^{13}C$ variability of macroalgae: Effects of water motion via baffling by seagrasses and mangroves. *Mar. Ecol. Progr. Ser.* **149**, 305–308.

Frank, K. T., Carscaddia, J. E., Legget, W. C. (1993): Causes of spatiotemporal variation in the patchiness of larval fish distributions: Differential mortality of behavior? *Fish Oceanogr.* **2**, 114–123.

Fréchette, M., Butman, C. A., Geyer, W. G. (1989): The importance of boundary-layer flows in supplying phytoplankton to the benthic suspension feeder, *Mytilus edulis* L. *Limnol. Oceanogr.* **34**, 19–36.

Freeman, G. H. (1977): A model relating numbers of dispersing insects to distance and time. *J. Appl. Ecol.* **14**, 477–487.

French, N. R. (1971): Simulation of dispersal in desert rodents. In *Statistical Ecology*, **3**, 367–375. Patil, G. P., Pielou, E. C., Waters, W. E. (eds.). University Park, PA: Penn. State Univ. Press.

French, D. P., Furnas, M. J., Smayda, T. J. (1983): Diel changes in nitrite concentration in the chlorophyll maximum in the Gulf of Mexico. *Deep-Sea Res. A.* **30**, 707–722.

Frenkiel, F. N. (1953): Turbulent diffusion: Mean concentration distribution in a flow field of homogeneous turbulence. *Adv. in Appl. Mech.* **3**, 61–107. New York: Academic Press.

Frenkiel, F. N., Munn, R. E. (eds.) (1974): *Turbulent Diffusion in Environmental Pollution. Adv. in Geophys.* **18A**. New York, Academic Press.

Fretwell, S. C. (1987): Food chain dynamics: The central theory of ecology? *Oikos* **50**, 291–301.

Frost, B. W. (1972): Effects of size and concentration of food particles on the feeding behaviour of the marine planktonic copepod *Calanus pacificus*. *Limnol. Oceanogr.* **17**, 805–815.

Frost, B. W. (1977): Feeding behaviour of *Calanus pacificus* in mixtures of food particles. Limnol. *Oceanogr.* **22**, 263–266.

Fujii, H., Minura, M., Nishiura, Y. (1982): A picture of the global bifurcation diagram in ecological interacting and diffusing systems. *Phys.* D **5**, no. 1, 1–42.

Fujimoto, M., Hirano, T. (1972): Study of the Kuroshio functioning as a means of transportation and diffusion of fish eggs and larvae. I. The results of drift bottle experiments. *Bull. Tokai Reg. Fish. Res. Lab.* **71**, 51–68 (Japanese with English abstract).

Furter, J., Grinfeld, M. (1989): Local vs. nonlocal interaction in population dynamics. *J. Math. Biol.* **27**, no. 1, 65–80.

Furusawa, J., Hill, A. V., Parkinson, J. L. (1928): The dynamics of "sprint" running. *Proc. Roy. Soc. London Ser. B*, **102**, 29–42.

Gallager, S. M., Davis, C. S., Epstein, A. W., Solow, A., Beardsley, R. C. (1997): High resolution observations of plankton spatial distributions correlated with hydrography in the Great South Channel, Georges Bank. *Deep-Sea Res. II* **43**, 1627–1663.

Gallucci, V. F. (1973): On the principles of thermodynamics in ecology. *Ann. Rev. Ecol. System.* **4**, 329–357.

Galton, F. (1981): *Macmillan's Mag., London* **23**, 353 (In Hamilton, W. D., 1971).

Gambi, M. C., Nowell, A. R. M., Jumars, P. A. (1990): Flume observations on flow dynamics in *Zostera marina* (eelgrass) beds. *Mar. Ecol. Progr. Ser.* **61**, 159–169.

Game, M. (1980): Best shape for nature reserves. *Nature* **287**, 630–632.

García-Moliner, G., Mason, D. M., Greene, C. H., Lobo, A., Li, B.-L., Wu, J., Bradshaw, G. A. (1993): Description and analysis of spatial patterns. In *Patch Dynamics. Lecture Notes in Biomathematics*, Vol. 96, 70–89. Levin, S. A. (ed.). Berlin: Springer-Verlag.

Gardner, L. R., Sharma, P., Moore, W. S. (1987): A regeneration model for the effect of bioturbation by fiddler crabs on ^{210}Pb profiles in salt marsh sediments. *J. Env. Radioact.* **5**, 25–36.

Garfield, P. C., Packard, T. T., Friederich, G. E., Codispoti, L. A. (1983): A sub surface particle maximum layer and enhanced microbial activity in the secondary nitrite maximum of the northeastern tropical Pacific Ocean. *J. Mar. Res.* **41**, 747–768.

Gary, N. E. (1973): The distribution and flight range of foraging honey bees (paper presented at the Entomol. Soc. Amer. meeting in Dallas, Texas, 26–30 Nov. 1973).

Gaylord, B., Denny, M. W. (1997): Flow and flexibility. I—Effects of size, shape and stiffness in determining wave forces on the stipitate kelps *Eisenia arborea* and *Pterygophora californica. J. Exp. Biol.* **200**, 3141–3164.

Geers, R., Puers, B., Goedseels, V., Wouters, P. (1997): *Electronic Identification, Monitoring and Tracking of Animals.* Wallingford, UK: CAB International.

George, D. G., Edwards, R. W. (1973): *Daphnia* distribution within Langmuir circulations. *Limnol. Oceanogr.* **18**, 798–800.

Georgii, B. (1980): Home range patterns of female red deer (Cervas elaphus L.) in the Alps. *Oecologia* **47**, 278–285.

Gibson, G. (1985): Swarming behavior of the mosquito Culex pipiens quinquefasciatus: A quantitative analysis. *Physiol. Entomol.* **10**, 283–296.

Gierer, A., Meinhardt, H. (1972): A theory of biological pattern formation. *Kybernetik* **12**, 30–39.

Gifford, F. (1959): Statistical properties of a fluctuating plume dispersion model. *Adv. in Geophys.* **6**, 117–137.

Gili, J. M., Coma, R. (1998): Benthic suspension feeders: Their paramount role in littoral marine food webs. *Trends Ecol. Evol.* **13**, 316–321.

Gillies, M. T. (1961): Studies on the dispersion and survival of *Anopheles gamviae* Giles in East Africa, by means of marking and release experiments. *Bull. Entomol. Res.* **52**, 99–127 (In Freeman, 1977).

Gillies, M. T., Wilkes, T. J. (1974): Evidence for downwind flights by host-seeking mosquitoes. *Nature* **252**, 388–389.

Gillis, J. (1956): Centrally biased discrete random walk. *Quart. J. Math.* (2) **7**, 144–152.

Gilpin, M. E. (1980): Subdivision of nature reserves and the maintenance of species diversity. *Nature* **285**, 567–568.

Glansdorff, P., Prigogine, I. (1971): *Thermodynamic Theory of Structure, Stability and Fluctuations*. New York: J. Wiley & Sons.

Glass, R. J., Parlange, J. Y., Steenhuis, T. S. (1989a): Wetting front instability. 1. Theoretical discussion and dimensional analysis. *Water Resour. Res.* **25**, 1187–1194.

Glass, R. J., Steenhuis, T., Parlange, J. Y. (1989b): Mechanism for finger persistence in homogeneous, unsaturated, porous media: Theory and verification. *Soil Sci.* **148**, 60–70.

Glick, P. A. (1939): The distribution of insects, spiders and mites in the air. *Tech. Bull. U.S. Dept. Agric.* No. 673.

Gmitro, J. I., Scriven, L. E. (1966): A physicochemical basis for pattern and rhythm. In *Intracellular Transport*, 221–255. Warren, K. B. (ed.). New York: Academic Press.

Godson, W. L. (1957): The diffusion of particulate matter from an elevated source. *Arch. Meteorol. Geophys. Biokl. A* **10**, 305–327.

Goel, N. S., Maitra, S. C., Montroll, E. W. (1971): On the Volterra and other nonlinear models of interacting populations. *Rev. Modern Phys.* **43**, 231–276.

Goel, N. S., Richter-Dyn, N. (1974): *Stochastic Models in Biology*. New York: Academic Press.

Goldstein, S. (1951): On diffusion by discontinuous movements, and on the telegraph equation. *Quart. J. Mech. Appl. Math.* **4**, 129–156.

Gonzalez, J. M., Suttle, C. A. (1993): Grazing by marine nonoflagellates on viruses and virus-sized particles: Ingestion and digestion. *Mar. Ecol. Progr. Ser.* **94**, 1–10.

Gopalsamy, K. (1976): On the asymptotic age distribution in dispersive population. *Math. Biosciences* **31**, 191–205.

Gopalsamy, K. (1977a): Competition, dispersion and coexistence. *Math. Biosciences* **33**, 25–33.

Gopalsamy, K. (1977b): Competition and coexistence in spatially heterogeneous environments. *Math. Biosciences* **36**, 229–242.

Gopalsamy, K. (1977c): Age-dependent population dispersion in linear habitats. *Ecol. Modelling* **3**, 119–132.

Gopalsamy, K. (1980): Pursuit-evasion wave trains in prey-predator systems with diffusionally coupled delays. *Bull. Math. Biol.* **42**, 871–877.

Gopalsamy, K., Aggarwala, B. D. (1981): The logistic equation with a diffusionally coupled delay. *Bull. Math. Biol.* **43**, no. 2, 125–140.

Gordon, A. L. (1986): Interocean exchange of thermocline water. *J. Geophys. Res.* **91**, 5037–5046.

Govoni, J. J., Grimes, C. B. (1992): The surface accumulation of larval fishes by hydrodynamic convergence within the Mississippi River plume front. *Cont. Shelf Res.* **12**, 1265–1276.

Grace, J. (1977): *Plant Response to Wind*. London, New York, San Francisco: Academic Press.

Graf, G. (1992): Benthic–pelagic coupling: A benthic view. *Oceanogr. Mar. Biol. Ann. Rev.* **30**, 149–190.

Gray, A. H., Caughey, T. K. (1965): A controversy in problems involving random parametric excitation. *J. Math. Physics* **44**, 288–296.

Greenbank, D. O., Schaefer, G. W., Rainey, R. C. (1980): Spruce budworm (Lepidoptera: Tortricidae) moth flight and dispersal: new understanding from canopy observations, radar, and aircraft. *Memoirs of the Entomological Society of Canada No. 110*. 49 pp. AN00137.

Greene, D. F., Johnson, E. A. (1989): A model of wind dispersal of winged or plumed seeds. *Ecology* **70**, 339–347.

Greene, D. F., Johnson, E. A. (1992): Can the variation in samara mass and terminal velocity on an individual plant affect the distribution of dispersal distances? *Amer. Nat.* **139**, 825–838.

Greene, D. F., Johnson, E. A. (1996): Wind dispersal of seeds from a forest into a clearing. *Ecology* **72**, 595–609.

Greenstone, M. H. (1990): Meteorological determinants of spider ballooning—the role of thermals vs. the vertical wind-speed gradient in becoming airborne. **84**, 164–168.

Gregory, P. H. (1961): *The Microbiology of the Atmosphere*. London: Leonard Hill Ltd.

Gregory, P. H. (1973): *The Microbiology of the Atmosphere*, 2nd ed. New York: J. Wiley & Sons.

Greig-Smith, P. (1964): *Quantitative Plant Ecology*, 2nd ed. Washington, DC: Butterworths.

Griffin, D. R., Hock, R. J. (1949): Airplane observations of homing birds. *Ecology* **30**, 176–198.

Grill, E. V. (1970): A mathematical model for the marine dissolved silicate cycle. *Deep-Sea Res.* **17**, 245–266.

Grimes, C. B., Able, K. W., Jones, R. S. (1986): Tilefish, *Lopholatilus chamaeleonti-ceps*, habitat, behavior and community structure in Mid-Atlantic and southern New England waters. *Env. Biol. Fish.* **15**, 273–292.

Grimm, V. (1999): Ten years of individual-based modelling in ecology: What have we learned and what could we learn in the future? *Ecol. Modelling* **115**, 129–148.

Grindrod, P. (1988): Models of individual aggregation or clustering in single and multi-species communities. *J. Math. Biol.* **26**, 651–660.

Grindrod, P. (1996): *The Theory and Applications of Reaction-Diffusion Equations (2nd ed.)*. Oxford: Clarendon Press.

Grizzle, R. E., Short, F. T., Newell, C. R., Hoven, H., Kindblom, L. (1996): Hydrodynamically induced synchronous waving of seagrasses: "Monami" and its possible effects on larval mussel settlement. *J. Exp. Mar. Biol. Ecol.* **206**, 165–177.

Gronell, A. M., Colin, P. L. (1985): A toroidal vortex for gamete dispersion in a marine fish, *Pygophlites diacanthus* (Pisces: Pomacanthidae). *Anim. Behav.* **33**, 1021–1040.

Groot, C., Margolis, L. (1991): *Pacific Salmon Life Histories*. Vancouver: UBC Press.

Grossart, H. P., Simon, M., Logan, B. E. (1997): Formation of macroscopic organic aggregates (lake snow) in a large lake: The significance of transparent exopolymer particles, phytoplankton, and zooplankton. *Limnol. Oceanogr.* **42**, 1651–1659.

Grünbaum, D. (1992): *Local Processes and Global Patterns: Biomathematical Models of Bryozoan Feeding Currents and Density-dependent Aggregations in Antarctic Krill. Ph.D. Thesis*, Ithaca, NY: Cornell University.

Grünbaum, D. (1994): Translating stochastic density-dependent individual behaviour with sensory constraints to an Eularian model of animal swarming. *J. Math. Biol.* **33**, 139–161.

Grünbaum, D. (1998): Schooling as a strategy for taxis in a noisy environment. *Evol. Ecol.* **12**, 503–522.

Grünbaum, D. (1998): Using spatially explicit models to characterize foraging performance in heterogeneous landscapes. *Amer. Nat.* **151**(2), 97–115.

Grünbaum, D. (1999): Advective-diffusive equations for generalized tactic searching behaviors. *J. Math. Biol.* **38**, 169–194.

Grunbaum, D. (2000): Advection-diffusion equations for internal state-mediated random walks. *SIAM J. Appl. Math.* **61**, 43–73.

Grünbaum, D. (2001): Population-level descriptions for random walks behaviors mediated by internal state dynamics. *SIAM J. Appl. Math.* (in press).

Grünbaum, D., Okubo, A. (1994): "Modelling social animal aggregations," In: *Frontiers of Theoretical Biology, Lecture Notes in Biomathematics*, Vol. 100 (ed. S. A. Levin). New York: Springer-Verlag.

Grundy, R. E., Peletier, L. A. (1987): Short time behavior of a singular solution to the heat equation with absorption. *Proc. Roy. Soc. Edinburgh* A **107**, no. 3–4, 271–288.

Gueron, S. (1998): The steady state distribution of coagulation–fragmentation processes. *J. Math. Biol.* **37**, 1–27.

Gueron, S., Levin, S. A. (1994): Self organization of front patterns in large wildebeest herds. *J. Theor. Biol.* **165**, 541–552.

Gueron, S., Levin, S. A. (1995): The dynamics of group formation. *Math. Biosci.* **128**, 243–264.

Gueron, S., Liron, N. (1989): A model of herd grazing as a travelling wave, chemotaxis and stability. *J. Math. Biol.* **27**, 595–608.

Gueron, S., Levin, S. A., Rubenstein, D. I. (1996): The dynamics of herds: From individual to aggregations. *J. Theor. Biol.* **182**, 85–98.

Guo, Y. H., Sperry, R., Cook, C. D. K., Cox, P. A. (1990): The pollination ecology of *Zannichelia palustris* L. (Zannicheliaceae). *Aquat. Bot.* **38**, 29–45.

Gupta, J. L., Deheri, G. M. (1990): A mathematical model of the effect of respiration on the gas concentration in an animal burrow. *Nat. Acad. Sci. Letters* **13**, 249–251.

Gurney, W. S. C., Nisbet, R. M. (1975): The regulation of inhomogeneous populations. *J. Theor. Biol.* **52**, 441–457.

Gurney, W. S. C., Nisbet, R. M. (1976): A note on non-linear population transport. *J. Theor. Biol.* **56**, 249–251.

Gurtin, M. E. (1973): A system of equations for age-dependent population diffusion. *J. Theor. Biol.* **40**, 389–392.

Gurtin, M. E. (1974): Some mathematical models for population dynamics that lead to segregation. *Quart. J. Appl. Math.* **32**, 1–9.

Gurtin, M. E., MacCamy, R. C. (1977): On the diffusion of biological populations. *Math. Biosciences* **33**, 35–49.

Gurtin, M. E., MacCamy, R. C. (1981): Diffusion models for age-structured populations. *Math. Biosci.* **54**, no. 1–2, 49–59.

Gurtin, M. E., Pipkin, A. C. (1984): On interacting populations that disperse to avoid crowding. *Q. Appl. Math.* **42**, 87–94.

Gyllenberg, M., Hanski, I., Hastings, A. M. (1997): Structured metapopulation models. In *Metapopulation Biology: Ecology, Genetics and Evolution*, 93–122. Hanski, I., Gilpin, M. (eds.). San Diego: Academic Press.

Haas, C. A. (1995): Dispersal and use of corridors by birds in wooded patches on an agricultural landscape. *Conservation Biol.* **9**, 845–854.

Haddad, N. M. (1999): Corridor use predicted from behaviors at habitat boundaries. *Am. Nat.* **153**, 215–227.

Hadeler, K. P., van der Heiden, U., Rothe, F. (1974): Nonhomogeneous spatial distributions of populations. *J. Math. Biol.* **1**, 165–176.

Hadeler, K. P., Rothe, F. (1975): Travelling fronts in nonlinear diffusion equations. *J. Math. Biol.* **2**, 251–263.

Haken, H. (1983): *Synergetics. An Introduction: Nonequilibrium Phase Transitions and Self-Organization in Physics, Chemistry, and Biology*, 3rd ed. Berlin: Springer-Verlag.

Hall, R. L. (1977): Amoeboid movement as a correlated random walk. *J. Math. Biol.* **4**, 327–335.

Hall, S. J., Wardle, C. S., MacLennan, D. N. (1986): Predator evasion in a fish school: test of a model of the fountain effect. *Marine Biology* **91**, 143–148.

Hallam, T. G. (1978/79): A temporal study of diffusion effects on a population modeled by a quadratic growth. *Nonlinear Anal.* **3**, no. 1, 123–133.

Halloway, C. F., Cowen, J. P. (1997): Development of a scanning confocal laser microscopic technique to examine the structure and composition of marine snow. *Limnol. Oceanogr.* **42**, 1340–1352.

Haltiner, G. H., Martin, F. L. (1957): *Dynamical and Physical Meteorology*. New York, Toronto, London: McGraw-Hill Book Co.

Hamilton, W. D. (1971): Geometry for the selfish herd. *J. Theor. Biol.* **31**, 295–311.

Hamilton, W. J., Gilbert, W. M., Heppner, F. H., Planck, R. J. (1967): Starling roost dispersal and a hypothetical mechanism regulating rhythmical animal movement to and from dispersal centers. *Ecology* **48**, 825–833.

Hamilton, W. J., Gilbert, W. M. (1969): Starling dispersal from a winter roost. *Ecology* **50**, 886–898.

Hamner, P., Hamner, W. M. (1977): Chemosensory tracking of scent trails by the planktonic shrimp *Acetes sibogae australis*. *Sci.* **195**, 886–888.

Hamner, W. M., Carleton, J. H. (1979): Copepod swarms: Attributes and role in coral reef ecosystems. *Limnol. Oceanog.* **24**, 1–14.

Hamner, W. M., Hauri, I. R. (1981): Long-distance horizontal migrations of zooplankton (Scyphomedusae: *Mastigias*). *Limnol. Oceanogr.* **26**, 414–423.

Hamner, W. M., Schneider, D. (1986): Regularly spaced rows of medusae in the Bering Sea: Role of Langmuir circulation. *Limnol. Oceanogr.* **31**, 171–177.

Hangartner, W. (1967): Spezifität und Inaktivierung des Spurpheromons von *Lasius fuliginosus* Latr. und Orientierung der Arbeiterinnen im Duftfeld. *Z. für Verg. Physiol.* **57**, 103–136.

Hansen, J. L. S., Kiørboe, T., Alldredge, A. L. (1996): Marine snow derived from abandoned larvacean houses: Sinking rates, particle content and mechanisms of aggregate formation. *Mar. Ecol. Progr. Ser.* **141**, 205–215.

Hanski, I. (1991): Single-species metapopulation dynamics: Concepts, models and observations. *Biol. J. Linnean Soc.* **42**, 17–38.

Hansteen, T. L., Andreassen, H. P., Ims, R. A. (1997): Effects of spatiotemporal scale on autocorrelation and home range estimators. *J. Wildl. Manage.* **61**, 280–290.

Hara, T. J. (1971): Chemoreception. In *Fish Physiology*, Vol. 5, 79–120. Hoar, W. S., Randall, D. J. (ed.). New York: Academic Press.

Hara, T. (1984): A stochastic-model and the moment dynamics of the growth and size distribution in plant-populations. *J. Theor. Biol.* **109**, 173–190.

Harada, K., Fukao, T. (1978): Coexistence of competing species over a linear habitat of finite length. *Math. Biosci.* **38**, 279–291.

Harder, L. D. (1998): Pollen-size comparisons among animal-pollinated angiosperms with different pollination characteristics. *Biol. J. Linnean Soc.* **64**, 513–525.

Hardy, K. R., Atlas, D., Glover, K. M. (1966): Multi-wave-length backscatter from the clear atmosphere. *J. Geophys. Res.* **71**, 1537–1552.

Harestad, A. S., Bunnell, F. L. (1979): Home range and body weight—a reevaluation. *Ecology* **60**, 389–402.

Hargrave, B. T. (1973): Coupling carbon flow through some pelagic and benthic communities. *J. Fish. Res. Board Can.* **30(9)**, 1317–1326.

Harper, J. L. (1977): *Population Biology of Plants.* London: Academic Press.

Harrison, L. G. (1987): What is the status of reaction-diffusion theory thirty-four years after Turing? *J. Theor. Biol.* **125**, 369–384.

Harrison, P. L., Babcock, R. C., Bull, G. D., Oliver, J. K., Wallace, C. C., Willis, B. L. (1984): Mass spawning in tropical reef corals. *Sci.* **223**, 1186–1189.

Harrison, W. G., Harris, L. R., Irwin, B. D. (1996): The kinetics of nitrogen utilization in the oceanic mixed layer: Nitrate and ammonium interactions of nanomolar concentrations. *Limnol. Oceanogr.* **41**, 16–32.

Hart, A., Lendrem, D. W. (1984): Vigilance and scanning patterns in birds. *Amer. Nat.* **32**, 1216–1224.

Hashimoto, H. (1974): Exact solution of a certain semilinear system of equations related to a migrating predation problem. *Proc. Japan Acad.* **50**, 622.

Hasler, A. D. (1971): Orientation and fish migration. In *Fish Physiology*, Vol. 6, 429–510. Hoar, W. S., Randall D. J. (eds.). New York: Academic Press.

Hassell, M. P., Comins, M. N., May, R. M. (1991): Spatial structure and chaos in insect population dynamics. *Nature* **353**, 255–258.

Haury, L. R., McGowan, J. A., Wiebe, P. H. (1978): Patterns and processes in the time-space scales of plankton distributions. In *Spatial Pattern in Plankton Communities*, 277–327. Steele, J. H. (ed.). New York: Plenum Press.

Havlin, S. (1990): Multifractals in diffusion and aggregation. *Physica* A **168**, 507–515.

Havlin, S., Araujo, M., Larralde, H., Stanley, H. E., Trunfio, P. (1992): Diffusion-controlled reaction $A + B \rightarrow C$ with initially separated reactants. *Physica* A **191**, 143–152.

Hawkes, C. (1972): The estimation of the dispersal rate of the adult cabbage root fly (*Erioischia brassicae* (Bouché)) in the presence of a brassica crop. *J. Appl. Ecol.* **9**, 617–632.

Hayashi, M., van der Kamp, G., Rudolph, D. L. (1998): Water and solute transfer between a prairie wetland and adjacent uplands. 2. Chloride cycle. *J. Hydrol.* **207**, 56–67.

Heaney, S. I., Eppley, R. W. (1981): Light, temperature and nitrogen as interacting factors affecting diel vertical migrations of dinoflagellates in culture. *J. Plankton Res.* **3**, 331–344.

Hebert, P. D. N., Good, A. G., Mort, M. A. (1980): Induced swarming in the predatory copepod *Heterocope septentrionalis. Limnol. Oceanogr.* **25**, 747–750.

Heiser, C. B., Jr. (1990): *Seed to Civilization: The Story of Food.* Cambridge: Harvard Univ. Press.

Herman, A. W. (1989): Vertical relationships between chlorophyll production and copepods in the eastern tropical Pacific. *J. Plankton Res.* **11**, 43–262.

Hida, T. (1975): *Brownian Motion.* Tokyo: Iwanami Shoten, Publ. Co. (Japanese) English translation to be published by Springer-Verlag, 1980.

Higgs and Usher. (1980): Should nature reserves be large or small. *Nature* **285**, 568–569.

Hilborn, R. (1979): Some longterm dynamics of predator-prey models with diffusion. *Ecol. Modelling* **6**, 23–30.

Hill, A. E. (1990): Pelagic dispersal of Norway lobster *Nephrops norvegicus* larvae examined using an advection–diffusion mortality model. *Mar. Ecol. Progr. Ser.* **64**, 217–226.

Hill, A. E. (1991): Advection–diffusion-mortality solutions for investigating pelagic larval dispersal. *Mar. Ecol. Progr. Ser.* **70**, 117–128.

Hill, C. J. (1995): Linear strips of rain forest vegetation as potential dispersal corridors for rain forest insects. *Conserv. Biol.* **9**, 1559–1566.

Hill, J. C. (1976): Homogeneous turbulent mixing with chemical reaction. *Ann. Rev. Fluid Mech.* **8**, 135–161.

Hill, P. S., Nowell, A. R. M., Jumars, P. A. (1992): Encounter rate by turbulent shear of particles similar in diameter to the Kolmogorov scale. *J. Mar. Res.* **50**, 643–668.

Hinckley, S., Bailey, K. M., Picquelle, S., Schumacher, J. D., Stabeno, P. J. (1991): Transport distribution and abundance of larval and juvenile walleye pollock (*Theragra chalcogramma*) in the western Gulf of Alaska. *Can. J. Fish. Aq. Sci.* **48**, 91–98.

Hinze, J. O. (1959): *Turbulence*. 2nd ed., 1975. New York: McGraw-Hill.

Hirano, T., Fujimoto, M. (1970): Preliminary results of investigation of the Kuroshio functioning as a mean of transportation and diffusion of fish eggs and larvae. In *The Kuroshio*, 405–416. Marr, J. C. (ed.). Honolulu: East-West Center Press.

Hobbs, R. J. (1992): The role of corridors in conservation: Solution or bandwagon? *Trends Ecol. Evol.* **11**, 398–392.

Holgate, P. (1967): The size of elephant herds. *Math. Gazette* **51**, 302–304.

Holgate, P. (1971): Random walk models for animal behavior. In *Statistical Ecology*, Vol. 2, 1–12. Patil, G. P., Pielou, E. C. Waters, W. E., (eds.). University Park, PA: Penn. State Univ. Press.

Holling, C. S. (1965): The functional response of predators to prey density and its role in mimicry and population regulation. *Mem. Entomol. Soc. Canada*, No. 45.

Holling, C. S. (1966): The functional response of invertebrate predators to prey density. *Mem. Entomol. Soc. Canada*, No. 58.

Holmes, E. (1993): Are diffusion models too simple? A comparison with telegraph models of invasion. *Amer. Nat.* **142**, 779–795.

Holmes, E. E., Lewis, M. A., Banks, J. E., Veit, R. R. (1993): Partial-differential equations in ecology—spatial interactions and population-dynamics. *Ecology* **75**, 17–29.

Hoppensteadt, F. (1975): *Mathematical Theories of Populations: Demographics, Genetics and Epidemics*. Philadelphia: Soc. for Industrial and Appl. Math.

Horenstein, W. (1945): On certain integrals in the theory of heat conductions. *Quart. J. Appl. Math.* **3**, 183–184.

Hori, J. (1977): *Langevin Equation*. Tokyo: Iwanami Shoten, Publ. Co. (Japanese).

Horwood, J. W. (1978): Observations on spatial heterogeneity of surface chlorophyll in one and two dimensions. *J. Mar. Biol. Assoc. UK* **58**, 487–502.

Hosono, Y. (1986): Traveling wave solutions for some density-dependent diffusion equations. Japan *J. Appl. Math.* **3**, 163–196.

Hosono, Y. (1987): Traveling waves for some biological systems with density dependent diffusion. *Japan J. Appl. Math.* **4**, 279–359.

Houde, E. D. (1997): Patterns and trends in larval-stage growth and mortality of teleost fish. *J. Fish. Biol.* **51**, 52–83.

Howard, L. N., Kopell, N. (1977): Slowly varying waves and shock structures in reaction–diffusion equations. *Studies in Appl. Math.* **56**, 95–145.

Howland, H. C. (1974): Optimal strategies for predator avoidance: The relative importance of speed and maneuverability. *J. Theor. Biol.* **47**, 333–350.

Humphrey, J. A. C. (1987): Fluid mechanical constraints on spider ballooning. *Oecologia* **73**, 469–477.

Humphrey, S. R. (1974): Zoogeography of the nine-banded armadillo (*Dasypus novemcinctus*) in the United States. *Bio. Sci.* **24**, 457–462.

Hundertmark, K. J. (1997): Home range, dispersal and migration. In *Ecology and Management of the North American Moose*, 303–335. Franzmann, A. W., Schwartz, C. C. (eds.). Washington, DC: Smithsonian Inst. Press.

Hunding, A., Sørenson, P. G. (1988): Size adaptation of Turing pre-patterns. *J. Math. Biol.* **26**, 27–39.

Hunter, J. R. (1966): Procedure for analysis of schooling behavior. *J. Fish. Res. Board, Canada* **23**, 547–562.

Hunter, J. R. (1969): Communication of velocity changes in jack mackeral (*Trachurus symmetricus*) schools. *Anim. Behav.* **17**, 507–514.

Hunter J., Nicholl, R. (1985): Visual threshold for schooling in northern anchovy Engraulis mordax. *Fishery Bulletin, U.S.* **83(3)**, 235–242.

Hurd, C. H., Stevens, C. L. (1997): Flow visualization around single- and multiple-bladed seaweeds with various morphologies. *J. Phycol.* **33**, 360–367.

Hurd, C. L., Harrison, P. J., Druehl, L. D. (1996): Effect of seawater velocity on inorganic nitrogen uptake by morphologically distinct forms of *Macrocystic integrifolia* from wave-sheltered and exposed sites. *Mar. Biol.* **126**, 205–214.

Hurlbert, S. H. (1990): Spatial distribution of the montane unicorn. *Oikos* **58**, 257–271.

Hurtt, G. C., Armstrong, R. A. (1996): A pelagic ecosystem model calibrated with BATS data. *Deep-Sea Res. II. Topical Studies in Oceanography* **43**, 653–684.

Huth, A., Wissel, C. (1990): The movement of fish schools: a simulation model. In: W. Alt, G. Hoffmann (Eds.) *Biological Motion, Lecture Notes in Biomathematics* **89**, 577–590. Berlin: Springer-Verlag.

Huth, A., Wissel, C. (1992): The simulation of the movement of fish schools. *J. Theor. Biol.* **156**, 365–385.

Ikawa, T., Okubo, A., Okabe, H., Cheng, L. (1998): Oceanic diffusion and the pelagic insects *Halobates* spp. (Gerridae:Hemiptera) *Marine Biol* **131**, 195–201.

Ilse, L. M., Hellgren, E. C. (1995): Spatial use and group dynamics of sympatric collared peccaries and feral hogs in southern Texas. *J. Mammal.* **76**, 993–1002.

Imai, I. (1970): *Fluid Mechanics*. Tokyo: Iwanami Shoten (Japanese).

Inagaki, T., Sakamoto, W., Aoki, I., Kuroki, T. (1976): Studies on the schooling behavior of fish. III. Mutual relationship between speed and form in schooling behavior. *Bull. Jap. Soc. Sci. Fish.* **42**, 629–635.

Incze, L. S. (1996): Small-scale biological-physical interactions. *Proc. GOM Ecosys. Dyn.* **97**, 105–116.

Inoue, E. (1950): On the turbulent diffusion in the atmosphere. I. *J. Meteorol. Soc. Japan* **28**, 441–456.

Inoue, E. (1951): On the turbulent diffusion in the atmosphere. II. *J. Meteorol. Soc. Japan* **29**, 246–253.

Inoue, E. (1955): Studies on the phenomena of waving plants ('Honami') caused by wind. I. Mechanism and characteristics of waving plants phenomena. *J. Agric. Meterol. Tokyo* **11**, 87–90.

Inoue, E. (1963): On the turbulent structure of airflow within crop canopies. *J. Meteorol. Soc. Japan* **41**, 317–326.

Inoue, E. (1965): On the CO_2-concentration profiles within crop canopies. *J. Agric. Meteorol. Tokyo* **21**, 137–140.

Inoue, E. (1971): Role of turbulence and diffusion in aerobiology. Preprint from *Proce. 1971 Fall Meeting of Meteorological Soc. of Japan*, paper no. 20.

Inoue, E. (1974): Current activities in aerobiology in Japan. *Boundary-Layer Meteorol.* **7**, 257–266.

Inoue, E., Tani, N., Imai, K., Isobe, S. (1958): The aerodynamic measurement of photosynthesis over the wheat field. *J. Agric. Meteorol. Tokyo* **13**, 121–125.

Inoue, E., Uchijima, Z., Udagawa, T., Horie, T., Kobayashi, K. (1968): Studies of energy and gas exchange within crop canopies (2): CO_2 flux within and above a corn plant canopy. *J. Agric. Meteorol. Tokyo* **23**, 165–176.

Inoue, T. (1972): Historical review of studies on animal movement patterns. *Kotaigun Seitai Gakkai Kaiho* **21**, 18–32 (Japanese).

Inoue, T. (1978): A new regression method for analyzing animal movement patterns. *Res. Popul. Ecol.* **19**, 141–163.

Irwin, M. E., Thresh, J. M. (1988): Long-range aerial dispersal of cereal aphids as virus vectors in North America. *Phil. Trans. Roy. Soc. London B* **321**, 421–446.

Isobe, S. (1972): A spectral analysis of turbulence in a corn canopy. *Bull. Nat. Inst. Agric. Sci. Japan. Series A*, no. **19**, 101–113.

Ito, K. (1944): Stochastic integral. *Proc. Imperial Acad. Tokyo* **20**, 519–524.

Ito, K. (1946): On a stochastic integral equation. *Proc. Japan Acad.* **1–4**, 32–35.

Ito, K. (1951): On a stochastic differential equation. *Mem. Amer. Math. Soc.* **4**, 51–89.

Ito, M. (1961): Some problems on a social process of insect population. *Jap. J. Ecol.* **11**, 202–208, 232–238 (Japanese).

Ito, Y. (1952): The growth form of populations in some aphids, with special reference to the relation between population density and movements. *Res. Popul. Ecol.* **1**, 36–48 (Japanese with English summary).

Ito, Y. (1963): *Introduction to Animal Ecology*. Tokyo: Kokin Shoin (Japanese).

Ito, Y. (1975): *Animal Ecology*, Vols. 1 and 2. Tokyo: Kokin Shoin (Japanese).

Ito, Y., Miyashita, K. (1961): Studies on the dispersal of leaf- and plant-hoppers. I. Dispersal of *Nephotettix cincticeps* Uhler on paddy fields at the flowering stage. *Jap. J. Ecol.* **11**, 181–186.

Iwao, S. (1968): A new regression method for analyzing the aggregation pattern of animal populations. *Res. Popul. Ecol.* **10**, 1–20.

Iwao, S., Kuno, E. (1971): An approach to the analysis of aggregation pattern in biological populations. In *Statistical Ecology*, Vol. 1, 461–513. Patil, G. P., Pielou, E. C., Waters, W. E. (eds.). University Park, PA: Penn. State Univ. Press.

Iwasa, Y., Teramoto, E. (1978): A mathematical model for the formation of a distributional pattern and an index of aggregation. *Jap. J. Ecol.* **27**, 117–124 (Japanese with English synopsis).

Izumi, T. (1973): Social behavior of the Norway rat (*R. norvegicus*) in their natural habitat: Especially on their group types. *Jap. J. Ecol.* **23**, 55–64 (Japanese with English synopsis).

Jackson, G. A. (1984): Internal wave attenuation by coastal kelp stands. *J. Phys. Oceanogr.* **14**, 1300–1306.

Jackson, G. A. (1988): Kelvin wave propagation in a high drag coastal environment. *J. Phys. Oceanogr.* **18**, 1733–1743.

Jackson, G. A. (1989): Simulation of bacterial attraction and adhesion to falling particles in an aquatic environment. *Limnol. Oceanogr.* **34**, 514–530.

Jackson, G. A. (1990): A model for the formation of marine algal flocs by physical coagulation processes. *Deep-Sea Res. I* **37**, 1197–1121.

Jackson, G. A. (1998): Currents in the high drag environment of a coastal kelp stand off California. *Cont. Shelf Res.* **17**, 1913–1928.

Jackson, G. A., Burd, A. B. (1998): Aggregation in the marine environment. *Environ. Sci. Technol.* **32**, 2805–2814.

Jackson, G. A., Lochmann, S. E. (1993): Modeling coagulation in marine ecosystems. In *Environmental Particles*, Vol. 2, 387–414. Buffle, J., van Leeuwen, H. P. (eds.). Boca Raton, FL: Lewis Publishers.

Jackson, G. A., Maffione, R., Costello, D. K., Alldredge, A. L., Logan, B. E., Dam, H. G. (1997): Particle size spectra between 1 μm and 1 cm at Monterey Bay determined using multiple instruments. *Deep-Sea Res. I* **44**, 1739–1767.

Jackson, G. A., Winant, C. D. (1983): Effects of kelp forest on coastal currents. *Cont. Shelf Res.* **2**, 75–80.

Jamart, B. M., Winter, D. F., Banse, K. (1979): Sensitivity analysis of a mathematical model of phytoplankton growth and nutrient distribution in the Pacific Ocean off the northwest U.S. coast. *J. Plankton Res.* **1**, 267–290.

Jamart, B. M., Winter, D. F., Banse, K., Anderson, G. C., Lam, R. K. (1977): A theoretical study of phytoplankton growth and nutrient distribution in the Pacific Ocean off the northwest U.S. coast. *Deep-Sea Res.* **24**, 753–773.

James, I. R. (1978): Estimation of the mixing proportion in a mixture of two normal distributions from simple, rapid measurements. *Biometrics* **34**, 265–275.

Jeans, J. H. (1952): *An Introduction to the Kinetic Theory of Gases.* Cambridge: Cambridge Univ. Press.

Jenkinson, I. R., Biddanda, B. A. (1995): Bulk-phase viscoelastic properties of seawater: Relationship with plankton components. *J. Plankton Res.* **17**, 2251–2274.

Jennrich, R. I., Turner, F. B. (1969): Measurement of non-circular home range. *J. Theor. Biol.* **22**, 227–237.

Jewell, P. A. (1966): The concept of home range in mammals. In *Play, Exploration and Territory in Mammals*, 85–109. Jewell, P. A., Loizos, C. (eds.). (*Symp. Zool. Soc. London*, 1966, no. 18). New York: Academic Press.

Johnson, B. K., Ager, A. A., Findholt, S. L., Wisdom, M. J., Marx, D. B., Kern, J. W., Bryant, L. D. (1998): Mitigating spatial differences in observation rate of automated telemetry systems. *J. Wildl. Manage.* **62**, 958–967.

Johnson, C. G. (1969): *Migration and Dispersal of Insects by Flight.* London; Methuen & Co. Ltd.

Johnson, C. P., Li, X., Logan, B. E. (1996): Settling velocities of fractal aggregates. *Environ. Sci. Technol.* **30**, 1911–1918.

Jones, R. (1959): A method of analysis of some tagged haddock return. *J. du Conseil int. Explor. Mer.* **25**, 58–72.

Jones, R. E. (1977): Movement pattern and egg distribution in cabbage butterflies. *J. Anim. Ecol.* **46**, 195–212.

Jorné, J. (1974): The effect of ionic migration on oscillations and pattern formation in chemical systems. *J. Theor. Biol.* **43**, 375–380.

Jorné, J. (1975): Negative ionic cross diffusion coefficients in electrolytic solutions. *J. Theor. Biol.* **55**, 529–532.

Jorné, J. (1977): The diffusive Lotka–Volterra oscillating system. *J. Theor. Biol.* **65**, 133–139.

Jorné, J., Carmi, S. (1977): Liapunov stability of the diffusive Lotka–Volterra equations. *Math. Biosciences* **37**, 51–61.

Joseph, J., Sendner, H. (1958): Über die horizontale Diffusion im Meere. *Dt. Hydrogr. Z.* **11**, 49–77.

Joyce, R. J. V. (1976): Insect flight in relation to problems of pest control. In *Insect Flight*, 135–155. Rainey, R. C. (ed.). New York: J. Wiley & Sons.

Jumars, P. A. (1993): *Concepts in Biological Oceanography: An Interdisciplinary Primer.* New York: Oxford Univ. Press.

Jury, W. A., Dyson, J. S., Butters, G. L. (1990): A transfer function model of field scale solute transport under transient water flow. *Soil. Sci. Soc. Amer. J.* **54**, 327–331.

Jury, W. A., Focht, D. D., Farmer, W. J. (1987): Evaluation of pesticide ground water pollution potential from standard indices of soil-chemical adsorption and biodegradation. *J. Environ. Qual.* **16**, 422–428.

Jury, W. A., Roth, K. (1990): Transfer functions and solute movement through soil: Theory and applications. Basel: Birkhauser.

Jury, W. A., Sposito, G., White, R. E. (1986): A transfer function model of solute movement through soil. 1. Fundamental concepts. *Water Resour. Res.* **22**, 243–247.

Kaae, R. S., Shorey, H. H. (1972): Sex pheromones of noctuid moths XXVII. Influence of wind velocity on sex pheromone releasing behavior of *Trichoplusia ni* females. *Ann. Entomol. Soc. Amer.* **65**, 436–440.

Kac, M. (1947): Random walk and the theory of Brownian motion. *Amer. Math. Monthly* **54**, 369–391.

Kaliappan, P. (1984): An exact solution for travelling waves of $u_t = Du_{xx} + u - u^k$. *Phys. D* **11**, no. 3, 368–374.

Kallen, A., Arcuri, P., Murray, J. D. (1985): A simple model for the spatial spread and control of rabies. *J. Theor. Biol.* **116**, 377–393.

Kametaka, Y. (1976): On the nonlinear diffusion equation of Kolmogorov–Petrovskii–Piskunov type. *Osaka J. Math.* **13**, 11–66.

Kametka, Y. (1977): *Nonlinear Partial Differential Equations.* Tokyo: Osho.

Kanehiro, H., Suzuki, M., Matuda, K. (1985): Characteristics of the schooling behavior by the group-size of rose bitterling in the experimental water tank. *Bull. Jpn. Soc. Sci. Fish.* **51(12)**, 1977–1982.

Kareiva, P. (1982): Experimental and mathematical analyses of herbivore movement. *Ecological Monographs* **52**, 261–282.

Kareiva, P. (1983): Local movements in herbivorous insects: Applying a passive diffusion model to mark-recapture field experiments. *Oecologia* **57**, 322–327.

Kareiva, P., Odell, P. (1987): Swarms of predators exhibit preytaxis if individual predators use area restricted search. *Amer. Nat.* **130**, 233–270.

Kareiva, P., Shigesada, N. (1983): Analyzing insect movement as a correlated random walk. *Oecologia* **56**, 234–238.

Karp-Boss, L., Boss, E., Jumars, P. A. (1996): Nutrient fluxes to planktonic osmotrophs in the presence of fluid motion. *Oceanogr. Mar. Biol.: An Ann. Rev.* **34**, 71–107.

Katona, S. K. (1973): Evidence for sex pheromones in planktonic copepods. *Limnol. Oceanogr.* **18**, 574–583.

Katul, G. G., Geron, C. D., Hsieh, C. I., Vidakovic, B., Guenther, A. B. (1998): Active turbulence and scalar transport near the forest–atmosphere interface. *J. Appl. Meteorol.* **37**, 1533–1546.

Kaufmann, S. A., Shymko, R., Trabert, K. (1978): Control of sequential compartment formation in *Drosophila*. *Science* **199**, 259–270.

Kawahara, T., Tanaka, M. (1983): Interaction of traveling fronts: an exact solution of a nonlinear diffusion equation. *Phys. Lett* A **97**, no. 8, 311–314.

Kawamura, A., Hirano, K. (1985): The spatial scale of surface swarms of Calanus plumchrus Marukawa observed from consecutive plankton net catches in the Northwestern North Pacific. *Bull. Mar. Sci.* **37**, 626–633.

Kawasaki, K. (1978): Diffusion and the formation of spatial distributions. *Math. Sciences* **16** (No. 183), 47–52 (Japanese).

Kawatani, T., Meroney, R. M. (1970): Turbulence and wind speed characteristics within a model canopy flow field. *Agric. Meteorol.* **7**, 143–158.

Keller, E. F., Segel, L. A. (1971a): Model for chemotaxis. *J. Theor. Biol.* **30**, 225–234.

Keller, E. F., Segel, L. A. (1971b): Traveling bands of chemotactic bacteria: A theoretical analysis. *J. Theor. Biol.* **30**, 235–248.

Kemble, E. C. (1937): *The Fundamental Principles of Quantum Mechanics*. New York: McGraw-Hill (also available at Dover Publ. Co., New York, 1958).

Kemp, W. M., Boynton, W. R. (1980): Influence of biological and physical processes on dissolved oxygen dynamics in an estuarine system: Implications for measurement of community metabolism. *Estuar. Coast. Mar. Sci.* **11**, 407–431.

Kendall, M. G. (1948): A form of wave propagation associated with the equation of heat conduction. *Proc. Cambridge Phil. Soc.* **44**, 591–593.

Kenney, B. C. (1993): Observations of coherent bands of algae in a surface shear layer. *Limnol. Oceanogr.* **38**, 1059–1067.

Kenward, R. E. (1985): Ranging behavior and population dynamics of grey squirrels. P. In *Behavioural Ecology*, 319–330. Sibley, R. M., Smith, R. H. (eds.). Oxford: Blackwell.

Keough, J. R., Hagley, C. A., Ruzycki, E., Sierszen, M. (1998): δ^{13}C composition of primary producers and the role of detritus in a freshwater coastal ecosystem. *Limnol. Oceanogr.* **43**, 734–740.

Kerner, E. H. (1959): Further considerations on the statistical mechanics of biological associations. *Bull. Math. Biophys.* **21**, 217–255.

Kessler, J. O. (1985): Hydrodynamic focusing of motile algal cells. *Nature* **313**, 218–220.

Kettle, D. S. (1951): The spatial distribution of *Culicoides impunctatus* Goet. under woodland and moorland conditions and its flight range through woodland. *Bull. Entomol. Res.* **42**, 239–291.

Kierstead, H., Slobodkin, L. B. (1953): The size of water masses containing plankton bloom. *J. Mar. Res.* **12**, 141–147.

Kiester, A. R., Slatkin, M. (1974): A strategy of movement and resource utilization. *Theor. Popul. Biol.* **6**, 1–20.

Kils, U. (1993): Formation of micropatches by zooplankton-driven microturbulences. *Bull. Mar. Sci.* **53**(1), 160–169.

Kim, S., Bang, B. (1990): Oceanic dispersion of larval fish and its implication for mortality estimates: Case study of walleye pollock in Shelikaf Strait. *Alask. Fish. Bull.* **88**, 303–311.

Kingsford, M. J., Choat, J. H. (1986): Influence of surface slicks on the distribution and onshore movements of small fish. *Mar. Biol.* **91**, 161–171.

Kiørboe, T. (2000): Colonization of marine snow aggregates by invertebrate zooplankton: Abundance, scaling, and possible role. *Limnol. Oceanogr.* **45**, 479–484.

Kiørboe, T., Hansen, J. L. S., Alldredge, A. L., Jackson, G. A., Passow, U., Dam, H. G., Drapeau, D. T., Waite, A., Garcia, C. M. (1996): Sedimentation of phytoplankton cells during a diatom bloom: Rates and mechanisms. *J. Mar. Res.* **54**, 1123–1148.

Kiørboe, T., Tiselius, P., Mitchell-Innes, B., Hansen, J. L. S., Visser, A. W., Mari, X. (1998): Intensive aggregate formation with low vertical flux during an upwelling-induced diatom bloom. *Limnol. Oceanogr.* **43**, 104–116.

Kirkpatrick, M. (1984): Demographic models based on size, not age, for organisms with indeterminate growth. *Ecology* **65**, 1874–1884.

Kishimoto, R. (1971): Long distance migrations of plant hoppers, *Sogetella furcifera* and *Nilaparvata lugens*, 201–216. *Proc. Symp. Rice Insects*, Tropical Agricultural Research Center.

Kishimoto, K. (1982): The diffusive Lotka-Volterra system with three species can have a stable, non-constant equilibrium solution. *J. Math. Biol.* **16**, 103–112.

Kishimoto, K., Mimura, M., Yoshida, K. (1983): Stable spatio-temporal oscillations of diffusive Lotka-Volterra system with 3 or more species. *J. Math. Biol.* **18**, 213–221.

Kishimoto, K., Weinberger, H. F. (1985): The spatial homogeneity of stable equilibria of some reaction-diffusion systems on convex domains. *J. Diff. Equations* **58**, 15–21.

Kitching, R. (1971): A simple simulation model of dispersal of animals among units of discrete habitats. *Oecologia* (Berl.) **7**, 95–116.

Kittredge, J. S., Takahasi, F. T., Lindsey, J., Lasker, R. (1974): Chemical signals in the sea: Marine allelochemics and evolution. *Fishery Bull.* **72**, 1–11.

Kleiber, M. (1932): Body size and metabolism. *Hilgardia* **6**, 315–353.

Kleiber, M. (1961): *The Fire of Life. An Introduction to Animal Energetics.* New York: J. Wiley & Sons.

Koch, A. L., Carr, A., Ehrenfeld, D. W. (1969): The problem of open-sea navigation: The migration of the green turtle to Ascension Island. *J. Theor. Biol.* **22**, 163–179.

Koch, E. W. (1994): Hydrodynamics, diffusion-boundary layers and photosynthesis of the seagrasses *Thalassia testudinum* and *Cymodocea nodosa. Mar. Biol.* **118**, 767–776.

Koch, E. W. (1996): Hydrodynamics of a shallow *Thalassia testudinum* beds in Florida. In *Seagrass Biology: Proceedings of an International Workshop*, 105–110. Kuo, J., Phillips, R. C., Walker, D. I., Kirkman, H. (eds.). Nedlands: Univ. Western Australia.

Koch, E. W., Gust, G. (1999): Water flow in tide and wave dominated beds of the seagrass *Thalassia testudinum. Mar. Ecol. Progr. Ser.* **184**, 63–72.

Koehl, M. A. R. (1986): Seaweeds in moving water: Form and mechanical function. In *On the Economy of Plant Form and Function*, 603–634. Givnish, T. J. (ed.). Cambridge: Cambridge Univ. Press.

Koehl, M. A. R., Alberte, R. S. (1988): Flow, flapping, and photosynthesis of *Nereocystis luetkeana*: A functional comparison of undulate and flat blade morphologies. *Mar. Biol.* **99**, 435–444.

Koehl, M. A. R., Powell, T. M. (1994): Turbulent transport of larvae near wave-swept rocky shores: Does water motion overwhelm larval sinking? In *Reproduction and development of marine invertebrates*, 261–274. Wilson, W. H., Jr., Stricker, S. A., Shinn, G. L. (eds.). Baltimore: Johns Hopkins Univ. Press.

Koehl, M. A. R., Powell, T. M., Dairiki, G. (1993): Measuring the fate of patches in the water: Larval dispersal. In *Patch Dynamics in Terrestrial, Marine, and Freshwater Ecosystems*, 50–60. Steele, J., Powell, T. M., Levin, S. A. (eds.). Berlin: Springer-Verlag.

Koehl, M. A. R., Strickler, J. R. (1981): Copepod feeding currents: Food capture at low Reynolds number. *Limnol. Oceanogr.* **26**, 1062–1073.

Koeppl, J. W., Slade, N. A., Harris, K. S., Hoffmann, R. S. (1977): A three-dimensional home range model. *J. Mammal.* **58**, 213–220.

Kolmogorov, A. N. (1941): The local structure of turbulence in incompressible viscous fluid for very large Reynolds numbers. *Comptes Rend. Acad. Sci., U.S.S.R.* **30**, 301–305.

Kolmogorov, A., Petrovsky, I., Piscounov, N. (1937): Étude de l'équation de la diffusion avec croissance de la quantité de matière et son application à un problème biologigue. *Moscow Univ. Bull. Ser. Internat. Sect. A*, **1**, 1–25.

Komar, P. D., Morse, A. P., Small, L. F., Fowler, S. W. (1981): Analysis of sinking rates of natural copepod and euphasid fecal pellets. *Limnol. Oceanogr.* **26**, 172–180.

Kondo, J., Akashi, S. (1976): Numerical studies on the two-dimensional flow in horizontally homogeneous canopy layers. *Boundary-Layer Meteorol.* **10**, 255–272.

Kono, T. (1952): Time-dispersion curve: Experimental studies on the dispersion of insects (2). *Res. Popul. Ecol.* **1**, 109–118 (Japanese with English summary).

Koopman, B. O. (1956): The theory of search: II Target detection. *Oper. Res.* **4**, 503–531.

Kopell, N., Howard, L. N. (1973): Plane wave solutions to reaction–diffusion equations. *Studies in Appl. Math.* **52**, 291–328.

Korhonen, K. (1980): Ventilation in the subnivean tunnels of the voles *Microtus agretis* and *M. oeconomus. Ann. Zool. Fennici* **17**, 1–4.

Koseff, J. R., Holen, J. K., Monismith, S. G., Cloern, J. E. (1993): Coupled effects of vertical mixing and benthic grazing on phytoplankton populations in shallow estuaries. *J. Mar. Res.* **51**, 843–868.

Kostitzin, V. A. (1939): *Mathematical Biology*. London: George G. Harrap & Co. Ltd. (Translated from the French book *Biologie mathématique*, 1937).

Kot, M., Lewis, M. A., van den Driessche, P. (1996): Dispersal data and the spread of invading organisms. *Ecology* **77**, 2027–2042.

Kot, M., Schaffer, W. M. (1986): Discrete-time growth dispersal models. *Math. Biosci.* **80**, 109–136.

Kowal, N. E. (1971): A rationale for modeling dynamic ecological systems. In *Systems Analysis and Simulation in Ecology*, Vol. 1, 123–196. Patten, B. C. (ed.). New York: Academic Press.

Koyama, J. (1962): The swarming behavior of *Fannia scalaris* Fabricuius, with special reference to the swarming individual number. *Jap. J. Ecol.* **12**, 11–16 (Japanese with English synopsis).

Koyama, J. (1974): The sexual and physiological character in the mating behavior of *Fannia scalaris* Fabricus. *Jap. J. Ecol.* **24**, 92–115 (Japanese with English synopsis).

Krakauer, D. C. (1995): Groups confuse predators by exploiting perceptual bottlenecks: A connectionist model of the confusion effect. *Behav. Ecol. Sociobiol.* **36**, 421–429.

Kramer, L. (1982): Absolute stability and transitions in ecosystems with a multiplicity of stable sates and dispersal. *J. Theoret. Biol.* **98**, no. 1, 91–108.

Kranck, K. (1980): Variability of particulate matter in a small coastal inlet. *Can. J. Fish. Aquat. Sci.* **37**, 1209–1215.

Krause, J., Treger, R. W. (1994): The mechanism of aggregation behaviour in fish shoals: Individuals minimize approach time to neighbors. *Anim. Behav.* **48**, 353–359.

Krembs, C., Juhl, A. R., Long, R. A., Azam, F. (1998): Nanoscale patchiness of bacteria in lake water studied with the spatial information preservation method. *Limnol. Oceanogr.* **43**, 307–314.

Kullenberg, G. (1978): Vertical processes and the vertical-horizontal coupling. In *Spatial Pattern in Plankton Communities*, 43–71. Steele, J. H. (ed.). New York: Plenum Press.

Kuno, E. (1968): Studies on the population dynamics of rice leafhoppers in a paddy field. *Bull. Kyushu Agricultural Experiment Station* **14**, 131–246, Chikugo-shi, Fukuoka Pref. Japan (Japanese with English summary).

Kuno, E. (1991): Sampling and analysis of insect populations. *Ann. Rev. Ent.* **36**, 285–304.

Labisky, R. F., Miller, K. E., Hartless, C. S. (1999): Effect of Hurricane Andrew on survival and movements of white-tailed deer in the Everglades. *J. Wildl. Manage.* **63**, 872–879.

Lacalli, T. C. (1981): Dissipative structures and morphogenetic patterns in unicellular algae. *Phil. Trans. Roy. Soc. B (London)* **294**, 547–588.

Lamare, M. D., Stewart, B. G. (1998): Mass spawning by the sea urchin *Evechinus chloroticus* (Echinodermata: Echinoidea) in a New Zealand fiord. *Mar. Biol.* **132**, 135–140.

Lamb, H. H. (1970): Volcanic dust in the atmosphere; with a chronology and assessment of its meteorological significance. *Phil. Trans. Roy. Soc. London* **266(1170)**, 425–533.

Lamp, K. P., Hassan, E., Scotter, D. R. (1970): Dispersal of scandium-46-labeled *Pantorhytes* weevils in papuan cacao plantations. *Ecology* **52**, 178–182.

Lampitt, R. S., Hillier, W. R., Challenor, P. G. (1993): Seasonal and diel variation in the upper ocean concentration of marine snow aggregates. *Nature* **362**, 737–739.

Land, M. F., Collett, T. S. (1974): Chasing behavior of houseflies (*Fannia canicularis*). A description and analysis. *J. Comp. Physiol.* **89**, 331–357.

Landa, J. T. (1998): Bioeconomics of schooling fishes: Selfish fish, quasi-free riders, and other fishy tales. *Envir. Biol. Fishes* **53**, 353–364.

Landahl, H. D. (1957): Population growth under the influence of random dispersal. *Bull. Math. Biophys.* **19**, 171–186.

Landahl, H. D. (1959): A note on population growth under random dispersal. *Bull. Math. Biophys.* **21**, 153–159.

Lande, R., Wood, A. M. (1987): Suspension times of particles in the upper ocean. *Deep-Sea Res. I* **34**, 61–72.

Landsberg, J. J., James, G. B. (1971): Wind profiles in plant canopies: Studies on an analytical model. *J. Appl. Ecol.* **8**, 729–741.

Lane, D. J. W., Beaumont, A. R., Hunter, J. R. (1985): Byssus drifting and the drifting threads of the young post-larval mussel *Mytilus edulis. Mar. Biol.* **84**, 301–308.

Langevin, P. (1908): Sur la théorie du mouvement brownien. *Comptes Rendus* **146**, 530–533.

Langlais, M. (1985): A nonlinear problem in age-dependent population diffusion. *Siam J. Math. Anal* **16**, no. 3, 510–529.

Langlais, M. (1987): Large time behavior in a nonlinear age-dependent population dynamics problem with spatial diffusion. *J. Math. Biol.* **26**, no. 3, 319–346.

Langlais, M. R. (1988): Stabilization in a nonlinear age dependent population model. In: *Biomathematics and related computational problems.* 337–344. Naples: Kluwer Acad. Publ.

Langmuir, I. (1938): Surface motion of water induced by wind. *Sci.* **87**, 119–123.

Lapidus, I. R. (1980): Pseudo-chemotaxis by microorganisms in an attractant gradient. *J. Theor. Biol.* **86**, 91–103.

Lapidus, I. R., Levandowsky, M. (1980): Modeling chemosensory responses of swimming protozoa. *Adv. Appl. Prob.* **12**, 568–568.

Lara-Ochoa, F. (1984): A generalized reaction-diffusion model for spatial structure formed by mobile cells. *Biosystems* **17**, 35–50.

Larson, R. J. (1992): Riding Langmuir circulations and swimming in circles: A novel form of clustering behavior by the scyphomedusa *Linuche unguiculata. Mar. Biol.* **112**, 229–235.

Lasker, H. R., Brazeau, D., Calderon, J., Coffroth, M. A., Coma, R., Kim, K. (1996): *In situ* rates of fertilization among broadcast spawning gorgonian corals. *Biol. Bull.* **190**, 45–55.

Laws, R. M., Parker, I. S. C., Johnstone, R. C. B. (1975): *Elephants and Their Habitats.* Oxford: Oxford Univ. Press.

Lazier, J. R. N., Mann, K. H. (1989): Turbulence and the diffusive layers around small organisms. *Deep-Sea Res.* **36**, 1721–1733.

Le Fevre, J., Frontier, S. (1988): Influence of temporal characteristics of physical phenomena on plankton dynamics as shown by Northwest European marine ecosystems. In *Towards a Theory on Biological-Physical Interactions in the World Ocean*, 245–272. Rothschild, B. J. (ed.). Dordrecht: Kluwer.

Leibovich, S. (1983): The form and dynamics of Langmuir circulations. *Ann. Rev. Fluid Mech.* **15**, 231–276.

Lekan, J. F., Wilson, R. E. (1978): Spatial variability of phytoplankton biomass in the surface water of Long Island. *Estuarine and Coastal Mar. Sci.* **6**, 239–250.

Leonard, L. A., Luther, M. E. (1995): Flow hydrodynamics in tidal marsh canopies. *Limnol. Oceanogr.* **40**, 1474–1484.

Lerman, A., Lal, D. (1977): Regeneration rates in the ocean. *Amer. J. Sci.* **277**, 238–258.

Les, D. H., Cleland, M. A., Waycott, M. (1997): Phylogenetic studies in the Alismatidae, II: Evolution of marine angiosperms (seagrasses) and hydrophily. *Syst. Bot.* **22**, 443–463.

Levandowsky, M., Childress, W. S., Spiegel, E. A., Hutner, S. H. (1975): A mathematical model of pattern formation by swimming microorganisms. *J. Protozool.* **22**, 296–306.

Levandowsky, M., White, B. S. (1977): Randomness: Time scales, and the evolution of bioloigical communities. In *Evolutionary Biology*, Vol. 10, 69–161. Hecht, M. K., Steere, W. C., Wallace, B. (eds.). New York: Plenum Press.

Levandowsky, M., Hauser, D. C. R. (1978): Chemosensory responses of swimming algae and protozoa. *Intl. Rev. Cytology* **53**, 145–210. New York: Academic Press.

Levin, L. A. (1983): Drift tube studies of bay-ocean water exchange and implications for larval dispersal. *Estuaries* **6**, 363–371.

Levin, L. A. (1990): A review of methods for labeling and tracking marine invertebrate larvae. *Ophelia* **32**, 115–144.

Levin, S. A. (1974): Dispersion and population interactions. *Amer. Naturalist* **108**, 207–228.

Levin, S. A. (1976a): Population dynamic models in heterogeneous environments. *Ann. Rev. Ecol. Systematics* **7**, 287–310.

Levin, S. A. (1976b): Spatial patterning and the structure of ecological communities. In *Some Mathematical Questions in Biology, 7. Lectures on Mathematics in the Life Science*, Vol. 8, 1–35. Levin, S. A. (ed.). Providence, RI: Amer. Math. Soc.

Levin, S. A. (1977): A more functional response to predator–prey stability. *Amer. Naturalist* **111**, 381–383.

Levin, S. A. (1978a): Population models and community structure in hetereogeneous environments. In *Mathematical Association of America Study in Mathematical Biology. Vol. II: Populations and Communities.* 439–476. Levin, S. A. (ed.). Washington: Math. Assoc. Amer.

Levin, S. A. (1978b): Pattern formation in ecological communities. In *Spatial Pattern in Plankton Communities*, 433–465. Steele, J. H. (ed.). New York: Plenum Press.

Levin, S. A., Paine, R. T. (1974): Disturbance, patch formation, and community structure. *Proc. Natl. Acad. Sci. USA* **71**, 2744–2747.

Levin, S. A., Paine, R. T. (1975): The role of disturbance in models of community structure. In *Ecosystem Analysis and Prediction*, 56–67. Levin, S. A. (ed.). Philadelphia: SIAM.

Levin, S. A., Segel, L. A. (1976): Hypothesis for origin of planktonic patchiness. *Nature* **259**, 659.

Levin, S. A., Segel, L. A. (1985): Pattern generation in space and aspect. *SIAM Rev.* **27**, 45–67.

Levins, R. (1968): *Evolution in Changing Environments*. Princeton: Princeton Univ. Press.

Levitan, D. R. (1991): Influence of body size and population density on fertilization success and reproductive output in a free-spawning invertebrate. *Biol. Bull.* **181**, 261–268.

Levitan, D. R. (1993): The importance of sperm limitation to the evolution of egg size in marine invertebrates. *Amer. Naturalist* **141**, 517–536.

Levitan, D. R. (1995): The ecology of fertilization in free-spawning invertebrates. In *The Ecology of Marine Invertebrate Larvae*, 123–156. McEdwards, L. R. (ed.). Boca Raton, FL: CRC Press.

Levitan, D. R. (1996): Effects of gamete traits on fertilization in the sea and the evolution of sexual dimorphism. *Nature* **382**, 153–155.

Levitan, D. R. (1998): Does Bateman's principle apply to broadcast-spawning organisms? Egg traits influence in situ fertilization rates among congeneric sea urchins. *Evol.* **52**, 1043–1056.

Levy II, H. Mahlman, J. D., Moxim, W. J. (1982): Tropospheric N_2O availability. *J. Geophys. Res.* **87**, 3061–3080.

Lewis, M. A. (1994): Spatial coupling of plant and herbivore dynamics: the contributions of herbivore dispersal to transient and persistent "waves" of damage. *Theor. Pop. Biol.* **45**, 277–312.

Lewis, M. A., van den Driessche, P. (1993): Waves of extinction from sterile insect release. *Math. Biosci.* **116**, 221–247.

Lewis, M. R., Harrison, W. G., Oakey, N. S., Hebert, D., Platt, T. (1986): Vertical nitrate fluxes in the oligotrophic ocean. *Sci.*, 870–873.

Lewis, M., Kareiva, P. (1993): Allee dynamics and the spread of invading organisms. *Theor. Popul. Biol.* **43**, 141–158.

Lewis, M., Schmitz, G., Kareiva, P., Trevors, J. (1996): Models to examine the containment and spread of genetically engineered microbes. *Molecular Ecol.* **5**, 165–175.

Li, Y., Ghodrati, M. (1994): Preferential transport of nitrate through soil columns containing root channels. *Soil Sci. Soc. Amer. J.* **58**, 653–659.

Li, Y., Ghodrati, M. (1995): Transport of nitrate in soils as affected by earthworm activities. *J. Env. Qual.* **24**, 432–438.

Li, X., Passow, U., Logan, B. E. (1998): Fractal dimensions of small (15–200 µm) particles in Eastern Pacific coastal waters. *Deep-Sea Res. I* **45**, 115–131.

Li, T.-Y., Yorke, J. A. (1975): Period three implies chaos. *Amer. Math. Monthly* **82**, 985–992.

Libelo, E. L., MacIntyre, W. G., Seitz, R. D., Libelo, L. F. (1994): Cycling of water through the sediment–water interface by passive ventilation of relict biological structures. *Mar. Geol.* **120**, 1–12.

Lidicker, W. Z., Caldwell, R. L. (1982): *Dispersal and Migration*. New York: Hutchinson Ross Pub. Co.

Linden, H. P., Midgley, J. (1996): Anemophilous plants select pollen from their own species from the air. *Oecologia* **108**, 85–87.

Lindstrom, F. T., Haque, R., Freed, V. H., Boersma, L. (1967): Theory on the movement of some herbicides in soils; linear diffusion and convection of chemicals in soils. *J. Environ. Sci. Tech.* **1**, 561–565.

Liss, P. S., Merlivat, L. (1986): Air–sea gas exchange rates: Introduction and synthesis. In *The Role of Air–Sea Exchange in Geochemical Cycling*. Buat-Menard, P. (ed.). Norwell, MA: D. Reidel.

Liu, Y., Steenhuis, T. S., Parlange, J. Y. (1994): Closed-form solution for finger width in sandy soils at different water contents. *Water Resour. Res.* **30**, 949–952.

Livshits, M. A., Gurija, G. T., Belintsev, B. N., Volkenstein, M. V. (1981): Positional differentiation as pattern-formation in reaction-diffusion systems with permeable boundaries. *J. Math. Biol.* **11**, 295–310.

Lloyd, A. L., May, R. M. (1996): Spatial heterogeneity in epidemic models. *J. Theor. Biol.* **179**, 1–11.

Lobel, P. S., Randall, J. E. (1986): Swarming behavior of the hyperiid amphipod, Anchylomera blossevilli. *Journal of Plankton Research* **8**, 253–262.

Loefer, J. B., Mefferd, R. B. (1952): Concerning pattern formation by free-swimming microorganisms. *Amer. Naturalist* **86**, 325–329.

Logan, B. E., Wilkinson, D. B. (1990): Fractal geometry of marine snow and other biological aggregates. *Limnol. Oceanogr.* **35**, 130–136.

Logan, J. D. (1994): *An Introduction to Nonlinear Partial Differential Equations*. New York: Wiley.

Long, G. E. (1977): Spatial dispersion in a biological control model for larch casebearer (*Coleophora laricella*). *Environmental Entomol.* **6**, 843–852.

Longhurst, A. R. (1991): Role of the marine biosphere in the global carbon cycle. *Limnol. Oceanogr.* **36**, 1507–1526.

Lopez, F., Garcia, M. (1997): Open-channel flow through simulated vegetation: Turbulence modeling and sediment transport. Technical Report WRP-CP-10. Washington: US Army Corp of Engineers.

Lopez-Gomez, J., Pardo, S. G., R. (1992): Coexistence in a simple food-chain with diffusion. *J. Math. Biol.* **30**, 655–668.

Lorenz, E. N. (1964): The problem of deducing the climate from the governing equations. *Tellus* **16**, 1–11.

Lotka, A. J. (1924): *Elements of Physical Biology*. Baltimore: Williams and Wilkins Co., Inc. (also *Elements of Mathematical Biology*. New York: Dover Publ., 1956).

Loukashkin, A. S. (1970): On the diet and feeding behavior of the northern anchovy, Engraulis mordax (Girard). *Proc. Calif. Acad. Sci.* **37**, 419–458.

Ludwig, D., Aronson, D. G., Weinberger, H. F. (1979): Spatial patterning of the spruce budworm. *J. Math. Biol.* (in press).

MacArthur, R. H., Wilson, E. O. (1967): *The Theory of Island Biogeography*. Princeton: Princeton Univ. Press.

MacArthur, R. H. (1972): *Geographical Ecology*. New York: Harper & Row.

MacCamy, R. C. (1981): A population model with nonlinear diffusion. *J. Differential Equations* **39**, no. 1, 52–72.

Machta, L. (1959): Transport in the stratosphere and through the tropopause. *Adv. Geophys.* **6**, 273–288.

Machtans, C. S., Villard, M., Hannon, S. J. (1996): Use of riparian buffer strips as movement corridors by forest birds. *Conserv. Biol.* **10**, 1366–1379.

Mackas, D. L., Boyd, C. M. (1979): Spectral analysis of zooplankton spatial heterogeneity. *Sci.* **204**, 62–64.

MacIntyre, S., Alldredge, A. L., Gotschalk, C. C. (1995): Accumulation of marine snow at density discontinuities in the water column. *Limnol. Oceanogr.* **40**, 449–468.

MacIsaac, J. J., Dugdale, R. C., Barber, R. T., Blasco, D., Packard, T. T. (1985): Primary production cycle in an upwelling center *Deep-Sea Res. Part A. Oceanogr. Res. Pap.* **32**, 503–529.

Mackenzie, B. R., Miller, J. T., Cyr, S., Leggett, W. C. (1994): Evidence for a dome-shaped relation between turbulence and larval fish ingestion rates. *Limnol. Oceanogr.* **39**, 1790–1791.

Madsen, S. D., Forbes, T. L., Forbes, V. E. (1997): Particle mixing by the polychaete Capitella species 1: Coupling fate and effect of a particle-bound organic contaminant (fluoranthene) in a marine sediment. *Mar. Ecol. Progr. Ser.* **147**, 129–142.

Maehr, D. S. (1997a): *The Florida Panther: Life and Death of a Vanishing Carnivore*. Washington, DC: Island Press.

Maehr, D. S. (1997b): The comparative ecology of bobcats, black bears and Florida panther in south Florida. *Bull. Florida Museum of Natural History* **40**, 1–176.

Maehr, D. S., Land, E. D., Roof, J. C., McCown, J. W. (1989): Early maternal behavior in the Florida panther (Felis concolor coryi). *Amer. Midl. Nat.* **122**, 34–43.

Maitland, D. P., Maitland, A. (1994): Significance of burrow-opening diameter as a flood-prevention mechanism for air-filled burrows of small intertidal arthropods. *Mar. Biol.* **119**, 221–225.

Maki, T. (1969): On zero-plane displacement and roughness length in the wind velocity profile equation over a corn canopy. *J. Agricul. Meteorol. Tokyo* **25**, 13–18 (Japanese with English summary).

Maki, T. (1976): Aerodynamic characteristics of wind within and above a plant canopy height. *Bull. National Inst. of Agricul. Sci., Tokyo, Ser. A.* no. **23**, 1–67.

Mangel M., Clark C. W. (1988): *Dynamic Modeling in Behavioral Ecology*. Princeton Univ. Press.

Mann, K., Lazier, J. R. N. (1991): *Dynamics of Marine Ecosystems: Biological-Physical Interactions in the Oceans*. Oxford: Blackwell.

Mann, K. H., Lazier, J. R. N. (1996): *Dynamics of Marine Ecosystems. Biological-Physical Interactions in the Ocean.* Cambridge, MA: Blackwell Science.

Manoranjan, V. S., van den Driessche, P. (1986): On a diffusion model for sterile insect release. *Math Biosci.* **79**, no. 2, 199–208.

Marsh, L. M., Jones, R. E. (1988): The form and consequences of random walk movement models. *J. Theor. Biol.* **133**, 113–131.

Marshall, E. J. P., Westlake, D. F. (1990): Water velocities around water plants in chalk streams. *Folia Geobot. Phytotax.* **25**, 279–289.

Marshall, T. J., Holmes, J. W. (1988): *Soil Physics*, 2nd ed. Cambridge: Cambridge Univ. Press.

Marshall, W. H. (1965): Ruffed grouse behavior (Biotelemetry). *Bio-Science* **15**, 92–94.

Martel, A., Chai, F. S. (1991): Foot-reaising behavior and active participation during the initial phase of post-metamorphic drifting in the gastropod *Lacuna spp. Mar. Ecol. Prog. Ser.* **72**, 247–254.

Martel, A., Chai, F. S. (1991): Drifting and dispersal of small bivalves and gastropods with direct development. *J. Exp. Mar. Biol. Ecol.* **150**, 131–147.

Martel, A., Diefenbach, T. (1993): Effects of body-size, water current, and microhabitat on mucous-thread drifting in post-metamorphic gastropods *Lacuna spp. Mar. Ecol. Prog. Ser.* **99**, 215–220.

Martin, W. R., Banta, G. T. (1992): The measurement of sediment irrigation rates: A comparison of the Br^- tracer and $^{222}Rn/^{226}Ra$ disequilibrium techniques. *J. Mar. Res.* **50**, 125–154.

Martin, J. H., Gordon, R. M., Fitzwater, S. E. (1991): The case for iron. *Limnol. Oceanogr.* **6**, 1793–1802.

Martin, J. H., Knauer, G. H., Karl, D. M., Broenkow, W. W. (1987): VERTEX: Carbon cycling in the northeast Pacific. *Deep-Sea Res.* **34**, 267–285.

Matlack, G. R. (1992): Influence of fruit size and weight on wind dispersal in *Betula lenta* a gap-colonizing tree species. *Amer. Midl. Nat.* **128**, 30–39.

Matsuda, H., Akamine, T. (1994): Simultaneous stimation of mortality and dispersal rated of an artificially released population. *Res. Popul. Ecol.* **30**, 73–78.

Maurer, B. A. (1994): *Geographical Population Analysis: Tools for the Analysis of Biodiversity. Methods in Ecology.* Lawton, J. H., Likens, G. E. (eds.). Oxford: Blackwell Scientific.

Maxey, M. R. (1987): The gravitational settling of aerosol-particles in homogeneous turbulence and random flow-fields. *J. Fluid Mech.* **174**, 441–465.

May, R. M. (1973): *Stability and Complexity in Model Ecosystems.* Princeton: Princeton Univ. Press (also, 2nd ed., 1974).

May, R. M. (1975): Biological populations obeying difference equations: Stable points, stable cycles, and chaos. *J. Theor. Biol.* **51**, 511–524.

May, R. M. (1976): Simple mathematical models with very complicated dynamics. *Nature* **261**, 459–467.

May, R. M. (ed.) (1976): *Theoretical Ecology: Principles and Applications.* Philadelphia, Toronto: W. B. Saunders Co.

May, R. M. (1977): Thresholds and breakpoints in ecosystems with a multiplicity of stable states. *Nature* **269**, 471–477.

May, R. M., Oster, G. F. (1976): Bifurcations and dynamic complexity in simple ecological models. *Amer. Naturalist* **110**, 573–599.

Maynard Smith, J. (1971): *Mathematical Ideas in Biology.* Cambridge, London, New York: Cambridge Univ. Press.

Maynard Smith, J. (1974): *Models in Ecology*. Cambridge, London, New York: Cambridge Univ. Press.

McCall, P. L., Soster, F. M. (1990): Benthos response to disturbance in western Lake Erie: Regional faunal surveys. *Can. J. Fish. Aquat. Sci.* **47**, 1996–2009.

McCave, I. N. (1984): Size spectra and aggregation of suspended particles in the deep ocean. *Deep-Sea Res. I* **31**, 329–352.

McGillicuddy, D. J., Robinson, A. R., Siegel, D. A., Jannasch, H. W., Johnson, R., Dickey, T. D., McNeil, J., Michaels, A. F., Knap, A. H. (1998): Influence of mesoscale eddies on new production in the Sargasso Sea. *Nature* **394**, 263–266.

McGurk, M. D. (1986): Natural mortality of marine pelagic fish eggs and larvae: Role of spatial patchiness. *Mar. Ecol. Progr. Ser.* **34**, 227–242.

McGurk, M. D. (1987): The spatial patchiness of Pacific herring larvae. *Environ. Biol. Fish.* **20**, 81–89.

McGurk, M. D. (1988): Advection diffusion and mortality of Pacific herring larvae *Clupea harengus pallasi* in Bamfield inlet, British Columbia, Canada. *Mar. Ecol. Progr. Ser.* **51**, 1–18.

McLeese, D. W. (1970): Detection of dissolved substances by the American lobster (*Homarus americanus*) and olfactory attraction between lobsters. *J. Fish. Res. Board Canada* **27**, 1371–1378.

McMurtrie, R. (1978): Persistence and stability of single-species and prey–predator systems in spatially heterogeneous environments. *Math. Biosciences* **39**, 11–51.

McNab, B. K. (1963): Bioenergetics and the determination of home range size. *Amer. Naturalist* **97**, 133–140.

McManus, M. L. (1973): A dispersal model for larvae of the gypsy moth *Porthetria dispar* L. In *Proc. Workshop/Conference II on Ecological Systems Approaches to Aerobiology II*. US/IBP Aerobiology Program Handbook No. 3, 129–138. Edmonds, R. L., Bennighoff, W. S. (eds.). Ann Arbor: Univ. Michigan Press.

Mead K. S., Denny, M. W. (1995): The effects of hydrodynamic shear stress on fertilization and early development of the purple sea urchin *Strongylocentrotus purpuratus*. *Biol. Bull.* **188**, 46–56.

Meadows, P. S., Meadows, A. (1991): *The Environmental Impact of Burrowing Animals & Animal Burrows*. New York: Oxford Univ. Press.

Meakin, P. (1986): A new model for biological pattern formation. *J. Theor. Biol.* **118**, 101–113.

Mellott, L. E., Berry, M. W., Comiskey, E. J., Gross, L. J. (1999): The design and implementation of an individual-based predator–prey model for a distributed computing environment. *Simulation Theory and Practice* **7**, 47–70.

Menge, B. A. (1992): Community regulation: Under what conditions are bottom-up factors important on rocky shores? *Ecol.* **73**, 755–765.

Mercer, J. W., Waddell, R. K. (1993): Contaminant transport in groundwater. In *Handbook of Hydrology*. Maidment, D. R. (ed.). New York: McGraw-Hill.

Meyers, D. E., O'Brien, E. E., Scott, L. R. (1978): Random advection of chemically reacting species. *J. Fluid Mech.* **85**, 233–240.

Meyers, M. B., Powell, E. N., Fossing, H. (1988): Movement of oxybiotic and thiobiotic meiofauna in response to changes in pore-water oxygen and sulfide gradients around macro-infaunal tubes. *Mar. Biol.* **98**, 395–414.

Midgley, J. J., Bond, W. J. (1991): How important is biotic pollination and dispersal to the success of the angiosperms? *Phil. Trans. Roy. Soc. Lond. B* **333**, 209–215.

Miller, G. L. (1984): Ballooning in Geolycosa-Turricola (Treat) and Geolycosa-Patellonigra (Wallce)—High dispersal frequencies in stable habitats. *Can. J. Zool.* **62**, 2110–2111.

Milton, K., May, M. L. (1976): Body weight, diet and home range area in primates. *Nature* **259**, 459–462.

Mimura, M. (1978): Asymptotic behavior of a parabolic system related to a planktonic prey and predatory system. *SIAM J. Appl. Math.*, Part C (in press).

Mimura, M., Nishiura, Y., Yamaguti, M. (1979): Some diffusive prey and predator systems and their bifurcation problem. Paper presented at Intl. Conf. Bifurcation Theory and Its Application to Scientific Disciplines, New York, Oct. 1977. *Ann. New York Academy of Sciences* **316**, 490–510.

Mimura, M., Murray, J. D. (1979): On a planktonic prey–predator model which exhibits patchiness. *J. Theor. Biol.* **75**, 249–262.

Mimura, M., Nishida, T. (1978): On a certain semilinear parabolic system related to Lotka–Volterra's ecological model. *Publ. Research Inst. Math. Sci., Kyoto Univ.* **14**, 269–282.

Mitchell, J. G., Fuhrman, J. A. (1989): Centimeter scale vertical heterogeneity in bacteria and chlorophyll a. *Mar. Ecol. Progr. Ser.* **54**, 141–148.

Mitchell, J. G., Martinez-Alonso, M-R., Lalucat, J., Esteve, I., Brown, S. (1991): Velocity changes, long runs and reversals in the *Chromatium minus* swimming response. *J. Bacteriol.* **173**, 997–1003.

Mitchell, J. G., Okubo, A., Fuhrman, J. A. (1985): Microzones surrounding phytoplankton form the basis for a stratified marine microbial ecosystem. *Nature* **316**, 58–59.

Mitchell, J. G., Okubo, A., Fuhrman, J. A. (1990): Gyrotaxis as a new mechanism for generating plankton heterogeneity on small scales. *Limnol. Oceanogr.* **35**, 123–130.

Mitchell, J. G., Pearson, L., Bonazinga, A., Dillon, S., Khouri, H., Paxinos, R. (1995): Long lag times and high velocities in the motility of natural assemblages of marine bacteria. *Appl. Environ. Microbiol.* **61**, 877–882.

Mitchell, R. G. (1979): Dispersal of early instars of the Douglas-fir tussock moth *Orgyia pseudotsugata*, a univoltine pest that defoliates *Pseudotsuga menziesii* and true fir, *Abies spp.*, in Western United States and Canada. *Ann. Entomol. Soc. Am.* **72**, 291–297.

Moen, R., Pastor, J., Cohen, Y. (1997): A spatially explicit model of moose foaging and energetics. *Ecology* **78**, 505–521.

Mogilner, A., Edelstein-Keshet, L. (1999): A non-local model for a swarm. *J. Math. Biol.* **38**, 534–570.

Mollison, D. (1977a): Long-distance dispersal of windborne organisms (manuscript).

Mollison, D. (1977b): Spatial contact model for ecological and epidemic spread. *J. Roy. Statist. Soc. B* **39**, 283–326.

Mollison, D. (1984): Simplifying simple epidemic models. *Nature* **310**, 224–225.

Moloney, K. A. (1993): Measuring process through pattern: Reality or fantasy. In *Patch Dynamics. Lecture Notes in Biomathematics*, Vol. 96, 61–69. Levin, S. A. (ed.). Berlin: Springer-Verlag.

Monk, P. B., Othmer, H. G. (1990): Wave propagation in aggregation fields of the cellular slime mould *Dictyostelium Discoideum. Proc. Roy. Soc. Lond.* B **240**, 555–589.

Monismith, S. G., Koseff, J. R., Thompson, J. K., O'Riordan, C. A., Nepf, H. M. (1990): A study of model bivalve siphonal currents. *Limnol. Oceanogr.* **35**, 680–696.

Monteith, J. L., Unsworth, M. H. (1990): *Principles of Environmental Physics*, 2nd ed. London: Edward Arnold.

Montgomery, R. B. (1939): Ein Versuch den vertikalen und seitlichen Austausch in der Tiefe der Sprungschicht im äquatorialen Atlantischen Ozean zu bestimmen. *Ann. d. Hydrogr. u. Mar. Meteor.* **67**, 242–246.

Montgomery, R. B., Palmén, E. (1940): Contribution to the question of the Equatorial Counter Current. *J. Mar. Res.* **3**, 112–133.

Montroll, E. W. (1967): On nonlinear processes involving population growth and diffusion. *J. Appl. Probability* **4**, 281–290.

Montroll, E. W. (1968): Lectures on nonlinear rate equations, especially those with quadratic nonlinearities. In *Lectures in Theoretical Physics*, Vol. X—A, 531–573. Barut, A. O., Brittin, W. E. (eds.).

Mooij, W. M., DeAngelis, D. L. (1999): Error propogation in spatially explicit population models: A reassessment. *Cons. Biol.* **13**, 930–933.

Moore, P. D. (1976): How far does pollen travel? *Nature* **260**, 388–389.

Moore, P. A., Fields, D. M., Yen, J. (1999): Physical constraints of chemoreception in foraging copepods. *Limnol. Oceangr.* **44**, 166–177.

Moran, V. C., Gunn, B. H., Walter, G. H. (1982): Wind dispersal and settling of 1st instar crawlers of the cochineal insect *Dactylopus austrinus. Ecol. Entomol.* **7**, 409–420.

Morgan, B. J. T. (1976): Stochastic models of grouping changes. *Adv. Appl. Prob.* **8**, 30–57.

Morisita, M. (1950): Dispersal and population density of a water-strider, *Gerris lacustris* L. *Contribut. Physiol. Ecol. Kyoto Univ.* No. 65, 1–149 (Japanese).

Morisita, M. (1952): Habitat preference and evaluation of environment of an animal: Experimental studies on the population density of an ant-lion, *Glenuroides japonicus* M'L. (1). *Physiol. and Ecol.* **5**, 1–16 (Japanese with English summary).

Morisita, M. (1954): Dispersion and population pressure: Experimental studies on the population density of an ant-lion, *Glenuroides japonicus M'L* (2). *Jap. J. Ecol.* **4**, 71–79 (Japanese with English synopsis).

Morisita, M. (1959): Measuring the dispersion of individuals and analysis of the distributional patterns. *Mem. Faculty Sci. Kyushu Univ. Series E (Biology)* **2**, 215–235.

Morisita, M. (1971): Measuring of habitat value by the "environmental density" method. In *Statistical Ecology*, Vol. 1, 379–401. Patil, G. P., Pielou, E. C., Waters, W. E. (eds.). University Park, PA: Penn. State Univ. Press.

Morowitz, H. J. (1972): *Entropy for Biologists.* New York: Academic Press.

Morris, W. (1993): Predicting the consequences of plant spacing and biased movement for pollen dispersal by honey bees. *Ecology* **74**, 493–500.

Mortensen, R. E. (1969): Mathematical problems of modeling stochastic nonlinear dynamic systems. *J. Statistical Phys.* **1**, 271–296.

Mullen, A. J. (1989): Aggregation of fish through variable diffusivity. *Fish Bull.* **87**, 353–362.

Munk, P., Larsson, P. O., Danielssen, D. S., Moksness, E. (1999): Variability in frontal zone formation and distribution of gadoid fish larvae at the shelf break in the northeastern North Sea. *MEPS.* **177**, 221–233.

Munk, W. H. (1966): Abyssal recipes. *Deep-Sea Res.* **13**, 707–730.

Murie, M. (1963): Homing and orientation of deer mice. *J. Mammal.* **44**, 338–349.

416   References

Murray, A. G., Jackson G. A. (1992): Viral dynamics: A model of the effects of size, shape, motion and abundance of single-celled planktonic organisms and other particles. *Mar. Ecol. Progr. Ser.* **89**, 103–116.

Murray, A. G., Jackson, G. A. (1993): Viral dynamics II: A model of the interaction of ultraviolet light and mixing processes on virus survival in seawater. *Mar. Ecol. Progr. Ser.* **102**, 105–114.

Murray, J. D. (1975): Non-existence of wave solutions for the class of reaction-diffusion equations given by the Volterra interacting-population equations with diffusion. *J. Theor. Biol.* **52**, 459–469.

Murray, J. D. (1976): Spatial structures in predator-prey communities – a non-linear time delay diffusional model. *Math. Biosci.* **31**, 73–85.

Murray, J. D. (1977): *Lectures on Nonlinear-Differential Equation Models in Biology.* Oxford: Oxford Univ. Press.

Murray, J. D. (1981): On pattern formation mechanisms for lepidopteran wing patterns and mammalian coat markings. *Phil. Trans. Roy. Soc. B (London)* **245**, 473–496.

Murray, J. D. (1989): *Mathematical Biology.* New York: Springer-Verlag.

Murray, J. D. (1993): *Mathematical Biology*, 2nd edition, (Biomathematics Texts, Vol. 19). Berlin: Springer-Verlag.

Murray, J. D., Stanley, E. A., Brown, D. L. (1986): On the spatial spread of rabies among foxes. *Proc. R. Soc. Lond.* B **229**, 111–150.

Murray, J. D. (1989): *Mathematical Biology.* Heidelberg: Springer-Verlag.

Murray, J., Stanley, E., Brown, D. (1986): On the spatial spread of rabies among foxes. *Proc. Roy. Soc. London B, Biol. Sciences* **229**, 111–150.

Murthy, C. R. (1976): Horizontal diffusion characteristics in Lake Ontario. *J. Phys. Oceanogr.* **6**, 76–84.

Murthy, C. R., Dunbar, D. S. (1981): Structure of the flow within the coastal boundary-layer of the Great Lakes. *J. Phys. Oceanogr.* **11**, 1567–1577.

Nakata, K., Ishikawa, K. (1975): Fluctuation of local phytoplankton abundance in coastal waters. *Jap. J. Ecol.* **25**, 201–205 (Japanese with English abstract).

Namba, T. (1980a): Asymptotic behavior of solutions of the diffusive Lotka-Volterra equations. *J. Math. Biol.* **10**, 295–303.

Namba, T. (1980b): Density-dependent dispersal and spatial distribution of a population. *J Theor. Biol.* **86**, 351–363.

Namba, T. (1989): Competition for space in a heterogeneous environment. *J. Math. Biol.* **27**, 1–16.

Neale, P. J., Talling, J. F., Heany, S. I., Reynolds, C. S., Lund, J. W. G. (1991): Long time series from the English Lake District: Irradiance dependent phytoplankton dynamics during the spring maximum. *Limnol. Oceanogr.* **36**, 751–760.

Nee, S., Read, A. F., Greenwood, J. J. D., Harvey, P. H. (1991): The relationship between abundance and body size in British birds. *Nature* **351**, 312–313.

Needham, D. J. (1992): A formal theory concerning the generation and propagation of travelling wave-fronts in reaction diffusion equations. *Quart. J. Mech. Appl. Math.* **45**, no. 3, 469–498.

Nelson, D. M., Treguer, P. (1992): Role of silicon as a limiting nutrient to Antarctic diatoms: Evidence from kinetic studies in the Ross Sea ice-edge zone. *Mar. Ecol. Progr. Ser.* **80**, 255–264.

Nelson, E. (1976): *Dynamical Theories of Brownian Motion.* Princeton: Princeton Univ. Press.

Nepf, H. M., Koch, E. W. (1999): Vertical secondary flows in stem arrays. *Limnol. Oceanogr.* **44**, 1072–1080.

Nepf, H. M. (1999): Drag, turbulence, and diffusion in flow through emergent vegetation. *Water Resour. Res.* **35**, 479–489.

Nepf, H. M., Sullivan, J. A., Zavistoski, R. A. (1997): A model for diffusion with emergent vegetation. *Limnol. Oceanogr.* **42**, 1735–1745.

Nepf, H. M., Vivoni, E. R. (1999): Turbulence structure in depth-limited, vegetative flow: Transition between emergent and submerged regimes. *Proc. XXVIII IAHR Congress*, Graz, Austria. Delft: IAHR.

Neushul, M., Benson, J., Harger, B. W. W., Charters, A. C. (1992): Macroalgal farming in the sea: Water motion and nitrate uptake. *J. Appl. Phycol.* **4**, 255–265.

Newman, W. I. (1980): Some exact solutions to a non-linear diffusion problem in population genetics and combustion. *J. Theor. Biol.* **85**, 325–334.

Newmark, W. D. (1987): A land-bridge island perspective on mammalian extinctions in western North American parks. *Nature* **325**, 430–432.

Nicol, S. (1986): Shape, size, and density of daytime surface swarms of the euphausid Meganyctiphanes norvegica in the Bay of Fundy. *J. Plankton Res.* **8**, 29–39.

Nicol, S., James, A., Pitcher, G. (1987): A first record of daytime surface swarming by Euphausia lucens in the southern Benguela region. *Mar. Biol.* **94**, 7–10.

Nicolis, G., Prigogine, I. (1977): *Self-Organization in Nonequilibrium Systems.* New York, London, Sydney, Toronto: J. Wiley & Sons.

Niklas, K. J. (1985): The aerodynamics of wind pollination. *Bot. Rev.* **51**, 328–386.

Niklas, K. J. (1992): *Plant Biomechanics.* Chicago: Univ. Chicago Press.

Niklas, K. J. (1997): *The Evolutionary Biology of Plants.* Chicago: Univ. Chicago Press.

Niklas, K. J., Buchmann, S. (1987): The aerodynamics of pollen capture in two sympatric *Ephedra* species. *Evol.* **41**, 104–123.

Niklas, K. J., Buchmann, S. (1988): Aerobiology and pollen capture in orchard-grown *Pistacia vera* (Anacardiaceae). *Amer. J. Bot.* **75**, 1813–1829.

Niklas, K. J., Paw U, K. T. (1983): Conifer ovulate cone morphology: Implications on pollen impaction patterns. *Amer. J. Bot.* **70**, 568–577.

Nishiki, S. (1966): On the aerial migration of spiders. *Acta Arachnologica* **20**, 24–34 (Japanese with English summary).

Nishizawa, S., Fukuda, M., Inoue, N. (1954): Photographic study of suspended mater and plankton in the sea. *Bull. Fac. Fisheries Hokkaido Univ.* **5**, 36–40.

Niwa, H. S. (1994): Self-organizing dynamic model of fish schooling, *J. Theor. Biol.* **171**, 123–136.

Niwa, H. S. (1996): Newtonian dynamical approach to fish schooling. *J. Theor. Biol.* **181**, 47–63.

Niwa, H. S. (1998a): Migration of fish schools in heterothermal environments. *J. Theor. Biol.* **193**, 215–231.

Niwa, H. S. (1998b): School size statistics of fish. *J. Theor. Biol.* **195**, 351–361.

Nixon, S. W., Oviatt, C. A. (1972): Preliminary measurements of midsummer metabolism in beds of eelgrass, *Zostera marina. Ecology* **53**, 150–153.

Noble, J. V. (1974): Geographic and temporal development of plagues. *Nature* **250**, 726–729.

Noble, P. S. (1983): *Biophysical Plant Physiology and Ecology.* New York: W.H. Freeman.

Nonacs, P., Smith, P. E., Bouskila, A., Lutteg, B. (1994): Modeling the behavior of the northern anchovy, *Engraulis mordax*, as a schooling predator exploiting patchy prey. *Deep-Sea Res. II* **41**(1), 147–169.

Nonacs, P., Smith, P. E., Mangel, M. (1998): Modeling foraging in the northern anchovy (*Engraulis mordax*): Individual behavior can predict school dynamics and population biology. *Can. J. Fish. Aquat. Sci.* **55**, 1179–1188.

Norcross, B. L., Shaw, R. F. (1984): Oceanic and estuarine transport of fish eggs and larvae: A review. *Trans. Amer. Fish. Soc.* **113**, 153–165.

Oaten, A., Murdoch, W. W. (1975): Functional response and stability in predator–prey systems. *Amer. Naturalist* **109**, 289–298.

Oaten, A., Murdoch, W. W. (1975): Switching, functional response, and stability in predatory–prey systems. *Amer. Naturalist* **109**, 299–318.

O'Brien, D. P. (1988): Direct observations of clustering (schooling and swarming) behaviour in mysids (Crustacea: Mysidacea). *Mar. Ecol. Progr. Ser.* **42**, 235–246.

O'Brien, D. P. (1989): Analysis of internal arrangement of individuals within crustacean aggregations (Euphausiacea, Mysidacea). *J. Exp. Biol. Mar. Ecol.* **128**, 1–30.

O'Brien, W. J., Evans, G. I., Howick, G. L. (1986): A new view of the predation cycle of a planktivorous fish, white crappie (Pomoxis annularis). *Can. J. Fish. Aquat. Sci.* **43**, 1894–1899.

Oda, T. (1963): Studies on the dispersion of the mulberry scale, *Pseudualacaspis pentagona*. *Jap. J. Ecol.* **13**, 41–46 (Japanese with English synopsis).

Odell, G. (1980): Biological waves. In *Mathematical models in molecular and cellular biology*. Segel, L. A. (ed.), 523–567. Cambridge: Cambridge Univ. Press.

Odendaal, F. J., Turchin, P., Stermitz, P. R. (1988): An incidental-effect hypothesis explaining aggregation of males in a population of Euphydryas anicia. *Amer. Naturalist* **132**, 735–749.

Odum, H. T. (1956): Primary production in flowing waters. *Limnol. Oceanogr.* **1**, 102–117.

Odum, H. T., Hoskin, C. M. (1958): Comparative studies in the metabolism of marine waters. *Texas University Marine Science Institute Contributions* **5**, 16–46.

Ogata, A., Banks, R. B. (1961): A solution of the differential equation of longitudinal dispersion in porous media. USGS. Prof. Paper 411-a.

Ogura, Y. (1952): The theory of turbulent diffusion in the atmosphere. I: *J. Meteorol. Soc. Japan* **30**, 23–28; II: *J. Meteorol. Soc. Japan* **30**, 53–58.

Ogura, Y. (1955): *Theory of Atmospheric Turbulence, Lectures in Meteorology*, Vol. 14. Tokyo: Chijin Shokan (Japanese).

Okubo, A. (1954): A note on the decomposition of sinking remains of plankton organisms and its relationship to nutriment liberation. *J. Oceanogr. Soc. Japan* **10**, 121–131.

Okubo, A. (1956): An additional note on the decomposition of sinking remains of organisms and its relationship to nutriment liberation. *J. Oceanogr. Soc. Japan* **12**, 45–57.

Okubo, A. (1962): A review of theoretical models of turbulent diffusion in the sea. *J. Oceanogr. Soc. Japan*, 20th Anniv., 286–320.

Okubo, A. (1968): Some remarks on the importance of the "shear effect" on horizontal diffusion. *J. Oceanogr. Soc. Japan* **24**, 60–69.

Okubo, A. (1970): Oceanic turbulence and diffusion. In *Physical Oceanography* I, 256–381. Tokyo: Tokai Univ. Press (Japanese).

Okubo, A. (1971a): Oceanic diffusion diagrams. *Deep-Sea Res.* **18**, 789–802.

Okubo, A. (1971b): Horizontal and vertical mixing in the sea. In *Impingement of Man on the Oceans*, 89–168. Hood, D. W. (ed.). New York: J. Wiley & Sons.

Okubo, A. (1972): A note on small organism diffusion around an attractive center: A mathematical model. *J. Oceanogr. Soc. Japan* **28**, 1–7.

Okubo, A. (1974): Diffusion-induced instability in model ecosystems. Chesapeake Bay Institute, The Johns Hopkins University, Baltimore, Tech. Rept. No. 86.

Okubo, A. (1978a): Advection-diffusion in the presence of surface convergence. In *Oceanic Fronts in Coastal Processes*, 23–28. Bowman, M. J., Esaias, W. E. (eds.). Berlin, Heidelberg, New York: Springer.

Okubo, A. (1978b): Horizontal dispersion and critical scales for phytoplankton patches. In *Spatial Pattern in Plankton Communities*, 21–42. Steele, J. H. (ed.). New York: Plenum Press.

Okubo, A. (1984): Critical patch size for plankton and patchiness. In *Symposium on Mathematical Ecology*, 456–477. Levin, S. A., Hallam, T. G. (eds.). Trieste (Italy), 1982. *Math. Ecol. Proc.* **54**.

Okubo, A. (1986): Dynamical aspects of animal grouping: swarms, schools, flocks, and herds. *Adv. Biophys.* **22**, 1–94.

Okubo, A. (1987): Fantastic voyage into the deep: Marine biofluid mechanics. In *Mathematical Topics in Population Biology, Morphogenesis and Neurosciences: Lecture Notes in Biomathematics* Vol. 71, 32–47. Teramoto, E., Yamaguti, M. (eds.). Berlin: Springer-Verlag.

Okubo, A. (1988): Biological vortex rings: Fertilization and dispersal in fish eggs. In *Mathematical Ecology*, 270–283. Hallam, G., Gross, L. J., Levin, S. A. (eds.). Singapore: World Scientific.

Okubo, A. (1994): The role of diffusion and related physical processes in dispersal and recruitment of marine populations. In *The Bio-Physics of Marine Larval Dispersal*, 5–32. Sammarco, P. W., Heron, M. L. (eds.). Washington: American Geophysical Union.

Okubo, A. (1994): The role of diffusion and related physical processes in dispersal and recruitment of marine populations. *Coast. Est. Stud.* **45**, 5–32.

Okubo, A., Anderson, J. J. (1984): Mathematical models for zooplankton swarms: Their formation and maintenance. *Eos* **65**(40), 731–732.

Okubo, A., Bray, D. J., Chiang, H. C. (1981): Use of shadows for studying the three-dimensional structure of insect swarms. *Ann. Entomol. Soc. Amer.* **74**, 48–50.

Okubo, A., Chiang, H. C. (1974): An analysis of the kinematics of swarming of *Anarete pritchardi* Kim (Diptera: Cecidomyiidae). *Res. Popul. Ecol.* **16**, 1–42.

Okubo, A., Chiang, C., Ebbesmeyer, C. C. (1977): Acceleration field of individual midges, *Anarete pritchardi* Kim within a swarm. *Canad. Entomol.* **109**, 149–156.

Okubo, A., Levin, S. A. (1989): A theoretical framework for data analysis of wind dispersal of seeds and pollen. *Ecology* **70**, 329–338.

Okubo, A., Sakamoto, W., Inagaki, T., Kuroki, T. (1977): Studies on the schooling behavior of fish. V. Note on the dynamics of fish schooling. *Bull. Jap. Soc. Sci. Fish* **43**, 1369–1377.

Olgivi, C. S., Dubois, A. B. (1981): The hydrodynamic drag of swimming bluefish (*Potomus salt atrix*) in different intensities of turbulence—variation with changes of buoyancy. *J. Exp. Biol.* **92**, 67–85.

Oliver, R., Kinnear, A., Ganf, G. (1981): Measurements of cell density of three freshwater phytoplankters by density gradient centrifugation. *Limnol. Oceanogr.* **26**, 285–294.

Oliver, R. L., Thomas, R. H., Reynolds, C. S., Walsby, A. E. (1985): The sedimentation of buoyant *Microcystis* colonies caused by precipitation with an iron-containing colloid. *Proc. Roy. Soc. London B* **223**, 511–528.

Olmstead, W. E. (1980): Diffusion systems reacting at the boundary. *Quart. Appl. Math.* **38**, no. 1, 51–59.

Olson, D. B., Backus, R. H. (1985): The concentrating of organisms at fronts: A cold-water fish and a warm-core Gulf Stream ring. *J. Mar. Res.* **43**, 113–137.

Oosawa, F., Nakaoka, Y. (1977): Behavior of micro-organisms as particles with internal state variables. *J. Theor. Biol.* **66**, 747–761.

O'Riordan, C. A., Monismith, S. G., Koseff, J. R. (1995): The effect of bivalve excurrent jet dynamics on mass transfer in the benthic boundary layer. *Limnol. Oceanogr.* **40**, 330–344.

Orr, R. T. (1970): *Animals in Migration.* New York: Macmillan Co.

Oschlies, A., Garcon, V. (1998): Eddy-induced enhancement of primary production in a model of the North Atlantic Ocean. *Nature* **394**, 266–269.

Oster, G. (1976): Internal variables in population dynamics. In *Some Mathematical Questions in Biology* 7. Levin, S. A. (ed.). Lectures on Mathematics in the Life Sciences, Vol. 8, 37–68. Providence, RI: Amer. Math. Soc.

Oster, G. (1978): The dynamics of nonlinear models with age structure. In *Mathematical Association of America Study in Mathematical Biology*, Levin, S. A. (ed.). Vol. II. Populations and Communities, 411–438, Washington: Math. Assoc. Amer.

Oster, G., Takahashi, Y. (1974): Models for age-specific interactions in a periodic environment. *Ecol. Monogr.* **44**, 483–501.

Ostro, L. E., Young, T. P., Silver, S. C., Koontz, F. W. (1999): A geographic information system method for estimating home range size. *J. Wildl. Manage.* **63**, 748–755.

Otake, A. (1970): *Animal Ecology.* Tokyo: Kyoritsu Publ. Co. (Japanese).

Othmer, H. G. (1969): *Interactions of reaction and diffusion in open systems.* (Ph.D. diss.) Dept. of Chemical Engin., Univ. of Minnesota.

Othmer, H. G. (1977): Current problems in pattern formation. In *Lectures on Mathematics in the Life Sciences*, Vol. 9, 57–85. Levin, S. A. (ed.). Providence, RI: Amer. Math. Soc.

Othmer, H. G., Dunbar, S. R., Alt, W. (1988): Models of dispersal in biological systems. *J. Math. Biol.* **26**, 263–298.

Othmer, H. G., Scriven, L. E. (1971): Instability and dynamic pattern in cellular networks. *J. Theor. Biol.* **32**, 507–537.

Otis, D. L., White, G. C. (1999): Autocorrelation of location estimates and the analysis of radiotracking data. *J. Wildl. Manage.* **63**, 1039–1044.

Owen, R. W. (1966): Small-scale, horizontal vortices in the surface layer of the sea. *J. Mar. Res.* **24**, 56–66.

Owens, N. J. P. (1993): Nitrate cycling in marine waters. In *Nitrate. Process, Patterns, and Management.* Burt, T. P., Heathwaite, A. L., Trudgill, S. T. (eds.). Chichester, UK: John Wiley & Sons.

Owen, R. W. (1989): Microscale and finescale variations of small plankton in coastal and pelagic environments. *J. Mar. Res.* **47**, 197–240.

Ozmidov, R. V. (1958): On the calculation of horizontal turbulent diffusion of the pollutant patches in the sea. *Doklady Akad. Nauk, SSSR* **120**, 761–763.

Pacala, S. W., Canham, C. W., Silander, J. A. (1993): Forest models defined by field measurements. I. The design of a northeastern forest simulator. *Canadian J. Forestry* **23**, 1980–1988.

Pacala, S. W., Roughgarden, J. (1982): Spatial heterogeneity and interspecific competition. *Theor. Pop. Biol.* **21**, 92–113.

Pacenka, S., Steenhuis, T. (1984): User's guide for the MOUSE computer program. Ithaca, NY: Cornell Univ. Press.

Paine, R. T., Levin, S. A. (1981): Intertidal landscapes: disturbance and the dynamics of pattern. *Ecol. Monog.* **51**, 145–178.

Pao, C. V. (1981): Coexistence and stability of a competition – diffusion system in population dynamics. *J. Math. Anal.* **83**, 54–76.

Papoulis, A. (1984): *Probability, Random Variables, and Stochastic Processes, 2nd ed.*, New York, McGraw-Hill, 1984.

Parr, A. E. (1927): A contribution to the theoretical analysis of the schooling behavior of fishes. Occasional Papers of the Bingham Oceanographic Collection, No. 1.

Parrington, J. R., Zoller, W. H., Aras, N. K. (1983): Asian dust: Seasonal transport to the Hawaiian islands. *Sci.* **220**, 195–197.

Parrish, J. K. (1992): Do predators "shape" fish schools: Interactions between predators and their schooling prey. *Neth. J. Zool.* **42**(2–3), 358–370.

Parrish, J. K., Hamner, W. M. (eds.) (1997): *Animal Groups in Three Dimensions.* Cambridge: Cambridge Univ. Press.

Parrish, J. K., Turchin, P. (1997): Individual decisions, traffic rules, and emergent pattern: A Lagrangian analysis. In *Animal Groups in Three Dimensions*, 126–142. Parrish, J. K., Hamner, W. M. (eds.). Cambridge: Cambridge Univ. Press.

Parsons, T. R., LeBrasseur, R. S., Fulton, J. D. (1967): Some observations on the dependence of zooplankton grazing on the cell size and concentration of phytoplankton blooms. *J. Oceanogr. Soc. Japan* **23**, 10–17.

Parsons, T. R., Takahashi, M. (1973): *Biological Oceanographic Processes.* Oxford, New York, Toronto, Sydney, Paris, Braunschweig: Pergamon Press (also 2nd ed., 1977).

Partridge, B. L. (1980): The effect of school size on the structure and dynamics of minnow schools. *Animal Behaviour* **28**, 68–77.

Partridge, B. L. (1982). The structure and function of fish schools. *Scientific American*, **246**, 90–99.

Pasquill, F. (1962): *Atmospheric Diffusion.* New York: D. Van Nostrand Co. Ltd. (also 2nd ed., 1976).

Patlak, C. S. (1953a): Random walk with persistence and external bias. *Bull. Math. Biophys.* **15**, 311–338.

Patlak, C. S. (1953b): A mathematical contribution to the study of orientation of organisms. *Bull. Math. Biophys.* **15**, 431–476.

Pattiaratchi, C. (1994): Physical oceanographic aspects of the dispersal of coral spawn slicks: A review. In *The Bio-Physics of Marine Larval Dispersal*, 89–105. Sammarco, P. W., Heron, M. L. (eds.). Washington: American Geophysical Union.

Pattle, R. E. (1959): Diffusion from an instantaneous point source with a concentration-dependent coefficient. *Quart. J. Mech. Appl. Math.* **12**, 407–409.

Pearcy, R. W., Ehleringer, J., Mooney, H. A., Rundel, P. W. (1989): *Plant Physiological Ecology: Field Methods and Instrumentation.* London: Chapman and Hall.

Pearson, K., Blakeman, J. (1906): Mathematical contributions to the theory of evolution—XV. A mathematical theory of random migration. Drapers' Company Research Mem. Biometric Series III, Dept. Appl. Math., Univ. College, Univ. London.

Pedley, T. J., Kessler, J. O. (1987): The orientation of spheroidal microorganisms swimming in a flow field. *Proc. Roy. Soc. London B* **231**, 47–69.

Peloquin, J. J., Olson, J. K. (1985): Observation on male swarms of Psorophora columbiae in Texas (USA) ricelands. *J. Amer. Mosquito Control Assn.* **1**(4), 482–488.

Pennington, J. J. (1985): The ecology of fertilization of echinoid eggs: The consequence of sperm dilution, adult aggregation, and synchronous spawning. *Biol. Bull.* **169**, 417–430.

Peters, R. H. (1983): *The Ecological Implications of Body Size.* Cambridge: Cambridge Univ. Press.

Pettitt, J. M. (1984): Aspects of flowering and pollination in marine angiosperms. *Oceanogr. Mar. Biol. Ann. Rev.* **22**, 315–342.

Pettitt, J., Ducker, S., Knox, B. (1981): Submarine pollination. *Sci. Amer.* **244**, 131–143.

Philbrick, C. T. (1988): Evolution of underwater outcrossing from aerial pollination systems: A hypothesis. *Ann. Missouri Bot. Gard.* **75**, 836–841.

Philip, J. R. (1957): Sociality and sparce populations. *Ecology* **38**, 107–111.

Phillips, O. M. (1966): *The Dynamics of the Upper Ocean.* Cambridge: Cambridge Univ. Press (also 2nd ed., 1977).

Pielou, E. C. (1969): *An Introduction to Mathematical Ecology.* New York: J. Wiley & Sons.

Pielou, R. C. (1975): *Ecological Diversity.* New York: J. Wiley & Sons.

Pielou, E. C. (1977): *Mathematical Ecology.* New York, London, Sydney, Toronto: J. Wiley & Sons.

Pilskaln, C. H., Churchill, J. H., Mayer, L. M. (1998): Resuspension of sediment by bottom trawling in the Gulf of Maine and potential geochemical consequences. *Cons. Biol.* **12**, 1223–1229.

Pitcher, T. J. (1973): The three-dimensional structure of schools in the minnow, *Phoxinus phoxinus* (L.). *Anim. Behav.* **21**, 673–686.

Pitcher, T. J. (1995): The impact of pelagic fish behaviour on fisheries. *Scientia Marina* **59**, 295–306.

Pitcher, T. J. (1998): Shoaling and schooling in fishes. In *Comparative Psychology: A Handbook*, 748–760. Greenberg, G., Haraway, M. (eds.). New York: Garland Publ.

Pitcher, T. J., Misund, O. A., Ferno, A., Totland, B., Melle, V. (1996): Adaptive behaviour of herring schools in the Norwegian Sea as revealed by high-resolution sonar. *ICES J. Mar. Sci.* **53**, 449–452.

Pitcher, T. J., Parrish, J. K. (1993): Functions of shoaling behaviour in teleosts. In *Behaviour of Teleost Fishes*, 2nd ed. Pitcher, T. J. (ed.). London: Chapman & Hall.

Pitcher T. J., Wyche C. J. (1983): Predator avoidance behaviour of sand-eel schools: why schools seldom split. pp. 193–204. In *Predators and Prey in Fishes.* Noakes D. L. G., Lindquist B. G., Helhman G. S., Ward J. A. (Eds) Netherlands, Junk, The Hague.

Plagens, M. J. (1986): Aerial dispersal of spiders (Araneae) in a Florida cornfield ecosystem. *Envir. Entomol.* **15**, 1225–1233.

Platt, J. R. (1961): "Bioconvection patterns" in cultures of free-swimming organisms. *Sci.* **133**, 1766–1767.

Platt, T. (1972): Local phytoplankton abundance and turbulence. *Deep-Sea Res.* **19**, 183–187.

Platt, T., Denman, K. L. (1975a): Spectral analysis in ecology. *Ann. Rev. Ecol. and Systematics* **6**, 189–210.

Platt, T., Denman, K. L. (1975b): A general equation for the mesoscale distribution of phytoplankton in the sea. *Mem. Soc. Roy. des Sci. Liege*, 6th serie, **7**, 31–42.

Platt, T., Denman, K. L. (1977): Organization in the pelagic ecosystem. Helgoländer wiss. Meeresunters **30**, 575–581.

Platt, T., Denman, K. L., Jassby, A. D. (1977): Modeling the productivity. In *The Sea*, Vol. 6 (Marine modeling), 807–856. Goldberg, E. D. et al. (eds.). New York: J. Wiley & Sons.

Plesset, M. S., Winet, H. (1974): Bioconvection patterns in swimming microorganism cultures as an example of Rayleigh-Taylor instability. *Nature* **248**, 441–443.

Plesset, M. S., Whipple, C. G., Winet, H. (1976): Rayleigh-Taylor instability of surface layers as the mechanism for bioconvection in cell cultures. *J. Theor. Biol.* **59**, 331–351.

Pointer, J., Samadi, S., Jarne, P., Delay, B. (1998): Introduction and spread of Thiara granifera in Martinique, French West Indies. *Biodiv. Cons.* **7**, 1277–1290.

Pollack, J. B. (1981): Measurements of the volcanic plumes of Mount St. Helens in the stratosphere and troposphere: Introduction. *Sci.* **211**, 815–816.

Pond, S., Pytkowicz, R. M., Hawley, J. E. (1971): Particle dissolution during settling in the oceans. *Deep-Sea Res.* **18**, 1135–1139.

Posedel, N., Faganeli, J. (1991): Nature and sedimentation of suspended particulate matter during density stratification in shallow coastal waters (Gulf of Trieste, northern Adriatic). *Mar. Ecol. Progr. Ser.* **77**, 135–145.

Possingham, H. P., Roughgarden, J. (1990): Spatial population dynamics of a marine organism with complex life cycle. *Ecol.* **71**, 973–985.

Powell, R. A., Zimmerman, J. W., Seaman, D. E. (1997): *Ecology and Behavior of North American Black Bears: Home Ranges, Habitat and Social Organization.* London: Chapman and Hall.

Powell, T. M., Okubo, A. (1994): Turbulence, diffusion and patchiness in the sea. *Phil. Trans. Roy. Soc. London B* **343**, 11–18.

Powell, T. M., Richerson, P. J. (1985): Temporal variation, spatial heterogeneity, and competition for resources in plankton systems: a theoretical model. *Amer. Natur.* **125**, 431–464.

Powell, T. M., Richerson, P. J., Dillon, T. M., Agee, B. A., Dozier, B. J., Godden, D. A., Myrup, L. O. (1975): Spatial scales of current speed and phytoplankton biomass fluctuations in Lake Tahoe. *Sci.* **189**, 1088–1090.

Prandtl, L., Tietjens, O. G. (1957): *Applied Hydro- and Aeromechanics* (translated from German). New York: Dover Publ. Inc.

Prezant, R. S., Chalermwat, K. (1984): Flotation of the bivalve *Corbicula fluminea* as a means of dispersal. *Science* **225**, 1491–1493.

Priede, I. G., Swift, S. M. (1992): *Wildlife Telemetry: Remote Monitoring and Tracking Animals.* New York: Ellis Horwood.

Pritchard, D. W., Carpenter, J. H. (1960): Measurements of turbulent diffusion in estuarine and inshore waters. *Bull. Intl. Assn. Sci. Hydrol.* **20**, 37–50.

Proctor, L. M., Fuhrman, J. A. (1990): Viral mortality of marine bacteria and cyanobacteria. *Nature* **343**, 60–62.

Proctor, M., Yeo, P., Lack, A. (1996): *Natural History of Pollination.* Portland: Timber Press.

Prospero, J. M. (1981): Aeolian transport to the world ocean. In *The Oceanic Lithosphere*, 801–874. Emiliani, C. (ed.). New York: Wiley-Interscience.

Prospero, J. M. (1990): Mineral aerosol transport to the North Atlantic and North Pacific: The impact of African and Asian sources. In *The Long-Range Atmospheric*

Transport of Natural and Contaminant Substances, 59–86. Knap, A. H. (ed.). Dordrecht, The Netherlands: Kluwer.

Prospero, J. M., Barrett, K., Church, T., Dentener, F., Duce, R. A., Galloway, J. N., Levy, II, H., Moody, J., Quinn, P. (1996): Atmospheric deposition of nutrients to the North Atlantic basin. In *Nitrogen Cycling in the North Atlantic Ocean and Its Watersheds*, 27–73. Howarth, R. W. (ed.). Dordrecht, The Netherlands: Kluwer.

Pudsey, C. J., King, P. (1997): Particle fluxes, benthic processes and the palaeo-environmental record in the Northern Weddell Sea. *Deep-Sea Res. Part I Oceanogr. Res. Pap.* **44**, 1841–1876.

Pugh, P. R. (1978): The application of particle counting to an understanding of the small-scale distribution of plankton. In *Spatial Pattern in Plankton Communities*, 111–129, Steele, J. H. (ed.). New York: Plenum Press.

Pulliam, H. R. (1973): On the advantages of flocking. *J. Theor. Biol.* **38**, 419–422.

Purcell, E. M. (1977): Life at low Reynolds number. *Amer. J. Phys.* **45**, 3–11.

Radakov, D. V. (1973): *Schooling in the Ecology of Fish* (translated from Russian text, 1972). New York: J. Wiley & Sons, and Israel Program for Scientific Translations.

Ragotzkie, R. A., Bryson, R. A. (1953): Correlation of currents with the distribution of adult *Daphnia* in Lake Mendota. *J. Mar. Res.* **12**, 157–172.

Railsback, S. F., Lamberson, R. H., Harvey, B. C., Duffy, W. E. (1999): Movement rules for individual-based models of stream fish. *Ecol. Modelling* **123**, 73–89.

Rainey, R. C. (1958): Some observations on flying locusts and atmospheric turbulence in eastern Africa. *Quart. J. Roy. Meteorol. Soc. London*, **84**, 334–354.

Rainey, R. C. (1959): Air current and the behavior of air-borne insects: Locusts. *Proc. Roy. Entomol. Soc. London* (C) **24**, 9.

Rainey, R. C. (1960): Applications of theoretical models to the study of flight-behavior in locusts and birds. In *Model and Analogues in Biology* (Symp. Soc. Exp. Biol. No. 14), 122–139. Cambridge: Cambridge Univ. Press.

Rainey, R. C. (1976): Flight behavior and features of the atmospheric environment. In *Insect Flight*, 75–112. Rainey, R. C. (ed.). New York: J. Wiley & Sons.

Rainey, R. C., Waloff, Z. (1951): Flying locusts and convection currents. *Anti-Locust Bull.* **9**, 51–72.

Rainey, R. C. (1989): *Migration and Meteorology: Flight Behaviour and the Atmospheric Environment of Locusts and Other Migrant Pests*. Oxford: Oxford Science.

Raman, R. K. (1977): Cellular aggregation towards steady point sources of attractant. *J. Theor. Biol.* **64**, 43–69.

Rapaport, D. C. (1985): Asymptotic properties of lattice trails. *J. Physics A—Math Gen* **18**, L475–L479.

Raupach, M. R., Antonia, A. R., Rajagopalan, S. (1991): Rough-wall turbulent boundary layers. *Appl. Mech. Rev.* **44**, 1–25.

Raupach, M. R., Finnigan, J. J., Brunet, Y. (1996): Coherent eddies and turbulence in vegetation canopies: The mixing-layer analogy. *Boundary-Layer Meteorol.* **78**, 351–382.

Raupach, M. R., Thom, P. G. (1981): Turbulence in and above plant canopies. *Ann. Rev. Fluid Mech.* **13**, 97–129.

Raven, P. H., Evert, R. F., Eichorn, S. E. (1999): *Biology of Plants*, 6th ed. New York: Freeman/Worth.

Ray, A. J., Aller, R. C. (1985): Physical irrigation of relict burrows: Implication for sediment chemistry. *Mar Geol.* **62**, 371–379.

Raynor, G. S., Ogden, E. C., Hayes, J. V. (1970): Dispersion and deposition of ragweed pollen from experimental source. *J. Appl. Meteorol.* **9**, 885–895.

Raynor, G. S., Ogden, E. C., Hayes, J. V. (1972): Dispersion and deposition of timothy pollen from experimental source. *Agric. Meteorol.* **9**, 347–366.

Raynor, G. S., Ogden, E. C., Hayes, J. V. (1973): Dispersion of pollens from low-level, crosswind line sources. *Agric. Meteorol.* **11**, 177–195.

Regal, P. J. (1982): Pollination by wind and animals: Ecology and geographic patterns. *Ann. Rev. Ecol. Syst.* **13**, 497–524.

Reid, C. (1899): *The Origin of the British Flora.* London: Dulaw (in Skellam, J. G., 1951a).

Renner, S. S., Ricklefs, R. E. (1995): Dioecy and its correlates in the flowering plants. *Amer. J. Bot.* **82**, 596–606.

Renshaw, E. (1982): The development of a spatial predator-prey process on inter-connected sites. *J. Theor. Biol.* **94**, 355–365.

Revelante, N., Gilmartin, M. (1990): Vertical water column resource partitioning by a ciliated protozoan population under stratified condition in the Northern Adriatic. *J. Plankton Res.* **12**, 89–108.

Richardson, L. F. (1926): Atmospheric diffusion shown on a distance-neighbor graph. *Proc. Roy. Soc. London A* **110**, 709–727.

Richardson, R. H. (1970): Models and analyses of dispersal patterns. In *Mathematical Topics in Population Genetics*, 79–103. Ken-ichi Kojima (ed.). New York: Springer.

Richardson, M. J., Weatherly, G. L., Gardner, W. D. (1993): Benthic storms in the Argentine basin. *Deep-Sea Res.* Part II. **40**, 975–987.

Richter, K. E. (1985a): Acoustic scattering at 1.2 MHz from individual zooplankters and copepod populations. *Deep-Sea Res.* **32**, 149–161.

Richter, K. E. (1985b): Acoustic determination of small-scale distributions of in-dividual zooplankters and zooplankton aggregations. *Deep-Sea Res.* **32**, 163–182.

Richter, K. E., Bennett, J. C., Smoth, Jr., K. L. (1985): Bottom-moored acoustic array to monitor density and vertical movement of deep-sea benthopelagic ani-mals. *IEEE J. Ocean Engineering* **OE-10**, 32–37.

Richmond, R. H., Hunter, C. L. (1990): Reproduction and recruitment of corals: Comparisons among the Caribbean, the Tropical Pacific, and the Red Sea. *Mar. Ecol. Progr. Ser.* **60**, 185–203.

Ridd, P. V. (1996): Flow through animal burrows in mangrove creeks. *Est. Coast. Shelf Sci.* **43**, 617–625.

Riebesell, U. (1991): Particle aggregation during a diatom bloom. I. Physical aspects. *Mar. Ecol. Progr. Ser.* **69**, 27–280.

Riley, G. A. (1951): Oxygen, phosphate, and nitrate in the Atlantic Ocean. *Bull. Bingham Oceanogr. Coll.* **12**, 1–126.

Riley, G. A. (1963): Organic aggregates in seawater and the dynamics of their formation and utilization. *Limnol. Oceanogr.* **8**, 372–381.

Riley, G. A., Van Hemert, D., Wangersky, P. J. (1965): Organic aggregates in surface and deep waters of the Sargasso Sea. *Limnol. Oceanogr.* **10**, 354–363.

Riley, G. A., Wangersky, P. J., Van Hemert, D. (1964): Organic aggregates in tropi-cal and subtropical surface waters of the North Atlantic Ocean. *Limnol. Oceanogr.* **9**, 46–550.

Riley, J. R. (1994): Flying insects in the field. In *Video Techniques in Animal Ecology and Behavior*, 1–15. Wratten, S. D. (ed.). London: Chapman Hall.

Riley, J. R., Smith, A. D., Bettany, B. W. (1990): The use of video equipment to record in three-dimensions the flight trajectories of Heliathis armigera and other moths at night. *Physiol. Entomol.* **15**, 73–80.

Riskin, H. (1984): *The Fokker-Planck Equation: Methods of Solution and Applications*. New York: Springer-Verlag.

Roberts, A. (1981): Hydrodynamics of protozoan swimming. In *Biochemistry and Physiology of Protozoa*, Vol. 4. Levandowsky, M., Hutner, S. H. (eds.). New York: Academic Press.

Robock, A., Matson, M. (1983): Circumglobal transport of the El Chichón volcanic dust cloud. *Sci.* **221**, 195–197.

Rogers, D. (1977): Study of a natural population of *Glossina fuscipes fuscipes* Newstead and a model of fly movement. *J. Anim. Ecol.* **46**, 309–330.

Rohlf, F. J., Davenport, D. (1969): Simulation of simple models of animal behavior with a digital computer. *J. Theor. Biol.* **23**, 400–424.

Roman, M. R., Yentsch, C. S., Gauzens, A. L., Phinney, D. A. (1986): Grazer control of the fine-scale distribution of phytoplankton in warm-core Gulf Stream rings. *J. Mar. Res.* **44**, 795–814.

Romano, J. C., Marquet, R. (1991): Occurrence frequencies of sea-surface slicks at long and short time-scales in relation to wind speed. *Est. Coast. Shelf Sci.* **33**, 445–458.

Rombakis, S. (1947): Über die Verbreitung von Pflanzensamen und Sporen durch turbulente Luftströmungen. *Z. Meteor.* **1**, 359–363.

Romey, W. L. (1996): Individual differences make a difference in the trajectories of simulated schools of fish. *Ecol Model* **92**(1), 65–77.

Rosen, R. (1971): *Dynamical System Theory in Biology*, Vol. 1. New York, London, Sydney, Toronto: J. Wiley & Sons.

Rosen, G. (1975): Nonexistence of dissipative structure solutions to Volterra many-species models. *J. Math. Phys.* **16**, 836.

Rosen, G. (1977): Effects of diffusion on the stability of the equilibrium in multi-species ecological systems. *Bull. Math. Biol.* **39**, 373–383.

Rosen, G. (1984): Brownian-motion correspondence method for obtaining approximate solutions to nonlinear reaction-diffusion equations. *Phys. Rev. Lett.* **53**, no. 4, 307–310.

Ross, R. M., Quentin, L. B. (1985): Depth distribution of developing *Euphausia superba* embryos, predicted from sinking rates. *Mar. Biol.* **79**, 47–53.

Rotenberg, M. (1972): Theory of population transport. *J. Theor. Biol.* **37**, 291–305.

Rotenberg, M. (1975): Equilibrium and stability in populations whose interactions are age-specific. *J. Theor. Biol.* **54**, 207–224.

Rotenberg, M. (1982): Diffusive logistic growth in deterministic and stochastic environments. *J. Theoret. Biol.* **94**, no. 2, 253–280.

Rothe, F. (1976): Convergence to the equilibrium state in the Volterra–Lotka diffusion equations. *J. Math. Biol.* **3**, 319–324.

Rothe, F. (1979): Some analytical result about a simple reaction-diffusion system for morphogenesis. *J. Math. Biol.* **7**, no. 4, 375–384.

Rothe, F. (1981): Global existence of branches of stationary solutions for a system of reaction diffusion equations from biology. *Nonlinear Anal.* **5**, no. 5, 487–489.

Rothe, F., de Mottoni, P. (1979): A simple system of reaction-diffusion equations describing morphogenesis: asymptotic behavior. *Ann. Mat. Pura Appl.* (4) **144**, 141–157.

Rothlisberg, P. C., Church, J. A., Forbes, A. M. G. (1983): Modelling the advection of penaeid shrimp larvae in the Gulf of Carpenteria, Australia. *J. Mar. Res.* **41**, 511–538.

Roughgarden, J., Gaines, S., Pacala, S. (1987): Supply-side ecology: The role of physical transport processes. *Proc. Br. Ecol. Soc. Symp.* **27**, 491–518.

Rubenfeld, L. A. (1979): A model bifurcation problem exhibiting the effects of slow passage through critical. *SIAM J. Appl. Math.* **37**, no. 2, 302–306.

Ruckleshaus, M. H. (1995): Estimation of outcrossing rates and of inbreeding depression in a population of the marine angiosperm, *Zostera marina.* *Mar. Biol.* **123**, 583–593.

Ruckelshaus, M., Hartway, C., Kareiva, P. (1997): Assessing the data requirements of spatially explicit dispersal models. *Cons. Biol.* **11**, 1298–1306.

Ruiz, J. (1997): What generates daily cycles of marine snow? *Deep-Sea Res. I* **44**, 1105–1126.

Runge, J. A., Ohman, M. D. (1982): Size fractionation of phytoplankton as an estimate of food available to herbivores. *Limnol. Oceanogr.* **27**, 570–576.

Russell, G., Marshall, B., Jarvis, P. G. (1989): *Plant Canopies: Their Growth, Form and Function.* Cambridge: Cambridge Univ. Press.

Rust, M. K., Bell, W. J. (1976): Chemo-anemotaxis: A behavioral response to sex pheromone in nonflying insects. *Proc. Natl. Acad. Sci. USA* **73**, 2524–2526.

Ruth, T. K., Logan, K. A., Sweanor, L. L., Temple, L. J. (1998): Evaluating cougar translocation in New Mexico. *J. Wildl. Manage.* **62**, 1264–1275.

Ryan, E. P. (1966): Pheromone: Evidence in a decapod crustacean. *Sci.* **151**, 340–341.

Saffmann, P. G. (1962): The effects of wind shear on horizontal spread from an instantaneous ground source. *Quart. J. Roy. Meteorol. Soc.* **88**, 382–393.

Saila, S. B., Shappy, R. A. (1963): Random movement and orientation in salmon migration. *J. Cons. Intl. Explor. Mer.* **28**, 153–166.

Saito, T., Nagai, Y., Isobe, S., Horibe, Y. (1970): An investigation of turbulence within a crop canopy. *J. Agric. Meteorol. Tokyo* **25**, 205–214.

Sakai, S. (1973): A model for group structure and its behavior. *Biophysics* **13**, 82–90 (Japanese).

Sakamoto, W., Aoki, I., Kuroki, T. (1975): Studies on the schooling behavior of fish-I. Spectral analysis of interaction between two individuals of fish in locomotion. *Bull. Japanese Soc. Scientific Fisheries* **41**, 945–952.

Salvatori, V., Skidmore, A. K., Corsi, F., van er Meer, F. (1999): Estimating temporal independence of radio-telemetry data on animal activity. *J. Theor. Biol.* **198**, 567–574.

Sanderson, G. C. (1966): The study of mammal movements—a review. *J. Wildl. Manage.* **30**, 215–235.

Sand-Jensen, K., Mebus, J. R. (1996): Fine-scale patterns of water velocity within macrophyte patches in streams. *Oikos* **76**, 169–180.

Sand-Jensen, K., Pedersen, O. (1999): Velocity gradients and turbulence around macrophyte stands in streams. *Freshwater Biol.* **42**, 315–328.

Sannomiya, N., Matuda, K. (1981): A mathematical model of fish behavior in a water tank, *IEEE Transactions on Systems Man and Cybernetics.* **14(1)**, 157–162.

Sarmiento, J. L., Slater, R. D., Fasham, M. J. R., Ducklow, H. W., Toggweiler, J. R., Evans, G. T. (1993): A seasonal three-dimensional ecosystem model of nitrogen cycling in the North Atlantic euphotic zone. *Global Biogeochem. Cycles,* **7**, 417–450.

Sayer, H. J. (1956): A photographic method for the study of insect migration. *Nature* **177**, 226.

Schaefer, G. W. (1976): Radar observations of insect flight. In *Insect Flight,* 157–197. Rainey, R. C. (ed.). (Symp. Roy. Entomol. Soc. London, No. 7). New York: J. Wiley & Sons.

Scheltema, R. S. (1971a): Dispersal of phytoplanktotrophic shipworm larvae (*Bivalvia: Teredinidae*) over long distances by ocean currents. *Mar. Biol.* **11**, 5–11.

Scheltema, R. S. (1971b): Larval dispersal as a means of genetic exchange between geographically separated populations of shallow-water benthic marine gastropods. *Biol. Bull.* **140**, 284–322.

Scheltema, R. S. (1971c): The dispersal of the larvae of shoal-water benthic invertebrate species over long distances by oceanic currents. In *Fourth European Marine Biology Symposium*, 7–28. Crisp, D. J. (ed.). Cambridge: Cambridge Univ. Press.

Scheltema, R. S. (1986): On dispersal and planktonic larve of benthic invertebrates: An eclectic overview and summary of problems. *Bull. Mar. Sci.* **39**, 290–322.

Scheltema, R. S. (1975): Relationship of larval dispersal, gene-flow, and natural selection to geographic variation of benthic invertebrates in estuaries and along coastal regions. *Estuar. Res.* **1**, 372–391.

Scheltema, R. S. (1995): The relevance of passive dispersal for the biogeography of Caribbean mollusks. *Am. Malacol. Bull.* **11**, 99–115.

Scheltema, R. S., Carlton, J. T. (1984): Methods of dispersal among fouling organisms and possible consequences for range extension and geographic variation. In *Marine Biodeterioration: An Interdisciplinary Study*, 311–315. Costlow, J. D., Tipper, R. C. (eds.). Naval Inst. Press.

Schlesinger, D. A., Molot, L. A., Shuter, B. J. (1981): Specific growth-rates of freshwater algae in relation to cell-size and light-intensity. *Can. J. Fish. Aquat. Sci.* **38**, 1052–1058.

Schmidt, W. (1925): Der Massenaustausch in freier Luft und verwandte Erscheinungen. *Probl. Kosm. Phys.* **7**, 1–118.

Schmidt-Nielsen, K. (1972): *How Animals Work*. Cambridge: Cambridge Univ. Press.

Schoener, T. W. (1968): Sizes of feeding territories among birds. *Ecology* **49**, 123–141.

Schoener, T. W. (1981): An empirically based estimate of home range. *Theor. Popul. Biol.* **20**, 281–325.

Schrödter, H. (1960): Dispersal by air and water—the flight and landing. In *Plant Pathology*, Vol. 3, 169–227. Horsfall, J. G., Dimond, A. E. (eds.). New York: Academic Press.

Schultz, C. (1998): Dispersal behavior and its implications for reserve design in a rare Oregon butterfly. *Cons. Biol.* **12**, 284–292.

Schwarz, M., Poland, D. (1975): Random walk with two interacting walkers. *J. Chem. Phys.* **63**, 557–568.

Schwink, I. (1954): Experimentelle Untersuchungen üder Geruchssinn und Strömungswahrnehmung in der Orientierung bei Nachtschmetterlingen. *Z. für vergleichende Physiol.* **37**, 19–56.

Scott, J. T., Meyer, G. E., Stewart, R., Walther, E. G. (1969): On the mechanism of Langmuir circulation and their role in epilimnion mixing. *Limnol. Oceanogr.* **14**, 493–503.

Seaman, D. E., Millspaugh, J. J., Kernohan, B. J., Brundige, G. C., Raedeke, K. J., Gitzen, R. A. (1999): Effects of sample size on kernel home range estimators. *J. Wildl. Manage.* **63**, 739–747.

Segel, L. A. (1984): *Modeling Dynamic Phenomena in Molecular and Cellular Biology*. Cambridge: Cambridge University Press.

Segel, L. A., Jackson, J. L. (1972): Dissipative structure: An explanation and an ecological example. *J. Theor. Biol.* **37**, 545–559.

Segel, L. A., Levin, S. A. (1976): Application of nonlinear stability theory to the study of the effects of diffusion on predator–prey interactoins. In *Topics in Statistical Mechanics and Biophysics: A Memorial to Julius L. Jackson*, AIP Conf. Proc. No. 27, 123–152. Piccirelli, R. A. (ed.). New York: Amer. Inst. Phys.

Segel, L. A., Stoeckly, B. (1972): Instability of a layer of chemotactic cells, attractant and degrading enzyme. *J. Theor. Bio.* **37**, 561–585.

Seginer, I., Mulhearn, P. J., Bradley, E. F., Finnigan, J. J. (1976): Turbulent flow in a model plant canopy. *Boundary-Layer Meteor.* **10**, 423–453.

Seinfeld, J. H. (1986): *Atmospheric Chemistry and Physics of Air Pollution.* New York, NY: Wiley and Sons.

Shaffer, G. (1996): Biogeochemical cycling in the global ocean 2. New production, Redfield ratios, and remineralization in the organic pump. *J. Geophys. Res. C.* **101**, 3723–3745.

Shaffer, G., Sarmiento, J. L. (1995): Biogeochemical cycling in the global ocean 1. A new analytical model with continuous vertical resolution and high-latitude dynamics. *J. Geophys. Res. C.* **100**, 2659–2672.

Shanks, A. L. (1986a): Tidal periodicity in the daily settlement of the intertidal barnacle larvae and a hypothesized mechanism for the cross-shelf transport of cyprids. *Biol. Bull.* **170**, 429–440.

Shanks, A. L. (1986b): Vertical migration and cross-shelf dispersal of larval *Cancer* spp. and *Randallia ornata* (Crustacea: Brachyura) off the coast of southern California. *Mar. Biol.* **92**, 189–199.

Shanks, A. L. (1995): Oriented swimming by megalopae of several eastern North Pacific crab species and its potential role in the onshore migration. *J. Exper. Mar. Biol. Ecol.* **186**, 1–16.

Shannon, C. E., Weaver, W. (1949): *The Mathematical Theory of Communication.* Urbana: Univ. Illinois Press.

Shapiro, A. M. (1987): Transport equations for fractured porous media. In *Advances in Transport Phenomena in Porous Media (NATO Adv. Stud. Inst. Ser. E 128)*, 407–471. Bear, J., Corapcioglu, M. Y. (eds.). Dordrecht: Martinus Nijhoff.

Sharov, A., Liebhold, A. (1998): Model of slowing the spread of gypsy moths with a barrier zone. *Ecol. Applic.* **8**, 1170–1179.

Shaw, E. (1978): Schooling fishes. *Amer. Scientist* **66**, 166–175.

Shaw, R. H., den Hartog G., King, K. M., Thurtell, G. W. (1974): Measurements of mean wind flow and three-dimensional turbulence intensity within a mature corn canopy. *Agric. Meteorol.* **13**, 419–425.

Shaw, R. H., Pereira, A. R. (1982): Aerodynamic roughness of a plant canopy: A numerical experiment. *Agric. Meteor.* **26**, 51–65.

Sheets, R. G., Linder, R. L., Dahlgren, R. B. (1971): Burrow systems of prairie dogs in South Dakota. *J. Mammal.* **52**, 451–453.

Shigesada, N., Kawasaki, K. (1997): *Biological Invasions: Theory and Practice.* New York: Oxford Univ. Press.

Shigesada, N., Kawasaki, K., Takeda, Y. (1995): Modeling stratified diffusion in biological invasions. *Amer. Naturalist* **146**, 229–251.

Shigesada, N., Kawasaki, K., Teramoto, E. (1979): Spatial segregation of interacting species. *J. Theor. Biol.* **79**, 83–99.

Shigesada, N., Roughgarden, J. (1982): The role of rapid dispersal in the population-dynamics of competition. *Theor. Pop. Biol.* **21**, 353–372.

Shigesada, N., Teramoto, E. (1978): A consideration on the theory of environmental density. *Jap. J. Ecol.* **28**, 1–8 (Japanese with English synopsis) (also see Shigesada et al., 1979).

Shimoyama N., Sano, M. (1996): Collective motion in a system of motile elements. *Phys. Rev. Lett.*, **76(20)**, 3870–3873.

Shinn, E. A., Long G. E. (1986): Technique for 3-D analysis of Cheumatopsyche pettiti swarms. *Environ. Entomol.* **15**, 355–359.

Shono, S. (1958): *Introduction to Meteorology. Lectures in Meteorology*, Vol. 1, Tokyo: Chijin Shokan (Japanese).

Shukla, J. B., Verma, S. (1981): Effects of convective and dispersive interactions on the stability of 2 species. *Bull. Math. Biol.* **43**, 593–610.

Shukla, V. P., Das, P. C. (1982): Effects of dispersion on stability of multi-species prey-predator systems. *Bull. Math. Biol.* **44**, 571–578.

Shukla, V. P., Shukla, J. B. (1982): Multi-species food webs with diffusion. *J. Math. Biol.* **13**, 339–344.

Shukla, V. P., Shukla, J. B., Das, P. C. (1981): Environmental Effects on the linear stability of a 3-species food chain. *Math. Biosci.* **57**, 35–58.

Shum, K. T. (1992): Wave-induced advective transport below a rippled water-sediment interface. *J. Geophys. Res.* **97**, 789–808.

Shum, K. T. (1993): The effects of wave-induced pore water circulation on the transport of reactive solutes below a rippled sediment bed. *J. Geophys. Res.* **98**, 10289–10301.

Siegel, D. A., Deuser, W. G. (1997): Trajectories of sinking particles in the Sargasso Sea: Modeling of statistical funnels above deep-ocean sediment traps. *Deep-Sea Res.* **44**, 1519–1541.

Siegel, D. A., Granata, T. C., Michaels, A. F., Dickey, T. D. (1990): Mesoscale eddy diffusion, particle sinking, and the interpretation of sediment trap data. *J. Geophys. Res.* **95**(C4), 5303–5311.

Siegfried, W. R., Underhill, L. G. (1975): Flocking as an anti-predator strategy in doves. *Animal Behaviour* **23**, 504–508.

Sigurdsson, J. B., Titman, C. W., Davies, P. A. (1976): The dispersal of young post-larval bivalve molluscs by byssus threads. *Nature* **262**, 386–387.

Silvert, W., Platt, T. (1978): Energy flux in the pelagic ecosystem: A time-dependent equation. *Limnol. Oceanogr.* **23**, 813–816.

Simberloff, D., Farr, J. A., Cox, J., Mehlman, D. W. (1992): Movement corridors: Conservation bargains or poor investments? *Conserv. Biol.* **6**, 493–504.

Sinclair, A. R. E. (1977): *The African Buffalo*. Chicago, London: Univ. Chicago Press.

Sinclair, M. (1988): *Marine Populations: An Essay on Population Regulation and Speciation*. Seattle: Univ. Washington Press.

Sinclair, M., Iles, T. D. (1989): Population regulation and speciation in the oceans. *J. Cons. Intl. Explor. Mer.* **45**, 65–175.

Silver, M. W., Coale, S. L., Pilskaln, C. H., Steinberg, D. R. (1998): Giant aggregates: Importance as microbial centers and agents of material flux in the meso-pelagic zone. *Limnol. Oceanogr.* **43**, 498–507.

Silver, M. W., Shanks, A. L., Trent, J. D. (1978): Marine Snow: Microplankton habitat and source of small-scale patchiness in pelagic populations. *Sci.* **201**, 371–373.

Sinclair, M. (1988): *Marine Populations: An Essay on Population Regulation and Speciation.* Seattle: Univ. Washington Press.

Siniff, D. B., Jessen, C. R. (1969): A simulation model of animal movement patterns. *Adv. in Ecol. Res.* **6**, 185–219.

Sinko, J. W., Streiffer, W. (1967): A new model for age-size structure of a population. *Ecology* **48**, 910–918.

Skellam, J. G. (1951a): Random dispersal in theoretical populations. *Biometrika* **38**, 196–218.

Skellam, J. G. (1951b): Gene dispersion in heterogeneous population. *Heredity* **5**, 433–435.

Skellam, J. G. (1955): The mathematical approach to population dynamics. In *The Number of Man and Animals*, 31–46. Cragg, J. B., Pirie, N. W. (eds.). London: Oliver and Boyd.

Skellam, J. G. (1972): Some philosophical aspects of mathematical modelling in empirical science with special reference to ecology. In *Mathematical Models in Ecology*, 13–28. Jeffers, J. N. R. (ed.). London: Blackwell Sci. Publ.

Skellam, J. G. (1973): The formulation and interpretation of mathematical models of diffusionary processes in population biology. In *The Mathematical Theory of the Dynamics of Biological Populations*, 63–85. Bartlett, M. S., Hiorns, R. W. (eds.). New York: Academic Press.

Slade, D. H. (ed.) (1968): *Meteorology and Atomic Energy 1968*. Washington, DC: U.S. Atomic energy Commission.

Slobodkin, L. B. (1953a): A possible initial condition for red tides on the coast of Florida. *J. Mar. Res.* **12**, 148–155.

Slobodkin, L. B. (1953b): An algebra of population growth. *Ecology* **34**, 513–519.

Slobodkin, L. B. (1954): Population dynamics in *Daphnia obtuse Kurz*. *Ecol. Monogr.* **24**, 69–89.

Smayda, T. J. (1970): The suspension and sinking of phytoplankton in the sea. *Oceanogr. Mar. Biol. Ann. Rev.* **8**, 353–414.

Smith, F. A., Walker, N. A. (1980): Photosynthesis by aquatic plants: Effects of unstirred layers in relation to assimilation of CO_2 and HCO_3^- and to carbon isotopic discrimination. *New Phytol.* **86**, 245–259.

Smith, J. L., Wollkind, D. J. (1983): Age structure in predator-prey systems: intraspecific carnivore interaction, passive diffusion, and the paradox of enrichment. *J. Math. Biol.* **17**, no. 3, 275–288.

Smoluchowski, M. (1916): Drei Vorträge über Diffusion, Brownsche Bewegung und Koagulation von Kolloidteilchen. *Physik. Z.* **17**, 557–585.

Sokal, R. R., Rohlf, F. J. (1969): *Biometry*. San Francisco: W. H. Freeman and Co.

Soong, T. T. (1973): *Random Differential Equations in Science and Engineering*. New York: Academic Press.

Soster, F. M., Harvey, D. T., Troksa, M. R., Grooms. T. (1992): The effect of tubificid oligochaetes on the uptake of zinc by Lake Erie sediments. *Hydrobiologia* **248**, 249–258.

South, A. (1999): Extrapolating from individual movement behaviour to population spacing patterns in a ranging mammal. *Ecol. Model.* **117**, 343–360.

Southern, W. E. (1965): Avian navigation (Biotelemetry). *BioScience* **15**, 87–88.

Spitzer, F. (1976): *Principles of Random Walk*, 2nd ed. New York, Heidelberg, Berlin: Springer.

Spigler, R., Zanette, D. (1992): Reaction-diffusion models from the Fokker-Planck formulation of chemical processes. *IMA J. Appl. Math.* **49**, no. 3, 217–229.

Springer, S. (1957): Some observations on the behavior of schools of fishes in the Gulf of Mexico and adjacent waters. *Ecology* **38**, 166–171.

Squires, K. D., Eaton, J. K. (1991): Lagrangian and Eulerian statistics obtained from direct numerical simulation of homogeneous turbulence. *Physics of Fluids A— Fluid Dynamics* **3**, 130–143.

Srinivasan, S., Ostling, J., Charlton, T., Denys, R., Takayama, K., Kjelleberg, S. (1998): Extracellular signal molecules involved in the carbon starvation response of marine vibrio sp. strain S14. *J. Bacteriol.* **180**, 201–209.

Srinivasan, S. K., Vasudevan, R. (1971): *Introduction to Random Differential Equations and Their Applications.* New York: Amer. Elsevier Publ. Co., Inc.

Stamhuis, E. J., Videler, J. J. (1998a): Burrow ventilation in the tube-dwelling shrimp *Callianassa subterranae* (Decapoda: Thalassinidae). I. Morphology and motion of the pleopods, uropods and telson. *J. Exp. Biol.* **201**, 2151–2158.

Stamhuis, E. J., Videler, J. J. (1998b): Burrow ventilation in the tube-dwelling shrimp *Callianassa subterranae* (Decapoda: Thalassinidae). II. The flow in the vicinity of the shrimp and the energetic advantages of a laminar non-pulsating ventilation current. *J. Exp. Biol.* **201**, 2159–2170.

Stamhuis, E. J., Videler, J. J. (1998c): Burrow ventilation in the tube-dwelling shrimp *Callianassa subterranae* (Decapoda: Thalassinidae). III. Hydrodynamic modelling and the energetics of pleopod pumping. *J. Exp. Biol.* **201**, 2171–2181.

Stander, G. H., Shannon, L. V., Campbell, J. A. (1969): Average velocities of some ocean currents as deduced from the recovery of plastic drift cards. *J. Mar. Res.* **27**, 293–300.

Stavn, R. H. (1971): The horizontal-vertical distribution hypothesis: Langmuir circulations and *Daphina* distributions. *Limnol. Oceanogr.* **16**, 453–466.

Steele, J. H. (1974a): *The Structure of Marine Ecosystems.* Cambridge: Harvard Univ. Press.

Steele, J. H. (1974b): Spatial heterogeneity and population stability. *Nature* **248**, 83.

Steele, J. H. (1976a): Patchiness. In *The Ecology of the Sea*, Cushing, D. H., Walsh, J. J. (eds). 98–115. Philadelphia: W. B. Saunders, Co.

Steele, J. H. (1976b): Application of theoretical models in ecology. *J. Theor. Biol.* **63**, 443–451.

Steele, J. H. (1977): Plankton patches in the northern North Sea. In *Fisheries Mathematics*, 1–19, Steele, J. H. (ed.). London, New York, San Francisco: Academic Press.

Steele, J. H. (1977): Structure of plankton communities. *Philos. T. Roy. Soc.* B **280**, 485–534.

Steele, J. H. (1978): Some comments on plankton patches. In *Spatial Pattern in Plankton Communities*, 1–20, Steele, J. H. (ed.). New York: Plenum Press.

Steele, J. H., Yentsch, C. S. (1960): The vertical distribution of chlorophyll. *J. Mar. Biol. Assoc. UK* **39**, 217–226.

Stepanov, K. M. (1935): Dissemination of infective diseases of plants by air currents. *Bull. Pl. Prot. Leningr. Ser. Z. Phytopathology*, No. 8, 1–68. (In Gregory, 1973).

Stephens, G. R., Aylor, D. E. (1978): Aerial dispersal of red pine scale, *Matsucoccus resinosae* (Homoptera: Margarodidae). *Envir. Entomol.* **7**, 556–563.

Stevens, C. L., Hurd, C. L. (1997): Boundary-layers around bladed aquatic macrophytes. *Hydrobiol.* **346**, 119–128.

Stigebrandt, A. (1991): Computations of oxygen fluxes through the sea surface and the net production of organic matter with application to the Baltic and adjacent seas. *Limnol. Oceanogr.* **36**, 444–454.

Stommel, H. (1949): Trajectories of small bodies sinking slowly through convection cells. *J. Mar. Res.* **8**, 24–29.

Stone, E. L. (1993): Soil burrowing and mixing by a crayfish. *Soil Sci. Soc. Amer. J.* **57**, 1096–1099.

Stratonovich, R. L. (1966): A new representation for stochastic integrals and equations. *J. SIAM Control* **4**, 362–371.

Strathmann, R. (1974): The spread of sibling larvae of sedentary marine invertebrates. *Amer. Naturalist* **108**, 29–44.

Streeter H. W., Phelps, E. B. (1925): A study of the pollution and natural purification of the Ohio River, III, Factors concerned in the phenomena of oxidation and reaeration. US Pub. Health Serv. Pub. Health Bull. 146. February 1925.

Streifer, W. (1974): Realistic models in population ecology. In *Adv. in Ecological Research*, Vol. 8, 199–266. Macfadyen, A. (ed.). New York, San Francisco, London: Academic Press.

Strickler, J. R. (1982): Calanoid copepods, feeding currents and the role of gravity. *Sci.* **218**, 158–160.

Strutton, P. G., Mitchell, J. G., Parslow, J. S. (1997): Using non-linear analysis to compare the spatial structure of chlorophyll with passive tracers. *J. Plankton Res.* **19**, 1553–1564.

Sture, K., Nordholm, J., Zwanzig, R. (1974): Strategies for fluctuation renormalization in nonlinear transport theory. *J. Statistical Phys.* **11**, 143–158.

Styan, C. A. (1998): Polyspermy, egg size, and the fertilization kinetics of free-spawning marine invertebrates. *Amer. Naturalist* **152**, 290–297.

Sugiura, Y. (1964): Some chemico-oceanographical properties of the Kuroshio and its adjacent regions. In *Recent Researches in the Fields of Hydrosphere, Atmosphere and Nuclear Geochemistry*, 49–63 (Editorial Committee of Sugawara Festival Volume). Tokyo: Maruzen Co. Ltd.

Sugiura, Y., Yoshimura, H. (1964): Distribution and mutual relation of dissolved oxygen and phosphate in the Oyashio and the northern part of Kuroshio regions. *J. Oceanogr. Soc. Japan* **20**, 14–23.

Sullivan, P. J. (1988): Effect of boundary conditions, region length, and diffusion rates on a spatially heterogeneous predator-prey system. *Ecol. Modelling* **43**, 235–249.

Sullivan, R. T. (1981): Insect swarming and mating. *Florida Entomologist*, **64**, 44–65.

Sundby, S. (1983): A one-dimensional model for the vertical distribution of pelagic fish eggs in the mixed layer. *Deep-Sea Res.* **30**, 645–661.

Sutcliffe, O. L., Thomas, C. D. (1996): Open corridors to facilitate dispersal by ringlet butterflies (Aphantopus hyperantus) between woodland clearings. *Conserv. Biol.* **10**, 1359–1365.

Suttle, C. A. (1992): Inhibition of photosynthesis by the submicron size fraction concentrated from seawater. *Mar. Ecol. Progr. Ser.* **87**, 105–112.

Suttle, C. A. (1994): The significance of viruses to mortality in aquatic microbial communities. *Microb. Ecol.* **28**, 237–243.

Suttle, C. A., Chan, A. M., Cottrell, M. T. (1990): Infection of viruses by phytoplankton and reduction of primary productivity. *Nature* **347**, 467–469.

Sutton, O. G. (1943): On the equation of diffusion in a turbulent medium. *Proc. Roy. Soc. A* **182**, 48–75.

Sutton, O. G. (1953): *Micrometeorology.* New York: McGraw-Hill.

Suzuki, R., Sakai, S. (1973): Movement of a group of animals. *Biophysics* **13**, 281–282 (Japanese).

Sverdrup, H. U., Johnson, M. W., Fleming, R. H. (1942): *The Oceans*. Englewood Cliffs, NJ: Prentice-Hall.

Svensson, U., Rahm, L. (1991): Toward a mathematical model of oxygen transfer to and within bottom sediments. *J. Geophys. Res.* **96**, 2777–2783.

Swaney, D. P. (1999): Analytical solution of Boudreau's equation for a tracer subject to food-feedback bioturbation. *Limnol. Oceanogr.* **44**, 697–698.

Swihart, R. K., Slade, N. A. (1985): Influence of sampling interval on estimates of home-range size. *J. Wildl. Manage.* **49**, 1019–1025.

Swingland, I. R., Greenwood, P. J. (1983): *The Ecology of Animal Movement*. Oxford: Oxford Univ. Press.

Symons, P. E. K. (1971): Spacing and density in schooling threespine sticklebacks (*Gasterosteus aculeatus*) and mummichog (*Fundulus heteroclitus*). *J. Fish. Res. Bd. Canada* **28**, 999–1004.

Syrjämäki, J. (1964): Swarming and mating behavior of *Allocironomus crassiforceps* Kieff. (Diptera: Chironomidae). *Annales Zoologici Fennici* **1**, 125–145.

Takahashi, K., Honjo, S. (1983): Radiolarian skeletons: Size, weight, sinking speed, and residence time in tropical pelagic oceans. *Deep-Sea Res. I* **30**, 543–568.

Takahashi, M., Nakai, T., Ishimaru, T., Hasumoto, H., Fujita, Y. (1985): Distribution of the subsurface chlorophyll maximum and its nutrient-light environment in and around the Japan current off Japan. *J. Oceanogr. Soc. Jpn.* **41**, 73–80.

Takeda, K. (1965): Turbulence in plant canopies (2). *J. Agric. Meteorol.* **21**, 11–14 (Japanese with English summary).

Takeda, K. (1966): On roughness length and zero-plane displacement in the wind profile of the lowest air layer. *J. Meteorol. Soc. Japan* **44**, 101–107.

Takeuchi, Y. (1989): Diffusion-mediated persistence in two-species competition Lotka-Volterra model. *Math. Biosci.* **95**, 65–83.

Talbot, J. W. (1974): Diffusion studies in fisheries biology. In *Sea Fisheries Research*, 31–54. Harden Jones, F. R. (ed.). New York: J. Wiley & Sons.

Talbot, J. W. (1977): The dispersal of plaice eggs and larvae in the Southern Bight of the North Sea. *J. Cons. Intl. Explor. Mer.* **37**, 221–248.

Tani, I. (1951): *Fluid Dynamics*. Tokyo: Iwanami Shoten (Japanese).

Taylor, G. I. (1921): Diffusion by continuous movements. *Proc. London Math. Soc.* **20**, 196–211.

Taylor, G. I. (1953): Dispersion of soluble matter in solvent flowing slowly through a tube. *Proc. Roy. Soc. London A* **219**, 186–203.

Taylor, G. I. (1954): The dispersal of matter in turbulent flow through a pipe. *Proc. Roy. Soc. London A* **223**, 446–468.

Taylor, L. R., Taylor, R. A. J. (1977): Aggregation, migration and population mechanics. *Nature* **265**, 415–421.

Taylor, L. R., Woiwod, I. P., Perry, J. N. (1978): The density-dependence of spatial behavior and the rarity of randomness. *J. Animal Ecol.* **47**, 383–406.

Taylor, R. A. J. (1978): The relationship between density and distance of dispersing insects. *Ecol. Entomol.* **3**, 63–70.

Temam, R. (1997): *Infinite-dimensional dynamical systems in mechanics and physics* 2nd ed. New York: Springer.

Tennekes, H., Lumley, J. L. (1972): *A First Course in Turbulence*. Cambridge, MA: The MIT Press.

Terada, T. (1933): Material and words. In Fukushima, H. (1942): *Fluctuation Pheromena in Physics*. Tokyo: Iwanami Shoten.

Teramoto, E. (1978): Mathematics of harvesting and predation: Mathematical doctrine for the morals of the biological society. *Math. Sciences* **16** (No. 183): 12–17, Science Sha, Tokyo (Japanese).

Thacker, W. C., Lavelle, J. W. (1978): Stability of settling of suspended sediments. *Phys. Fluids* **21**, 291–292.

Therriault, J.-C., Platt, T. (1981): Environmental control of phytoplankton patchiness. *Can. J. Fish. Aquat. Sci.* **38**, 638–641.

Thibodeaux, L. J. (1996): *Environmental Chemodynamics: Environmental Movement of Chemicals in Air, Water and Soil*, 2nd ed. New York: John Wiley & Sons.

Thom, A. S. (1971) (in Thom, A. S. 1975).

Thom, A. S. (1975): Momentum, mass and heat exchange of plant communities. In *Vegetation and the Atmosphere*, Vol. 1, 57–109. Monteith, J. L. (ed.) New York: Academic Press.

Thom, R. (1975): *Structural Stability and Morphogenesis.* (Translated from French edition, 1972.) Reading, MA: W. A. Benjamin, Inc.

Thomann, R. V., Mueller, J. A. (1987): *Principles of Surface Water Quality Modeling and Control.* New York: Harper and Row.

Thomas, F. I. M. (1994): Physical properties of gametes in three sea urchin species. *J. Exp. Biol.* **194**, 263–284.

Thompson, R. W., Cauley, D. A. (1979): A population balance model for fish population dynamics. *JTB* **81**, 289–307.

Thompson, W. A., Vertinsky, I., Krebs, J. R. (1974): The survival value of flocking in birds: a simulation model. *The Journal of Animal Ecology*, **43**, 785–820.

Tiebout, H. M., Anderson, R. A. (1997): A comparison of corridors and intrinsic connectivity to promote dispersal in transient successional landscapes. *Conservation Biology* **11**, 620–627.

Tilman, D. (1982): *Resource Competition and Community Structure.* Princeton: Princeton University Press.

Tilman, D., Kareiva, P. (1997): *Spatial Ecology: The Role of Space in Population Dynamics and Interspecific Interactions.* Princeton: Princeton Univ. Press.

Tilman, D., Kilham, S. S., Kilham, P. (1982): Phytoplankton community ecology: The role of limiting nutrients. *Ann. Rev. Ecol. Syst.* **1**, 349–372.

Timm, U., Okubo, A. (1994): Gyrotaxis: A plume model for self-focusing microorganisms. *Bull. Math. Biol.* **56**, 187–206.

Timm, U., Okubo, A. (1995): Gyrotaxis: Interactions between algae and flagellates. *Bull. Math. Biol.* **57**, 631–650.

Titman, D., Kilham, P. (1976): Sinking in freshwater phytoplankton: Some ecological implications of cell nutrient status and physical mixing process. *Limnol. Oceanogr.* **21**, 409–417.

Tolbert, W. W. (1977): Aerial dispersal behavior of two orb weaving spiders. *Psyche* **84**, 13–27.

Tonolli, V., Tonolli, L. (1960): Irregularities of distribution of plankton communities: Considerations and methods. In *Perspectives in Marine Biology*, 137–143. Buzzati-Traverso, A. A. (ed.). Univ. California Press.

Toride, N., Leij, F. J., van Genuchten, M. Th. (1993): A comprehensive set of analytical solutions for nonequilibrium solute transport with first-order decay and zero-order production. *Water Resour. Res.* **29**, 2167–2182.

Townsend, D. W., Cucci, T. L. Berman, T. (1984): Subsurface chlorophyll maxima and vertical distribution of zooplankton in the Gulf of Maine USA Canada. *J. Plankton Res.* **6**, 793–802.

Tranquillo, R. T., Alt, W. (1990): Glossary of terms concerning oriented movement. pp. 564–565 in W. Alt and G. Hoffmann, eds. *Biological Motion.* Lecture Notes in Biomathematics, Vol. 89. Berlin: Springer.

Treger, R. W., Krause, J. (1995): Density dependence and numerosity in fright stimulated aggregation behaviour of shoaling fish. *Phil Trans Roy. Soc. Lond. B*, **350**, 381–390.

Treherne, J. E., Foster, W. A. (1980): The effects of group-size on predator avoidance in a marine insect. *Anim. Behav.* **28**, 1119–1122.

Treherne, J. E., Foster, W. A. (1981): Group transmission of predator avoidance-behavior in a marine insect—the Trafalgar effect. *Anim. Behav.* **29**, 911–917.

Treisman, M. (1975a): Predation and the evolution of gregariousness. I. Models for concealment and evasion. *Anim. Behav.* **23**, 779–800.

Treisman, M. (1975b): Predation and the evolution of gregariousness. II. An economic model for predator–prey interaction. *Anim. Behav.* **23**, 801–825.

Trucco, E. (1965a): Mathematical models for cellular systems. The von Foerster equation. Part 1. *Bull. Math. Biophys.* **27**, 285–304.

Trucco, E. (1965b): Mathematical models for cellular systems. The von Foerster equation. Part 2. *Bull. Math. Biophys.* **27**, 449–471.

Truscott, J. E., Brindley, J. (1994): Ocean plankton populations as excitable media. *Bull. Math. Biol.* **56**, 981–998.

Tsuda, A., Furuya, K., Nemoto, T. (1989): Feeding of microzooplankton and macrozooplankton at the subsurface chlorophyll maximum in the subtropical North Pacific. *J. Exp. Mar. Biol. Ecol.* **132**, 41–52.

Tsunogai, S. (1972a): An estimate of the vertical diffusivity of the deep water. *J. Oceanogr. Soc. Japan* **28**, 145–152.

Tsunogai, S. (1972b): An estimate of the rate of decomposition of organic matter in the deep water of the Pacific Ocean. In *Biological Oceanography of the Northern North Pacific Ocean* (dedicated to Sigeru Motoda), 517–533. Takenouti, A. Y. (ed.). Tokyo: Idemitsu Shoten.

Tufto, J., Anderson, R., Linnell, J. (1996): Habitat use and ecological correlates of home range size in a small cervid: The roe deer. *J. Anim. Ecol.* **65**, 715–724.

Turchin, P. (1987): The role of aggregation in the response of Mexican bean beetles to host-plant density. *Oecologia* **71**, 577–582.

Turchin, P. (1989): Population consequences of aggregative movement. *J. Anim. Ecol.* **58**, 75–100.

Turchin, P. (1991): Translating foraging movements in heterogeneous environments into the spatial distribution of foragers. *Ecology* **72**(4), 1253–1266.

Turchin, P. (1997): Quantitative analysis of animal movements in congregations. In *Animal Groups in Three Dimensions*, 107–112. Parrish, J. K., Hamner W. M. (eds.). Cambridge: Cambridge Univ. Press.

Turchin, P. (1998): *Quantitative Analysis of Movement.* Sunderland, MA: Sinauer Press.

Turchin, P., Kareiva, P. (1989): Aggregation in Aphis varians: An effective strategy for reducing predation risk. *Ecology* **70**, 1008–1016.

Turchin, P., Simmons, G. (1997): Movements of animals in congregations: An Eulerian analysis of bark beetle swarming. In *Animal Groups in Three Dimensions*, 113–125. Parrish, J. K., Hamner, W. M. (eds.). Cambridge: Cambridge Univ. Press.

Turing, A. M. (1952): The chemical basis of morphogenesis. *Phil. Trans. Roy. Soc. London B* **237**, 37–72.

Turner, F. B., Jennrich, R. I., Weintraub, J. D. (1969): Home ranges and body size of lizards. *Ecology* **50**, 1076–1081.

Uchijima, Z. (1970): Carbon dioxide environment and flux within a corn crop canopy. In *Prediction and Measurement of Photosynthetic Productivity*, 179–196. Centre for Agricultural Publishing and Documentation. Wegeningen, The Netherlands.

Uchijima, Z., Udagawa, T., Horie, T., Kobayashi, K. (1970): Studies of energy and gas exchange within crop canopies (8): Turbulent transfer coefficient and foliage exchange velocity within a corn canopy. *J. Agric. Meterorol. Tokyo* **25**, 215–228.

Uchijima, Z., Inoue, K. (1970): Studies of energy and gas exchange within crop canopies (9): Simulation of CO_2 environment within a canopy. *J. Agric. Meteorol. Tokyo* **25**, 5–18.

Uhlenbeck, G. E., Ornstein, L. S. (1930): On the theory of the Brownian motion. *Phy. Rev.* **36**, 823–841.

United States Congress, Office of Technology Assessment. (1993): *Harmful Nonindigenous Species in the United States*. Washington, DC.

Vabo, R., Nottestad, L. (1997): An individual based model of fish school reactions: Predicting antipredator behaviour as observed in nature. *Fish Oceanogr* **6**(3), 155–171.

Vail, S. G. (1990): *Experimental and theoretical investigations of plant-animal interactions: ecological and evolutionary consequences of plant quality variation.* (Ph.D. diss.) Univ. of Calif., Davis.

Vail, S. G. (1993): Scale-dependent responses to resource spatial pattern in simple models of consumer movement. *Amer. Natur.* **141**, 199–216.

Van Genuchten, M. Th., Alves, W. J. (1982): Analytical solutions of the one-dimensional convective-dispersive solute transport equations. U.S. Dept of Agriculture, Technical Bulletin 1661.

Van der Heijde, P., Bachmat, Y., Bredehoeft, J., Andrews, B., Holtz, D., Sebastian, S. (1985): Groundwater management: The use of numerical models. *AGU Water Resources Monogr. 5*. Washington: Amer. Geophys. Union.

van der Pijl, L. (1972): *Principles of Dispersion in Higher Plants*, Berlin: Springer-Verlag.

Van Kirk, R. W. Lewis, M. A. (1997): Integrodifference models for persistence in fragmented habitats. *Bull. Math. Biol.* **59**, 107–137.

Van Lawick-Goodall, J. (1968): The behavior of free-living chimpanzees in the Gombe Stream Reserve. *Anim. Behav. Monographs.* **1**, 161–301 (In Morgan, 1976).

Van Olst, J. C., Hunter, J. R. (1970): Some aspects of the organization of fish schools. *J. Fish. Res. Bd. Canada* **27**, 1225–1238.

Vandevelde, T., Legendre, L., Therriault, J. C., Demers, S., Bah, A. (1987): Subsurface chlorophyll maximum and hydrodynamics of the water column. *J. Mar. Res.* **45**, 377–396.

Veit, R., Lewis, M. (1996): Dispersal, population growth and the Allee effect: dynamics of the house finch invasion of eastern North America. *Amer. Nat.* **148**, 255–274.

Venrick, E. L. (1982): Phytoplankton in an oligotrophic ocean: Observations and questions. *Ecol. Monogr.* **52**, 129–154.

Verduin, J. J., Walker, D. I., Kuo, J. (1996): *In situ* submarine pollination in the seagrass *Amphibolis antartica*: Research notes. *Mar. Ecol. Progr. Ser.* **133**, 307–309.

Vetter, Y. A., Deming, J. W., Jumars, P. A., Kriegerbrockett, B. B. (1998): A predictive model of bacterial foraging by means of freely released extracellular enzymes. *Microb. Ecol.* **36**(1), 75–92.

Vickers, N. J., Baker, T. C. (1994): Reiterative responses to single strands of odor promote sustained upwind flight and odor source location by moths. *Proc. Natl. Acad. Sci. USA* **191**, 5756–5760.

Villareal, T. A., Altabet, M. A., Culverrymsza, K. (1993): Nitrogen transport by vertically migrating diatom mats in the North Pacific Ocean. *Nature* **363**, 709–712.

Villareal, T. A., Pilskaln, C., Brzezinski, M., Lipschultz F., Dennett, M., Gardner, G. B. (1999): Upward transport of oceanic nitrate by migrating diatom mats. *Nature* **397**, 423–425.

Vine, I. (1971): Risk of visual detection and pursuit by a predator and the selective advantage of flocking behavior. *J. Theor. Biol.* **30**, 405–422.

Visser, A. (1997): Using random walk models to simulate the vertical distribution of particles in a turbulent water column. *Mar. Ecol. Progr. Ser.* **158**, 275–281.

Vogel, H., Czihak, G., Chang, P., Wolf, W. (1982): Fertilization kinetics of sea urchin eggs. *Math. Biosci.* **58**, 189–216.

Vogel, S. (1977a): Flows in organisms induced by movements of the external medium. In *Scale Effects in Animal Locomotion*, 285–297. Pedley, T. J. (ed.). London, New York, San Francisco: Academic Press.

Vogel, S. (1977b): Current-induced flow through living sponges in nature. *Proc. Nat. Acad. Sci. USA* **74**, 2069–2071.

Vogel, S. (1994): *Life in Moving Fluids*, 2nd ed. Princeton: Princeton Univ. Press.

Vogel, S., Bretz, W. L. (1972): Interfacial organisms: Passive ventilation in the velocity gradients near surfaces. *Sci.* **175**, 210–211.

Vogel, S., Ellington, C. P., Kilgore, D. L. (1973): Wind-induced ventilation of the burrow of the prairie-dog, *Cynomys ludovicianus. J. Comp. Physiol.* **85**, 1–14.

Volterra, V. (1926): Variazioni e fluttuazioni del numero d'individui in specie animali conviventi. *Memor. Accad. Lincei*, Ser. 6, **2**, 31–113 (partially translated in "Variations and fluctuations of the number of individuals in animal species living together." *J. du Cons. Int. Explor. Mer.* **3**, 1–51, 1928).

Von Foerster, H. (1959): Some remarks on changing populations. In *The Kinetics of Cellular Proliferation*, 382–407. Stohlman, F. (ed.). New York: Grune and Stratton.

Voorhees, B. H. (1982): Dissipative structures associated to certain reaction-diffusion equations in mathematical ecology. *Bull. Math. Biol.* **44**, 339–348.

Vugts, H. F., van Wingerden, W. K. R. E. (1976): Meteorological aspects of aeronautic behavior of spiders. *Oikos* **27**, 433–444.

Wagenet, R. J., Hutson, J. L. (1989): LEACHM: A finite difference model for simulating water, salt and pesticide movement in the plant root zone. Version 2.0. Continuum. Vol. 2. NY State Water Resour. Inst., Ithaca, NY: Cornell Univ. Press.

Wagenet, R. J., Rao, P. S. C. (1990): Modeling pesticide fate in soils. In *Pesticides in the Soil Environment: Processes, Impacts, and Modeling*. Cheng, H. H. (ed.). Madison, WI: Soil Sci. Soc. Am.

Wallace, A. R. (1876): *The Geographical Distribution of Animals*, Vol. 1. New York: Harper.

Waloff, Z. (1958): The behavior of locusts in migrating swarms. *Proc. X Intl. Congr. Entomol.* (Montreal, 1956) **2**, 567–570.

Waloff, Z. (1972): Orientation of flying locusts, *Schistocerca gregaria* (Forsk.) in migrating swarms. *Bull. Entomol. Res.* **62**, 1–73.

Walsh, P. D. (1996): Area-restricted search and the scale dependence of patch quality discrimination. *J. Theor. Biol.* **183**, 351–361.

Waltman, P. (1974): *Deterministic threshold models in the theory of epidemics. Lecture Notes in Biomathematics.* No. 1. Berlin: Springer-Verlag.

Wang, M. C., Uhlenbeck, G. E. (1945): On the theory of the Brownian motion II. *Rev. Modern Phys.* **17**, 323–342.

Wang, X., Matisoff, G. (1997): Solute transport in sediments by a large freshwater oligochaete, *Branchiura sowerbyi. Environ. Sci. & Tech.* **31**, 1926–1933.

Wanninkhof, R. (1992): Relationship between wind speed and gas exchange over the ocean. *J. Geophys. Res.* **97**, 7373–7382.

Warburton K., Lazarus J. (1991): Tendency distance models of social cohesion in animal groups. *J. Theor Biol* **150(4)**, 473–488.

Ward-Smith, A. J. (1984): *Biophysical Aerodynamics and the Natural Environment.* Chichester: Wiley.

Washington, J. W. (1996): Gas partitioning of dissolved volatile organic compounds in the vadose zone: Principles, temperature effects and literature review. *Groundwater* **34**, 709–718.

Watanabe, S., Utida, S., Yosida, T. (1952): Dispersion of insect and change of distribution type in its process: Experimental studies on the dispersion of insects (1). *Res. Popul. Ecol.* **1**, 94–108 (Japanese with English summary).

Watkins, J. L., Morris, D. J., Ricketts, C., Priddle, J. (1986): Differences between swarms of Antarctic krill and some implications for sampling populations. *Mar. Biol.* **93**, 137–146.

Watt, K. E. F. (1966): The nature of systems analysis (pp. 1–14) and Ecology in the future, 247–253. In *Systems Analysis in Ecology.* New York: Academic Press.

Watt, K. E. F. (1968): *Ecology and Resource Management.* New York: McGraw-Hill.

Waycott, M., Sampson, J. F. (1997): The mating system of a hydrophilous angiosperm *Posidonia australis* (Posidoniaceae). *Amer. J. Bot.* **84**, 621–665.

Wax, N. (1954): *Selected Papers on Noise and Stochastic Processes.* New York: Dover Publ.

Webb, G. F. (1981): A reaction-diffusion model for a deterministic diffusive epidemic. *Jour. of Math. Anal. and Applications* **84**, 150–161.

Weber, P., Greenberg, J. M. (1985): Can spores survive in interstellar space? *Nature* **316**, 403–407.

Webster, D. R., Weissburg, M. J. (2001): Chemosensory guidance cues in a chemical odor plume. *Limnol. Oceanogr.* **46**, 1013–1026.

Webster, D. R., Rahman, S., Dasi, L. P. (2001): On the usefulness of bilateral comparison to tracking turbulent chemical odor plumes. *Limnol. Oceanogr.* **46**, 1026–1032.

Weihs, D. (1973): Hydromechanics of fish schooling. *Nature* **241**, 290–291.

Weinbauer, M. G., Wilhelm, S. W., Suttle, C. A., Garza, D. R. (1997): Photoreactivation compensates for UV damage and restores infectivity to natural marine virus communities. *Appl. Environ. Microbiol.* **63**, 2200–2205.

Weiland, R. H., Fessas, Y. P., Ramarao, B. V. (1984): On instabilities arising during sedimentation of two-component mixtures of solids. *J. Fluid Mech.* **142**, 383–389.

Weilenmann, U., O'Melia, C. R., Stumm, W. (1989): Particle transport in lakes: Models and measurements. *Limnol. Oceanogr.* **34**, 3–18.

Weir, J. S. (1973): Air flow, evaporation and mineral accumulation in mounds of *Macrotermes subhyalinus* (Rambur). *J. Animal Ecol.* **42**, 509–520.

Weiss, G. H. (1968): Equation for the age structure of growing populations. *Bull. Math. Biophys.* **30**, 427–435.

Weissburg, M. J. (1997): Chemo- and mechanosensory orientation by crustaceans in laminar and turbulent flows: From odor trails to vortex streets. In *Orientation and Communication in Arthropods*, 215–246. Lehrer, M. (ed). Basel: Birkhauser Verlag.

Weissburg, M. J. (2000): The hydrodynamic context of olfactory mediated behavior. *Biol. Bull.* **198**, 188–202.

Weissburg, M. J., Doall, M. H., Yen, J. (1998): Following the invisible trail: Mechanisms of chemosensory mate tracking by the copepod *Temora*. *Phil. Trans. Roy. Soc. London B* **353**, 701–712.

Weller, R. A., Dean, J. P., Marra, J., Price, J. F., Francis, E. A., Boardman, D. C. (1985): Three-dimensional flow in the upper ocean. *Sci.* **227**, 1552–1556.

Weller, R. A., Price, J. F. (1988): Langmuir circulation within the oceanic mixed layer. *Deep-Sea Res.* **35**, 711–747.

West, B. J. (1976): Nonlinear models of population growth with diffusion. *Collect. Phenom.* **2**, no. 3, 111–118.

Wetzler, R., Risch, S. (1984): Experimental studies of beetle diffusion in simple and complex crop habitats. *J. Anim. Ecol.* **53**, 1–19.

Wheeler, W. N. (1980): Effect of boundary layer transport on the fixation of carbon by the giant kelp *Macrocystis pyrifera*. *Mar. Biol.* **56**, 103–110.

Wheeler, W. N. (1988): Algal productivity and hydrodynamics—a synthesis. In *Progress in Phycological Research*, 23–58. Round, F. E., Chapman, D. J. (eds.). Bristol: Biopress.

White, G. C., Garrott, R. A. (1990): *Analysis of Wildlife Radio Tracking Data*. San Diego: Academic Press.

Whitehead, D. R. (1983): Wind pollination: Some ecological and evolutionary perspectives. In *Pollination Biology*, 97–108. Real, L. (ed.). New York: Academic Press.

Widder, D. V. (1975): *The Heat Equation*. New York, San Francisco, London: Academic Press.

Wiebe, P. H. (1970): Small-scale spatial distribution in oceanic zooplankton. *Limnol. Oceanogr.* **15**, 205–217.

Wiens, J. A., Crawford, C. S., Gosz, J. R. (1985): Boundary dynamics: A conceptual framework for studying landscape ecosystems. *Oikos* **45**, 421–427.

Wierenga, P. J., Hills, R. G., Hudson, D. B. (1991): The Las Cruces Trench site: Characterization, experimental results, and one-dimensional flow predictions. *Water Resour.* **27**, 2695–2705.

Wildish, D. J., Kristmanson, D. D. (1997): *Benthic Suspension Feeders and Flow*. Cambridge: Cambridge Univ. Press.

Wilhelm, H. E. (1972): Nonlinear initial-boundary-value problem for convection, diffusion, ionization, and recombination processes. *J. Math. Phys.* **13**, 252–256.

Wilhelm, S. W., Weinbauer, M. G., Suttle, C. A., Jeffrey, W. H. (1998): The role of sunlight in the removal and repair of viruses in the sea. *Limnol. Oceanogr.* **43**, 586–592.

Wilkinson, D. H. (1952): The random elements in bird "navigation." *J. Exp. Biol.* **29**, 532–560.

Williams, E. J. (1961): The distribution of larvae of randomly moving insects. *Aust. J. Biol. Sci.* **12**, 598–604.

Williams, G. C. (1964): *Mich. St. Univ. Mus. Publ. Biol. Serv.* **2**, 351 (In W. D. Hamilton, 1971).

Williams, G. (1966): Measurements of consociation among fishes and comments on the evolution of schooling. *Pub. Michigan State Univ. Ser.* **2**, 349–384.

Willson, M. F. (1992): The ecology of seed dispersal. In *Seeds: The Ecology of Regeneration in Plant Communities*, 61–85. Fenner, M. (ed.). Wallingford: CAB International.

Wilson, D. E., Findley, J. S. (1972): Randomness in bat homing. *Amer. Naturalist* **106**, 418–424.

Wilson, E. O. (1958): A chemical release of alarm and digging behavior in the ant *Pogonomyrmex badius* (Latreille). *Psyche* **65**, 41–51.

Wilson, E. O. (1962): Chemical communication among workers of the fire ant *Solenopsis saevissima* (Fr. Smith) 1. The organization of mass-foraging. *Animal Beh.* **10**, 134–147.

Wilson, E. O. (1971): *The Insect Societies.* Cambridge: Harvard Univ. Press.

Wilson, K. J., Kilgore, D. L. (1978): The effects of location and design on the diffusion of respiratory gases in mammal burrows. *J. Theor. Biol.* **71**, 73–101.

Wilson, R. E., Esaias, W. E., Okubo, A. (1979): A note on time-dependent spectra for chlorophyll variance. *J. Mar. Res.* **37**, 485–491.

Winet, H., Jahn, T. L. (1972): Effect of CO_2 and NH_3 on bioconvection: Patterns in *Tetrahymena* cultures. *Exp. Cell Res.* **71**, 356–360.

Withers, P. C. (1978): Models of diffusion-mediated gas exchange in animal burrows. *Amer. Naturalist* **112**, 1101–1112.

Wolf, K. U., Woods, J. D. (1988): Lagrangian simulation of primary production in the physical environment—the deep chlorophyll maximum and nutricline. In *Towards a Theory on Biological-Physical Interactions in the World Ocean*, 51–70. Rothschild, B. J. (ed.). Dordrecht: Kluwer.

Wolfe, G. V. (2000): The chemical defense ecology of marine unicellular plankton: Constraints, mechanisms, and impacts. *Biol. Bull.* **198**, 225–244.

Wolfenbarger, D. O. (1946): Dispersion of small organisms. *Amer. Midland Naturalist* **35**, 1–152.

Wolfenbarger, D. O. (1959): Dispersion of small organisms, incidence of viruses and pollen; dispersion of fungus, spores and insects. *Lloydia* **22**, 1–106.

Wolfenbarger, D. O. (1975): *Factors Affecting Dispersal Distances of Small Organisms.* Hicksville, NY: Exposition Press.

Wolfram, S. (1985): Origins of randomness in physical systems. *Phys. Rev. Lett.* **55**, 449–452.

Wolfram, S. (ed.) (1986): *Theory and Application of Cellular Automata.* Singapore: World Scientific.

Wollkind, D. J., Manoranjan, V. S., Zhang, L. (1994): Weakly nnlinear stability analyses of prototype reaction-diffusion model equations. *SIAM Rev.* **36**, 176–214.

Worcester, S. E. (1995): Effects of eelgrass beds on advection and turbulent mixing in low current and low shoot density environments. *Mar. Ecol. Progr. Ser.* **126**, 223–232.

Worton, B. J. (1995): Using Monte Carlo simulation to evaluate kernel-based home range estimators. *J. Wildl. Manage.* **59**, 794–800.

Wright, R. H. (1964): *The Science of Smell.* New York: Basic Books, Inc.

Wright, S. (1968): Dispersion of *Drosophila pseudoobscura*. *Amer. Naturalist* **102**, 81–84.

Wroblewski, J. S., O'Brien, J. J., Platt, T. (1975): On the physical and biological scales of phytoplankton patchiness in the ocean. *Mem. Soc. Roy. des Sci. Liege*, 6th Ser., Vol., 7, 43–57.

Wu, Y. S., Kool, J. B., Huyakorn, P. S., Saleem, Z. A. (1997): An analytical model for nonlinear adsorptive transport through layered soils. *Water Resour. Res.* **33**, 21–29.

Wyatt, T. D., Foster, W. A. (1991): Intertidal invaders: Burrow design in marine beetles. *Symp. Zool. Soc. London* **63**, 281–296.

Wynne-Edwards, V. C. (1962): *Animal Dispersion in Relation to Social Behaviour.* New York: Hafner Publ. Co., Inc.

Wyrtki, K. (1962): The oxygen minima in relation to ocean circulation. *Deep-Sea Res.* **9**, 11–23.

Yahel, G., Post, A. F., Fabricus, K., Marie, D., Vaulot, D., Genin, A. (1998): Phytoplankton distribution and grazing near coral reefs. *Limnol. Oceanogr.* **43**, 551–563.

Yamaguti, M. (1976): Un systeme des equations pour la competition et la migration. Exposé à Paris VI.

Yamaguti, M. (1978): Chaos with special reference to mathematical ecology. *Math. Sciences*, **16** (No. 183), 5–11 (Japanese).

Yamazaki, H. and Okubo, A. (1995): A simulation of grouping: an aggregating random walk. *Ecological Modelling*, **79**, 159–165.

Yan, J. G. G. (1993): Some characteristic features in the pattern of bifurcating steady states in coupled reaction-diffusion systems. *Internat. J. Bifur. Chaos Appl. Sci. Engrg.* **2**, no. 2, 285–293.

Yasuda, N. (1975): The random walk model of human migration. *Theor. Popul. Biol.* **7**, 156–167.

Yen, J., Bundock, E. A. (1997): Aggregate behaviour in zooplankton: Phototactic swarming in four developmental stages of *Coullana canadenssis* (Copepoda, Harpacticoida). In *Animal Groups in Three Dimensions.* Parrish, J. K., Hamner, W. M. (eds.). Cambridge: Cambridge Univ. Press.

Yen, J., Weissburg, M. J., Doall, M. H. (1998): The fluid physics of signal perception by a mate-tracking copepod. *Phil. Trans. Roy. Soc. London B* **353**, 787–804.

Yokoyama, N. (1960): Empirical formulas of atmospheric diffusion and their applications. *Meteorol. Research Note*, **11**(5), 355–369, Meteorol. Soc. Japan (Japanese).

Yoshikawa, A., Yamaguti, M. (1974): On some further properties of solutions to a certain semilinear system of partial differential equations. *Publ. Research Inst. Math. Sci., Kyoto Univ.* **9**, 577.

Yoshizawa, S. (1970): Population growth process described by a semilinear parabolic equation. *Math. Biosci.* **7**, 291–303.

Zeeman, E. C. (1976): Catastrophe theory. *Scientific Amer.* 65–83, April 1976.

Zeil, J. (1986): The territorial flight of male houseflies (Fonia canicuiris L.). *Behav. Ecol. Sociobiol.* **19**, 213–219.

Zeldis, J. R., Jillett, J. B. (1982): Aggregation of pelagic *Munida gregaria* (Fabricius) (Decapoda, Anomura) by coastal fronts and internal waves. *J. Plankton Res.* **4**, 839–857.

Zhang, J., Huang, W. W., Liu, S. M., Liu, M. G., Yu, Q., Wang, J. H. (1992): Transport of particulate heavy metals towards the China Sea: A preliminary study and comparison. *Mar. Chem.* **40**, 161–178.

Zhou, L., Pao, C. V. (1982): Asymptotic behavior of a competition – diffusion system in population dynamics. *Non Lin. Anal.* **6**, 1163–1184.

Zimmer-Faust, R. K., Finelli, C. M., Pentcheff, D., Wethey, D. S. (1995): Odor plumes and animal navigation in turbulent water flow: A field study. *Biol. Bull.* **188**, 111–116.

Author Index

Subject Index

462   Subject Index

Interdisciplinary Applied Mathematics